A Textbook of
SOIL
Chemical Analysis

A Textbook of
SOIL
Chemical Analysis

P. R. HESSE Ph.D., F.R.I.C.

CHEMICAL PUBLISHING CO., INC.
NEW YORK

Printed in Great Britain

First American Edition 1972

CHEMICAL PUBLISHING CO., INC.

Preface

The author was trained as a 'pure' chemist and for many years revelled in the niceties of exactitude obtainable in the analysis of true solutions or homogeneous solids. Soil, if it was ever considered at all, was something in which to grow cabbages. Not until force of circumstance precipitated him into basic soil research did he appreciate the enormous challenge of attempting a meaningful analysis of a substance having a constantly changing chemical and biological equilibrium.

Nearly every university now has a Soil Science department and many confer degrees or diplomas specifically in that subject. All such degree courses include, or should include, practical instruction in the chemical analysis of soil. In many cases students merely work through a series of routine analyses using methods established many years ago and, although outmoded, perpetuated in lecture notes. There is no generally accepted textbook on the subject as exists for pure chemical analysis.

Several factors have to be considered when teaching, or learning, soil analysis; the practical manipulation and often the improvization of apparatus, the chemical theory of the analysis in question, the theory underlying the reason for making the analysis and the interpretation of the result. Most courses rely upon the student attending chemistry lectures to absorb the necessary chemical theory, but although training in chemistry is essential, many theories applying to chemical systems are not applicable to complex biological systems as exist in soil.

One of the prime aims of this textbook is to narrow the gap between the theory of soil chemistry and the manuals of soil analysis. A manual is incorporated, however, thus freeing the student from additional expenditure, and as it is based upon the preceding theoretical discussion it should be more meaningful than the 'cook-book' type of manual.

The subject of soil analysis is relatively young and procedures are constantly being modified, improved or superseded. The only way of competently performing many analyses has been by reference to articles in scientific journals which are not always readily available and not all analysts are up to date in their reading. There is often controversy over the relative merits of different analytical procedures and it is even not unknown for one soil laboratory to use method B because another is using method A. At every soils conference there is a plea and even a decision to standardize methods but this, like many conference decisions, has failed to materialize. It must be remembered of course that different soils sometimes need different methods for the same analysis.

The author has attempted to bring together the findings and opinions of soil analysts from all over the world during the last hundred years. The book may, in a way, be considered as a concise history of soil analysis. The scope of the book is such as to include non-routine methods of analyzing soils and to discuss special techniques and apparatus. For example, the routine procedures for determining nitrate and ammonium in soils are of limited value to a modern student of nitrification; also needed is discussion of techniques such as percolation and respiration.

World food shortage has focused attention on the reclamation and use of previously ignored land such as saline swamps. The cultivation and production of rice has boomed with concomitant increased research into waterlogged soils. Soils which are permanently or periodically wet need special consideration with regard to analytical techniques and interpretation of analytical results. Part of the textbook has been expressly designed to meet the problem of analyzing wet soils—a problem hitherto not considered except in isolated technical papers.

Each chapter commences with a brief résumé of the theoretical background of the particular analysis or group of analyses as related to natural soil conditions. An attempt is made to explain the reasons for making the analysis and although the subject is not exhaustively discussed, sufficient facts are presented to ensure that the procedures are not followed merely to obtain a numerical answer. A list of relevant references from which the analyst can select further reading is given in chapter order at the end of the book.

The chemistry of certain soil constituents such as nitrogen and phosphorus has received limited attention as this forms in its own right the subject matter of whole books. Other substances, such as sulphur, which have been rather neglected by soil scientists, have been discussed more fully. Substances such as yttrium and uranium compounds and other fission products are mentioned briefly but sufficient has been said to make the reader aware of their importance in modern soil chemistry.

Specialized analytical techniques such as spectrography, X-ray diffraction, polarography and chromatography are not detailed although they are commonly used in soil analysis. This is to avoid duplication as experimental details can be found elsewhere; where such techniques can be applied is indicated with relevant references.

Recommended analytical methods have been chosen with the facilities of the average soil laboratory in mind. Other factors being equal, preference has been given to procedures involving simple apparatus and commonly available reagents. More intricate apparatus usually intended for specialized research has not been ignored but a 'do-it-yourself' attitude is encouraged. The author has had long experience of laboratories in out-of-the-way parts of the world where delivery of supplies may take over one year and which have operated under a stringent budget. For those who can afford to contact expensive laboratory furnishers the textbook will give ample scope and at the same time will serve its purpose in student education. When it comes to recommended procedures, however, the student himself has been catered for; not every analyst has a scintillator-spectrometer for measuring potassium and so, although the method is mentioned, more emphasis is placed upon simple flame photometry.

Nearly every chapter has a section dealing with the determination of plant-available nutrients but a recommended procedure is not necessarily given. The myth of soil analysis being the answer to the farmer's problem is not kept up. The reader will repeatedly be reminded that field experiments are essential.

In order to conform with the present international trend, SI units have been used throughout this book. Thus °C represents degrees Celsius and not degrees Centigrade, although in practice no conversion factor is involved. The Celsius scale is being retained until such time as the Kelvin scale has been more generally accepted.

No doubt many readers (like the author) will, for a while, be mildly exasperated at having to translate mentally 'cm^3' into the more familiar 'ml' and in making standard solutions with x g dm^{-3} instead of with x g/l. More drastic changes, though, exist for the soil analyst; for example the replacement of mmhos/cm by mS cm^{-1} (millisiemens per cm) and the use of the pascal (N m^{-2} or newton per square metre) instead of atmospheres. In such cases therefore, the more commonly understood unit is given in parentheses after the SI unit. Non-SI units have been retained when quoting the results of previously published work and the corresponding SI units put in parentheses. The unit used in the special field of ion exchange has been retained as the milliequivalent. Some of the more frequently used symbols are explained in Appendix III.

P.R.H.

Lahore
West Pakistan

Acknowledgements

The author has drawn not only upon his own experience but upon that of many colleagues and he wishes to express his gratitude for their permission to do so. Numerous published works, books and articles in periodicals have been consulted and these are acknowledged in the text. Certain diagrams and tables have been reproduced with the kind permission of their authors and publishers and these too are acknowledged in the text. Certain of the photographs have been provided by institutions and commercial firms as acknowledged in the captions. Special thanks are due to colleagues, students and assistants who have helped in the practical investigations necessary and to those who have criticized the text, in particular Dr P. B. Tinker of Oxford University, and Drs S. J. Tinsley and J. W. Parsons of Aberdeen University.

Contents

Plates

1

Introduction

1:1 ORIGIN AND NATURE OF SOILS

Of the great many different kinds of soil on earth, each has its own special collective characteristics upon which its behaviour will depend; consequently, any definition of the term 'soil' must be comprehensive in context. It is necessary to distinguish between soil as a substance and soil as a natural body. In the laboratory there is an unfortunate, yet persistent, tendency to regard soil as a material in a bottle, but in the field due notice must be taken of the fact that soils have shape, volume, boundaries and so on. A general definition of soil is given in the *Soil Survey Manual* of the United States Department of Agriculture (1951):

'Soil is the collection of natural bodies occupying portions of the earth's surface that support plants and that have properties due to the integrated effects of climate and living matter acting upon parent material, as conditioned by relief, over periods of time.'

This definition takes into account the landscape nature of soils and the many varied combinations of genetic factors.

The origin of soil is the earth's crust, the primary rocks of which have weathered to produce what is known as 'soil parent material'. The rocks from which parent material is formed are known as 'parent rocks' and these may in some cases be the primary rocks themselves, or secondary rocks formed by weathering.

Four main classes of soil parent material are now recognized:

i. Those formed *in situ* from soft rocks such as chalk and volcanic ash.

ii. Those formed *in situ* from hard rocks which may be igneous (e.g. granite and basalt), sedimentary (e.g. limestone and sandstone) or metamorphic (e.g. gneiss and marble).

iii. Those transported from their place of origin before weathering to form soils. This is the most important of the parent material classes and includes all material moved by water (such as alluvium, colluvium, lacrustine deposits and marine and beach sediments), by wind (such as loess and eolian sands), and by glacial phenomena such as glacial drift and till.

iv. Those formed from organic material such as peat.

Rocks are weathered mechanically by the action of frost, water and wind, and chemically by such processes as oxidation, hydration, hydrolysis and solution in carbonic acid. Similar or even identical rocks can produce quite different materials according to the kind of weathering.

The products of the weathered rock, that is, the soil parent material, then undergo further change to produce soil as we know it. This process does not necessarily occur at the same time or place as production of parent material, nor on the other hand is a parent material necessarily found between a soil and its parent rock; for example, the soil-forming processes may have proceeded concomitantly with rock weathering. In many soils as now found, more than one parent material may have been involved in their formation.

The process of soil formation from parent material can be divided into two main parts. One involves addition and admixture of organic matter and the other involves solution and translocation of certain constituents and is known as 'eluviation'. Eluviation itself can be further divided into mechanical eluviation (movement of finer particles, usually downwards by washing) and chemical eluviation, which involves the partial decomposition of colloidal matter with subsequent movement of the decomposition products.

Either after or during the formation of a soil, its nature can be modified by otherwise unrelated effects such as volcanic action, floods, erosion and human activities.

The eventual soil is a chemically, physically and biologically complex, dynamic system, the constituents of which are constantly undergoing change. Hall (1949)* likens soil to a 'three-phase system of solid, liquid and gaseous components with constantly shifting equilibria'.

The reader is referred to standard works on soil genesis and constitution (e.g. Hall, 1949; Russell, 1961) for a full discussion of the origin of soils, but from the analyst's point of view it is apparent that soil, even as a material in a bottle, provides a stimulating challenge to his skill and capabilities.

1:2 PURPOSE OF SOIL ANALYSIS

Any particular soil may be analyzed for such varied reasons as to predict its behaviour if planted with wheat, if used to construct a road, if drained, if fertilized and so on; or it may be analyzed to discover its origin or to see if a certain constituent is present or to investigate the effects of a changing environment.

In the study of soil as a nutrition factor, for example, the analyses may be of a routine nature for advisory purposes, or highly specific as in basic research. For the classification of soils as in a soil survey, again the analyses may be general in nature, but if a soil map is required illustrating a particular soil property such as salinity, then specific analyses are called

---

* All references are listed in chapter order on pp. 485–512.

for. In this book emphasis is upon the analysis of soil in its capacity as a medium for plant growth but the reader should realize that factors other than the chemistry of a soil influence plant growth. Such factors include climate, microbiological conditions, structure and other physical aspects. These factors are but briefly touched upon here but their importance in limiting the usefulness of soil chemical analysis must be emphasized.

For whatever purpose a soil is to be analyzed, plans should be made in advance as to what analyses are required, why they are required and what level of accuracy is needed. This is particularly important when an investigator has the analyses done for him rather than performing them himself, as all too often a field or research worker will demand an excessive number of analyses chosen from the book rather than from the requirements of the investigation.

1:3 PRECISION AND ACCURACY

The degree of accuracy with which a determination is carried out is very often a neglected factor in soil analysis and a particular procedure is merely followed from written instructions without reference to the purpose of analysis. As a consequence some measurements are made with unnecessary accuracy, often at the expense of considerable time. For example, there is no point in using a lengthy gasometric method to measure the carbonate content of a soil to three decimal places when all that is called for is a statement as to whether or not the soil is calcareous. The farcicality of such a measurement is more apparent if one considers that perhaps 1 g of a 2-mm sample will have been used in the experiment, this 1 g having been taken from about a kilogramme sample, itself selected from a much larger bulk of soil which may or may not have been representatively sampled in the field. The variability of calcium carbonate distribution in a soil could be such that the sampling error is enormous.

Another and oft-quoted example is that of pH determination. It is well known that the pH of a soil may vary by as much as a whole unit over a relatively small area, and thus it is pointless (as a routine) to measure pH values with greater accuracy than 0·2 or even 0·5 of a unit.

Conversely, some analyses are made with insufficient attention to accuracy. For example in a specific experiment concerning changes in pH of a particular soil sample with time or with certain treatments such as dilution of suspension, it may be necessary to obtain results much closer than to within 0·2 of a unit. This is the case for example when measuring exchange acidity by Brown's method (section 4:3). Similarly great accuracy would be demanded in ammonium or nitrate determination when investigating the effect of certain treatments upon nitrification in a soil, whereas it would be unnecessary to achieve such accuracy when analyzing as a routine for the inorganic nitrogen content of a soil.

Another common malpractice is to achieve an apparent accuracy by applying mathematical techniques to an experimental result out of proportion to the accuracy with which that result was obtained. An extreme example of this was once encountered by the author who found a soil chemist reporting fertilizer responses to seven decimal places using yield figures obtained in lb/acre to the nearest 5 lb. His explanation was that the calculating machine gave the answer to seven decimal places. While such examples are fortunately rare, less obvious cases are extremely common and it should be borne in mind that in order to be significant, only the last reported digit may be uncertain.

The chief errors involved in a chemical analysis have been adequately discussed by, *inter al.*, Vogel (1962). Additional remarks of particular relevance to soil analysis can be made regarding personal errors, sampling errors and errors of method.

Personal errors, apart from those due to bad technique and carelessness, are due to personal characteristics which influence results in a standard manner, resulting in bias. These errors can be revealed by having the same analysis performed by more than one person but in practice such a procedure is not justified as a routine. It is, however, an advisable precaution when testing new or modified procedures.

Errors of method, although common, are difficult to detect. Good agreement of replicate analyses is meaningless if the method of analysis itself does not yield the correct result. As an example we can consider the Kjeldahl procedure for measuring total nitrogen in a material which may contain much nitrate; unless the necessary modifications to the method are introduced the nitrate-nitrogen will remain undetected. Whenever possible, but always when introducing a new method, independent methods of measuring the same quantity should be used. A control analysis using material of known composition should always be carried out and helps detect errors of method. Standardization of method permits the comparison of analyses of different soils made in different places. It is far more important to use a standard method of analysis than to modify it for what may be a slight increase in accuracy. Not that such modifications are to be discouraged, but their adoption should be made by all if results are to be comparable. Regression analysis can be used to compare statistically a new method of analysis with a standard method and an example of this is given by Pantony (1961). All proposed new methods of soil analysis should have a statement of the standard deviation and of the number of degrees of freedom involved in its calculation.

Sampling errors are perhaps the most common errors encountered in soil analysis owing to the extreme heterogeneity of material. In the laboratory such errors can be minimized by grinding and mixing, but the real sampling error occurs in the field. Again, procedures exist for reducing these, but interpretation of soil analyses must always be made keeping in mind the probabilities of field sampling errors.

As a general means of reducing errors all experiments should include a blank determination which will reveal errors of reagent, a control which reveals errors of method, reagent and instrument, and all experiments should be replicated. Good agreement of replicate analyses, however, is useless without blank and control analyses as a constant error may be involved.

Gross errors can often be detected by using certain interrelationships, some of which are listed:

i. pH and carbonates: If free carbonates are present in a soil the pH value will usually exceed 7.

ii. pH and base saturation: A pH of over 6 indicates a percent base saturation of between 60 and 100.

iii. pH and sodium carbonate: Soluble carbonates usually are not present in a soil unless the pH exceeds 9·5.

iv. Base exchange capacity, organic matter and mechanical analysis: In general a high cation exchange capacity in a soil is associated with high organic matter content and to a lesser degree with high clay content. The relationship is of most value when a large number of samples of the same kind of soil are being analyzed, when any discrepancy shows up at once.

v. Cations and anions: The sum of the cations should agree with the sum of the anions when both are expressed as milliequivalents per 100 g.

In each of the detailed methods of analysis described in this book the errors most likely to occur are indicated and guidance is given as to the limits of accuracy that can be obtained.

The 'accuracy' of an analysis refers to its correctness and is determined by the absolute error involved, the 'absolute error' being the difference between the observed value and the true value. 'Accuracy' must not be confused with 'precision' which refers to the reproducibility of an analytical result. Thus a series of analyses performed with a high degree of precision are not necessarily accurate; the analyst may consistently be making the same mistake (using a wrong reagent for example) and although his results are reproducible, they will be quite wrong. Measures of precision are the 'mean deviation', the 'relative mean deviation' and the 'coefficient of variation' which can be computed statistically.

The 'standard deviation' is a measure of the spread of values and is calculated from analyses of standard samples by the expression,

$$s = \sqrt{\frac{\sum d^2}{(n-1)}} \qquad [1:1$$

where d is the deviation of individual samples from the arithmetical mean value and n is the number of sample values. Where soil samples are concerned, Vermeulen (1953) recommended duplicate analyses and application of the expression,

$$s = \sqrt{\frac{\sum y^2}{2n}} \qquad [1:2$$

where y is the difference between duplicates and n is the number of samples.

Detailed discussions of the statistical treatment of analytical results to estimate their reliability can be found in several books and monographs (e.g. Fisher, 1958; *Agronomy 9*, 1965) and will not be repeated here. An excellent and simple introduction to statistical treatment of analytical data is given by Pantony (1961).

Vermeulen (1953), during a study of the reliability of soil analyses made by his staff, found large variations in standard errors according to the season of the year. Errors were minimal after the summer holidays, high during Christmas and very high at the beginning of Spring. Vermeulen pointed out that his analysts were all young ladies. This, of course, only shows that Vermeulen was guilty of bias in his experiments.

1:4 RECORDING AND PRESENTATION OF RESULTS

1:4:1 Immediate recording of final results

All analytical results such as titration figures, volumes and weights should be recorded at the time of obtaining in order to avoid uncertainty or error later. Memory should not be relied upon, even for a few minutes, as an unexpected distraction could result in the waste of hours of previous work. The results should not be jotted down on odd scraps of paper which may be mislaid, but in a notebook kept for the purpose. These notebooks should not be destroyed as very often it is necessary to refer back to some figures.

1:4:2 Permanent recording of final results

When all calculations have been made and the result of an analysis appears to be correct, the figures should be entered into the laboratory ledger or on printed standard forms which are then filed. The layout of the ledger or of the printed sheets will depend upon the purpose of analysis. Results sheets used for a soil survey may differ from those used in fertility trials or in a particular research problem; for a large laboratory dealing with many samples for different purposes it may be most convenient to have forms permitting the entry of every analysis that may be made in the laboratory, although such forms can become unwieldy.

Whatever the nature of the final permanent records, they should never be allowed to leave the laboratory and should be kept in a fire-proof cabinet. If necessary, copies can be made for issue or loan.

1:4:3 Reporting of results

The way in which analytical results are issued from a laboratory will depend not only upon the purpose of the analyses but upon the knowledge and training of the person receiving the results. It would be of little use, for example, to send the average farmer a list of cation exchange data expressed as m.e./100 g, whereas such a list would suffice for a research worker dealing with clay mineralogy. For laboratories providing an agri-

1. Riffle sampler for reducing the bulk of soil samples (*see* p. 15)

2. Battery-operated conductivity bridge for field work (*see* p. 69)

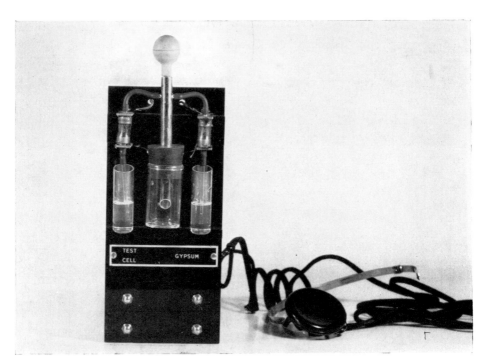

3. Detail of battery-operated bridge showing conductivity cells

4. Bureau of Soils cup and holder for measuring the resistance of saturated soil paste (*see* p. 76)

5. Apparatus for the preparation of soil saturation extracts (*see* p. 81)

6. Portable vacuum filtration apparatus for saturation extracts

7. Pressure membrane apparatus for extracting soil solutions (*see* p. 81)

(Photo: Soil Moisture Equipment Corp., Santa Barbara, Calif., USA).

9. Detail of pipette-type conductivity cell

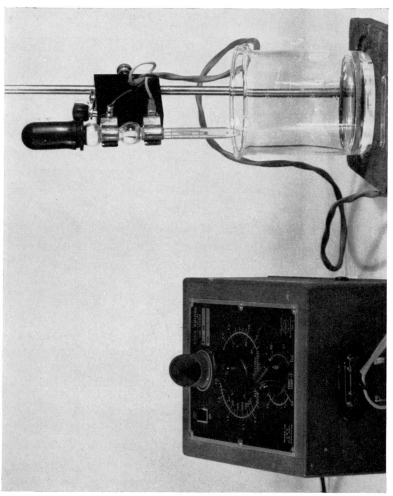

8. Pipette-type conductivity cell and Solu bridge (*see* p. 82)

10. Semi-micro Kjeldahl flasks on electric digestion stand with glass fume exhaust (*see* p. 197)

(Photo: A. Gallenkamp & Co. Ltd., London.)

11. Jenkin mud sampler (*see* p. 468)

(Photo: Freshwater Biological Assoc.

cultural advice service, analytical results must be translated into practical information for the layman, and usually an expert on this subject is attached to the laboratory staff.

1:4:4 Checking analytical results

All results should first be checked by the analyst himself with respect to arithmetic and feasibility. All doubtful results should be checked by repeating the analysis. If there is a chemist in charge of the analyst he should re-check all results before they are entered in the permanent record and if necessary he should have an experiment repeated.

The procedure upon finding a doubtful or unusual result should be:

i. Check if the result has been correctly copied from the original notebook.

ii. Using the original analytical results from the notebook, check the arithmetic and calculations.

iii. Check on the method of analysis used, the apparatus, the chemicals, the quantities of reagents and the sampling procedure. It may be that the method is not quite suitable for the particular determination under existing conditions; for example, when determining exchangeable cations the calcareous nature of a soil may have been overlooked. A piece of apparatus may have been incorrectly set up, leaks may be occurring, voltage may be fluctuating and so on. Reagents may be contaminated or it may even be that their limits of purity are not suitable for the particular analysis; for example, the content of sulphur in a sodium carbonate reagent may be sufficiently low for most purposes but not when very small quantities of sulphur are being measured. Sampling errors and their subsequent weighing errors are very common in soil analysis and exceptional care has to be taken to minimize them.

iv. If the above routine fails to explain the unusual result then the experiment must be repeated.

v. Having repeated the experiment with the utmost care, if the same result is obtained this should be commented on in any reports made as the matter will almost certainly provide material for research, either upon the soil in question or upon methodology.

1:4:5 Expression of results

In nearly every case of soil analysis it is preferable to express results on an oven-dry basis even though analysis of an oven-dry soil will be rare. This permits comparison of results and provided that the moisture content is also recorded, results can, if required, be calculated to air-dry or even field moisture contents, although the latter would be very arbitrary. A possible exception to this general rule is the expression of a moisture content. Sometimes it is easier to visualize a soil with 60% moisture on a wet basis; that is, 40 g of soil mixed with 60 g of water, rather than as with 150% moisture on a dry basis. This, however, is a matter of individual

2+T.S.C.A.

preference and both methods of expression are commonly encountered; the important thing is to state clearly which method is being followed.

1:5 INTERPRETATION OF RESULTS

Before the results of an analysis can be considered in relation to purpose they have to be interpreted correctly in relation to the method by which they were obtained. For example, the meaning of a pH determination will depend upon the soil–water ratio during measurement and whether or not potassium chloride has been added. With some soils the treatment given prior to measurement can profoundly affect the results. With certain soils of East Pakistan, for example, a routine measurement using air-dry soil or a field test made during the dry season leads to an acid soil classification. In fact during the monsoon and growing season, the soils are neutral to alkaline. Any fertilizer or crop recommendations for these soils would be incorrect if the drying effect upon pH had not been considered. Another and common example of interpretation in relation to method is instanced by the meaning of 'available' phosphate, methods of measuring which are legion.

Some years ago (1960) it was published that the rice plant was capable of fixing atmospheric nitrogen. Subsequent investigation by other workers (Hart and Roberts, 1961) demonstrated that this spectacular discovery was nothing more than a misinterpretation of results in relation to method. Nitrogen measured by the unmodified Kjeldahl process had been taken as total nitrogen and the considerable quantities of nitrate-nitrogen involved had been neglected.

The limitations imposed upon interpretation by any particular method of analysis are discussed in the appropriate sections of this book.

Once an analysis had been completed the result has to be considered in relation to the purpose for which the analysis has been made. The correct interpretation of analytical data is of fundamental importance and even if the ultimate interpretation is to be made by someone else, the analyst himself should be sufficiently aware of the factors involved to make an intelligent appraisal of the figures.

As a simple example of interpretation in relation to purpose, consider the determination of nitrate in a soil. The relative importance and the meaning of the result would vary considerably according as to whether one was measuring the salt content for classification purposes, for reclamation purposes, for fertility purposes or for special research on nitrogen changes in that soil.

The most difficult interpretations are those pertaining to fertility experiments. Many factors other than the analysis in question have to be taken into account. The so-called 'available' nutrients in a soil are those commonly measured and it is important to realize the great need of their correct interpretation (e.g. Hesse, 1968). No one element should be considered by itself but in conjunction with other relevant factors known to influence it, such as other elements or groups, soil conditions, the use to which the land is to be put and so on.

An analysis may indicate an adequacy of, say, calcium or magnesium in a soil for a particular crop, yet, if the soil should be potassium deficient necessitating fertilization with potash, this could render it calcium or magnesium deficient due to ionic antagonism. The methods of determining available nutrients normally apply only to a particular soil in a particular place and intended for a particular crop. It is essential to calibrate all such laboratory measurements by field experiments and even this does not allow for seasonal changes in the soil or changes due to land management.

The rate of release of nutrient elements is one of the principal factors in the soil–nutrient–crop relationship. The processes governing rate of nutrient release include ionic diffusion, mass flow, reaction surface area and mineralization of organic matter. The rate of organic matter mineralization will be dealt with extensively in Chapter 11 but the other factors are deemed to be outside the scope of this book. They should, however, be considered when interpreting the results of 'quantity' analyses.

2

Preparation of soil samples for analysis

2:1 RECEPTION AT THE LABORATORY

Each batch of soil samples arriving at the laboratory should be accompanied or preceded by an explanatory letter or a standard form giving relevant field data and indicating the analyses required. If the analyst himself collected the samples, then adequate field notes will have been kept. Every individual sample should be labelled unambiguously with its field number; all unidentifiable samples should be discarded.

Information regarding the samples should be entered in a ledger and each sample given a laboratory number. It is more convenient to use laboratory numbers rather than field numbers, which often are complicated and may even be duplicated if more than one source of samples is involved. A suggested ledger format with an example entry is given in Table 2:1.

Once the samples have been received and recorded the first consideration is the urgency with which analysis, or perhaps one particular determination, is required. It may happen, for example, that a certain measurement such as pH is required before the soil is handled further. There will occasionally be instances where a very recently sampled soil needs immediate attention, such as a redox potential measurement or iron(II) determination. Such cases however are special; they involve particular sampling and transporting techniques and will not be encountered in the normal routine.

2:2 DRYING OF SOILS

2:2:1 Methods

The soils are emptied from their containers and spread to dry on flat trays. The trays are commonly about 50 by 60 cm, 2·5 cm deep and made from wood, zinc, fibreglass or plastic.

It is most important to maintain the identity of each sample at all stages of preparation. During the drying stage the trays can be numbered but it is advisable to place a plastic tag in with the soil as well. It is perhaps superfluous to say that the tag should be removed before the soil is prepared, but a case actually occurred where plastic tags tipped with a soil into an electric grinding machine were intimately incorporated with the 2-mm sample.

Table 2:1

Suggested format for Soil Reception Ledger with specimen entries

Lab. no.	Field no.	Origin of sample	Depth cm	Field notes*	Sampled by	Analyses required	Date received	Date completed
1	Nx/4/G	Goodmans Farm Anywhere	0–10	Under oats showing Mn deficiency symptoms. Suspected calcareous.	A. Smith, Agric. Div. Adv. Service	pH, CaCO₃, C, N, ex. cat., mech., Mn	4/7/67	15/8/67
2								
3								
...								
117	SS1	Unknown Plains Exshire	0–5	From Profile 14. Bleached layer at 5 cm. Mottling at 15–30 cm. Becomes dark and blueish at 35 cm.	P. Jones, Party II	As completely as possible. Profile to be preserved for exhibition and tuition.	22/9/67	
118	SS2		5–7					
119	SS3		7–15					
120	SS4		15–25					
121	SS5		25–35					
122	SS6		35–60					
123								

* The entry under Field Notes should be relevant to the required analyses. Comprehensive field notes will be kept by the person who sampled the soil.

The soils are allowed to dry in the air, the trays being placed in special racks—ideally in a cabinet with warm air circulating. The temperature should not be allowed to exceed 35°C and the relative humidity should be between 30 and 70%.

2:2:2 Effects, chemical and physical

Drying causes changes in the chemical and physical characteristics of a soil, the chemical changes often being the result of microbiological changes. The degree to which such changes occur varies with the temperature and time of drying, and for this reason the procedure for drying soils should be standardized as far as possible.

For a true evaluation of a particular constituent in a soil as encountered in the field, special techniques must be used whereby the soil is sampled and analyzed immediately without further treatment. Examples of such techniques will be found in subsequent chapters where applicable, but the analysis only gives the position at that particular time and results would differ with time of sampling. Consequently except for specific purposes such analyses are of little value. For example, the nitrate content of a soil could change over a relatively short period of time due to natural causes such as rainfall, drainage or sunshine, and in order to compare soils it is essential to standardize by air-drying even though the results may be different from those which would obtain in the field.

Oven-drying a soil can cause such profound changes that it is not recommended as a preparatory procedure in spite of its convenience and reproducibility.

In recent years a considerable amount of work has been done on the effects of drying soils and the reader is referred particularly to the papers of Birch (1958; 1959; 1960) and Griffiths (1961). Some of the more important effects are listed and discussed below.

Chemical effects

NITROGEN

As far as measurement of total nitrogen is concerned, drying a soil before analysis has little effect. On the other hand, if water-soluble nitrogen fractions are to be measured it must be borne in mind that water-soluble organic material in a soil increases with time and temperature of drying and this effect is greatly enhanced if the soil is oven-dried.

Similarly, drying profoundly affects the ammonifying and nitrifying organisms in a soil. The longer a soil remains dry, the greater is the amount of organic nitrogen mineralized when the soil is re-wetted. For example, Birch (1960) found that with one soil that had remained dry for 9 weeks the extra nitrogen mineralized on re-wetting (that is, in addition to that mineralized when the soil was not dried) was equivalent to over 1 ton of ammonium sulphate per acre 6-in. (more than 251 kg 10^3 m^2 per 15 cm).

The explanation for this phenomenon involves the fact that drying a soil destroys some micro-organisms as well as liberating soluble organic matter, and on re-wetting there occurs a high microbial activity due to the developing population. It is apparent that this drying effect must be taken into account if a soil is to be used for experiments involving mineralization of nitrogen.

Air-drying a soil seems to have little effect upon the nitrifying organisms, and consequently on re-wetting the increased mineral nitrogen fraction will appear as nitrate, whereas if the soil had been dried at higher temperatures the nitrifiers would have been killed and on re-wetting there would be an accumulation of ammonium-nitrogen.

CARBON

Again, air-drying a soil will not affect a total carbon analysis, but owing to the increased solubility of organic matter and destruction of micro-organisms referred to above, the time and temperature of drying must be considered in experiments concerning oxidation of organic matter. Oven-drying may cause loss of carbon due to oxidation of organic matter.

POTASSIUM

It was shown by Attoe (1946) that as soils are dried potassium may be released or fixed depending upon the original level of exchangeable potassium. Generally the exchangeable potassium is increased on drying a soil if originally it was less than about 1 milliequivalent per 100 grammes, and is rendered non-exchangeable if the original level exceeded that value. In this respect the critical levels of exchangeable potassium present before drying a soil vary with the kind of soil and appear to be dependent upon the kind of clay mineral present. The importance of the drying effect in routine analysis is difficult to assess as the original exchangeable potassium content would have to be measured upon a freshly sampled soil for comparison. For experiments involving soils which may have been fertilized with potash the possibility of misleading exchangeable potassium results being obtained should be considered.

PHOSPHORUS

Air-drying an acid soil leads to an increase in water-soluble or dilute acid-soluble phosphorus, and if the soil is dried at high temperatures this increase can be as much as 100%. Dry soils with high pH values, on the other hand, release less phosphorus than if not dried. With some soils drying can affect the phosphorus fixation capacity and this is probably linked with aluminium and iron changes.

SULPHUR

Air-drying some soils has been found (Barrow, 1961; Freney, 1958) to release more sulphate to extracting solutions than if the corresponding fresh soils were extracted.

pH

The effect of air-drying upon the pH value of certain soils has been mentioned in Chapter 1 (1:5) and is further discussed in Chapter 3 (3:1) and in Chapter 13 (13:1). Some soils, and particularly those containing sulphur compounds, can have their reaction changed drastically by drying. Sulphide- or elemental sulphur-containing soils which are neutral when wet develop a pH value of less than 2 when air-dried, the effect being chemical and biological.

MANGANESE

Exchangeable manganese increases in amount in a soil when that soil is air-dried. The effect appears to have biological causes and is discussed in greater detail in Chapter 14 (14:3:2).

Physical effects

An investigation into the effects upon the physical characteristics of a soil when it is dried was made by van Schuylenborgh (1954). Using a tropical mountain forest soil which was subject to excessive rainfall, van Schuylenborgh found that drying affected the results of mechanical and consistency analyses.

Mechanical analyses were made with and without prior oxidation of organic matter on wet and air-dried soil samples. The overall effect was to change a clay soil to a sandy soil and the same effect was shown by differences in water-holding capacity and plasticity; the effect was not one of drying upon the organic matter. It was suggested that the effect is one of irreversible dehydration causing cementation of clay particles which are then not dispersed. In arid climates where soils are normally dry in any case, the matter is of little importance.

Similar results were obtained by Schalscha and co-workers (1965) with a volcanic ash soil. Drying caused irreversible changes resulting in clay textures being analyzed as sandy textures.

Nevo and Hagin (1966) found that changes occurring in a soil during the first 3 months of dry storage were independent of micro-organisms; the major factor involved was a change in the physical structure of the organic fraction.

2:3 GRINDING AND SIEVING

When a soil is air-dry—and the time for this will vary according to the procedure and amount of soil—it is taken to the preparation room for further treatment. The preparation room should be quite separate from the drying room, and particularly from the laboratory, and should have some means of dust extraction to prevent contamination of other samples.

In some instances it may be desired to keep a part of the sample as a reference, or for setting up an exhibit or monolith, and such a sub-sample should be taken before any further treatment is given to the soil.

Stones and pieces of macro-organic matter are picked out and, if necessary, weighed, and the remainder of the sample is crushed and sieved. Large lumps are broken up by hand and then the soil is ground by rolling gently with a wooden roller. Some laboratories employ mechanical grinders for this purpose, with rotating wooden rollers mounted above a sieve, or an electrically driven rubber pestle and mortar.

After grinding, the soil is screened through a 2-mm (10-mesh) sieve to give what is commonly referred to as the 2-mm fine earth and which is used for most routine analyses.* Approximately 1 kilogramme of fine earth is selected by quartering or by using a riffle sampler. The greater than 2-mm soil retained on the sieve is weighed and may or may not be analyzed separately.

In 'quartering', the soil is spread uniformly over a sheet of brown paper or polythene and divided into four equal portions as illustrated in Figure 2:1. The portions marked 1 and 4 are discarded and the remaining portions mixed together, spread out again and reduced to half by the same

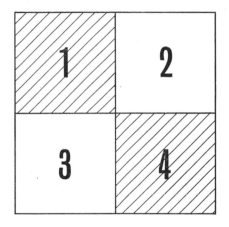

Figure 2:1 Reducing a soil sample by quartering

procedure. This process is repeated until a sample of the required bulk is obtained.

The riffle sampler is depicted in Plate 1; a soil sample tipped in at the top is automatically halved by a series of chutes and the process is repeated as many times as necessary.

For certain analyses, in particular those carried out at micro or semi-micro level, a sample much finer than the 2 mm is required. In some laboratories the preparation of such a sub-sample is carried out as a routine in continuation of the initial preparation. It may, however, be

* In this book all references to mesh size use the US Standard Sieve dimensions and not the IMM. A table showing both systems with the corresponding mesh openings in millimetres is given in Appendix III.

more desirable to prepare the sub-sample as and when required. In either case the whole of the 2-mm sample is tipped out on to a sheet of glazed paper or plastic and sampled by quartering until the required amount is obtained (usually about 100 g). The selected sub-sample is then ground in an agate mortar, either by hand or electrically, to pass a 0·5-mm (40-mesh) sieve or, for certain specific purposes, finer sieves. The complete sub-sample must be ground to pass the sieve and none is discarded. The author has occasionally come across the practice of tipping the 2-mm sample into a finer sieve and collecting what comes through; this is most inexact and must not be done.

The composition of the grinding and sieving apparatus can be important especially if trace elements are to be determined. If minor elements are to be measured, extra precautions against contamination of the sample must be taken throughout its preparation.

2:4 STORAGE OF SOIL SAMPLES

The ideal container for a soil sample is a screw-capped glass jar which should be clearly labelled with the laboratory number and the degree of fineness of the sample: for example, 4631—0·5 mm. If a large number of samples is to be kept it will be found more convenient to store the bottles on racks in strict numerical order rather than to attempt some other system, such as classifying according to district or kind of analyses required. As labels are liable to be lost or defaced, each bottle should contain a plastic disc bearing the laboratory number.

Owing to expense many laboratories find it impossible to use glass jars and samples are stored in wooden boxes, waxed cartons or paper bags. Such containers, however, should be avoided if at all possible in order to reduce the risk of contamination. Of recent years plastic containers have become popular for storing soil samples, being cheaper and less breakable than glass. However, investigations made in the USA (J. M. Coleman, private communication) have revealed that collection and storage of soil samples in plastic containers can result in certain fundamental changes in the clay minerals. Samples were collected from Mississippi in plastic bags and in waxed cartons; subsequent X-ray analysis showed significant differences in the clay minerals according to the type of container used. The matter was investigated and the effect of the plastic containers confirmed. It is thought that an organic complex passes from the plastic into the clay mineral—specifically montmorillonitic clays in the soils examined. Soil samples collected in plastic bags in East Pakistan have given X-ray pictures which have proved unidentifiable. The subject has, at the time of writing, yet to be more fully investigated, but the possible effects of such a phenomenon upon properties like cation exchange could be important.

Whatever the container, soils will undergo changes during storage and some of these changes are important when considering analytical results.

This has been partly discussed in section 2:2 dealing with the drying of soils, but so far unconsidered are the changes which occur when soils are stored in a moist state. In practice this concerns the time elapsing between sampling and preparing a soil, as moist storage for long periods is done only for specific reasons.

A change with perhaps the most important effects is that in the microbial population. Stotzky and co-workers (1962) found that the greatest microbial changes in a moist soil occur in that sampled from the top 15 cm when all the organisms except actinomycetes considerably decrease in number. In soils sampled from deeper horizons (15–45 cm) similar changes occur, except that the denitrifyers and fungi increase in number. The most striking find of Stotzky, however, was the increase with time of the actinomycetes and the increased contribution of that group to the total microflora. As actinomycetes are active in the production of antibiotics, their increase may be responsible for the fungistatic materials found to occur in stored, moist soils.

Stotzky's samples were sieved and stored in plastic bags. Thus, as most of the macro-organic matter had been removed, once the available substrate had been used up only those organisms capable of forming spores or existing at low metabolic rates could persist, and hence the decrease in bacteria. Moreover, the increase in spore-forming bacteria and of bacteria capable of development in carbon dioxide suggested that the changes may have been brought about by restricted gas exchange during storage. It was realized that the polythene bags were permeable to oxygen and carbon dioxide, but nevertheless it was considered that toxic levels of carbon dioxide may have occurred due to a permeability differential.

Concomitant with microbial change will be changes between the organic and the mineral constituents of a moist soil. Storing a soil in a moist state is, in effect, incubating it, but without temperature or moisture control and possibly with a build-up of carbon dioxide at the expense of oxygen. As such treatment will result in many complicated reactions it is most undesirable to keep a soil in a moist state for any length of time for the purpose of general analysis.

2:5 DETERMINATION OF MOISTURE CONTENT

Except for special purposes, a knowledge of the moisture content of an air-dried soil is of little interest in itself, but the determination is necessary for the calculation of most other analytical results.

Various methods exist for determining soil moisture, but for the soil analytical chemist the standard procedure of determining the loss in weight when a sample is oven-dried is the most suitable. The determination is best made using small, non-corrosible metal tins and the temperature control of the oven should be as accurate as possible.

For special, non-routine determination of moisture content, temperature control of drying becomes very important. Some soils lose weight on heating through decomposition of organic matter but this is not normally a serious source of error. Ideally the soil sample should be dried to constant weight, but in practice it is sufficient to dry for a definite period of time, which should not be too long.

Approximately 5 g of soil are accurately weighed into a tared tin and then dried in an electric oven at a temperature of 105–110°C for 8 hours. After cooling in a desiccator, the loss in weight is determined.

The results can be expressed either on a wet basis or a dry basis as discussed in section 1:4:5.

If a soil is stored for some time before being further analyzed it is advisable to redetermine its moisture content. When analyzing a fresh, wet soil, it is imperative to weigh the sample for moisture determination at the same time as for the analysis in question.

3

Soil reaction (pH value) and lime potential

3:1 INTRODUCTION

When 'soil pH value' is referred to, what is meant is the pH value of a soil–water system the composition of which is variable according to circumstances.

Sørenson's (1909) definition of pH as the negative logarithm of the hydrogen ion concentration—that is:

$$pH = \lg \frac{1}{a_{H^+}} = -\lg a_{H^+} \qquad [3:1$$

where a_{H^+} represents the activity of hydrogen ions—refers strictly to a true solution in which the ions are completely dissociated and where there exists a large volume compared to molecular dimensions.

A soil–water system does not comply with these limitations and consequently the pH value of a soil cannot be defined as precisely as that of a solution. Nevertheless, the pH value remains the most important single measurement made upon soils.

The main value of a pH measurement is not that it shows a soil to be acid or alkaline but the information it gives about associated soil properties such as phosphorus availability, base status and so on.

Most agricultural soils have pH values lying between 4 and 8. More acid soils are usually peaty in nature and often contain a high proportion of sulphur or aluminium. For example, the discovery that certain soils in the West Indies were too acid to support crops led to the establishment of a highly profitable bauxite industry. Strongly alkaline soils are found in the arid regions of the world. Certain soils exhibit ambivalency with regard to their pH value. For example in East Pakistan the basin soils near Dacca have a pH of 5 or less when sampled during the dry season or when dried before analysis, whereas the same soils sampled during the monsoon season, and not dried before measuring pH, have values of 7 or more. A routine pH determination of the air-dry soil in this case would lead to an acid soil classification, whereas in fact during the growing season the land is neutral or alkaline. Thus great care has to be taken in interpreting pH measurements of such soils since the results influence soil classification, crop recommendation and so on.

The more acid a soil the more mobile will become such elements as iron, manganese, zinc, copper and other minor elements. Thus at very low pH values a soil may contain toxic quantities of certain elements and it is considered that this induced toxicity is responsible for poor plant growth rather than acidity *per se*. The most common example is aluminium toxicity which restricts plant growth by causing, *inter al.*, stunted and deformed roots. This will be discussed more fully in Chapter 4, but the usual method of counteracting aluminium toxicity is to raise the pH value of the soil by liming. Conversely, too high a pH value of a soil can induce minor element deficiencies.

The degree of acidity is critical when considering the phosphorus fixation capacity of a soil. The rapidity with which phosphorus is fixed by a very acid soil is due to the secondary effects of iron and aluminium liberation. In acid soils the availability of phosphorus to crops has been shown (Birch, 1953) to be significantly related to the percent base saturation which in turn is related to the pH value.

Another important soil property depending upon pH status is that of microbial activity. Many micro-organisms, and in particular the nitrifyers, are inhibited by acidity; others, such as *Thiobacillus thio-oxidans*, require a low pH in order to function effectively.

Soils as buffer systems

The pH value of a soil is a measure only of intensity of acidity and not of the amount of acid present. Certain soils such as heavy clays and peats have greater reserves of acidity than, say, sands, and such soils with large reserves of either acidity or alkalinity are said to be 'well buffered'.

Buffer action is the resistance to changes in hydrogen ion concentration and the most common buffers consist of a mixture of a weak acid and its corresponding salt made with a strong base; for example, acetic acid and sodium acetate. Such systems are common in soils containing organic acids. Aluminium is partly responsible for the buffering action of soils because as the pH value rises aluminium dissociates hydrogen ions from co-ordinated water molecules in the clay.

3:2 SOIL pH VALUES

3:2:1 Background and theory

If the water content of a soil falls below wilting point the pH fluctuates and the electrical resistance of a dry soil is so great that a pH meter may give incorrect readings; a soil must not be too dry when its pH is measured.

Thus, as previously mentioned, it is really the pH value of a soil–water system that is measured and the relative proportions of the components of this system will affect the pH results. Generally, the higher the proportion of water the higher the pH value will tend to be. Huberty and Haas (1940) quote one soil as rising from pH 6·45 at 6·3% moisture to 8·60 when a 1:5 suspension was measured.

As soils normally contain salts to greater or less extent, the suspension is strictly a system of soil particles in salt solution and one effect of diluting the system is to reduce the salt concentration. If much carbon dioxide is absorbed in the soil, dilution would raise the pH value and a considerable rise would occur if the soil contained much exchangeable sodium.

A further complication in measuring soil pH is the fact that the pH value of a soil paste will be different from that of its supernatant liquid, due to a hydrogen ion gradient. If a small charged particle of soil is in water suspension, its charge will be neutralized by a diffuse cloud of exchangeable ions around it (the electric double layer). As salts are also present their ions will mingle with the exchangeable ions and tend to reduce the double layer. The higher the valency of the salt ions and the greater their concentration the more will the double layer be reduced, and the pH at the particle surface will approach that of the bulk of solution. Thus one way of obtaining the same pH value in the supernatant liquid as for the soil paste is to add an excess of salt, usually potassium chloride, to destroy the electric double layer.

Although steps can be taken to standardize the variables of water content and salt concentration, the measurement of soil pH remains somewhat arbitrary. Different analysts have adopted different soil–water ratios from 1:1 to 1:10 but the most convenient ratio is that obtained in the saturated soil paste. This takes into account the textural differences between soils; for example, a peaty soil needs a greater amount of water added in order to measure its pH than does a sandy soil. Moreover, the saturated soil paste must often be prepared in any case for other measurements, either on the paste itself or on its extract.

According to the purpose of measurement it may be useful to obtain the pH value of a soil in potassium chloride solution as well as in water. As explained, this standardizes the salt effect and it may also help to minimize errors due to the liquid junction potential if electrometric methods are being used (see section 3:2:2). It is possible to reduce the salt effect by leaching the soil before measuring pH but, apart from the tediousness of the procedure, the result would be artificial and would not correspond to field conditions. For reasons explained later (section 3:3:2) it is considered preferable to use dilute calcium chloride solution rather than potassium chloride to standardize the salt effect in some soils.

Very occasionally it is desirable to eliminate the effect of carbon dioxide absorbed in a soil when measuring the pH. With soils of pH less than 6 the carbon dioxide concentration is of little consequence, but neutral, calcareous soils are very sensitive to changes in carbon dioxide content. As pointed out by Reed and Cummings (1945) the necessity for considering the carbon dioxide effect depends upon the purpose of analysis. In plant nutrition studies the effect of carbon dioxide upon pH values is important and should be taken into account, whereas for soil classification it is preferable to standardize the carbon dioxide concentration for reproducibility. The fluctuation of pH values of soils in the field is due partly to changes in carbon dioxide concentration, particularly in poorly buffered soils.

Whitney and Gardner (1943) found that pH is a straight line function of the logarithm of carbon dioxide pressure at constant moisture and that at constant carbon dioxide pressure, pH values decrease with dilution of suspension. These workers studied the effect of carbon dioxide upon soil pH measurement by aspirating the soil suspension with a gas of constant carbon dioxide content. Known amounts of carbon dioxide were passed into the tank of an air compressor which was then filled with air to the desired pressure. Gas mixtures were analyzed before use and the carbon dioxide pressure calculated; the method of doing this was that of Johnston and Walker (1925). Soil suspensions were equilibrated at 25°C by bubbling the gas mixture through water and then through the suspension. At high carbon dioxide pressures equilibrium was obtained within 2–3 days and with low carbon dioxide pressures, 5–7 days were required.

3:2:2 Electrometric measurements of soil pH

Methods

pH values of soils have, in the past, been measured using the ultimate standard of the hydrogen electrode (e.g. Crowther, 1925) but this is an inconvenient method apart from being sensitive to oxidizing and reducing substances, catalytic poisons and a whole variety of ions.

The quinhydrone electrode was introduced for soil analysis by Biilman (1924). A cell was prepared containing a Veibel buffer (KCl/HCl) at pH 2·03 to which some quinhydrone was added and in which a platinum electrode dipped. The soil suspension mixed with quinhydrone carried a second platinum electrode and an agar–potassium chloride salt bridge connected the two systems. The platinum electrodes were connected to a potentiometer and the pH of the soil suspension calculated from the expression:

$$pH = 2{\cdot}03 + \frac{\text{observed e.m.f. in volts}}{RT} \qquad [3{:}2$$

the function RT being corrected for the temperature of the suspension from tables.

The quinhydrone electrode is not suitable for measuring the pH of soils containing manganese oxides as the oxides become reduced and the pH drifts. Furthermore, at pH values greater than 8·5 quinhydrone dissociates to hydroquinone which is acidic.

Nowadays the glass electrode in conjunction with a standard half-cell is employed almost universally.

The glass electrode has many advantages over others. It is unaffected by oxidizing and reducing substances, it can be used in the presence of protein or in coloured or opaque solutions, it does not liberate dissolved gases from the system and is convenient and inexpensive. Glass electrodes are made in a variety of shapes that permit their use in very small quantities of material or in soils of stiff texture which would break the conventional electrode.

The ordinary glass electrode is capable of measuring pH values from 0 to 9 but, owing to possible transfer of ions other than hydrogen, particularly of sodium, in order to measure more alkaline suspensions a special glass electrode must be used in which the sodium of the glass has been replaced by lithium. Such electrodes are commercially available.

Before use, glass electrodes are soaked in dilute acid and thereafter should not be allowed to dry out. When not in use they are kept immersed in distilled water. The presence of water is essential to maintain the pH function of a glass electrode, this function being impaired even if the bulb is immersed in a dehydrating fluid such as ethanol.

For a detailed discussion of the theory of the function of glass electrodes the reader is referred to standard physical chemistry textbooks. Briefly, the electrode consists of an extremely thin-walled glass bulb filled with dilute hydrochloric acid into which dips a silver wire coated with silver chloride and connected to the electrode lead. The thin glass wall of the bulb permits transfer of hydrogen ions and thus the electrode is sensitive to changes in hydrogen ion concentration. It is used in conjunction with a calomel or other standard cell.

On every occasion, before use the pH meter must be calibrated using buffer solutions. This is because an asymmetric potential develops across the glass of the electrode even when it is immersed in a solution with a hydrogen ion concentration identical to that inside the bulb, due to a difference in strain between the inside and outside of the bulb. Moreover, this asymmetric potential is not constant and varies from time to time as well as being influenced by previous use of the electrode in strong acid or alkali.

pH METERS

A pH meter must be an electronic instrument if a glass electrode, which has a very high resistance, is to be used. There are two main kinds of meter, the potentiometric and the direct reading. The direct reading type of instrument, although possibly less accurate than the potentiometric, is used almost exclusively in modern soil laboratories. The e.m.f. of the glass electrode–calomel electrode cell is applied across a resistance and the resulting current, after amplification, is passed through an ammeter causing deflection of a pointer across a scale marked in pH units.

These instruments are available to operate either on mains a.c. current or by means of batteries, the latter type being of particular use for field measurements. In most pH meters some form of automatic temperature control is provided and usually a device can be incorporated for the continuous recording of pH.

Operational details differ for each commercial instrument but such details are always supplied by the manufacturer and should be rigorously followed. However, calibration with buffer solutions is a procedure that must be carried out whatever make of instrument is used.

STANDARD BUFFER SOLUTIONS

At least two buffer solutions must be used, one at low pH and one at high pH. The buffer solution most commonly used for low pH values is a 0·05 M solution of potassium hydrogen phthalate (10·21 g dm^{-3}) which has a pH value of 4·001 at 20°C. The pH of this buffer varies with temperature from 4·000 at 15°C to 4·020 at 35°C. The reagent must be of the highest purity and the water used for its solution should be de-ionized or freshly distilled and free from any carbon dioxide. The buffer can be stored in well-sealed Pyrex or polythene bottles but preferably should be freshly prepared every 2 weeks.

For a high pH buffer solution a 0·01 M solution of borax (3·81 g dm^{-3} $Na_2B_4O_7.10H_2O$) is convenient and at 20°C has a pH of 9·22. The same precautions should be taken for its preparation as for the phthalate buffer and provided that it is protected from atmospheric carbon dioxide it can be kept for about a month.

Standard, certified buffer tablets are available for a whole range of pH values and these are most convenient. One tablet dissolved in a specified amount of water, usually 50 or 100 cm^3, provides a solution of known pH to within about 0·02 of a unit.

Sources of error

LIQUID JUNCTION POTENTIAL

The liquid junction potential is the most important source of error when using the glass electrode–calomel electrode system.

When two solutions of different strength or composition come into contact the more concentrated solution will diffuse into the more dilute. If the ions of the diffusing solution move at different speeds the dilute solution will assume an electric charge with respect to the concentrated solution corresponding to that of the faster moving ion. For example, if the diffusing anions move more quickly than the cations they will cause the dilute solution to become negative with respect to the concentrated solution. The resulting difference in potential across the interface of the solutions is called the 'liquid junction potential' (E_j) and either adds to or subtracts from the electric potential.

Such a potential is likely to arise at the liquid junction between a soil suspension and the salt bridge of the calomel electrode. The presence of colloids or suspensions has a marked effect on liquid junction potentials and hence this error may be more important in soil pH measurement than when using pure solutions.

Attempts have been made to allow for liquid junction potentials by calculation. The formulae, however, involve knowledge of activity co-efficients and even for true solutions have proved of little use and would be quite impossible to derive for soil suspensions.

One procedure to minimize the liquid junction potential is to use saturated potassium chloride solution as the salt bridge. It is the relative mobilities of the oppositely charged ions at the interface that decide the potential gradient and thus it is desirable to equate these mobilities as far as possible. Potassium chloride is used as potassium ions and chloride ions have about the same mobility, and if the concentration of the salt is much greater than that of other electrolytes present it will be responsible for transferring almost all the current across the liquid junction.

Peech *et al.* (1953), in a study of liquid junction potentials, concluded that for calcium-clays and soils the effect of short circuiting the small potential drop across the diffuse double layer is offset by a diffusion potential due to calcium chloride formed just ahead of the potassium chloride front and which decreases the ion activity of the suspension. This view was strongly criticized by Marshall (1953) who considered that an alkaline suspension effect was involved.

In pH measurements the salt bridge is incorporated in the calomel cell and saturated potassium chloride solution forms an impeded-flow junction via a porous plug or ground-glass sleeve. As the bridge solution is denser than the test solution the latter tends to travel up the tube by convection and causes a drifting potential. This is prevented by having a head of saturated potassium chloride solution giving a small downward flow.

As hydrogen and hydroxyl ions are very mobile, if either is introduced in large amount, the liquid junction potential will increase; consequently if the pH of the solution under test is very different from that of the buffer, the junction potential increases and a standardization error is introduced.

SOLUBLE SALTS

The presence of neutral salts affects pH readings by influencing ionic activities and gives rise to what are known as 'activity errors'. These errors can be significant if a small amount of dilute solution is being measured when potassium chloride from the salt bridge causes changes in activity. The same effect will be obtained when dealing with saline soils.

The salt effect is overcome, or rather standardized, by taking measurements in potassium chloride solution rather than in water. Usually pH values in potassium chloride solution are lower than in water; but for certain kinds of soil, for example strongly weathered oxic horizons, the converse is true and the effect is one of the diagnostics for these soils (see *Soil Classification*, 7th approximation, 1960, p. 53).

ELECTRODES

If the calomel electrode is open to the air it absorbs oxygen which affects its e.m.f. The resultant error is minimized by using a saturated potassium chloride solution–solid potassium chloride system in the electrode and it is important to check repeatedly that this system is maintained.

The electrodes should not be removed from solution when the instrument is set for pH measurement as they are likely to become polarized. The instrument should always be switched to the check or rest position before removing and washing the electrodes.

Limits of accuracy

The accuracy of pH measurements using commercially available glass electrodes and electrometers rarely exceeds 0·02 pH unit. Although equipment capable of greater accuracy is obtainable, it is not usually needed for soil analysis.

3:2:3 Colorimetric measurement of soil pH

Methods

Colorimetric measurement of pH is of greatest value for field testing where it is more convenient and more rapid than electrometric methods. It is not recommended as a laboratory procedure; however, experimental details are given (section 3:4:3) for a student exercise or in case it should be impossible to use a pH meter.

The method depends upon a range of organic chemicals that have different colours at different pH values. The colour changes are due to a reaction between the dyestuff and the ions in solution, the actual colour being determined by hydrogen ion activity.

In the Brönsted concept the colour of an indicator is determined by the ratio of concentrations of acidic and basic forms, that is by In_A/In_B. By using activity coefficients it can be deduced that:

$$pH = pK\,In + \lg \frac{[In_B]}{[In_A]} \qquad\qquad [3:3$$

where $pK\,In$ is the apparent indicator constant, $[In_A]$ is the concentration of the acidic form and $[In_B]$ the concentration of the basic form.

A characteristic of pH indicators is that the change from acid colour to alkaline colour is not sharp but takes place over a pH range known as the 'colour change interval'. The position of this interval in the pH range varies for different indicators.

Choice of a suitable indicator can be difficult. It must give well-defined colour changes over a small pH range and be relatively unaffected by all substances other than hydrogen and hydroxyl ions. The colours should be stable and quickly formed.

The methods of using indicators to measure the pH of soils are very arbitrary and have been developed more from experience than from theoretical considerations.

It is of little use to add an indicator solution by itself to a soil as the colour is nearly always absorbed and masked by the colour of the soil. It is equally useless to make a water extract of a soil and to measure its pH as the extract is not in equilibrium with the soil. Possibly measurements made upon the extracted soil solution would be feasible but this is too tedious as a routine.

The procedures differ for the field and the laboratory and are compromises of the above-mentioned methods. In the laboratory the soil is shaken with water, the indicator added and the suspension allowed to settle, when the colour of the supernatant liquid is matched with standards. The whole procedure is fraught with theoretical objections but in practice gives reasonable results. Steps must be taken to prevent turbidity and this is best done by adding barium sulphate as a flocculant and centrifuging. The extracts are often coloured by organic matter which involves viewing the standards through similar extracts.

Soil extracts will be only slightly buffered and thus susceptible to pH changes. Consequently the amount of indicator added has to be kept to a minimum or else the resulting colour will represent merely the equilibrium pH of the two solutions. It is possible to overcome to some extent the errors due to poor buffering by using the isohydric indicator method in which three portions of the same indicator are adjusted to three different pH values. These indicators are then used to determine the pH of the soil extract and the reading which coincides most nearly to the pH of one of the indicators is taken as the true value. This method will find little application in soil analysis, however.

It is more practical to keep the amount of added indicator in small proportion. This can be assisted by viewing a deep layer of solution in a Nessler glass; standard Nessler tubes can be prepared from buffer mixtures or standard pH discs of coloured glass can be used in a comparator. The amount of indicator solution added to 50 cm^3 of test solution should be about 0·2 cm^3.

Temperature can affect the colorimetric readings in as much as it affects the activity of the ions in solution as well as affecting the colours of some indicators. The nature and concentration of soluble salts also affect colorimetric pH readings by changing the activity coefficient. Ideal agreement between test solution and standard will occur only if the salt concentrations are the same.

Sources of error

The principal sources of error when determining the pH of soils colorimetrically are:

i. Comparison of unequal depths of liquid.

ii. Measuring at the extreme range of the indicator. A pH value which corresponds to one or other of the indicator limits is untrustworthy and another indicator should be chosen so as to obtain a value in the middle of its scale.

iii. Use of an inappropriate indicator, which is sometimes difficult to avoid. Initial use of a wide-range 'universal' indicator will assist in choosing the most suitable reagent.

iv. Poor buffering capacity of soil extracts. This has been discussed and it is recommended that the proportion of indicator be kept small in relation to the test solution.

v. Temperature and salt effects. The errors due to temperature are not likely to be of importance in soil analysis but the errors due to soluble salts cannot easily be avoided, except by prior removal of salts which gives an artificial result. For comparative purposes the salt effect can be standardized by adding a known excess of potassium chloride.

vi. Coloured solutions and turbidity. It is essential to attempt some control over these conditions; otherwise the results will be meaningless.

Limits of accuracy

As far as the measurement of solutions is concerned, a universal indicator can be made to give results accurate to within about 0·5 of a pH unit. Using the most appropriate indicator and taking all precautions against error, comparison by eye of a test solution with a series of buffers cannot be relied upon to give values more accurate than to within 0·2 of a pH unit. The makers of standard pH colour discs claim a similar degree of accuracy. Greater accuracy of comparison between test solutions and buffers can be obtained photoelectrically but such accuracy would be out of proportion to that with which the colour of the test solution reflects the pH of the soil.

Field colorimetric measurement of soil pH

Numerous pH colorimetric test kits are available commercially, nearly all of which involve mixing indicator solutions with a small quantity of soil on a white tile. Sometimes the excess indicator solution is separated from the soil for colour comparison with standard colours printed on a card, as in the LaMotte–Hester procedure for example, and sometimes a neutral, white powder (such as barium sulphate) is dusted on to the mixture in order to reveal the colour (Hellige–Truog method). The Hellige–Truog method will be found convenient, rapid and sufficiently accurate for field pH measurement and an accuracy of 0·1 to 0·2 of a pH unit is claimed if the method is very carefully followed. As it stands the Hellige–Truog method is not too satisfactory for the determination of pH values above 8 and in such cases the LaMotte procedure which uses several different indicators should be followed.

Often the pH of waterlogged soils cannot be determined colorimetrically as reducing conditions affect the indicators.

3:3 LIME POTENTIALS

As discussed in section 3:1, pH values of soil suspensions cannot be precisely defined; they are affected by soil–water ratios and salt concentration and are difficult to interpret.

If, instead of single ion activity which varies in a soil suspension, we consider ionic activity products or ratios, these are found to be relatively constant and characteristic for a soil.

Consider a cation C with valency v. In dilute solution in equilibrium with a soil the activity of the hydrogen ions divided by the activity of the C ions will be constant (Schofield and Taylor, 1955). That is,

$$\frac{a_{H^+}}{(a_C)^{1/v}} = k \qquad\qquad [3\!:\!4$$

The activity function depends upon the valency of the ion concerned. Thus for a soil in equilibrium with calcium chloride solution,

$$\frac{a_{H^+}}{(a_{Ca^{2+}})^{\frac{1}{2}}} = \frac{a_{H^+}}{\sqrt{a_{Ca^{2+}}}} = k \qquad\qquad [3\!:\!5$$

Schofield and Taylor suggested using the negative logarithm of the function, that is:

$$-\lg \frac{a_{H^+}}{\sqrt{a_{Ca^{2+}}}} \quad \text{or} \quad pH - \tfrac{1}{2}pCa = k' \qquad\qquad [3\!:\!6$$

and they called this the 'lime potential'.

The lime potential of a soil is independent of the calcium–magnesium ratio, is less affected by salt concentration and moisture than is pH and is more characteristic of that soil.

In certain cases the lime potential can be used to estimate the degree of base saturation. Turner and Nichol (1962) found that in clays saturated with aluminium and titrated with calcium hydroxide, the lime potential was directly related to the percent base saturation regardless of the type of clay. Similar results were obtained when the clays were saturated with iron or with other trivalent cations, providing that the respective hydroxides were insoluble. Turner and Nichol further found that the magnitude of the lime potential at any given degree of percent base saturation depended upon the solubility of the hydroxide of the metal used, and that in some cases the potential can be used to determine whether the clay has any appreciable fraction of its charge neutralized by exchangeable iron or aluminium. Clark and Hill (1964), however, demonstrated that the lime potential is poorly correlated with the degree of base saturation in acid soils. Turner and Clark (1967) then derived what they called a 'corrected lime potential' or C.L.P., which is based upon the exchange equilibrium,

$$2AlX_3 + 3Ca(OH)_2 = 2Al(OH)_3 + 3CaX_2 \qquad\qquad [3\!:\!7$$

The C.L.P. in contrast to the lime potential is governed by a ratio of activities and can differ from the lime potential by absolute values greater than $+0\!\cdot\!4$ and $-1\!\cdot\!0$.

Another possible use of the lime potential is to determine the carbon dioxide concentration in the soil air of calcareous soils (Ulrich, 1961).

By using a calcium chloride solution of known strength, the activity of the calcium ions can be calculated and by experiment it was found that a $0\!\cdot\!01$ M solution was convenient; the concentration of the calcium ions in this solution is relatively unaffected by soils. At a concentration of

0·01 M, the activity of calcium ions is 1·14 and hence the lime potential of a soil measured in 0·01 M calcium chloride solution will be given by pH − 1·14. In practice it is more convenient to consider calcium and magnesium as a single ionic species and to define lime potential as:

$$pH - \tfrac{1}{2}p[Ca + Mg].$$

3:4 RECOMMENDED METHODS: pH

3:4:1 Electrometric measurement of soil pH in the laboratory

Preparation

Soil

Spread the 2-mm fine earth on a sheet of polythene and snatch-sample to obtain a representative sample of the required size. Prepare the saturated soil paste as described in Chapter 6 (section 6:3:1I) and allow to stand for 1 hour. For measurement of pH in potassium chloride solution, snatch-sample 10 g of soil and mix in a small beaker with 25 cm³ of 1 M potassium chloride solution (75 g dm⁻³ KCl); allow the mixture to stand for 1 hour before measuring.

Instruments

Switch the pH meter on and follow the instructions provided by the manufacturer; meters need several minutes to half an hour to warm up.

Wash the glass electrode, which should have been prepared according to the directions provided with it, in a stream of distilled water and remove excess water by gently touching the bulb with clean filter paper. Check the calomel electrode to make sure that it is full of saturated potassium chloride solution, that it contains solid potassium chloride and that the porous plug is not blocked.

Set the temperature control according to the soil suspension temperature or switch into circuit the automatic temperature control and dip the electrodes into the low pH buffer solution. Adjust the pointer or dial to show the theoretical pH value. Switch to the rest position and wash the electrodes. Repeat this process using a high pH buffer and then re-check with the low buffer.

Measurement

Procedure

Insert the electrodes into the soil paste* or suspension; the bulb-type glass electrode will usually enter the paste easily but if much resistance is offered use a spear-type electrode. Move the electrodes in the paste to ensure good contact and read the pH. Again move the electrodes and repeat the

---

* Note that if the electrical resistance of the saturated soil paste is to be measured also, this should be done before measuring pH as traces of potassium chloride from the calomel electrode will affect the resistance.

reading. When a constant reading has been obtained record it and wash the electrodes with water before measuring the next sample.

Most instruments have a 'check' position by means of which it is possible to check the calibration between each reading without having to use buffer solutions. When all the readings have been taken, wash the electrodes and leave the glass electrode dipping into distilled water and the calomel electrode dipping into saturated potassium chloride solution.

Drifting pH *readings*

Occasionally a soil will exhibit pH 'drift'; that is, the pH will slowly but continuously increase or decrease, and it is difficult to decide upon the true value. There is no hard and fast rule for dealing with this problem. Some workers recommend allowing the soil paste to stand for an arbitrary period of time, say, 15 minutes or 1 hour, with the electrodes in place and the instrument on, and to accept the reading obtained. Whatever is done, it is obvious that a single figure will have little significance and it is best to record that the pH is drifting and to give the limits over a certain period of time. The most important result of the measurement is that the pH does drift and in which direction.

Continuous recording of pH

For some experiments it may be useful to measure the pH of a soil over a period of time. One example could be an investigation of a soil exhibiting pH drift as just described. Certain sulphur-containing soils become extremely acid as they slowly dry over a period of time and it would be convenient to have a continuous record of pH changes.

Most pH meters can be equipped so as to make a continuous recording of pH values. Output terminals can be connected to a chart-recorder of almost any type, voltage or current operated. A most convenient instrument is a portable recording milliammeter with a moving coil scaled at 0–14 pH corresponding to 0–1·4 milliamps d.c.

No special technique is involved other than to follow the manufacturer's instructions for setting up the recorder; the pH meter is connected directly to the recorder by the appropriate terminals after removing the shorting strap. If the average chart is constantly moved at 1 inch per hour, approximately a month's continuous recording can be obtained.

Although the fundamental range of a pH meter is 0–14, by suitably adjusting the sensitivity the range of the recorder can be varied.

3:4:2 Electrometric measurement of soil pH in the field

Soil sampling

The soil should be sampled according to the purpose of analysis. The soil can be measured just as it is provided that it is sufficiently moist, or it can be made into a saturated paste using the field equipment for that purpose. Alternatively the soil can be air-dried and re-wetted as desired. It should be remembered that the mere sampling of a soil will alter such a factor as carbon dioxide concentration.

Measurement

Apparatus and technique

The equipment used is very similar, and may even be identical, to that used in the laboratory, except that batteries provide the power.

Buffer solutions should be prepared in advance and carried to the field. Most portable pH meters are built into a case suitably provided for carrying the various accessories. Transport of electrodes—and more than one should always be taken in case of breakage—is best done by fitting them through rubber bungs into test-tubes containing distilled water.

The makers' instructions are followed to set up and calibrate the instrument and soil pH values are measured exactly as described for the laboratory determination (section 3:4:1).

Compact, so-called 'pocket' pH meters are available but although convenient, are not always sufficiently reliable. They are powered with dry cells and are used with a single electrode unit in which the glass and reference electrode are combined. Some such single units include yet a third electrode (platinum) for redox potentials.

It is again emphasized that pH values of soils in the field are variable and depend upon moisture content, aeration and temperature and thus upon season and climate; they can also vary significantly over a relatively small area. Accurate measurement of soil pH in the field by electrometric methods is not justified except for specific purposes.

3:4:3 Colorimetric measurement of soil pH in the laboratory (after Kuhn, 1930)

Reagents

The following range of indicator solutions are prepared by grinding 0·1 g of reagent in an agate mortar with the indicated amount of 0·1 M sodium hydroxide solution and diluting to 250 cm^3.

Indicator	pH range of colour change	Colour change	cm^3 of 0·1 M NaOH used to prepare 250 cm^3 solution
Thymol blue*	1·2–2·8	red–yellow	2·15
Bromphenol blue	3·0–4·6	yellow–blue	1·49
Bromcresol green	3·8–5·4	yellow–blue	1·43
Bromcresol purple	5·2–6·8	yellow–purple	1·85
Bromthymol blue	6·0–7·6	yellow–blue	1·60
Phenol red	6·8–8·4	yellow–red	2·82
Cresol red	7·2–8·8	yellow–red	2·62
Thymol blue*	8·0–9·6	yellow–blue	2·15

* Thymol blue has two different colour changes for different pH ranges.

Also required is a 'universal' indicator which covers a wide range of pH; these are available commercially to cover from pH 3·0 to pH 11·0 but the analyst can prepare a universal indicator as follows:

Methyl yellow	60 mg
Methyl red	40 mg
Bromthymol blue	80 mg
Thymol blue	100 mg
Phenolphthalein	20 mg

Dissolve in 100 cm^3 of ethanol and add 0·1 M NaOH solution dropwise until yellow.

The approximate pH values are indicated by the following colours:

pH	Colour
1–2	red
3–4	reddish-orange
5	orange
6	yellow
7	yellow-green
8	green
9	blue-green
10	blue

Procedure

Find the approximate pH of each kind of soil to be measured using the universal indicator and a spot-tile. From these values choose the most appropriate indicator solution.

Place about 1 cm depth of pure barium sulphate powder in a series of 15-cm^3 conical centrifuge tubes. Add about a 3-cm depth of soil and a 10-cm depth of carbon dioxide-free water. Add 0·1 cm^3 of the chosen indicator solution and fill the tubes with water. Shake for 1 minute and centrifuge at 250 rev/s for 5 minutes. If the supernatant liquid is not clear, add a little more barium sulphate, shake the tube and re-centrifuge for a longer time. For each kind of soil prepare a blank with the indicator solution omitted. Compare the indicator colours in a comparator with standard pH discs, placing the blank behind the disc, in order to find the pH values.

3:4:4 Colorimetric measurement of soil pH in the field

The apparatus for field pH determination is that provided in a Hellige–Truog pH kit and consists merely of a white porcelain spot tile, a bottle of mixed indicator solution and a bottle of purified barium sulphate which is shaken out through a fine mesh screen.

Procedure

Place exactly 3 drops of indicator solution in one of the tile cavities and add just enough soil to absorb the liquid. Mix with a spatula and move the wet mass to one side of the cavity; on tilting the tile no free solution should drain from the soil. If the mixture is too wet, mix in a little more soil, or if too dry, add a little more solution. As soon as possible after mixing, sprinkle barium sulphate powder to cover the soil completely and prevent its colour from showing. After 1–2 minutes compare the colour produced with the standard colours printed in the instruction booklet. These standard colours vary from a purplish-red through blue and green to yellow and cover a pH range of from 8·5 to 4·0.

The test works best when air-dry soil is used; when wet soil is used the amount of indicator should be reduced, whereas for laterites or peaty soil a little more solution may be required. Periodically check the indicator solution for neutrality by viewing a thin film in the dropping tube and comparing it with the colour chart. When replacing the barium sulphate use only that which has been specially prepared for the purpose. Samples of so-called 'pure' barium sulphate are often slightly alkaline and this can lead to erroneous results.

For soils with pH values exceeding 8 it is best to use the LaMotte pH kit which is supplied with full instructions.

3:5 RECOMMENDED METHOD: LIME POTENTIAL

Reagent

Calcium chloride solution, 0·01 M: About 1·3 g $CaCl_2$ (anhyd.) is dissolved in water and the solution diluted to 1 dm^3. Standardize with EDTA solution (section 8:4:3) and dilute to exactly 0·01 M.

Procedure

Shake 10 g of 2-mm soil with 20 cm^3 of calcium chloride solution for 30 minutes and measure the pH of the suspension as described in section 3:4:1.

$$\text{Lime potential} = \text{Measured pH} - 1·14 \qquad [3:8$$

4

Titrateable acidity, exchangeable hydrogen, and lime requirement of soil

4:1 INTRODUCTION

The pH as measured in a soil is an expression of the hydrogen ion activity and does not take into account the activities of other ions such as aluminium, iron and manganese which can influence the acidity of a soil. Consequently, if a base-unsaturated soil is leached with a neutral salt solution the leachate will exhibit a greater degree of acidity than would be inferred from the soil pH measurement. For a more precise determination of soil acidity, buffer curves are prepared by titrating the soil suspension with alkali whilst simultaneously measuring change in pH.

In potentiometric titration of acid soils the influence of the cation of the added base is quite pronounced (Peech and Bradfield, 1948). Most hydrogen clays give potentiometric titration curves resembling those of weak monobasic organic acids, except that there is a more indefinite end-point. This is partly due to the fact that the pH value for any given degree of neutralization varies with the cations according to a lyotropic series, and partly to continued reaction between clay and added base. The titrateable hydrogen seems to occur in several different forms: electrovalent, hydrolytic and a form intermediate between electrovalent and covalent. The electrovalent and intermediate forms are exchangeable and the hydrolytic form is produced by hydrolysis of aluminium and iron salts.

4:2 TITRATEABLE ACIDITY

Titrateable acidity, in effect, is measured as exchangeable hydrogen by most of the methods described in section 4:3 unless the contribution of aluminium is specifically eliminated.

Conductometric titration of acid soils can be used to determine the proportion of titrateable hydrogen to titrateable aluminium (Harward and Coleman, 1954). After leaching the soil with acidified barium chloride solution, which displaces exchangeable hydrogen and aluminium ions, determination of the barium held by the soil is a measure of total exchangeable acidity (Russell, 1961).

4:3 EXCHANGEABLE HYDROGEN

The difference between cation exchange capacity and total exchangeable bases is sometimes taken as exchangeable hydrogen, but as extractable aluminium contributes largely to the acidity of a soil such an arbitrary method is unsatisfactory except as a broad approximation. Furthermore, as aluminium salts hydrolyze to give hydrogen ions, titration methods are unreliable as an accurate estimation of exchangeable hydrogen.

Conventional methods of analysis employ solutions of neutral salts to leach out exchangeable hydrogen, subsequent measurements being made on the leachate. The principle underlying this method is that if the solution is well buffered at pH 7 with respect to the soil then the pH of the soil extract will decrease linearly with exchangeable hydrogen.

The extractant most commonly used is neutral, 1 M ammonium acetate solution. This is well buffered at pH 7 and effective for the purpose; it is also convenient as it is widely used to extract exchangeable bases from soils and thus the same extract can be used to determine exchangeable hydrogen.

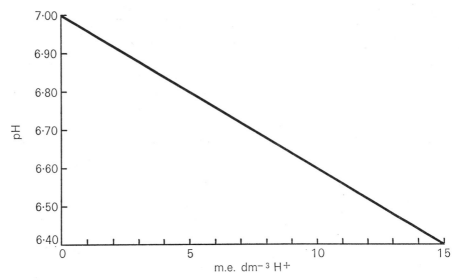

Figure 4:1 Variation in pH of ammonium acetate solution with exchangeable hydrogen (Brown; *Soil Science* © 1943)

Brown (1943) titrated 100 cm³ of neutral, 1 M ammonium acetate solution with 0·02 M acetic acid and plotted pH against milliequivalents of acid. He obtained the curve shown in Figure 4:1 and estimated exchangeable hydrogen in ammonium acetate suspensions of soils by measuring the pH.

In another procedure a standard graph is prepared in which milliequivalents of exchangeable hydrogen are plotted against millivolts as obtained by measuring a series of ammonium acetate solutions containing different amounts of acetic acid using a quinhydrone electrode (Schollenberger and Dreibelis, 1930). The technique is to add 0·0, 0·05, 0·10 and 0·15 cm³ of 1 M acetic acid to 10 cm³ aliquots of ammonium acetate solution; this gives standards corresponding to 0, 5, 10 and 15 milliequivalents of exchangeable hydrogen. Approximately 0·05 g of quinhydrone is added to each solution (most conveniently as an acetone solution) and after standing a few minutes the e.m.f. is read in millivolts. Similar measurements are made on ammonium acetate leachates from soils and permit direct estimation of exchangeable hydrogen.

Bradfield and Allison (1933) presented a method in which soil is brought into equilibrium with excess calcium carbonate at the partial pressure of carbon dioxide in the air; residual calcium carbonate is then determined.

Mehlich (1938; 1948) used barium chloride solution buffered at pH 8·1 with triethanolamine to extract exchangeable hydrogen and the procedure was slightly modified by Peech and co-workers (1962). Parker (1929) used barium acetate as extracting solution and Jackson (1958) modified this by adjusting the pH to 8·1 with barium hydroxide. The leachate is subsequently titrated with barium hydroxide solution using phenolphthalein as indicator. A blank titration is made on the extracting solution and milliequivalents exchangeable hydrogen calculated.

Woodruff (1948) developed a buffer solution consisting of calcium acetate, magnesium oxide and *p*-nitrophenol which had a p*K* value of 7·1 and was adjusted to pH 7. The change in pH of this solution due to leaching of exchangeable hydrogen gives a direct measure, over a certain range, of that exchangeable hydrogen.

Yuan (1959) pointed out that even when using salt solutions buffered at pH 7, the results can be influenced by hydrolysis of aluminium salts formed by extraction of aluminium. Consequently Yuan recommended the use of 1 M potassium chloride solution as extractant with the subsequent addition of sodium fluoride to the leachate to eliminate the effect of aluminium by formation of sodium fluo-aluminate,

$$AlCl_3 + 6NaF = Na_3AlF_6 + 3NaCl \qquad [4:1$$

The fluo-aluminate complex is neutral to phenolphthalein and thus the treated extract can be titrated with sodium hydroxide using that indicator to give a measure of exchangeable hydrogen. Yuan's procedure has been found, however, to be not generally applicable, possibly due to the fact that the extractant is a salt of a strong acid. The principle of suppressing aluminium with sodium fluoride can be applied to ammonium acetate extracts of soil with subsequent estimation of exchangeable hydrogen from pH measurement. For example, a certain soil from East Pakistan gave no exchange acidity when measured by potassium chloride extraction although the pH of the soil was 4·2. By ammonium acetate extraction and pH measurement an exchange acidity of 13·2 m.e./100 g was obtained. Addition of sodium fluoride made no difference to the pH of the soil extract, indicating that aluminium was probably not involved. To the same soil was added aluminium sulphate in amount equivalent to 5 m.e./100 g and the experiments were repeated. In this case the pH of the extract was 6·60 instead of 6·70 and was equivalent to an exchange acidity of 18·2 m.e./100 g, that is quantitatively including the added aluminium. After adding sodium fluoride the pH rose to 6·70 again showing complete suppression of the aluminium effect.

Middleton and Westgarth (1964) estimated exchangeable hydrogen in leached acid soils of the humid tropics by a method in which equilibrium determinations were made with solutions of barium nitrate and barium acetate. The results were then extrapolated to pH 7, the basis for the extrapolation being a linear relationship between pH and exchangeable hydrogen for the soils examined.

4:4 LIME REQUIREMENT OF SOILS

Addition of lime to a base unsaturated soil raises the pH and the lime potential; it increases the proportion of calcium on the exchange complex and can immobilize by precipitation certain elements such as iron, manganese and aluminium which may be present in a soil to excess.

The optimum degree of acidity of a soil will vary considerably for different crops on different soils, and it is considered that the most important contribution made to the fertility of a soil by liming is the increase in calcium supply (Albrecht and Smith, 1952).

The true meaning of the term 'lime requirement' reflects the amount of lime needed for maximum economic return from a particular crop on a particular soil. The term does not necessarily refer to the amount of lime needed to neutralize an acid soil, although most of the methods for determining lime requirement are based upon such a consequence.

Work in New Zealand has indicated that the most efficient chemical method of determining lime requirement is a complete cation exchange analysis (Metson, 1961). Having determined the cation exchange capacity and the percent base saturation of a soil, lime is added in such amount as to bring the percent base saturation to sixty.

Early methods of determining lime requirement involved titration of the soil with calcium hydroxide solution until pH 6–7 was reached. Puri and Anand (1934) employed an antimony electrode assembly to measure pH during titration. One electrode was placed in a buffer solution at pH 6·5 or 7·0 and the other was placed in the soil suspension. Titration was then carried out with calcium or barium hydroxide solution to zero deflection of the galvanometer. Dunn (1943) placed portions of the soil in a series of flasks and added different amounts of 0·02 M calcium hydroxide solution. After dilution and protection from microbiological action with chloroform, the mixtures were allowed to stand for 4 days before measurement of pH. A graph was prepared by plotting pH against lime added and was used to estimate lime requirement.

Woodruff (1948) recommended the determination of exchangeable hydrogen as being the best measure of lime requirement and he developed a buffer solution for this purpose (section 4:3). Using 5 g of 0·5-mm soil mixed with 5 cm^3 of water and 10 cm^3 of buffer solution, Woodruff measured the change in pH over 30 minutes. Calculations were based upon a reduction of 0·1 pH unit being equivalent to 1130 kg per 10^4 m^2 (1000 lb/acre) of limestone. McLean and co-workers (1960) found that Woodruff's method indicated only half the acidity as measured by incubation with calcium carbonate, except in partially limed soils when too high an acidity was indicated. A weaker buffer solution devised by Shoemaker *et al.* (1961), however, gave the correct acidity of soils as indicated by incubation values. Similar results were obtained by Keeney and Covey (1963) who found a high correlation between incubation results and those obtained using the Shoemaker buffer, whereas the Woodruff buffer was less satisfactory. As a means of comparison Keeney and Covey incubated a variety of soils with different amounts of dolomitic limestone for 12 months at field moisture capacity. Organic soils needed larger amounts of lime than did mineral soils. Final pH values were plotted against added lime and the lime requirement was taken as that necessary to give a pH of 6·5. Organic matter was found to be correlated with the lime requirement of soils.

Pratt and Bair (1962) incubated 400 g of soil with various amounts of calcium carbonate for 4 months. When no further increase in pH occurred, the lime needed to raise the pH to 6·5 was estimated from a plot of pH against added carbonate. Results by this incubation technique were compared with those obtained using Shoemaker's buffer solution and were found to agree fairly well, except that for sandy soils the proportion of soil to buffer had to be increased five times to obtain satisfactory results.

McLean and co-workers (1960) investigated the buffer solutions of Mehlich and of Woodruff as a means of determining lime requirement. Soils of different acidity and with different amounts of extractable aluminium were treated with increasing amounts of calcium carbonate and incubated for seventeen months. Although Woodruff's buffer was found unsatisfactory, that of Mehlich showed correct acidities provided that in partially limed soils extractable aluminium was taken into account. It appeared that the high equilibrium pH of these buffer solutions preserved the aluminium in exchangeable positions so that it could not react. The buffer solution of Shoemaker gave correct results regardless of aluminium content.

Adams and Evans (1962) found Woodruff's method unsatisfactory for soils with low exchange capacity as the buffer was insufficiently sensitive when small changes in exchange acidity affect pH. Adams and Evans were particularly concerned with determining the lime requirement of soils in which the need for lime was very small, yet definite; that is, too much lime

3 + T.S.C.A.

would cause damage. For a particular group of soils (red–yellow podsolic) a curve was plotted of soil pH measured in 1:1 water suspension against base unsaturation. It was shown that base unsaturation could be satisfactorily estimated from the change in pH when the soil was mixed with buffer solution rather than with water. The curve was then used to calculate the acidity to be neutralized and tables were composed showing lime requirement in lb/acre to adjust soil pH to desired values. Adams and Evans developed a new buffer solution for use with soils of low exchange capacity. This solution contained 20 g *p*-nitrophenol, 15 g boric acid, 74 g potassium chloride and 10·5 g potassium hydroxide in 2 dm³. The solution had a pH of 8·0 ± 0·1 and each 0·08 m.e. of acid resulted in a pH change of 0·10 units in 20 cm³.

Markert (1961) and Peters and Markert (1961) utilized the close correlation that exists between the buffering of soils and their methylene blue sorption. Laboratory application of lime to soils containing from 2% to 30% organic matter in amounts calculated from pH and methylene blue sorption values gave the desired final pH. The authors presented their results in tabular form in relation to initial pH value (measured in 0·1 M KCl) and to desired final pH values of from 4 to 6.

4:5 RECOMMENDED METHODS

4:5:1 Titrateable acidity by extraction with ammonium acetate solution (Brown)

Brown's equilibrium method determines total exchange acidity and is rapid and convenient when large numbers of soils are to be analyzed. The procedure is required when classifying soils by the 7th approximation.

Reagents

Ammonium acetate solution, neutral, 1·0 M: 288 cm³ glacial acetic acid and 275 cm³ conc. NH₄OH diluted to 5 dm³. Adjust to pH 7 with either 2 M acetic acid or 2 M NH₄OH. The final pH should be checked with a pH meter and must be 7·00 ± 0·05.

Acetic acid, 0·02 M: Dilute a standardized, approximately 0·1 M solution prepared from 6 cm³ dm⁻³ glacial acetic acid.

Procedure

i. Preparation of standard curve. Titrate potentiometrically 100 cm³ of the ammonium acetate solution with 0·02 M acetic acid with simultaneous measurement of pH. Any of the commercially available potentiometers can be used but the simple apparatus depicted in Figure 4:2 is adequate. The reference electrode is a calomel half-cell and the indicator electrode is a standard glass electrode; the electrodes are connected to a pH meter. Stir the solution magnetically whilst titrating and record the pH after

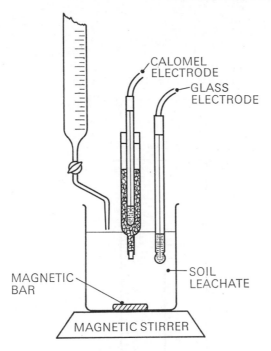

Figure 4:2 Apparatus for potentiometric titration of ammonium acetate leachates of soil

about 30 seconds for each addition of acid. Plot pH values as ordinate and m.e. dm^{-3} of acid as abscissa; Figure 4:1 shows a typical curve obtained in this manner.

ii. Measurement of titrateable acidity. Mix 2·5 g of soil with 25 cm^3 of ammonium acetate solution in a small beaker. After agitating, allow the mixture to stand for 1 hour with occasional stirring. Measure the pH of the suspension to within $\pm 0\cdot02$ of a unit and read the exchange acidity from the graph in m.e. dm^{-3}; the result is then converted to m.e./100 g soil.

The method can be applied to the ammonium acetate extract obtained for measuring exchangeable cations (section 7:3:2) provided that the ammonium acetate solution has a pH of 7·00 and provided that any difference in soil–solution ratio is taken into account.

4:5:2 Exchangeable hydrogen (Brown; Yuan)

This procedure applies Yuan's sodium fluoride treatment, for eliminating the effect of aluminium, to the method of Brown.

Reagents

Ammonium acetate solution, 1·0 M, pH 7·00: As in section 4:5:1.
Sodium fluoride solution: 4% w/v, aqueous.

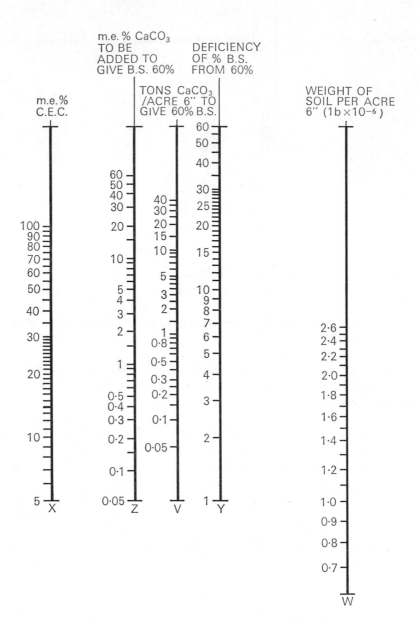

Figure 4:3 Nomogram for estimation of lime requirement of soil from cation exchange data (New Zealand DSIR). *Note:* The factors for converting to SI units are: 1 ton = 1016 kg; 1 lb × 10^6 = 0·45 × 10^6 kg; 1 acre 6-in = 4 × 10^3 m^2 15-cm

Procedure

Prepare an ammonium acetate extract of soil as described in section 7:3:2 or as in section 4:5:1, and filter off the soil. Take a suitable aliquot of extract (about 50 cm³) and add 10 cm³ of sodium fluoride solution. Stir for 30 seconds and measure the pH to within ± 0.02 of a unit. The result is converted to m.e. dm^{-3}, as in section 4:5:1, and expressed as m.e. H$^+$/100 g soil. If much soluble iron is present this must be determined, expressed in m.e. and deducted from the exchangeable hydrogen.

4:5:3 Lime requirement of soil

From cation exchange data (Metson)

This method depends upon an ammonium acetate extraction of soil to measure cation exchange capacity and per cent base saturation as described in section 7:3:2. The lime requirement is calculated as the calcium carbonate necessary to raise the base saturation to 60%; this is most conveniently done using the nomogram given by Metson (Figure 4:3 to which he appended the example:

C.E.C. = 40 m.e./100 g
% B.S. = 35
2.2×10^6 lb of soil to the acre 6-in
   (i.e. 2.4×10^6 kg per 10^4 m² per 15 cm).
Deficiency of the % B.S. from 60% = 25.

Align the number 40 of scale X and the number 25 of scale Y and read m.e. % CaCO$_3$ on scale Z (= 10). Align the number 10 on scale Z with 2·2 on scale W and read the lime requirement in tons/acre 6-in on scale V (= 5).

Buffer method (Shoemaker *et al.*)

Reagent

Buffer solution:
   p-Nitrophenol, 1·8 g
   Triethanolamine, 2·5 cm³
   Potassium chromate, 3·0 g
   Calcium acetate, 2·0 g
   Calcium chloride, 40·0 g (CaCl$_2$.2H$_2$O)

Dissolve in 800 cm³ of distilled water and adjust to pH 7·5 with dilute HCl or NaOH. Dilute to 1 dm³.

Procedure

Mix 10 g of soil (or 50 g if very sandy) with 20 cm³ of buffer solution and shake for 10 minutes. Read the pH of the suspension and estimate the lime

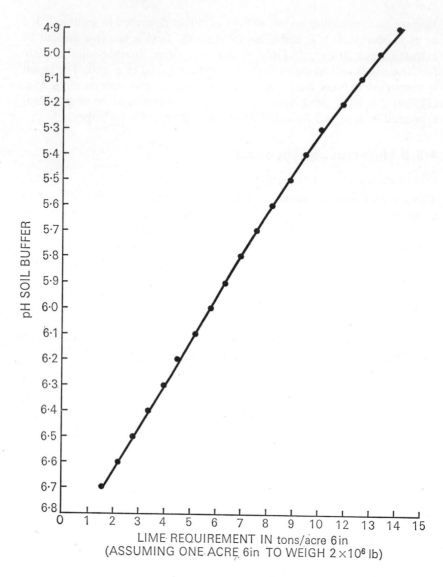

Figure 4:4 Lime requirement of soil by the method of Shoemaker *et al.*

requirement from the observed decrease in pH of the buffer using the curve shown in Figure 4:4 which has been prepared from the data of Shoemaker. If 50 g of soil were taken the result should be divided by 5.

5

Insoluble carbonates and sulphur requirement of soil

5:1 INTRODUCTION

Calcium carbonate occurs in the form of calcite in some crystalline rocks but is more generally deposited as a secondary mineral from water. Lime-containing complex silicates are weathered by carbon dioxide-charged water to liberate soluble hydrogen carbonates which eventually re-precipitate as carbonates. Calcareous soils develop particularly in arid regions where carbonate is precipitated at various depths depending upon the effective depth of leaching. Chernozems, for example, are characterized by an accumulation of calcium carbonate at a depth of about 2 metres and which, under slightly more arid conditions, may be concretionary. In more arid zones Chestnut soils occur with an accumulation of calcium carbonate nearer to the surface and, in general, the more arid the climate the nearer to the surface are the calcium carbonate accumulations. Increasing temperature has the effect of thickening the carbonate layer which sometimes appears as a pan upon the soil surface. Podsolization of rendzinas is associated with deep distribution of carbonate. Under very arid conditions gypsum can accumulate beneath the carbonate.

Magnesium carbonate occurs as magnesite and dolomite which is a compound carbonate of magnesium and calcium.

In calcareous soils the carbonate occurs mostly in the silt fraction thus affecting the soil texture. Deb (1963) found that the clay and silt fractions of soils near Aswan contained more than 75% of the calcium carbonate present. Addition of calcium carbonate to clay soils usually improves the tilth, although some conflicting results have been obtained.

Chemically, the alkaline-earth carbonates yield soluble calcium and magnesium which are able to replace exchangeable sodium on the soil colloids. The volatilization of ammonia from soils has been correlated more closely with the calcium carbonate content than with pH, and even in soils with low calcium carbonate content, loss of ammonia is appreciable (Lehr and Wesemael, 1961).

Calcium and magnesium carbonates are not usually found in soils if the pH is less than 7 but if the pH exceeds this value carbonates should always be tested for. If no sodium carbonate is present the pH of a calcareous soil will seldom exceed 8·5 although, as discussed by Russell (1961), higher pH values can occur under certain conditions.

An excess of calcium carbonate can introduce too much calcium on to the soil complex and can lead also to deficiency of minor elements. The analysis of calcareous soils for exchangeable cations presents a special problem and is discussed in Chapter 7. In humid climates particularly, calcium and magnesium carbonates are lost as hydrogen carbonates through leaching action; the rate of loss depends largely upon the particle size of the carbonates.

Insoluble carbonates in a soil are usually reported as calcium carbonate but if appreciable amounts of magnesium carbonate are present the analytical procedures must allow for this even if the final result is still reported as calcium carbonate. This is because magnesium carbonate reacts more slowly with acid than does calcium carbonate and time of reaction and acid strength should be increased. Fine grinding of the sample also facilitates reactions. Sometimes it is necessary to determine magnesium carbonate as such and methods of doing this are described.

5:2 INSOLUBLE CARBONATES

5:2:1 Total insoluble carbonates

Methods of determining calcium and magnesium carbonates in soil can be divided into classes according to the accuracy required. At the bottom of the accuracy scale is the qualitative test whereby dilute acid is added to a moistened soil and carbonate down to 0·5% is detected by effervescence.

A rapid, routine method of limited accuracy is to decompose the carbonates with excess dilute acid and to titrate the unused acid with standard alkali (Piper, 1950); the result is referred to as the 'calcium carbonate equivalent'. Greater accuracy is obtained by absorbing liberated carbon dioxide in alkali which is then titrated with standard acid (Hutchinson and MacLennan, 1914; Tinsley, Taylor and Moore, 1951). Schollenberger (1945) liberated carbon dioxide *in vacuo* and absorbed it in barium hydroxide, but his technique is rather involved to be adopted for routine analysis.

Gravimetric analysis of carbonates can be made by absorbing evolved carbon dioxide in soda-lime and measuring the increase in weight (Vogel, 1962) or by determining the loss of weight of a sample when the carbonate is decomposed. For the latter, indirect gravimetric method, a weighed quantity of soil is added in small portions to dilute acid in a small flask, the combined weight of which is known. After decomposition is complete, carbon dioxide in the flask is swept out with a stream of air and the new weight found. The weight of carbon dioxide lost is calculated and expressed as per cent calcium carbonate; the accuracy of the method depends largely upon the balance used. In the direct gravimetric analysis water vapour has to be absorbed separately from carbon dioxide and this is best done with magnesium perchlorate ('anhydrone'). Vogel recommends the use of syrupy phosphoric acid to decompose the carbonates as it is relatively non-volatile and permits greater control of reaction. The method, although it can be very accurate, is not really suitable for routine soil analysis, but

Erikson and Gieseking (1947) have described a compact form of the apparatus (Figure 5:6) for so determining soil carbonates. These authors recommend trichloracetic acid for decomposing the carbonates.

The most convenient accurate methods of carbonate analysis are manometric when the volume of evolved carbon dioxide is measured. Horton and Newson (1953) have reduced this method to a simple form in which the carbon dioxide is liberated in a vessel connected to a water reservoir when it displaces water into a measuring cylinder (Figure 5:1).

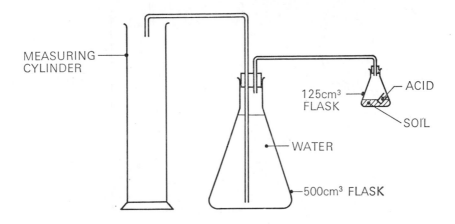

Figure 5:1 Apparatus for the determination of calcium carbonate equivalent (Horton, 1953; Soil Science Society of America, *Proceedings*)

Williams (1948) devised a procedure in which carbon dioxide is liberated by dilute acid from soil contained in a 250-g screw-capped bottle. A manometer is then connected by plunging an attached hypodermic needle through a rubber gasket which seals the bottle. The apparatus (Figure 5:2) is calibrated with pure calcium carbonate and future manometer readings are converted to calcium carbonate percentages from a graph. The method would be suitable for large numbers of samples provided that a series of bottles could be obtained of exactly the same volume; alternatively each bottle of a series could be individually calibrated. The calcimeter (Figure 5:3) designed by Collins (1906) gives accurate results within a reasonable time once the apparatus has been set up, although it is a tedious method when large numbers of samples are involved. Bascomb (1961) devised an apparatus (Figure 5:7) permitting the determination of up to 1 gramme of calcium carbonate instead of less than 0·1 g as in Collins's method. Thus ten times the weight of soil as used in Collins's method can be taken and in most cases this eliminates the need for grinding and fine sieving of the sample. Some soils, particularly if manganese is present, can give carbon dioxide by oxidation of organic matter during reaction. This source of error is avoided by adding dilute (3% w/v) iron(II) chloride solution to

3*

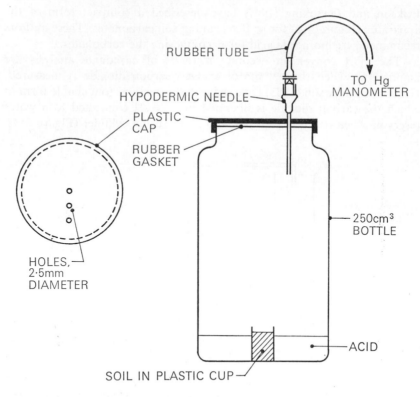

Figure 5:2 Apparatus for manometric determination of carbonate in soil (Williams, 1948; Soil Science Society of America, *Proceedings*)

the acid and by heating under reduced pressure to avoid high temperatures (Schollenberger, 1930).

Sokolovich (1964) treated from 4 to 10 g of soil depending upon carbonate content, with 80 cm³ of water and 5 g of sodium fluoride. The mixture was shaken and boiled for 5 minutes then cooled to 25°C in the presence of phenolphthalein. The suspension was finally titrated with normal hydrochloric acid; 1 cm³ of this acid is equivalent to 100 mg of calcium carbonate. Sokolovich claimed that the method had an accuracy greater than that of the calcimeter.

5:2:2 Magnesium carbonate

The separate determination of magnesium carbonate has presented analytical problems. Bruin (1938) suggested that use could be made of the fact that whereas both calcium and magnesium carbonates dissolve in hydrochloric acid, only calcium carbonate dissolves in oxalic acid. Weissman and Diehl (1953) used EDTA to separate the two carbonates but this procedure has been found unsatisfactory. X-ray diffraction is an efficient method but tedious (Tenant and Berger, 1957).

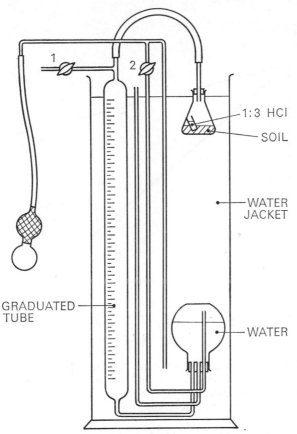

Figure 5:3 Collins's calcimeter for determining carbonate in soil.

With both taps open the closed reaction vessel is immersed in the water jar. Tap 2 is closed and the water stirred by pumping air. Tap 2 is then opened and tap 1 closed and the bulb squeezed to level the burette at zero. On releasing the bulb the level sinks to the bottom in the central tube. The reaction flask is removed, tilted to spill the acid, shaken and returned to the water jacket. The level is adjusted with the bulb, tap 2 being closed, and the gas volume measured. Shaking and levelling is repeated until constant readings are obtained. Knowing the volume of the reaction flask, the volume of acid, the weight of soil, the temperature of the water and the barometric pressure, the weight of gas in mg can be calculated and converted to per cent calcium carbonate. A special slide rule is available to facilitate the calculations.

The most promising method appears to be that of Skinner and co-workers (1959) and is based upon the difference in rates of solution of the two carbonates. Carbon dioxide is liberated from the soil with dilute acid and is manometrically measured at frequent intervals of time until reaction ceases. The logarithm of the carbon dioxide equivalent to unreacted carbonate is plotted against time. Calcium carbonate reacts fairly quickly and then the slope of the curve becomes smaller and temporarily linear during magnesium carbonate decomposition. This linear portion of the curve is extrapolated to zero time to obtain the carbon dioxide equiva-lent to the total magnesium carbonate present, and the carbon dioxide corresponding to the calcium carbonate is obtained by difference from the

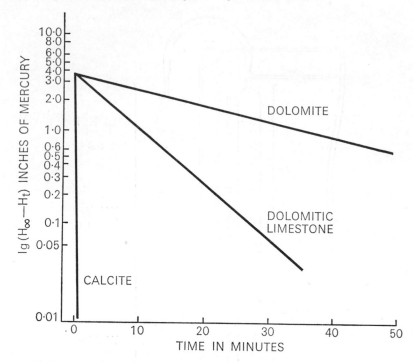

Figure 5:4 Relative rates of solution in acid of calcite, dolomite and dolomitic limestone (Skinner *et al.*, 1959; *Can. J. Soil Sci.*). *Note:* The SI unit of pressure is the pascal ($N\,m^{-2}$) and the conversion factor here is that 1 in of Hg corresponds to $3 \cdot 32\ kN\ m^{-2}$

total carbon dioxide. In Figure 5:4 is shown the results obtained by Skinner *et al.* using dolomite, calcite and dolomitic limestone. The curves reveal the marked difference in rates of solution. Other carbonates naturally would interfere but in soils this factor can usually be ignored. The degree of fineness of the sample is important and must be standardized. Skinner's results agreed well with those obtained by X-ray measurements and the accuracy is such as to give a standard deviation of $\pm 0 \cdot 005$ g carbon dioxide.

In a critical examination of the method Turner (1960) pointed out that the value $\lg (H_\infty - H_t)$ (see section 5:4:1) used in the plot against time is proportional to the volume, V, of dolomite at any time, t, and can be replaced by this volume on the graph. Turner queried whether the plot of $\lg V$ against t really is linear as required by the method. That it is experimentally found to be linear may be, it was considered, because departure from linearity is very small or because the experimental error in measuring V may be sufficiently large to mask the non-linearity.

A mathematical analysis showed that the plot may be concave either upwards or downwards depending upon the size distribution of crystals. The deviations from linearity during the time needed to dissolve 50% of the smallest crystal are within limits.

The question was followed up by Turner and Skinner (1960) who concluded that for accuracy the method requires the calcite in a sample to be quite dissolved in sixty seconds. This will be so if the effective diameter of the calcite crystals is not greater than 5.6×10^{-2} cm and so it is essential for the sample to pass a 0.5-mm sieve. If the smallest dolomite crystal has an effective diameter greater than 1.7×10^{-3} cm the method is fundamentally sound, but if any smaller crystals are present the method is uncertain.

Petersen *et al.* (1966) suspected that erroneous results would be obtained by Skinner's procedure as the rate of carbon dioxide release is dependent upon particle size and exposed surface area. Petersen thus developed a procedure based upon the difference in solubilities of the two carbonates in sodium citrate–sodium dithionite solution. Calcium carbonate solution is favoured by the acidic and chelating nature of the citrate and the dithionite destroys iron oxide coatings. The method consists of a total carbonate determination by the approximate back-titration method (section 5:4:1I), selective dissolution of the calcium carbonate in the citrate solution and calculation of magnesium carbonate by difference. The quantity of calcium carbonate, and if applicable of magnesium carbonate, dissolved by the citrate is calculated from the calcium and magnesium extracted and thus exchangeable calcium and magnesium must be removed first. The solution used was 0.30 M with respect to sodium citrate and 0.14 M with respect to sodium dithionite and the potassium-saturated soil sample was warmed to $80°C$ with the solution for 15 minutes. After centrifuging the extract was analyzed with a flame photometer. By assuming that the calcium–magnesium ratio of dolomite is 1, the calculations were made thus:

m.e. calcite–calcium   = m.e. (citrate-soluble Ca–citrate-soluble Mg) [5:1

m.e. calcite–carbonate = m.e. calcite-calcium   [5:2

m.e. dolomite–calcium = m.e. (total carbonate–calcite-carbonate)   [5:3

Wright and co-workers (1960) heated samples to $1030°C$ in steps of $300°C$ per hour in an atmosphere of carbon dioxide in a thermobalance. The pyrolysis curves obtained were used for the quantitative estimation of total carbonates and to differentiate calcite and dolomite.

5:3 SULPHUR REQUIREMENT OF SOILS

If a soil contains exchangeable sodium in appreciable quantity its physical condition deteriorates. The impermeability of alkali soils, for example, is related to the percentage saturation of the exchange complex with sodium. As an approximation, a soil with more than 15% exchangeable sodium can be expected to have poor physical properties. Such soils are commonly found in arid regions and are usually, although not necessarily, saline.

The basis of reclamation of alkali soils is to replace the exchangeable sodium with calcium. Simple liming can be effective but only if the pH of the soil is not too high, owing to the low solubility of calcium carbonate at high pH.

Often an alkali soil is also calcareous, and in this case the exchangeable sodium can be replaced by treatment with soluble calcium salts such as the chloride or sulphate or by liberating the calcium already present by acidifying the soil. Acidification can be done by adding elemental sulphur, iron sulphate, aluminium sulphate or by adding sulphuric acid directly.

If the soil is not calcareous, treatment with acid or acid-forming chemicals can render it excessively acid and for such soils it is preferable to use a soluble calcium salt unless the acid is added together with lime.

Sulphuric acid and the acid-forming soil amendments, with the exception of elemental sulphur, are rapid in action and calcium sulphate (gypsum) is also fairly quick-acting, its rate of reaction being limited by its rate of solution or by its particle size. The oxidation of elemental sulphur is slow because of the microbiological action involved, but even so, provided that moisture and temperature conditions are suitable, elemental sulphur oxidizes fast enough for practical purposes.

The analyst is thus called upon sometimes for the 'sulphur requirement' and sometimes for the 'gypsum requirement' of a soil.

Various experimental methods have been advanced to determine sulphur and gypsum requirement. As calcium exchanges for sodium Schoonover (1952) suggested shaking the soil with standard gypsum solution and determining the loss of calcium from solution. This method is widely used but can be criticized in that exchangeable potassium is also replaced by calcium, giving high results.

Another method depends upon a knowledge of the cation exchange capacity and the exchangeable sodium content of a soil (section 5:4:2).

Pratt and Bair (1962) incubated 200-g samples of soil with various amounts of elemental sulphur until there was no further change in pH. Soluble salts were then leached out and the pH again measured. The sulphur requirement was estimated as that needed to bring the pH of the saturated paste of a non-saline soil to 5·2. Pratt and Bair compared this incubation technique with that of a buffer method in which the change in pH of a buffer solution after shaking with a soil is used to calculate the sulphur requirement. The method was found satisfactory for coarse-textured soils but it greatly underestimated the sulphur needs of clay and organic soils.

5:4 RECOMMENDED METHODS

5:4:1 Carbonates

Total insoluble carbonates

I / RAPID TITRATION METHOD (PIPER)

This method is suitable for the routine analysis of a large number of samples where an accuracy of about 1% is sufficient. The results are referred to as 'calcium carbonate equivalent'. In all methods any water-soluble carbonate present must be allowed for.

Reagents

Hydrochloric acid, 1·0 M: 88 cm³ dm⁻³ conc. HCl. Standardize with
  sodium borate and adjust to exactly 1·0 M.
Sodium hydroxide solution, 1·0 M: 40 g dm⁻³ NaOH. Standardize with
  the HCl and dilute until exactly 1·0 M.
Bromthymol blue indicator: 0·1 g of reagent ground with 1·60 cm³ of
  0·1 M NaOH solution and diluted to 250 cm³.

Procedure

Weigh 5 g of 2-mm soil into a 250-cm³ tall-form beaker and add 100 cm³
of 1·0 M hydrochloric acid slowly from a burette. Cover the beaker with

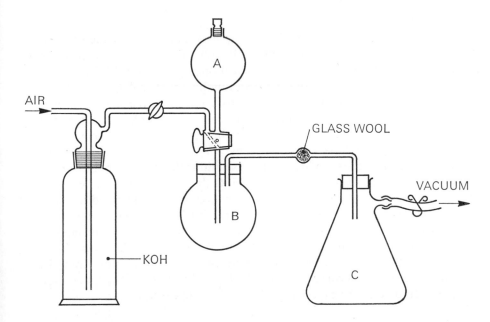

Figure 5:5 Apparatus for determining carbonates by the method of
Hutchinson and MacLennan (*J. agric. Sci.*, 1941)

a Speedy-vap clock glass (Vogel, 1962) and allow to stand with occasional
stirring for 1 hour. When the soil has settled after its final agitation, pipette
20 cm³ of the supernatant liquid into a conical flask and titrate with 1 M
sodium hydroxide solution using 6 drops of indicator solution. As a
blank, titrate 20 cm³ of the original acid.

Calculation

$$\% \text{ CaCO}_3 \text{ equivalent} = (\text{Blank titre} - \text{Actual titre}) \times 5 \qquad [5:4$$

II / VOLUMETRIC METHOD (HUTCHINSON AND MACLENNAN)

Apparatus

This is depicted in Figure 5:5 and is slightly different from that originally described.

Reagents

Hydrochloric acid: 200 cm³ dm⁻³ conc. HCl.

Sodium hydroxide solution, 0·1 M: 4 g dm⁻³ NaOH and standardize.

Barium chloride solution: 150 g dm⁻³ $BaCl_2 . 2H_2O$.

Potassium hydroxide solution: 40 g dm⁻³ KOH.

Phenolphthalein indicator: 0·5 g reagent/50 cm³ ethanol diluted to 100 cm³ with water.

Procedure

Weigh into the reaction flask B 5–10 g, according to carbonate content, of 0·5-mm (40-mesh) soil and moisten it with a little distilled water. Place 50 cm³ of dilute hydrochloric acid in the dropping funnel A. In the absorption flask C place 50 cm³ of 0·1 M sodium hydroxide solution and a few drops of indicator and stopper the flask. Evacuate the apparatus and cautiously add the dilute acid from the funnel. Gently shake the reaction flask while adding the acid, then leave for 30 minutes with occasional shaking. Slowly release the vacuum by allowing carbon dioxide-free air to enter via the potassium hydroxide wash bottle; this will sweep the evolved carbon dioxide from the reaction vessel into the absorption flask. Leave the apparatus for another 20 minutes with occasional shaking, disconnect the absorption flask, rinse the stopper into it (using CO_2-free water) and add an excess (about 10 cm³) of barium chloride solution to precipitate the carbonate. Titrate the free alkali with 0·1 M hydrochloric acid using phenolphthalein as indicator. The presence of the indicator from the beginning in the absorption flask will show whether or not too much soil has been used in the experiment. Also do a blank titration.

Calculation

$$\% \text{ CaCO}_3 = \frac{(\text{blank titre} - \text{actual titre})}{\text{weight of soil}} \times 5 \qquad [5:5$$

III / GRAVIMETRIC METHOD (ERIKSON AND GIESEKING)

Apparatus

The apparatus is shown in Figure 5:6.

Reagents

Trichloroacetic acid: Mix 15 g of reagent with 1 cm³ of water.
Magnesium perchlorate.

Procedure

Place a suitable weight of 0·15-mm (100-mesh) soil in the reaction flask A,
and moisten it with a little water; place 5 cm³ of trichloroacetic acid in the
tube B. Close the stopcocks and weigh the apparatus. Attach the two
U-tubes containing magnesium perchlorate, open tap 1 and tilt the reaction
flask to mix the acid with the soil. Leave the apparatus standing overnight,

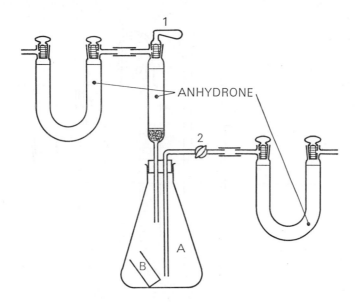

Figure 5:6 Apparatus for determining carbonates by the method of
Erikson and Gieseking (*Soil Science* © 1947)

open tap 2 and displace the carbon dioxide by drawing air through for
about 10 minutes. Finally, close both taps, disconnect the U-tubes and
re-weigh the apparatus.

Calculation

$$\% \ CaCO_3 = \frac{\text{(initial weight} - \text{final weight)}}{\text{weight of soil}} \times 227 \cdot 4 \qquad [5{:}6$$

IV / MANOMETRIC METHOD (BASCOMB)

Apparatus

The apparatus is shown in Figure 5:7. The acid and water reservoirs have
a capacity of 1 dm³ and the reaction flask a capacity of 250 cm³. The wide

arm of the U-tube is graduated in 0·5 cm³ for the range 0–10 cm³ and in 1·0 cm³ for the range 10–250 cm³.

Reagents

Dilute hydrochloric acid: 1:3, v/v.

The dilute acid and the water used in the reservoirs are partially saturated with carbon dioxide to reduce the amount of gas dissolved during the experiment. This is done by decomposing calcium carbonate with acid and with the taps A and B closed; the apparatus is left for 30 minutes.

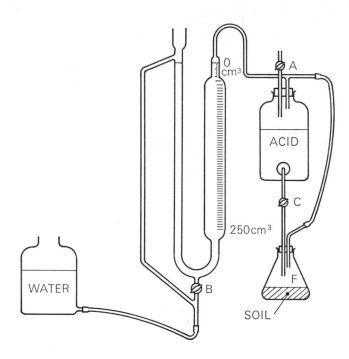

Figure 5:7 Apparatus for determining carbonates by the method of Bascomb (*Chem. Ind.*, 1961)

Procedure

The weight of soil taken will depend upon its approximate carbonate content, thus:

20 g for soil containing less than					5% CaCO₃
10 g ,, ,,	,,	,,	,,		5–10%
5 g ,, ,,	,,	,,	,,		10–20%
2 g ,, ,,	,,	,,	,,		20–50%
1 g ,, ,,	,,	,,	more than		50%

Place the weighed 2-mm soil sample in the reaction vessel F. Open tap A and insert the moistened bung fitting the acid reservoir through tap C into F. Open tap B and raise the water reservoir until the level in both tubes is just above zero. Close tap B and place the water reservoir upon the bench. Close tap A and open C to add acid sufficient to make the soil sample fluid; shake the flask and allow it to stand for several minutes. When reaction ceases open tap B until the water in the left-hand tube becomes level with that in the graduated tube; then close B. Add more acid and shake the flask. Repeat the process until constant level is obtained in the two tubes. Read the volume of gas, the air temperature and the atmospheric pressure.

Calculations

$$\% \ CaCO_3 = \frac{cm^3 \ CO_2}{wt. \ of \ sample} \times \frac{Pressure \ (kilopascal \ or \ kN \ m^{-2})}{(°C + 273)} \times 1·22$$

$$[5:7]$$

or in non-SI units of pressure,

$$\% \ CaCO_3 = \frac{cm^3 \ CO_2}{wt. \ of \ sample} \times \frac{Pressure \ (mmHg)}{(°C + 273)} \times 0·16 \quad [5:8]$$

Simultaneous determination of calcium and magnesium carbonates (Skinner)

Apparatus

Small wax-paper cups. 700-cm³ wide-mouth bottle fitted with a two-hole rubber stopper carrying a thermometer and a glass tube. The glass tube is connected to a mercury manometer and the bottle is fitted to a wrist-action shaking machine and immersed in a water-bath at constant temperature.

Reagent

Hydrochloric acid, 4 M: 350 cm³ dm⁻³ conc. HCl.

Procedure

Weigh a suitable amount, depending upon carbonate content, of soil which has passed a 0·5-mm sieve into a paper cup. Float the cup in 30 cm³ of dilute acid in the bottle and, after checking the zero of the manometer, commence the shaking. At regular intervals of time, say every 10 or 15 seconds, record the manometer reading until no further change occurs.

Calculations

Subtract the manometer readings for each time (H_t) from the final reading (H_∞) and plot time in seconds against lg ($H_\infty - H_t$) as in Figure 5:8. The straight line portion of the graph usually begins after about 1 minute; extrapolate this line to zero time to obtain the intercept value H_D.

H_D represents the carbon dioxide evolved from magnesium carbonate and thus $(H_\infty - H_D)$ represents the carbon dioxide from calcium carbonate. These figures are converted to g CO_2 and per cent carbonate by curves prepared from results obtained using known amounts of carbonate (also ground to pass a 0·5-mm sieve).

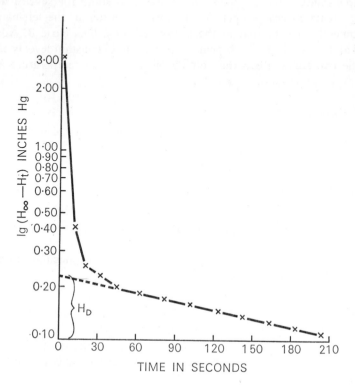

Figure 5:8 Estimation of calcium and magnesium carbonates from rate of solution data (Skinner *et al.*, 1959; *Can. J. Soil Sci.*). *Note:* The SI unit of pressure is the pascal (N m^{-2}) and the conversion factor here is that 1 in of Hg corresponds to 3·32 kN m^{-2}

Field test for carbonates

Place a small portion of soil on a watch-glass and moisten it with distilled water in order to remove air bubbles, which are sometimes confused with evolving carbon dioxide. Add a few drops of 5 M hydrochloric acid and observe the effervescence. From the degree of effervescence, carbonate content is estimated as small, medium or high; with practice, carbonate contents of down to 0·5% can be estimated.

For the field detection of magnesium carbonate the acid should be warmed and it has been found that placing a small beaker of acid on the engine of a car is sufficient to heat the acid to the required temperature.

5:4:2 Sulphur requirement of soil

From cation exchange data

Calculate the exchangeable sodium percentage (E.S.P.) from cation exchange capacity and exchangeable sodium (determined as in sections 7:3:1 and 9:3:5).

$$E.S.P. = \frac{ex.\ Na\ m.e./100\ g}{C.E.C.\ m.e./100\ g} \times 100 \qquad [5:9$$

Decide by how much it is desired to reduce the E.S.P. and insert the new E.S.P. value into the above equation, the C.E.C. remaining constant. This gives the desired content of exchangeable sodium and hence the amount of sodium to be replaced. Express this exchangeable sodium as milliequivalents per 100 g soil, when it will be equivalent to the m.e./100 g of sulphur to be added.

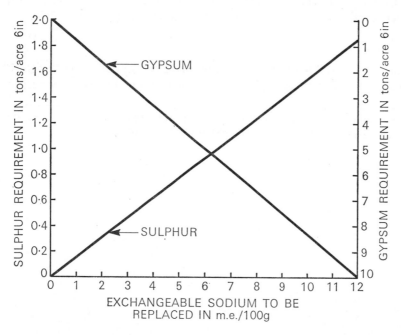

Figure 5:9 Gypsum and sulphur requirements of soil by calculation from cation exchange data and assuming an acre 6-in of soil to weigh $2·2 \times 10^6$ lb (from data in USDA Handbook No. 60; US Salinity Laboratory). *Note:* The corresponding SI factor is that a hectare (10^4 m²) 15-cm of soil weighs $2·22 \times 10^6$ kg

Figure 5:9 shows the quantities of gypsum and of sulphur in tons/acre 6-in and kg/ha 15-cm required to replace indicated amounts of exchangeable sodium assuming that an acre 6-in of soil weighs $2·2 \times 10^6$ lb (a hectare, or 10^4 m², 15-cm weighs $2·22 \times 10^6$ kg). For soils where this assumption cannot be made (highly organic soils for instance) the figures can be converted using Metson's nomogram (Figure 4:3).

Buffer method (Pratt and Bair)
This method is applicable to sandy soils only.

Reagents
Buffer solution:

> Calcium acetate, 2·0 g
> p-Nitrophenol, 1·8 g
> Potassium chromate, 3·0 g
> Potassium hydrogen phthalate, 4·7 g
> Formic acid, 88%, 1·0 cm³

Dissolve the reagents in about 800 cm³ of water and titrate to pH 2·00 with dilute hydrochloric acid; dilute to 1 dm³. When 10·0 cm³ of this buffer solution are diluted with 70 cm³ of water the pH will be 2·50.

Procedure

Shake 40 g of soil with 35 cm³ of water and 5 cm³ of buffer solution for 10 minutes. Allow to stand for 20 minutes and measure the pH of the suspension.

The sulphur requirement of the soil is then calculated as 113 kg per hectare or (10^4 m²) for each 0·10 pH unit above 3·70.

6

Soluble salts

6:1 INTRODUCTION

The main source of the soluble salts which are present to some extent in all soils is the primary minerals. Of the soluble anions, sulphates and chlorides predominate, but exchangeable cations in equilibrium with carbonic acid in the soil give rise to soluble carbonates and hydrogen carbonates. Nitrates, phosphates and silicates occur in varying amounts, often negligible but sometimes predominant. The cations involved are usually calcium, magnesium, potassium and sodium, although other cations become more important in certain kinds of soil—aluminium, iron and manganese in acid soils, for example.

When a soil has excess soluble salts it is said to be saline. The term 'excess soluble salt', however, is a vague criterion and attempts have been made to make salinity a more specific soil property. Some workers consider a saline soil as one in which the salt content is sufficiently high to interfere with plant growth, but this depends upon other soil conditions and the nature of the plant. In the USDA handbook (No. 60) on salinity a saline soil is defined as one having a conductivity (of saturation extract) greater than 4 mS cm^{-1} (4 millimhos/cm) at 25°C, a soluble sodium content of less than half the total soluble cations and a pH value usually less than 8·5. If the sodium content increases the soil becomes a saline alkali soil, the pH remaining below 8·5. If salts are leached from a saline alkali soil it reverts to a non-saline alkali soil, whereas leaching a saline soil converts it to a normal soil.

Although the ultimate origin of soluble salts is primary minerals, saline soils usually accumulate the excess salts by drainage and seepage from other areas, unless they are coastal soils where the origin of salts is more likely to be oceanic. Yaalon (1963) reported that in Israel the contribution to salinity of weathering primary minerals is relatively small except in areas where soils are formed from gypseous lake deposits. The coastal soils of Israel had oceanogenic salinity and the desert salinity was caused by wind-borne salts. Apart from coastal regions, saline soils usually occur in arid and semi-arid parts with inadequate drainage and hardly any leaching. Poor drainage of these soils may be due to a high water table or to low permeability or to both. Very often, as such soils occur in regions of low rainfall the land is irrigated and unless, as is seldom the case, the available water is low in salt content, irrigation aggravates the salinity problem. Owing to the lowered vapour pressure of salt solutions saline soils do not evaporate water from their surface as readily as non-saline soils and consequently salt solutions can move towards the surface for relatively long periods, thus increasing surface salt patches (Russell, 1961).

The major portion of the world's saline soils tend to have neutral to alkaline reaction but acid saline soils also exist. Acid saline soils are commonly coastal, or have been under marine influence, and contain large amounts of sulphur. When such soils are dried and thus oxidized, sulphates of iron and aluminium are produced and may even form a crust on the surface of the soil. These acid sulphur soils are discussed further in Chapter 13 (section 13:1).

In certain soils of the tropics a seasonal accumulation of nitrate has been observed (Griffith, 1950; Simpson, 1960; Wetselaar, 1960) and this is due partly to upward movement of nitrate-containing water in the dry seasons followed by leaching in the wet seasons. Simpson, however, by adding chloride to a topsoil towards the end of a rainy season and using it as a tracer for future analyses, concluded that the upward movement of salts was insufficient to account for the quantities of nitrate found. The excess nitrate appears to be formed microbiologically and probably the wetting–drying cycle is involved (see sections 2:2 and 10:2:5).

Plants vary greatly in their tolerance to salinity, and injury due to salts is dependent partly upon soil conditions other than salt content. Wadleigh and co-workers (1947) devised a method of measuring the salinity tolerance of plants by growing them in columns of soil in which the salt content increased with depth. Yaron and Mokady (1962) decided that pot tests for investigating saline soils need special techniques. Satisfactory interpretation of results depends upon the concentration profile of salt solution and upon a uniform, constant moisture profile in the pots. Yaron and Mokady thus developed a technique based on a principle of simultaneously leaching accumulated salts and removing excess water. The harmful effects of excess salts upon plant growth is due partly to a specific effect of certain ions, for example of Cl^- upon citrus, and partly to the increase caused in osmotic pressure around the plant roots which inhibits water uptake. Heimann (1958) investigated the influence of osmotic pressure exerted by saline water upon plant growth. He introduced the concept of 'balanced ionic environment' combining the effects of nutrients, interfering salts and hydrogen ions. Heimann emphasized that the suitability of saline water for irrigation cannot be judged by its analysis alone; the circumstances accompanying its application must be considered also. A very broad generalization of plant susceptibility to soluble salts is given in the USDA salinity handbook and is here reproduced. It is based upon the conductivity of soil saturation extracts.

$\sigma \times 10^3$ millisiemens, mS (or mmhos) cm^{-1}:

0–2	Salinity effects negligible
2–4	Very sensitive crops affected (citrus, beans)
4–8	Many crops affected

8–16 Only salt-tolerant crops yield satisfactorily (wheat, grapes, olives)

> 16 Only very few salt-tolerant crops yield satisfactorily (dates, barley, sugar-beet).

As shown by Baeyens and Appelmans (1947) and by Themlitz and Ishakohln (1962), high salt concentration ($> 0.2 \%$) in soils reduces nitrification and at higher concentrations nitrification is completely inhibited and some nitrogen immobilization occurs. Birch (1959) pointed out that soluble salts in a soil depress the solubility of organic matter and thus a salt-free, dry soil will have a greater quantity of organic carbon oxidized on re-wetting than will a dry saline soil. Birch further postulated that this does not necessarily mean a concomitant reduction of nitrogen oxidation in saline soils.

Much nitrate in a soil depresses plant uptake of chloride and can be used to counteract chloride injury (Buchner, 1951). Soluble salts also affect the plant uptake of phosphorus; sodium, potassium and especially calcium depress the solubility of phosphate in soil, and chloride and nitrate have a similar effect; sulphate however, particularly in high concentration, increases phosphate solubility (van Wesemael and Lehr, 1954).

6:2 DETERMINATION: BACKGROUND AND THEORY

The determination of soluble salts in a soil falls into two main categories, the measurement of total salinity and the determination of individual soluble ions.

6:2:1 Total salinity

Total salinity can be appraised by several methods, the choice of which will depend upon the level of precision required. For rapid routine determinations, such as would be made in a large-scale soil survey, measurement of the electrical resistance of a soil is most convenient. Electrical resistance is measured in ohms and is defined as E (volts)/I (amps). The method is an approximation only and it is assumed that the concentration of salts constitutes the electrolyte concentration of which resistance is a function. Other factors are involved, however, such as temperature, moisture content, specific ions and degree of dissociation. Magistrad and co-workers (1945) found that different types of soil, even with the same salt content, had different values of electrical resistance. In spite of this failing the method remains excellent for rapid, routine salinity appraisal and no way of improving the technique without destroying its simplicity has yet been devised. To standardize the method as far as possible the resistance of the soil saturated paste is measured rather than that of an arbitrary mixture of soil and water, and results are corrected to a standard temperature of 16°C. The temperature correction is made by using the tables of Whitney and Means (1897), who used the Fahrenheit scale and corrected to 60°F. These tables converted to the Celsius scale are reproduced in Table 6:1. Alternatively the nomogram (Figure 6:5) published in the USDA *Soil Survey Manual* can be used.

The saturated soil paste is made by manually mixing soil with water until certain specified conditions are satisfied (section 6:3:1ɪ). The method is liable to error and reproducibility may be poor. Thus Longenecker and Lyerly (1964) sent representative samples of the same group of soils, including four different textural classes, to four different laboratories. The soils were made into saturated pastes and the quantities of water used were determined. The range of water used for the coarse soil was ± 17 cm^3 per 200 g of soil and for the fine-textured soil the range was ± 58 cm^3. As a result of these findings Longenecker and Lyerly proposed a new method of preparing saturated soil pastes utilizing capillary absorption instead of manual mixing. The apparatus consists of an outer container holding distilled water at constant level and an inner container filled with medium or fine pure quartz sand. The inner container is suspended in the water and has a perforated base covered with a fine cloth screen. The surface of the sand is kept smooth and level and is covered with a sheet of blotting paper. Dry, sieved, 200 g soil samples are first weighed and then placed in cups made from Whatman No. 52, 15-cm filter paper. The cups have flat bottoms and sloping sides and are made by stapling the paper. The soil-filled cups are then placed on the sand table and left until a state of capillary saturation is achieved. Different soils need different times of contact to become saturated and a period of 18 hours is recommended as a routine. The method has the advantage of enabling several soils to be saturated simultaneously and possibly of eliminating human error, but is obviously of no use for field work. Indeed, it was designed primarily to obtain saturation extracts and not saturated pastes as an end-product. Comparison of results shows that the capillary saturation method is not in fact greatly superior to hand mixing and in the case of very fine-textured soils is less accurate.

The resistance of soil paste is measured with a Wheatstone bridge of which several portable models are available for field use. The paste is placed inside a specially designed conductivity cell known as the 'Bureau of Soils cup' (Plate 2). The cup must be filled completely with soil paste, care being taken to exclude air bubbles, and it is then placed in a rack having spring contacts for the two electrodes which are built into the sides of the cup. The US Salinity Laboratory has now designed a modified cup having electrodes at the top and bottom.

To interpret readings of electrical resistance a relationship is required between ohms and salt concentration. Davis and Bryan (1910) used four non-saline soils representative of the textural classes sand, loam, clay-loam and clay, and measured their electrical resistance after adding known amounts of chloride and sulphate. A rough relationship was thus obtained between salt concentration and resistance and the data of Davis and

Bryan are still used (Table 6:2). Davis and Bryan also gave figures based upon the presence of carbonates and recommended that if the soil paste gave an alkaline reaction to phenolphthalein the relevant column of figures be used. These figures are still given in some books but the USDA salinity handbook considers that the procedure is unreliable owing to unconsidered cation exchange reactions.

Whilst measurement of the resistance of a saturated soil paste is a conveniently rapid procedure to establish salinity, it is unreliable as an index of soil solution concentration (Reitemeier and Wilcox, 1946). When a reasonable degree of accuracy is required the solution should be extracted from the saturated paste and the conductivity of that solution measured.

Conductance, G, is the reciprocal of resistance and is measured in siemens (mhos). The conductance of a solution varies with the salt concentration and the 'conductivity' of a solution is the conductance of that solution at 25°C between electrodes 1 cm square and 1 cm apart. For soil solutions the siemens is rather unwieldy and it is more usual to express results of conductivity as millisiemens per cm and refer to it as $\sigma \times 10^3$. Again the apparatus for measuring conductivity is essentially a Wheatstone bridge and the rheostat can be calibrated to read conductance.

Conductivity values of soil extracts will depend, *inter alia*, upon the ratio of soil to water, and as a standard procedure measurements are made on the extract from the saturated soil paste. Temperature affects the readings, the conductivity increasing approximately 2% per degree Celsius, and thus the results must be reduced to a standard temperature which is usually taken as 25°C. On some instruments provision is made to correct automatically for temperature, otherwise the correction can be made by reference to tables (Table 6:5).

The saturation extract of a soil is obtained simply from the paste by suction using a Buchner or Richards (1949) funnel. For routine work it is convenient to set up several funnels for simultaneous filtration by means of a manifold (Plate 3). Leo (1963), to avoid the tedious process of cleaning the funnels, devised an apparatus for filtering soil pastes using ordinary glass funnels. Leo's apparatus (Figure 6:1) consists of a flat suction plate made from a perforated metal or plastic disc sealed with plastic cement on to the top of a 75-mm glass funnel. The funnel is mounted in a filter flask and dips into a 35-cm^3 glass vial supported on a foam rubber pad. A filter paper is placed on top of the plate and soil paste spread over its surface; suction is applied to about 4 kN m^{-2} (30 in). Plastic-tipped, straightened tongs are used to remove the vial and several units can be connected to a manifold suction source.

For field work a convenient portable filtering device is obtainable (Richards, 1955a); the apparatus (Plate 4 and Figure 6:2) consists of a vacuum jar with a rubber stopper to hold the funnel, connected to a hand-operated vacuum pump worked by a rubber bulb. A glass vial is placed inside the jar to collect the extract; the pump gives a vacuum better than 68 kN m^{-2} ($\frac{2}{3}$ atmosphere).

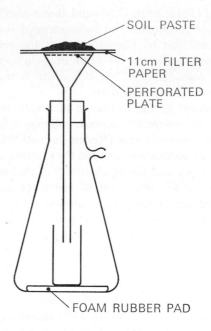

Figure 6:1 Filtration apparatus for preparing saturation extracts (Leo; *Soil Science* © 1963)

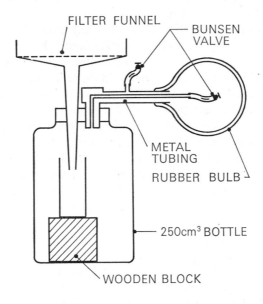

Figure 6:2 Portable vacuum filter for preparing saturation extracts (Richards). See also Plate 4

For interpretation of results the percentage of water needed to prepare the saturated paste is required and this is called the 'saturation percentage' (S.P.). If possible, for example in the laboratory, the saturation percentage is best obtained in the conventional manner by drying a weighed portion of the paste in an oven at 110°C and recording the loss in weight. Alternatively the volume of water added is recorded and the saturation percentage calculated from a knowledge of the original moisture content. Wilcox (1951) proposed a method for determination of saturation percentage which depends upon the weight of a known volume of paste. The procedure is to fill the cup, the volume and weight of which are predetermined, with saturated paste and weigh it. The saturation percentage is then calculated from the expression,

$$\text{S.P.} = \frac{100 \times (2 \cdot 65V - W)}{2 \cdot 65(W - V)} \qquad [6:1$$

where V is the volume of soil paste in cm^3 and W is the weight of V cm^3 of paste. It is assumed that the soil particles have a density of $2 \cdot 65$ g cm^{-3} and thus the method can be used as an approximation only.

The saturation percentage is related directly to the field moisture range; that is, from wilting point to a moisture content of about twice the wilting point moisture, and thus measurements of the conductivity of the saturation extract are particularly relevant. However, for some purposes the salt concentration of the actual soil solution is required and this is experimentally more difficult to obtain than a saturation extract. Several methods have been proposed for extracting soil solutions, the most obvious of which is suction, although this is not particularly effective. Centrifuging or, better, ultra-centrifuging, can be used but the most commonly employed procedure is that of displacement (Parker, 1921). In the displacement method a liquid such as ethanol or even water is allowed to percolate through a column of soil when it pushes before it the displaced soil solution. Burd and Martin (1923) used water as a displacing liquid and speeded up the percolation by applying air pressure. Jackson (1958) used very dilute potassium thiocyanate solution and periodically tested the leachate with iron(III) to determine when the flow of solution ceased.

Probably the most convenient routine procedure for extracting soil solutions is that using a pressure membrane (Richards, 1947; Reitemeier and Richards, 1944). The modern version of the apparatus is shown in Plate 5 and Figure 6:3 and is essentially a semipermeable membrane on which the soil is placed and to which gas pressure can be applied to force out the moisture. The membrane is of cellophane sheet and pressure is obtained by pumping in nitrogen to avoid any chemical effects of oxygen or carbon dioxide. The soil is placed in a special retaining cylinder which fits into the extractor.

In some circumstances it is necessary to measure the salinity of more dilute soil extracts, for example a 1:1 or a 1:2 soil–water extract (Jackson, 1958), and Metson (1961) suggests a 1:5 extract. The results naturally will

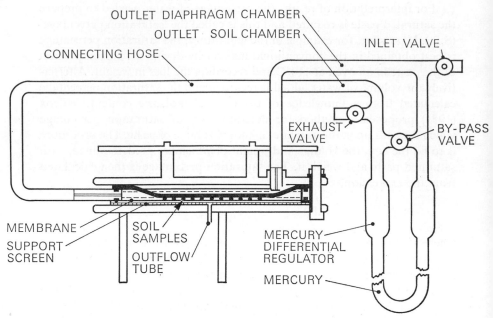

OUTLET DIAPHRAGM CHAMBER
OUTLET SOIL CHAMBER
INLET VALVE
CONNECTING HOSE
EXHAUST VALVE
BY-PASS VALVE
MEMBRANE
SUPPORT SCREEN
SOIL SAMPLES
OUTFLOW TUBE
MERCURY DIFFERENTIAL REGULATOR
MERCURY

Figure 6:3 Pressure membrane apparatus for extracting soil solutions. (See also Plate 7)

vary according to the ratio of soil to water and such arbitrary extracts are to be used for specific purposes only. In every case measured conductivities should be calibrated with concentrations of the particular salts predominating in the type of soil examined. Such extracts, which are more easily obtained than soil solutions or saturation extracts, are particularly useful if the same soil is to be analyzed repeatedly over a period of time. In gypseous soils the more dilute the extract the more calcium sulphate will be dissolved and this constitutes a difficulty in itself. It has to be decided whether the current concentration of soluble gypsum is to be measured or the total, potentially soluble gypsum; to obtain the total gypsum content great dilution may be needed. Metson (1961) when preparing the 1:5 extract shakes the mixture for 30 minutes and points out that calcium sulphate in excess of 15 m.e./100 g will not all be dissolved.

To measure the conductivity of a soil extract some type of conductivity cell is required and a dip-type is convenient if sufficient liquid is available, from a 1:5 extract for example. Where only small quantities are available, for example soil solutions or saturation extracts, a micro-dip-type is commercially available but the US Salinity Laboratory have devised a particularly convenient pipette-type cell (Plates 8 and 9) which has a capacity of 2–3 cm³. The cell has platinized platinum electrodes and its constant (k) must be known.

With modern apparatus an automatic temperature compensation device is incorporated, but if not, then corrections have to be made to reduce the conductivity readings to 25°C. The conductivity bridge devised to go with the pipette-type cell is the most convenient and a portable, battery-operated version exists for field work (Plates 2 and 3). The bridge scale has been calibrated to read directly from 0·15 to 15 mS cm^{-1} and the instrument has a cathode ray tube null indicator; the field model (Richards, 1955b) has a telephone detector instead of a cathode ray tube, the null point being that at which the volume of sound is minimal.

Interpretation of conductivities in terms of salt concentration can be made more precisely than with paste resistance values. Even so, as the salts in a soil extract vary in composition, the relationship between conductivity and salt concentration can be only approximate. The USDA salinity handbook gives a curve (Figure 6:4) relating salt concentration

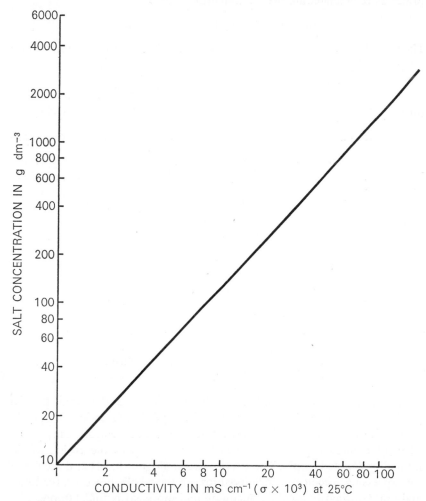

Figure 6:4 Relationship of concentration of saturated extracts of soils with electrical conductivity (USDA Handbook No. 60, US Salinity Laboratory)

with the conductivity of saturation extracts. More accurate results will be obtained from curves prepared for the soil and salts in question. With more dilute extracts the reliability of such curves depends more upon the type of salt present and this is particularly so if calcium sulphate or soluble carbonates are present. Thus the use of dilute extracts to determine total salinity is not recommended for general analysis; it will be of greatest value when the same soil is to be repeatedly analyzed. Piper (1942) gives the following relationship between electrical conductivity and salt concentration for a 1:5 extract,

$$K_{25°} \times 336 \simeq \% \text{ total soluble salts} \qquad [6:2$$

and Metson (1961), for New Zealand soils in particular, quotes Hurst and Miller as recommending the relationship,

$$K_{25°} \times 350 \simeq \% \text{ total soluble salts} \qquad [6:3$$

The USDA salinity handbook converts conductivities to salt concentration by the following equations,

$$K_{25°} \text{ mS cm}^{-1} \times 10 = \text{salts m.e./dm}^3 \qquad [6:4$$

thus

$$K_{25°} \text{ mS cm}^{-1} \times 5 = \text{salts m.e./100 g for a 1:5 extract} \qquad [6:5$$

Hyperbolic relationships were shown by Delver and Kadry (1960) between saturation moisture content or salinity level and conductivity of saturated paste extract and a method was given for estimating the conductivity of saturation extracts from the conductivity of the saturated paste.

On the basis of conductivities soil samples can be assigned to so-called 'salinity classes', thus

Conductivity of saturation extract	Salinity class
0–4 mS cm^{-1} at 25°C	Free
4–8 ,,	Slight
8–15 ,,	Moderate
> 15 ,,	Strong

Whittles and Schofield-Palmer (1951) introduced the term pC for conductivity and pS for salt content where pC is the co-log of specific conductivity in S cm^{-1} and pS is the co-log of salt concentration in g cm^{-3}. The salt content expressed in m.e./dm$^3 \times 10^{-3}$ gives the co-log of the normality denoted by pN. These terms are analogous to pH and are useful for expressing specific interrelationships; on the whole, however, the terms and units ordinarily used are adequate and more easily understood.

As might be expected, the salt content of a soil extract affects the osmotic pressure of the extract and hence electrical conductivity is related to osmotic pressure. This relationship is sometimes of value, particularly for plant salt-tolerance experiments.

Another method of determining the total salinity of a soil is by chemical analysis. The basis of the method is to extract the soluble salts with water, evaporate off the solvent and to determine the total salts gravimetrically. As in the case of electrical measurements, the first problem is to decide upon the extraction procedure. In order to approximate to field conditions the saturation extract can be used but, as explained, this will not give a true analysis of gypseous soils other than showing the approximate composition of the soil solution, and thus as a general routine it may be preferable to employ a more dilute extract. Metson (1961) recommends a 1:5 extract and Merkle and Dunkle (1948) used a 1:2 extract. The second problem is to obtain a clear extract from the soil. With saline soils this may not be a problem and centrifuging or simple filtration will usually give a clear extract. With other soils, however, it is sometimes extremely difficult to clarify the extract. A filter candle is useful but is not really convenient to use; Metson suggests that a bacterial membrane filter may be essential. The addition to the filtered extract of a few drops of glacial acetic acid sometimes is sufficient to coagulate the dispersed soil particles which can then be filtered off, but there is danger of loss of soluble carbonate.

The next question to be decided is the temperature at which the extracted salts should be dried before weighing. At high temperatures chloride may be lost and at low temperatures water of crystallization may be retained. Calcium sulphate loses its associated water only at temperatures above 180°C. Metson recommends the relatively high temperature of 470–500°C in order to dehydrate salts and destroy organic matter; the loss of chloride is corrected for by determining its concentration before and after heating. Richards and co-workers (1947) dried salts at 110°C and reported the result as 'total dissolved solids'. Jackson (1958) also prefers the lower temperature. On the whole the low temperature of 110°C is to be recommended as not only is it a quicker and more simple procedure than heating to 500°C but the double determination of chloride is obviated. The justification for the easier procedure is that the results will be sufficiently accurate for the determination of 'total salinity' which, in any case, is estimated more conveniently from conductivity measurements.

Tabikh and Russell (1961) used cation exchange resins to evaluate soil salinity. Portions of saturation extract are added to tapering columns of Dowex–50 resin and after 5 minutes the columns are flushed with water. The displaced hydrogen ions are then titrated with dilute sodium hydroxide solution using bromthymol blue as indicator; if carbonates are present the titration is made to excess with barium hydroxide solution followed by back-titration with hydrochloric acid.

The areometer, or hydrometer, has successfully been used to determine chloride plus sulphate and sulphate concentrations in soil (e.g. Dashevskii, 1959). The method is claimed to be convenient for finding the approximate content of physiologically injurious salts under field conditions.

4+T.S.C.A.

6:2:2 Individual soluble ions

If the total soluble salt content of a soil is less than 2·0 m.e./100 g (or 0·15%), there is little point in measuring individual ions except for some specific purpose. The summation of individual ions provides another method of estimating total salinity and if expressed as m.e./100 g the sum of the anions should equal the sum of the cations which provides a check on the accuracy of the determination.

The individual ions normally determined are calcium, magnesium, potassium, sodium, chloride, sulphate, hydrogen carbonate and carbonate. For specific purposes any ion which is soluble in water may be individually determined. Of the anions sulphate and chloride are usually the most important and carbonate and hydrogen carbonate usually occur in small amounts unless the pH is high. Nitrate, phosphate and silicate vary considerably in water extracts of soil, being usually in small concentration except in certain kinds of soil.

The individual determination of the cations and of nitrate, phosphate and silicate is discussed in appropriate later chapters and sulphate determination is described at length in Chapter 13 except, however, for the specific determination of gypsum as such. In this chapter is discussed the determination of soluble chloride, carbonate, hydrogen carbonate and gypsum.

Chloride

Volumetric methods for estimating chloride in solution include titration with silver nitrate using potassium chromate as indicator (Mohr titration), titration with silver nitrate using an absorption indicator, or electrometrically, treatment with excess silver nitrate and back-titration of reagent (Volhard's method) and titration with mercury(II) nitrate using diphenylcarbazone as indicator.

Gravimetric and colorimetric procedures for chloride determination are not very suitable for soil extracts.

The Mohr titration with silver nitrate is the most commonly used procedure and depends upon the formation of sparingly-soluble red silver chromate at the end-point. The theory of the titration can be found in most textbooks of chemical analysis but it should be noted that the solution titrated must be neutral or slightly alkaline. If an aliquot of soil extract is analyzed for carbonate or hydrogen carbonate first, then at the end of the determination the solution will be just neutral to methyl orange and chloride can be measured directly in the same aliquot. Metson (1961) recommends the addition of 1% solid sodium hydrogen carbonate before titration and Vogel (1962) recommends calcium carbonate to be added with a blank titration of a suspension of calcium carbonate. Bower and Wilcox (1965) suggest that the end-point of the titration is more easily detected under yellow light.

By using dichlorofluorescein as an absorption indicator the chloride solution may be neutral or slightly acid; the end-point is reached when the white precipitate suddenly assumes a reddish tint and the titration is facilitated by the addition of a little dextrin solution to offset the coagulating effect of calcium ions (Vogel, 1962).

If the chloride ion alone is to be determined, this is conveniently done by electrometric titration of a soil suspension using a silver–silver chloride electrode (Best, 1929; 1950). For such titrations the pH should be adjusted to 8·2 (just colourless to phenolphthalein); the electrode is a silver wire coated with silver chloride and used in conjunction with a quinhydrone electrode connected by an agar–potassium nitrate salt bridge. 4 g of soil are suspended in 50 cm³ of water and the deflection of a galvanometer needle noted. The suspension is then titrated with silver nitrate until the needle reverses its direction.

Titration of chloride with mercury(II) nitrate is described in several books but offers no advantages over the silver chloride titration. Wang (1953) described a method for determining chloride in saline soils in which the soil is extracted with 0·2% calcium nitrate solution and the extract titrated with silver nitrate.

Carbonate and hydrogen carbonate

Soluble carbonates are unlikely to occur in a soil if the pH is less than 9·5. The usual method of determination in soil extracts is by titration with dilute acid, firstly to the hydrogen carbonate stage using phenolphthalein as indicator,

$$CO_3^{2-} + H^+ = HCO_3^- \qquad [6:6$$

and then, using the same aliquot, to the carbonate stage using methyl orange or other suitable indicator,

$$HCO_3^- + H^+ = H_2CO_3 \rightleftharpoons CO_2 + H_2O \qquad [6:7$$

The theory of the titration is in most textbooks (Vogel, 1962; Metson, 1961) and a variety of indicators has been proposed (Piper, 1942; Tinsley *et al.*, 1951) to increase the accuracy. The first titration should be made slowly and the flask agitated between each addition of acid. If the carbonate content is high Tinsley and co-workers (1951) recommend adding the acid below the surface of the liquid which is protected by a layer of petroleum ether. For carbonate–hydrogen carbonate mixtures containing relatively large amounts of carbonate Vogel recommends using two separate aliquots of extract if enough is available. In this case the first titration is a measure of total alkali using methyl orange as indicator and in the second titration a measured excess of standard 0·1 M sodium hydroxide solution is added to transform hydrogen carbonate to carbonate; excess barium chloride is then added to precipitate the carbonate and the excess alkali titrated. Such a refinement, however, will seldom be needed for soil analysis.

When the soil extract is strongly coloured, rendering a colorimetric titration impractical, an electrometric titration can be made using a glass electrode; hydrogen carbonate is titrated to pH 4·5 and carbonate to pH 8·2 (Bower and Wilcox, 1965).

6:2:3 Gypsum

Measurement of the gypsum in a soil saturation extract only shows the concentration of calcium sulphate soluble at any one time under approximate field conditions; in order to measure the total gypsum content of a soil a more dilute extract is required. As a routine a 1:5 extract is used but if the gypsum content as measured approaches or exceeds 15 m.e./100 g, the extraction must be repeated at greater dilution.

At one time the determination of gypsum in soil extracts was made by the obvious method of estimating separately calcium and sulphate ions but, as pointed out by Reitemeier (1946), calcium and sulphate ions can be dissolved from sources other than gypsum. Furthermore, the dissolved calcium enters into cation exchange reactions with sodium, magnesium and so on.

Bower and Huss (1948) presented a rapid conductance method in which calcium sulphate is precipitated from the soil extract with acetone, redissolved in water, and the conductivity of its solution measured. A standard curve is prepared showing the relationship between conductivity and concentration of calcium sulphate. Lagerwerff *et al.* (1965) modified the method of Bower and Huss to avoid exchange error by determining gypsum on the basis of sulphate associated with calcium; this minimizes errors due to acetone and to occluded ions. The US Salinity Laboratory estimated gypsum by the difference between the amounts of calcium and magnesium brought into solution in a dilute extract and the amounts present in a saturation extract.

The US Salinity Laboratory method and that of Bower and Huss were critically examined by Deb (1963) with particular reference to saline–alkali soils and were found to have limitations. As a result Deb proposed another method whereby calcium and sulphate ions are estimated in the saturation extract and in an extract sufficiently dilute to dissolve all the gypsum. Gypsum is then calculated as follows,

Gypsum (m.e./100 g) = Sulphate or calcium (whichever is   [6:8] less) in the saturation extract plus the additional amount of sulphate brought into the dilute solution.

Deb's method is precise and is unaffected by exchange reactions or the presence of sources other than gypsum of calcium or sulphate; it will not be precise for soils rich in magnesium or sodium sulphate.

The choice of method really depends upon the required degree of accuracy and for many purposes the rapid procedure of Bower and Huss will be found adequate.

6:3 RECOMMENDED METHODS

6:3:1 Determination of total salinity

Approximate total salinity by the electrical resistance of saturated soil paste

I / PREPARATION OF SATURATED SOIL PASTE

Place about 200 g of 2-mm soil in a suitable container and add distilled water whilst stirring with a spatula. Occasionally tap the container on the bench to consolidate the mass. Initially add sufficient water to bring the soil to near saturation as this gives better results than a slow, gradual addition. At saturation the soil surface glistens, the soil flows slightly if the container is tipped and it easily slides off the spatula. Allow the paste to stand for one hour and repeat the saturation tests; there should be no free water on the paste surface. If too much water has been added a little more dry soil can be mixed in. Note that peaty soils, if dry, will need soaking overnight before the paste can be prepared satisfactorily. With very fine textured soils it is preferable to add the initial water with a minimum of stirring to avoid puddling.

II / DETERMINATION OF SATURATION PERCENTAGE

Transfer a portion of the saturated soil paste to a tared tin and weigh. Dry the paste in an oven at 110°C, cool in a desiccator and re-weigh. Calculate the saturation percentage from the loss in weight,

$$\text{S.P.} = \frac{\text{loss in weight} \times 100}{\text{oven-dry weight}} \qquad [6:9$$

If the air-dry moisture content of the soil is known and the water added to prepare the paste has been measured then the saturation percentage can be calculated from,

$$\text{Wt. of oven-dry soil} = \frac{\text{wt. air-dry soil} \times 100}{(100 + \% \text{ air-dry moisture})} \qquad [6:10$$

$$\text{Total water} = \text{water added} + (\text{wt. air-dry soil} - \text{wt. oven-dry soil}) \qquad [6:11$$

$$\text{S.P.} = \frac{\text{total water} \times 100}{\text{wt. oven-dry soil}} \qquad [6:12$$

The soil cup is weighed and its capacity determined. After filling the cup with paste, weigh again. The saturation percentage can then be calculated from,

$$\text{S.P.} = \frac{100 \times (2 \cdot 65V - W)}{2 \cdot 65 \times (W - V)} \qquad [6:13$$

This method is of principal use in the field and is an approximation only.

III / MEASUREMENT OF SATURATED SOIL PASTE RESISTANCE

Apparatus

A Bureau of Soils cup should be used together with its special holder (Plate 4); the cell constant of the cup is a known factor (0·25).

Several models of bridge are available for measuring resistance and precise experimental details will depend upon the model used. Basically, however, all the instruments are Wheatstone bridges and their differences are minor. The resistance in ohms is measured when the paste-filled cup is placed in circuit and a current passed. The bridge pointer, usually a dial, is moved round the scale with certain standard resistances in circuit and the null point is indicated by zero deflection of a galvanometer needle, minimum sound in earphones or maximum opening of a cathode ray 'magic eye', according to type of instrument.

Procedure

Fill the cup completely with soil paste taking care to exclude all air bubbles and strike off level with a spatula. Clean and dry the outside of the cup and insert it into the spring-clips of the holder. After measuring the resistance, record the temperature of the paste and test a small portion qualitatively for carbonate with phenolphthalein.

Note: If the saturated soil paste is to be used also for pH measurement, this must be done after measuring resistance to avoid interference from potassium chloride transferred from the calomel electrode.

Calculations

First correct the soil paste resistance to a standard temperature of 16°C (60°F). This can be done by using the tables of Whitney and Means (Table 6:1) and as an example consider a measured resistance of 1247 ohms at 27°C. Using Table 6:1,

Resistance at 27°C $\longrightarrow$	Resistance at 16°C
1000	1310
200	262
40	52
7	9
___	___
1247	1633

As an alternative to using tables a nomogram (Figure 6:5) can be used, or, if the number of samples is sufficiently large to warrant it, a special slide-rule can be made.

To translate resistance in ohms to salt concentration, reference is made to Table 6:2 (from *Soil Survey Manual*, p. 349). More accurate results can be obtained if the saturation percentage is known by using Table 6:3

(from *Soil Survey Manual*, p. 351); a nomogram (Figure 6:5) can be prepared for this calculation also.

Table 6.1

Bureau of Soils data for reducing soil-paste resistance readings to values at 16°C (after Whitney and Means)

°C	Ohms								
	1 000	2 000	3 000	4 000	5 000	6 000	7 000	8 000	9 000
4	735	1 470	2 205	2 940	3 675	4 410	5 145	5 880	6 615
6	763	1 526	2 289	3 052	3 815	4 578	5 341	6 104	6 867
7	788	1 576	2 364	3 152	3 940	4 728	5 516	6 304	7 029
8	814	1 628	2 442	3 256	4 070	4 884	5 698	6 512	7 326
9	843	1 686	2 529	3 372	4 215	5 058	5 901	6 744	7 587
10	867	1 734	2 601	3 468	4 335	5 202	6 069	6 936	7 803
11	893	1 786	2 679	3 572	4 465	5 358	6 251	7 114	8 037
12	917	1 834	2 751	3 668	4 585	5 502	6 419	7 336	8 253
13	947	1 894	2 841	3 780	4 735	5 682	6 629	7 576	8 523
14	974	1 948	2 922	3 896	4 870	5 844	6 818	7 792	8 766
16	1 000	2 000	3 000	4 000	5 000	6 000	7 000	8 000	9 000
17	1 027	2 054	3 081	4 108	5 135	6 162	7 189	8 216	9 243
18	1 054	2 108	3 162	4 216	5 270	6 324	7 378	8 432	9 486
19	1 081	2 162	3 243	4 324	5 405	6 486	7 567	8 648	9 729
20	1 110	2 220	3 330	4 440	5 550	6 660	7 770	8 880	9 990
21	1 140	2 280	3 420	4 560	5 700	6 840	7 980	9 120	10 260
22	1 170	2 340	3 510	4 680	5 850	7 020	8 190	9 360	10 530
23	1 201	2 402	3 603	4 804	6 005	7 206	8 407	9 608	10 809
24	1 230	2 460	3 690	4 920	6 150	7 380	8 610	9 840	11 070
26	1 261	2 522	3 783	5 044	6 305	7 566	8 827	10 088	11 349
27	1 294	2 598	3 880	5 176	6 470	7 764	9 058	10 352	11 646
28	1 327	2 654	3 981	5 308	6 635	7 962	9 289	10 616	11 943
29	1 359	2 718	4 077	5 436	6 795	8 154	9 513	10 872	12 231
30	1 393	2 786	4 179	5 572	6 965	8 358	9 751	11 144	12 537
31	1 427	2 854	4 281	5 708	7 135	8 562	9 989	11 416	12 843
32	1 460	2 920	4 380	5 840	7 300	8 760	10 220	11 680	13 140
33	1 495	2 990	4 485	5 980	7 475	8 970	10 465	11 960	13 455
34	1 532	3 064	4 596	6 128	7 660	9 192	10 724	12 256	13 788
36	1 570	3 140	4 710	6 280	7 850	9 420	10 990	12 560	14 130
37	1 611	3 222	4 833	6 444	8 055	9 666	11 277	12 888	14 499

Table 6:2

Approximate amount of salts in soils containing predominantly sulphates and chlorides with given saturation paste resistances (adapted from the data of Davis and Bryan)

Resistance at 16°C ohms	Approximate salt content of soil in μg g^{-1}			
	Sand	Loam	Clay–Loam	Clay
18	30000	30000	—	—
19	24000	24000	30000	—
20	22000	24000	28000	30000
25	15000	17000	19400	22000
30	12400	13400	14600	15800
35	10400	11400	12200	13200
40	8600	9400	10400	11400
45	7500	7800	8800	9800
50	6700	7100	7700	8600
55	6000	6400	6900	7700
60	5500	5800	6300	7000
65	5100	5400	5700	6300
70	4800	5000	5300	5900
75	4500	4700	5000	5500
80	4200	4400	4700	5100
85	3900	4200	4400	4800
90	3700	3900	4100	4500
95	3500	3700	3900	4200
100	3 300	3 500	3 700	3 900
105	3 100	3 300	3 500	3 700
110	3 000	3 200	3 300	3 500
115	2 800	2 900	3 100	3 300
120	2 700	2 800	2 900	3 200
125	2 500	2 600	2 800	3 000
130	2 400	2 500	2 600	2 800
135	2 300	2 400	2 500	2 700
140	2 200	2 300	2 400	2 600
145	2 100	2 200	2 300	2 500
150	2 100	2 100	2 200	2 400
155	2 000	2 100	2 100	2 300
160	2 000	2 000	2 100	2 200
165	1 900	2 000	2 000	2 100
170	1 900	1 900	2 000	2 000

4* **Table 6:3**

Saline soils containing mixed neutral salts; conductivity of saturation extract; ohms resistance of saturated soil paste at 16°C for stated moisture percentage; percent salt in saturated soil paste for stated soil moisture (reproduced from Soil Survey Manual)

Conductivity of saturation extract Millisiemens cm^{-1} at 25°C ($\sigma \times 10^3$)	\multicolumn{18}{c}{Moisture percentage of soil paste}																
	20		30		40		50		60		70		80		90		100	
	R Ohms	S %	R Ohms	S %	R Ohms	S %	R Ohms	S %	R Ohms	S %	R Ohms	S %	R Ohms	S %	R Ohms	S %	R Ohms	S %
3	380	0·05	295	0·07	255	0·09	230	0·11	210	0·14	200	0·16	190	0·18	185	0·20	180	0·23
4	295	0·06	230	0·09	195	0·12	180	0·16	165	0·19	155	0·22	145	0·25	140	0·28	140	0·31
5	245	0·08	190	0·12	160	0·16	145	0·20	135	0·24	125	0·28	120	0·32	115	0·36	115	0·39
6	205	0·10	160	0·14	140	0·19	125	0·24	115	0·29	110	0·34	105	0·39	100	0·43	96	0·48
7	180	0·11	140	0·17	120	0·23	110	0·28	99	0·34	93	0·40	89	0·46	86	0·51	83	0·57
8	160	0·13	125	0·20	105	0·26	96	0·33	88	0·40	83	0·46	80	0·53	77	0·59	74	0·66
9	145	0·15	110	0·22	96	0·30	87	0·37	81	0·45	76	0·52	73	0·60	70	0·67	68	0·75
10	130	0·17	100	0·25	88	0·34	80	0·42	74	0·50	69	0·59	66	0·67	64	0·76	62	0·84
12	110	0·20	87	0·31	75	0·41	67	0·51	62	0·61	59	0·72	56	0·82	54	0·92	52	1·00
15	92	0·26	71	0·39	61	0·52	55	0·65	51	0·78	48	0·91	46	1·05	44	1·15	43	1·30
20	71	0·36	55	0·53	47	0·71	43	0·89	40	1·05	37	1·25	35	1·40	34	1·60	33	1·80
25	60	0·45	45	0·68	39	0·91	35	1·15	32	1·35	30	1·60	29	1·80	28	2·05	27	2·25
30	50	0·55	39	0·83	33	1·10	30	1·40	28	1·65	26	1·95	25	2·20	24	2·50	23	2·75
35	43	0·65	34	0·98	29	1·30	26	1·65	24	1·95	23	2·30	22	2·60	21	2·95	20	3·25
40	39	0·76	30	1·10	26	1·50	23	1·90	21	2·25	20	2·65	19	3·00	18	3·40	18	3·80
45	35	0·86	27	1·30	23	1·70	21	2·15	20	2·55	18·5	3·00	17·5	3·45	17	3·85	16·5	4·30
50	32	0·96	25	1·45	21	1·90	19	2·40	17·5	2·90	16·5	3·35	16	3·85	15	4·35	15·0	4·80

R = Resistance at 16°C for designated moisture percentage; S = Salt at designated moisture percentage.
Ohms resistance is given to the nearest 5 ohms above 100 and to nearest 0·5 ohms below 20. Percentage salt in the soil above 1·00 is given to the nearest 0·05%.

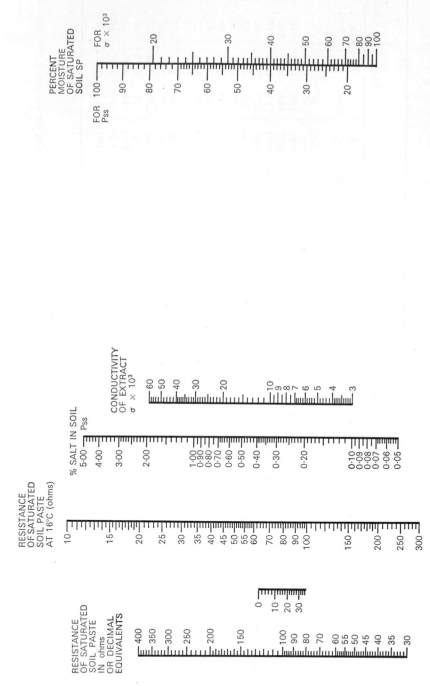

Figure 6:5 Nomogram for converting resistance of saturated soil paste to a standard temperature and for estimating soluble salt content of soils (*Soil Survey Manual*)

Total salinity by electrical conductivity of soil extracts

I / PREPARATION OF SATURATION EXTRACT

Prepare the saturated soil paste as in section 6:31, spread it on a filter paper in a Buchner, Richards or Leo funnel and filter by suction. If gypsum is known to be present, allow the saturated paste to stand overnight before filtering. Unless carbonate and hydrogen carbonate are to be determined immediately and if it is intended to measure them later, treat a portion of the extract with a solution of 100 μg cm^{-3} sodium hexametaphosphate at the rate of 1 drop per 25 cm^3; this prevents precipitation of calcium carbonate. If boron is to be measured do not use Pyrex glassware.

II / PREPARATION OF A 1:5 EXTRACT

Weigh 20 g of 2-mm soil into a suitable shaking bottle and add 100 cm^3 of distilled water. Stopper the bottle and shake for 30 minutes, then allow to stand for 15 minutes to let the bulk of soil settle. Centrifuge and filter the supernatant liquid; centrifuging first will remove much of the suspended matter. If the soil contains relatively little salt it may be highly dispersed and difficult to clarify; in such cases a filter candle or a bacterial membrane filter will help. In most cases however, soils with such low salt content will be reported as such without need of further analysis.

III / EXTRACTION OF SOIL SOLUTIONS (US SALINITY HANDBOOK NO. 60)

The apparatus is the pressure membrane (Plate 7 and Figure 6:3) with a cylinder 5 cm or 10 cm high. Cellophane (No. 600) sheets are used for the membranes. First soak the cellophane sheets in distilled water, with frequent changes of water, to remove excess electrolyte; conductivity measurements of the wash liquid indicate when this is complete. Store the membranes under water until required.

Bring the soil samples directly from the field in sealed tins or, if dry, re-wet to the desired moisture content using a fine spray of distilled water. Artificially wetted soils should be stored in airtight tins for 2 weeks at constant (approximately room) temperature before extracting.

Set up the apparatus and fit a partially dried membrane. Pack the soil to a suitable depth, depending upon the height of cylinder used, on the membrane and firmly close the chamber. Pump in washed nitrogen gas at a pressure of 1520 kN m^{-2} (15 atmospheres) using a mercury differential regulator with the by-pass valve open. When the required pressure is obtained close the by-pass valve and open the needle valve until the mercury gurgles. Mercury in the U-tube of the regulator is pushed up by the gas until nitrogen bubbles through and thus the pressure in the soil chamber is always lower than in the diaphragm chamber by the height of the mercury column. This pressure differential allows the diaphragm to keep the soil in close contact with the membrane and is applied only when several hours of extraction have elapsed and the sample has become firm.

Collect the extract in several fractions and discard the first. Conductivity measurements on all fractions will determine the uniformity of extracts. The process is slow and for some soils may take several days.

IV / MEASUREMENT OF ELECTRICAL CONDUCTIVITY OF SOIL EXTRACTS

Apparatus

The apparatus recommended is the 'Solu bridge' designed by the US Salinity Laboratory together with the micro-pipette conductivity cell (Plates 8 and 9). The line-operated bridge incorporates an automatic temperature correction device and the scale is calibrated to read directly in mS cm^{-1} at 25°C, the range being from 0·15 to 15 mS cm^{-1}. The conductivity cell has a constant of 0·5 cm^{-1} which is taken care of in the scale calibration. If a different cell is used or a different bridge, the cell constant must be determined and the readings corrected accordingly.

Determination of cell constant

Using double-distilled water prepare a 0·01 M solution of potassium chloride (0·7456 g dm^{-3} KCl) and measure its resistance at known temperature. Calculate the cell constant (k) from,

$$k = \text{conductivity (S cm}^{-1}) \times \text{resistance (ohms)} \qquad [6:14$$

The conductivity values at different temperatures of 0·01 M potassium chloride solution are obtained from Table 6:4, prepared from the data of Kohlrausch quoted by Metson (1961).

Table 6 : 4

Conductivity of 0·01 M potassium chloride solution at different temperatures (Kohlrausch)

Temperature °C	Conductivity mS cm^{-1}
10	0·001 020
15	0·001 147
16	0·001 173
17	0·001 199
18	0·001 225
19	0·001 251
20	0·001 278
21	0·001 305
22	0·001 332
23	0·001 359
24	0·001 386
25	0·001 413*
30	0·001 552

The bridge commonly used in the field is battery operated and has twin cells for comparing the test solution with a saturated gypsum solution (Plates 8 and 9). When using this type of bridge rinse both cells and fill with saturated gypsum solution. The bridge should then give a balance at 2·2 mS cm^{-1}. Provided that both cells are at the same temperature, temperature effects are automatically taken care of.

Procedure for measurement of conductivity

Measure the temperature of the soil extract and set the temperature compensation dial of the Solu bridge accordingly. If no compensation dial is present, note the temperature for future corrections. Rinse the pipette-cell with extract and then fill it. If a dip-type cell is being used follow the supplier's instructions as to the depth to which it should be inserted. Close the contact switch and balance the instrument. The point of balance is indicated by maximum opening of the 'magic eye' or minimum sound in earphones. Read the conductivity and record it in mS cm^{-1} at 25°C. If necessary, correct for temperature by the figures in Table 6:5. If the solution is too dilute or too concentrated to lie in the range of the bridge the instrument will not balance. For too concentrated solutions an aliquot should be diluted ten times and the obtained reading multiplied by ten.

Calculations

For many purposes once the conductivity of the saturation extract is known the soil sample can be assigned to a salinity class (section 6:2:1); if the conductivity values are to be translated into salt concentration then Figure 6:4 should be used. For a 1:5 extract an approximation of salt concentration can be obtained using the equations given in section 6:2:1.

More accurate results will be obtained from relationships formed by calibrating conductivity with known concentrations of the particular salts occurring in the particular soil type examined.

6:3:2 Soluble cations

Methods of determining the individual soluble cations extracted from soil are given in the relevant chapters.

6:3:3 Soluble anions

For the determination of anions other than chloride, carbonate, hydrogen carbonate and gypsum-sulphate, see the relevant chapters.

Water-soluble carbonate and hydrogen carbonate

Reagents

Sulphuric acid, 0·005 M: accurately standardized.
Phenolphthalein solution: 1% w/v in 50% ethanol.
Methyl orange: 0·05% w/v of free acid in water.

Table 6 : 5

Temperature factors (f) for correcting conductivity data on soil extracts to the standard temperature of 25°C (from USDA Handbook No. 60)

°C	f	°C	f	°C	f
3·0	1·709	22·0	1·064	29·0	0·925
4·0	1·660	22·2	1·060	29·2	0·921
5·0	1·613	22·4	1·055	29·4	0·918
6·0	1·569	22·6	1·051	29·6	0·914
7·0	1·528	22·8	1·047	29·8	0·911
8·0	1·488	23·0	1·043	30·0	0·907
9·0	1·448	23·2	1·038	30·2	0·904
10·0	1·411	23·4	1·034	30·4	0·901
11·0	1·375	23·6	1·029	30·6	0·897
12·0	1·341	23·8	1·025	30·8	0·894
13·0	1·309	24·0	1·020	31·0	0·890
14·0	1·277	24·2	1·016	31·2	0·887
15·0	1·247	24·4	1·012	31·4	0·884
16·0	1·218	24·6	1·008	31·6	0·880
17·0	1·189	24·8	1·004	31·8	0·877
18·0	1·163	25·0	1·000	32·0	0·873
18·2	1·157	25·2	0·996	32·2	0·870
18·4	1·152	25·4	0·992	32·4	0·867
18·6	1·147	25·6	0·988	32·6	0·864
18·8	1·142	25·8	0·983	32·8	0·861
19·0	1·136	26·0	0·979	33·0	0·858
19·2	1·131	26·2	0·975	34·0	0·843
19·4	1·127	26·4	0·971	35·0	0·829
19·6	1·122	26·6	0·967	36·0	0·815
19·8	1·117	26·8	0·964	37·0	0·801
20·0	1·112	27·0	0·960	38·0	0·788
20·2	1·107	27·2	0·956	39·0	0·775
20·4	1·102	27·4	0·953	40·0	0·763
20·6	1·097	27·6	0·950	41·0	0·750
20·8	1·092	27·8	0·947	42·0	0·739
21·0	1·087	28·0	0·943	43·0	0·727
21·2	1·082	28·2	0·940	44·0	0·716
21·4	1·078	28·4	0·936	45·0	0·705
21·6	1·073	28·6	0·932	46·0	0·694
21·8	1·068	28·8	0·929	47·0	0·683

Procedure

Pipette a suitable aliquot of extract into a white porcelain basin and add a few drops of phenolphthalein solution. Titrate with standard sulphuric acid until the pink colour disappears and note the titre (y cm³). Add 3–4 drops of methyl orange solution and continue the titration until the indicator turns red. Note the titre (z cm³) and keep the titrated solution for determination of chloride.

Calculation

$$HCO_3^- \text{ m.e. dm}^{-3} = \frac{2(z - 2y) \times \text{concn. in mol of acid}}{\text{cm}^3 \text{ in aliquot}} \times 1000 \quad [6:15$$

$$CO_3^{2-} \text{ m.e. dm}^{-3} = \frac{4y \times \text{concn. in mol of acid}}{\text{cm}^3 \text{ in aliquot}} \times 1000 \quad [6:16$$

These results can, if required, be converted to m.e./100 g soil:
For saturation extracts, multiply by saturation percentage/1000. For 1:5 extracts, divide by 2. The accuracy is ± 0.1 m.e.

Water-soluble chloride

Reagents

Silver nitrate solution, 0·02 M: 3·398 g dm^{-3} AgNO$_3$ and store in a brown glass bottle.
Potassium chromate solution: 5 g K$_2$Cr$_2$O$_4$ dissolved in about 50 cm^3 of water. Add silver nitrate solution until a permanent red precipitate is formed, filter and dilute to 100 cm^3.

Procedure

Titrate the aliquot of extract used for carbonate determination with silver nitrate solution using 1 cm^3 of potassium chromate solution as indicator. If a different aliquot of extract is used it should be made neutral or slightly alkaline with sodium hydrogen carbonate. The end-point is indicated by a permanent reddish-brown precipitate. A blank solution containing a suspension of calcium carbonate and indicator assists in matching colours.

Calculation

$$Cl^- \text{ m.e. dm}^{-3} = \frac{\text{cm}^3 \text{ 0·02 M AgNO}_3}{\text{cm}^3 \text{ in aliquot}} \times 20 \quad [6:17$$

This can be converted to m.e./100 g soil and the accuracy is ± 0.05 m.e.

6:3:4 Determination of gypsum

Conductance method of Bower and Huss

Shake 5 g of 2-mm air-dry soil with 100 cm^3 of distilled water for 30 minutes and filter until a clear extract is obtained. Pipette 20 cm^3 of extract into a 50-cm^3 centrifuge tube and add 20 cm^3 of acetone. Allow the tube to stand until the precipitate settles (about 10 minutes) and then centrifuge. Decant and discard the supernatant liquid and invert the tube on a pad of filter paper to drain for 5 minutes. Re-disperse the precipitate of calcium sulphate with 10 cm^3 of acetone blown from a pipette and repeat the centrifuging and draining.

Add exactly 40 cm^3 of water to the precipitate in the tube (using a burette), stopper the tube and shake until the precipitate has dissolved.

Measure the conductivity of the solution as described in section 6:3:1iv and correct if necessary to 25°C.

Calculation

Gypsum as m.e. $CaSO_4/100$ g soil is obtained from the conductivity using the curves in Figure 6:6.

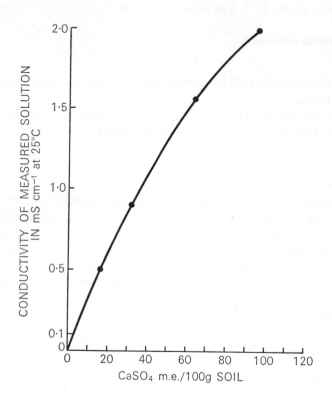

Figure 6:6:1 Relationship between concentration of gypsum in a soil and the conductivity in mS cm^{-1} at 25°C when 20 cm^3 of a 1:20 extract is taken and 40 cm^3 of water are used in the final solution measured by the method of Bower and Huss

Analytical method of Deb

Prepare a saturation extract of soil (as in section 6:3:1ii) and a 1:20 water extract as in the method of Bower and Huss (6:3:4).

Determine the calcium and sulphate contents of the saturation extract and the sulphate content of the dilute extract. For calcium determination see section 8:4:3, and sulphate is best determined by the rapid gravimetric method, section 13:3:2i.

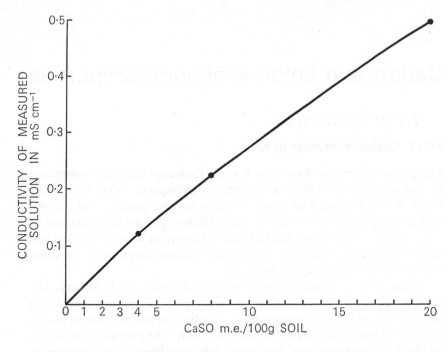

Figure 6:6:2 Relationship between concentration of gypsum in a soil and the conductivity in mS cm^{-1} at 25°C. As for Figure 6:6:1 but covering the range 0–20 m.e./100 g gypsum (1–5 m.e. dm^{-3}) and valid when 20 cm^3 of a 1:20 extract is taken and the final solution diluted to 40 cm^3

Calculation

Express all results as m.e./100 g soil.

> Gypsum m.e./100 g = m.e. of sulphate or calcium, whichever is less, in the saturation extract plus the extra amount of sulphate (m.e.) brought into dilute solution. [6:18

Example:
m.e. Ca/100 g in saturation extract = 0·41
m.e. SO_4/100 g in saturation extract = 0·98
m.e. SO_4/100 g in dilute extract = 6·80

Then, m.e. gypsum/100 g = 0·41 + (6·80 − 0·98)
= 6·23

If required, the result can be converted to percentage gypsum by multiplying by 0·0861 (i.e. equivalent weight/1000).

7

Cation and anion exchange properties

7:1 INTRODUCTION

7:1:1 Cation exchange in soils

The phenomenon now known as Cation Exchange has been recognized by soil scientists for well over a century (Thompson, 1850; Way, 1850, 1852). A soil leached with a salt solution has the power to adsorb the cation of the percolating solution and to liberate an equivalent amount of other cations. Thus a soil leached with ammonium acetate solution will adsorb some ammonium ions and liberate calcium, magnesium and other ions which will appear in the leachate.

Different cations have different powers of replacement; for example, calcium is more active than sodium in this respect and unless a percolating solution has at least three equivalents of sodium to one of calcium, practically no sodium will be adsorbed (Kelley, 1964). The predominant cations involved in exchange are hydrogen, calcium, magnesium, potassium, sodium and ammonium and such cations which can replace, or be replaced by, others in this manner are termed 'exchangeable'. Elements such as zinc, manganese and copper also occur in exchangeable forms but in small amounts.

It is seldom, if ever, that a cation in a soil exists wholly, or even largely, in the exchangeable form, but the exchangeable form is the most important source of immediately-available plant nutrient; in general, 'available' cations can be considered as 'exchangeable' cations. The total amount of exchangeable cations that can be held by a soil is known, rather arbitrarily, as its Cation Exchange Capacity.

The term cation exchange replaced the original term, base exchange, when it was realized that hydrogen ions are effective in replacing other ions. This is one reason for the acidity of soils in humid climates as they are leached with rain water charged with carbon dioxide.

The ability of a soil to hold cations in exchangeable forms is a property of its fine mineral particles and of its humus component. In mineral soils the clay fraction is largely responsible for cation exchange properties and this was recognized by Way (1850), but not until the 1930s, when X-ray measurements revealed the crystalline nature of clay minerals, was it realized that different clay minerals affected the cation exchange capacity of soils (Hendricks and Fry, 1930; Kelley, 1931). A concise review of the different clay minerals and their effect upon the cation exchange capacities of soils has been given by Metson (1961) and more detailed discussions can be found in several books, notably that of Kelley (1948).

In highly organic soils such as peats the influence of clay minerals is negligible compared to that of humus, and such soils usually have higher exchange capacities than do mineral soils. Even in soils with low organic matter content, the humus may be an important factor in cation exchange if the clay content is also low (Kamprath and Welch, 1962). Organic compounds hold bases by negative charges originating from such reactions as dissociation of hydrogen ions from carboxyl and hydroxyl groups. Attempts have been made to differentiate analytically between the exchange capacity due to organic matter and that due to the clay in a soil. Thus Olson and Bray (1938) determined the exchange capacity of a soil before and after treatment with hydrogen peroxide, and Tedrow and Gillam (1941) destroyed organic matter by ignition in a similar experiment. There is no clear evidence, however, to show that such treatments do not affect the exchange capacities of the mineral fraction of soils. Schnitzer (1965) using soils from which the total organic matter could be extracted found that the exchange capacity due to organic matter was very much higher than that found by the 'difference' method. Soils containing much hydrated amorphous mineral have been shown (Kanshiro and Sherman, 1960) to lose a lot of their cation exchange capacity merely on air-drying.

Apart from providing a source of plant-available cations, exchangeable cations affect the physical properties of a soil. Thus exchangeable calcium promotes good tilth whereas exchangeable sodium destroys soil structure, causing impermeability and all the consequent effects of poor drainage. This aspect of cation exchange is more fully discussed in the chapter dealing with sodium (section 9:2:1). Magnesium differs little from calcium in its replacing activity but in the exchangeable form its effect upon a soil is similar to that of sodium (see section 8:3:1).

Aluminium and iron(III) are special cases when considering cation exchange, due to the ease with which their salts hydrolyze and to the insolubility and weak basic character of their hydroxides. The still debatable question as to whether or not aluminium is an exchangeable ion has been intensively investigated. Marshall (1949) gives a summary of the work done to that date and it was concluded that aluminium is not much involved in exchange reactions but that it is ionized and brought into solution from acid clays by neutral salts. It is preferable to refer to 'extractable aluminium' rather than to 'exchangeable'. What applies to aluminium probably applies equally to iron(III) but this has been relatively little investigated.

The determination of cation exchange capacity and the individual exchangeable cations not only helps to evaluate the fertility of a soil but also to classify it. It is important to know the distribution of exchangeable cations in a profile as a measure of weathering. Basic cations will be leached by acidic waters and be either lost in drainage or taken up by plants; their replacement from primary minerals by weathering is a slow process except in some tropical climates. de Endredy and Quagraine (1960) used so-called 'Exchange Capacity Ratios' as a soil characteristic, that is, the capacity found by using ammonium as exchanging ion over that found by using barium; but, as discussed in section 7:2:1, this may be based upon a fallacy. Cation exchange capacity determinations furthermore provide one of the most efficient ways of assessing the lime requirement of a soil (section 4:4).

7:1:2 Anion exchange in soils

Anion exchange is analogous to cation exchange and is really a measure of the number of positive charges carried by a soil. Mehlich (1953) made anion exchange determinations using the phosphate ion, which is particularly suitable for the purpose. Sulphate is often involved in anion exchange reactions but chloride and nitrate have negligible effects. Anion exchange is dependent upon pH but can be quantitative under acid conditions. Haan and Bolt (1963) determined anion exchange capacities of clays using radio-nuclides and pointed out that when interpreting the experimentally observed interaction between anions and clays, two aspects should be recognized. That is, the negative adsorption of anions by negatively charged sites and the positive adsorption on other sites. A generalized equation was developed by Haan and Bolt describing the contribution of negative adsorption.

Negative adsorption refers to a deficit of anions near the clay particle in comparison with the equilibrium solution and is due to repulsion between anions and the negative charges on the particles. Very often negative adsorption masks the presence of any positive adsorption, and to interpret anion adsorption values found experimentally it is necessary to calculate the negative adsorption. Such calculations were first introduced by Schofield (1947) and were developed by Bolt and Warkentin (1958). The calculations at that time were confined to a consideration of one anion only; the calculations introduced by Haan and Bolt include more than one anion.

For the full mathematical derivation and discussion of the equations the reader is referred to the original paper of Haan and Bolt, but briefly, the equation dealing with the negative adsorption of monovalent anions is,

$$\frac{\Gamma}{(1 - f^{2-}) N_0} = \frac{\sqrt{2}}{\sqrt{\beta N_0}} \cdot Q'(f^{2+}, f^{2-}) - \delta \qquad [7:1$$

where, in a system of mono- and divalent cations and anions of total salt concentration N_0 eq. dm^{-3},

f^{2+} is the equivalent fraction of divalent cations
f^{2-} is the equivalent fraction of divalent anions
β is a constant of double layer theory and equal to $1 \cdot 06 \times 10^{15}$ at 25°C
Γ is the negative adsorption of monovalent anions in m.e. cm^{-2} surface
δ is a distance value obtained from the relationship, $\delta \simeq 4/z^+ \beta \Gamma$, in

which z^+ is the valency of the dominant cation and Γ is the surface charge density of the colloid in m.e. cm^{-2}

and the Q' function of f^{2+} and f^{2-} is,

$$Q' = \frac{1}{\sqrt{f^{2-}}} \cdot \left[\text{arg sinh} \frac{\sqrt{f^{2-}} \sqrt{f^{2+}} + 2 + f^{2-}}{\sqrt{1 - f^{2+} f^{2-}}} - \text{arg sinh} \frac{\sqrt{f^{2-}} \sqrt{f^{2+}}}{\sqrt{1 - f^{2+} f^{2-}}} \right]$$

[7:2

A similar equation was derived for the divalent anions.

7:2 DETERMINATION: BACKGROUND AND THEORY

7:2:1 Cation exchange capacity

The conventional method for determining cation exchange capacity is by saturating the soil complex with an index cation, washing out the excess and determining the amount of cation retained.

Originally (Gedroiz, 1924; Hissink, 1923; Kelley and Brown, 1924) cation exchange capacity was measured as the quantity of ammonium held in exchangeable form after leaching a soil with ammonium chloride solution. Later, ammonium acetate was more generally used (e.g. Schollenberger and Simon, 1945). It was soon recognized that the ammonium so determined was not always equal to the sum of the individually measured cations in the leachate owing to extraction of soluble, as distinct from exchangeable, cations.

Carlson (1962) investigated the anomaly of obtaining more equivalents of exchangeable cations than expected from the cation exchange capacity and found this to be due largely to errors in procedure. Adsorption of complex alkaline-earth metal cations contributed to the supersaturation of the soil complex, and in soils having much solid-phase phosphate, errors were introduced by incomplete removal of sparingly soluble polyvalent phosphate-associated cations during washing. Carlson concluded that ammonium is not a suitable cation with which to measure exchange capacity as its weak basic character makes it prone to hydrolysis when washing. He recommended sodium as a preferable index ion. Chapman (1965) points out that for vermiculite clay soils, whereas ammonium readily displaces other cations and is thus suitable for determining total exchangeable bases, it is not itself fully displaced and thus is not suitable for determining exchange capacity.

It has been found that in general, barium, strontium, calcium and magnesium used as index cations result in higher values of cation exchange capacity than when ammonium or sodium are used and that pH also affects the results. Mehlich (1942) reported that the cation exchange capacity of a soil is constant between pH 8 and pH 9 and consequently recommended a barium chloride–triethanolamine buffer solution at pH 8·2.

The complicating factor of different cation exchange capacities being shown in the same soil by divalent as opposed to monovalent index cations was for a long time accepted as a necessary and unexplainable evil. Then it was thought that the conflicting results may be due to methodology and that errors may occur, particularly during the washing out of excess index ion. However, de Endredy and Quagraine (1960) concluded that the difference between results obtained with mono- and divalent cations is real and not dependent upon methodology. Helmy (1963) found that according to concepts which follow directly from the double layer theory, cation exchange reactions in most instances are not stoichiometric. This, Helmy considered, explained why cation exchange capacities obtained with divalent cations are higher than those found using monovalent cations.

Okazaki and co-workers (1964) challenged the findings of de Endredy and Quagraine. These workers pointed out that in each step of the determination of cation exchange capacity errors are likely to occur, but will be most serious during the washing step. Most washing procedures involve, for convenience of routine, a fixed amount of wash liquid, and as soils differ in capacity to retain salts, some soils will retain excess index cation even after washing. On the other hand, with some soils hydrolysis of the index cation will occur during washing. Consequently if an accurate assessment of cation exchange capacity is wanted it is necessary to eliminate or balance salt-retention and hydrolysis errors. Okazaki and co-workers (1963) determined salt-retention and hydrolysis errors as suggested by Bower *et al.* (1952). Each portion of a wash liquid was analyzed for cation and anion and washing was continued until chloride, present in the index saturating salt, was completely removed. The anion analysis measured the removal of excess saturating salt and the cation analysis measured the removal of excess saturating salt plus the adsorbed cation removed by hydrolysis. The difference between them was a measure of index cation hydrolysis. Salt retention was calculated from measurements of anion removal by subsequent washings. The net error at any stage was measured by the imbalance between hydrolysis and salt retention. The conclusion was that for the particular soils investigated, hydrolysis errors were unimportant but that this would not necessarily be true for all soils.

In their subsequent investigations Okazaki *et al.* (1964) examined the results of de Endredy and Quagraine with a view to possible errors in methodology. Four soils were analyzed using sodium and barium as index cations and the values for cation exchange capacity were found to be the same with barium saturation as with sodium saturation at the same pH value. As the only difference in procedure from that used by de Endredy and Quagraine was elimination of the washing step, Okazaki claimed that de Endredy's results were due to errors during the washing step.

In a second experiment Okazaki and co-workers again analyzed the soils for cation exchange capacity, this time including a washing step and using the centrifuge technique. Barium acetate–barium chloride was used as saturating solution with subsequent water washing and barium acetate and sodium acetate were used with ethanol washing (80% as used by de Endredy). Water washing was continued until all the chloride had been removed or until the soil dispersed. The results showed that choice of index cation did affect the exchange capacities but it was pointed out that it is the apparent, and not the real, exchange capacity that is affected.

One of the soils used by Okazaki was very difficult to re-suspend after centrifuging; this retarded salt removal and gave high apparent exchange capacities. The curves in Figure 7:1 were obtained with this particular

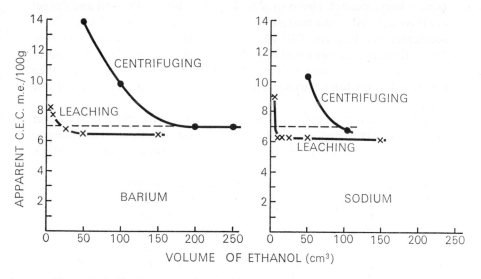

Figure 7:1 Comparison of centrifuging and leaching upon salt retention and hydrolysis of index ion using barium and sodium as index ions (Okazaki *et al.*; *Soil Science* © 1964)

soil and employing a washing step; it can be seen that with barium as index cation particularly large errors resulted. Complete suspension was obtained only by very careful shaking. The inference from this experiment is that regardless of whether or not a washing step is included, complete suspension of a soil must be obtained as this particular source of error will be present during saturation and displacement as well as during washing.

As a general conclusion from their work Okazaki *et al.* strongly oppose the inclusion of a washing step in cation exchange determinations. An attempt to use the controlled washing technique of Bower *et al.* (1960), in which washing is continued until the conductivity of the washings reaches 40 μS cm^{-1} (micro-mhos/cm), failed because the soils dispersed before washing was complete. Similarly some soils disperse before all tracer, for example chloride, ions can be washed out. Rich (1962) removed most of the excess saturating salt (CaCl$_2$) by washing with water and methanol and then calculated the exchange capacity from the calcium and chloride extractable by another salt solution (magnesium acetate). Hydrolysis errors were assumed to be small.

Ensminger (1944) proposed a method that depended upon washing out excess index solution with a more dilute solution of the same salt. In an example of the method Ensminger assumes barium acetate as index solution. The soil is weighed in a beaker, 25 cm³ of normal barium acetate solution added, the suspension stirred during 1 hour and filtered using a weighed funnel and paper. The soil is then leached with 250 cm³ of 0·5 M barium acetate solution containing 0·5% n-butyl alcohol during 24 hours and then with 250 cm³ of 0·05 M barium acetate solution containing butyl alcohol, also over a period of 24 hours. The soil and funnel are then weighed to the nearest 0·1 g. The net weight minus dry weight of soil represents the excess 0·05 M acetate solution. The dilute acetate solution is then leached out together with adsorbed barium with 0·2 M hydrochloric acid and the barium determined. Cation exchange capacity is then calculated from total barium less the amount present as excess dilute acetate.

Another source of error in washing procedures was reported by Baker (1962). The error was specifically that of traces of ethanol remaining to contaminate the final extract, thus causing errors when determining sodium with a flame photometer.

For the routine analysis of many samples, techniques involving a washing step may still be preferred as more rapid, convenient and sufficiently accurate. In such cases sodium is recommended as index ion. Chapman (1965) recommended the use of isopropyl alcohol to wash out excess index salt but Smith *et al.* (1966) found this less efficient than methanol or ethanol and it gives a possibility of precipitation of sodium acetate. Smith concluded that in general, errors due to salt retention are more important than those caused by hydrolysis and that for most soils the salts should be washed out and hydrolysis ignored. When particularly accurate results are required alternative methods of analysis must be followed.

Bascomb (1964) utilized the phenomenon of 'compulsive exchange' which occurs between a barium soil and magnesium sulphate solution and during which barium ions are removed by precipitation as sulphate. An EDTA titration is used to measure the amount of magnesium exchanged from a standard solution by a barium soil and the cation exchange capacity of the soil calculated. The method is widely applicable, even to calcareous soils or those rich in organic matter. It has the further advantages of speed, constant pH conditions, avoidance of excessive washing and complete replacement of saturating cation in one operation.

Saline, calcareous and alkali soils need special techniques for measurement of exchangeable cations and exchange capacities. Some workers remove soluble salts from saline soils by leaching prior to extraction of exchangeable ions (e.g. Gill and Sherman, 1952). This procedure is not recommended as even with alcohol washing, hydrolysis may occur. The most acceptable procedure is to make a separate determination of water-soluble cations and to subtract them from the soluble plus exchangeable. However, such a procedure raises the question of the conditions under which soluble salts are to be extracted. The commonly accepted method is to determine the cations in the saturation extract, although it has been pointed out by Nijensohn (1960) that such determinations are not reproducible and that it would be preferable to measure the cations in the displaced soil solution at field capacity. This too is difficult in practice and thus Nijensohn proposed a method whereby soil solutions are displaced at moisture contents equivalent to the soil–water retention stress which is equivalent to 1 cm water pressure. Nijensohn's procedure is to measure the moisture equivalent to 1 hectopascal (0·001 atmosphere) tension by a modification of Piper's (1942) method for determining water-holding capacity. The soluble cations are then extracted by a somewhat tedious procedure and the soil leached with calcium sulphate solution to extract exchangeable sodium, potassium and magnesium. The exchange capacity is obtained by placing the calcium-saturated soil in contact with ammonium oxalate solution and determining the changes in concentration of ammonium and of oxalate. The method, although ingenious, will find little application in routine analysis for which the reproducibility of the saturation extract method is sufficiently accurate.

Soils which are saline–alkali may prove difficult to leach owing to poor permeability. This difficulty is best overcome by centrifuging instead of leaching, although great care must be taken to ensure complete suspension of the sample for each new shaking period. An alternative procedure is to mix the soil with a weighed amount of pure sand to facilitate leaching. The ammonium ion can be fixed under moist conditions by some saline–alkali soils, and although this is immaterial if using ammonium acetate to displace exchangeable cations, it leads to low exchange capacity results when this quantity is measured by determination of displaced ammonium. Thus for cation exchange capacity determinations on such soils it is preferable to use a different index ion, for example sodium, which is not fixed. If a saline or saline–alkali soil also contains gypsum, additional difficulties arise with certain extractants. Barium cannot be used as an index ion and it is difficult to replace calcium as it is continually released from the gypsum (Bower *et al.*, 1952).

Shaw and McIntire (1935) found that cold ammonium chloride solution as saturant could not be used to determine the exchange capacity of calcareous soil as it dissolves some calcium carbonate. They thus investigated the use of boiling ammonium chloride solution and found that it effected complete replacement of exchangeable cations and complete disintegration of carbonates. Their final procedure was to distil the soil in ammonium chloride solution, collecting evolved ammonia in standard acid. After cooling, dilute ammonium hydroxide solution was added to the suspension which was then allowed to stand. The soil was filtered off, washed with ammonium chloride solution and leached with ethanol until chloride-free. A second distillation was made in the presence of magnesium oxide and the exchange capacity calculated.

Dell'Agnola and Maggioni (1964) investigated several extractants and techniques for determining the exchange capacity of calcareous soil. Barium chloride was found well suited as a saturant as it does not react with calcium carbonate and does not disperse clay or extract organic matter. Yaalon and co-workers (1960) working with calcareous and saline soils used lithium chloride solution buffered to pH 8·2 with lithium acetate as saturant, and the soil was then immediately leached with calcium acetate at pH 8·2. Soluble cations were determined separately and there was no need to wash out excess saturant if it was assessed as chloride. Amer (1960) found that for calcareous soils the sodium acetate method gave high values for exchange capacity, due to the high pH effects, and found ammonium acetate more accurate.

Blume and Smith (1954) improved the methods of determining exchangeable calcium and cation exchange capacity of calcareous soils by using radioactive calcium. Their method depends upon equilibration of the soil with a dilute solution containing known amounts of ^{40}Ca and ^{45}Ca the concentrations of which are followed by analysis. By this means the decrease in specific activity of calcium is measured in the solution and the calcium entering from the soil into equilibrium with the calcium in the solution is calculated.

Peter and Markert (1960) measured cation exchange capacities of soils by methylene blue sorption. Schönberg (1963) obtained a significant correlation of cation exchange capacity and methylene blue absorption, particularly with subsoils. The methylene blue sorption of organic components was, however, lower than their cation exchange capacities and therefore the ratio of absorption to cation exchange capacity will depend upon the humus content of a soil and will be wider as the proportion of humus increases.

Sieling (1941) proposed a rapid colorimetric procedure for determining exchange capacity based upon the sorption of copper from a standard solution of copper acetate in dilute acetic acid. The decrease in copper concentration in a measured volume of solution is determined using a weighed quantity of soil. Sieling claimed that the method is sufficiently accurate for quick tests and for soil classification, particularly in profile examination. Sieling further found that the variations in exchange capacity caused by drying a soil were greater than the variations in results obtained by copper sorption as opposed to standard leaching procedures.

Baker and Burns (1964) used nickel as the displacing ion as a 0·125 M neutral solution of nickel chloride. They found this to give higher values for cation exchange capacity than ammonium acetate and the values were consistently reproduced. The divalent nickel ion does not disperse the inorganic fraction of the soil as do ammonium, sodium and calcium acetates; neither does it hydrolyze the organic fraction. As nickel is normally not present to any great extent in soils, the analysis is more simple than that of the more conventional reagents and, more imporant, enables the washing step to be omitted entirely. The total nickel finally determined is that displaced from the exchange sites plus the excess from the nickel chloride solution. Chloride analysis gives the chloride from the excess nickel chloride, enabling the non-exchanged nickel to be calculated instead of washed out.

Street (1963) explained that a linear relationship exists between cation exchange capacity and the iso-conductivity value defined for that concentration of electrolyte at which the relative conductivity (K/K_1) remains unity as clay concentration varies. The relationship is due to a linear relationship between exchange capacity and surface area and to an approximate linear relationship between conductivity and concentration over a limited concentration range.

Ragland (1962) calculated cation exchange capacities of soils from thermometric titration curves and his results compared well with those obtained by the barium chloride–triethanolamine method. Ragland used apparatus equipped with an adiabatic titration cell and a thermistor for monitoring temperature changes. A Wheatstone bridge was balanced against the resistance of the thermistor at the beginning of the titration.

Simonson and Axley (1963) studied cation exchange in small samples of clay and soil by high frequency titration in 1:1 water–ethanol suspension with 0·05 M magnesium sulphate solution. The samples were pre-treated with sodium carbonate solution and, after washing out the carbonate, with barium chloride solution. Peech (1945) described a micro-method for determining exchange capacity in which small quantities of soil are leached on a small Buchner funnel with ammonium acetate, washed with ethanol, leached with sodium chloride solution and the displaced ammonia determined by Nesslerization. Bergseth and co-workers (1963) developed a method for measuring the exchange capacity of very small amounts of material. In such cases the determination of the very small quantities of displaced cations is difficult and so Bergseth flocculated the dispersed soil colloid with radio-strontium ($^{89}SrCl_2$). The flocculate is then washed free from chloride and the strontium activity determined. Such a procedure has been used also for routine analyses but, as pointed out by Halevy and Tzur (1964), no method involving strontium can be used for calcareous soils due to precipitation of strontium carbonate.

Maywald (1962) described a rapid procedure for determining the exchange capacity of a mineral soil in which 10 g of soil are shaken for 2 hours with 200 cm³ of 0·04 M hydrochloric acid. After standing for one day, 100 cm³ of supernatant liquid are titrated with 0·1 M sodium hydroxide solution. To the soil is added 20 cm³ of a 2·5% solution of barium chloride and 12 drops of 2% phenolphthalein solution and the suspension titrated with 0·1 M sodium hydroxide solution until, with repeated shaking, the foam remains pink. The difference between the two titres is taken as a measure of exchange capacity.

Middleton and Westgarth (1964) gave a rapid method for determining exchangeable hydrogen and exchange capacity in which equilibrium solutions are made with barium nitrate and barium acetate. The results are extrapolated to pH 7 as cation exchange capacity and pH have a linear relationship as demonstrated by the authors. The precision of the method was claimed to be as good as that obtained by standard methods.

Chapman (1965) considers that for acid and non-calcareous soils the best measure of cation exchange capacity is obtained by summing the total exchangeable bases and the exchangeable hydrogen.

Edwards (1967) introduced an entirely new technique for measuring the cation exchange capacity of soils which avoids the washing step. Edwards proposed shaking the soil with water and an insoluble cation exchange resin saturated with ammonium ions. The resin is then filtered off and the soil sample steam distilled with sodium chloride and magnesium oxide. As presented by Edwards the procedure was on a semi-micro scale (0·2–0·5 g soil) but can probably be adapted for larger quantities of soil. By suitable leaching of the resin individual exchangeable bases can be brought into solution for determination.

7:2:2 Total exchangeable cations

As far as extraction and determination of the individual exchangeable cations are concerned, ammonium acetate is to be preferred over all other extractants, mainly for analytical convenience, although it has the further advantage of ammonium ions usually being present naturally in soil in small amount. The ammonium acetate leachate is easily worked up for subsequent measurement of cations by digestion with nitric and hydrochloric acids to destroy the organic fraction. Excess ammonium is driven off by heating and the residue taken up in dilute hydrochloric acid. Metson (1961) gives a conveniently rapid procedure whereby the leachate is ignited to leave the bases as carbonates and oxides; simple titration of the residue with standard acid then gives an approximation of total exchangeable bases.

Mitchell (1936) developed the Lundegardh flame spectrographic method for determining the individual bases in an ammonium acetate leachate of soil, but his procedure necessitated large quantities of solution and covered a small range of cation content. Black and Smith (1950) considered Mitchell's procedure unsuitable for routine and modified it to cover a greater range and to use less reagent. In the modification of Black and Smith 10 g of soil are leached with 100 cm^3 of neutral, 1 M ammonium acetate solution and before digesting the leachate with acids, its pH is measured. This pH measurement gives useful and easily obtained information about the soil. An acid leachate indicates an unsaturated soil, a neutral reaction indicates a saturated soil and an alkaline reaction shows the presence of free carbonates. For unsaturated soils measurement of exchangeable hydrogen and total exchangeable bases will give an approximation of exchange capacity and percent base saturation.

Black and Smith tested the efficiency of the 1:10 extraction ratio and found that for calcium a 96% recovery was obtained, for manganese the recovery was 91% and potassium and magnesium were 100% extracted. The results in general agreed well with those obtained with Mitchell's original method. Full experimental details are given by Black and Smith for the extraction of soil, treatment of leachate and use of the Hilger medium quartz spectrograph including subsequent photometry. An ingenious leaching apparatus was devised by Black incorporating an easily prepared constant level device (Figure 7:2:1); this was later modified by Legg (1963) as shown in Figure 7:2:2. Gillingham (1965) described an apparatus for leaching soil in which clear leachates are obtained regardless of the soil texture (Figure 7:3).

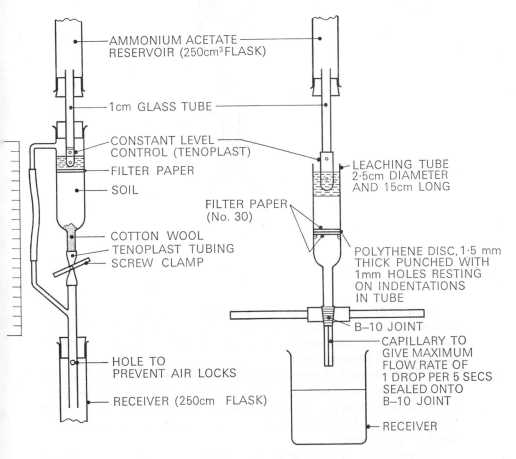

Figure 7:2:1 Apparatus for leaching soils for cation exchange measurements (Black and Smith, 1950; *J. Sci. Fd. Agric.*)

Figure 7:2:2 Modification of the leaching apparatus of Black and Smith (Legg; *Soil Science* © 1963)

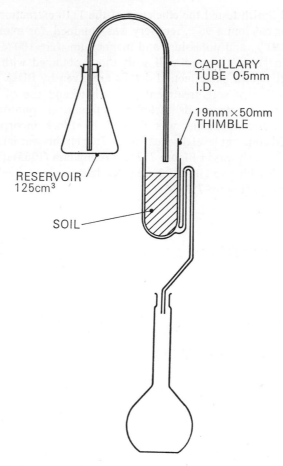

CAPILLARY
TUBE 0·5mm
I.D.

19mm × 50mm
THIMBLE

RESERVOIR
125cm³

SOIL

Figure 7:3 Apparatus for leaching soil for cation exchange measurements (Gillingham, 1965; *Can. J. Soil Sci.*)

Total exchangeable bases can be estimated by titration of the ignited residue of leachate or alternatively the soil is extracted with 1 M acetic acid and the total exchangeable bases calculated from the change in pH of the solution (Brown, 1943).

The measurement of exchangeable hydrogen is more difficult than that of exchangeable bases and the results are more dependent upon methodology. It is often approximately estimated by subtracting the total exchangeable bases from the exchange capacity. Its measurement has been discussed in Chapter 4.

Determination of other individual exchangeable cations is discussed and described in the chapters dealing with those elements.

Usually all cation exchange properties of a soil are expressed in terms of milliequivalents per 100 gramme of soil rather than as percentages. Hissink (1923) introduced the terms 'S-value', meaning total exchangeable bases as m.e./100 g soil and '*T*-value', meaning cation exchange capacity in m.e./100 g soil. Thus

$$T = S + H^+_{ex} \qquad\qquad [7:3$$

Apart from individual exchangeable cations, total exchangeable bases and exchange capacity, the 'per cent saturation' of a soil is often reported and, using the notation of Hissink,

$$\% \text{ saturation} = S/T \times 100. \qquad\qquad [7:4$$

Sometimes it is required to calculate the m.e. % of each cation as a percentage of the exchange capacity; for example,

$$\% \text{ Na saturation} = \frac{\text{m.e. Na/100 g}}{\text{C.E.C.}} \times 100 \qquad\qquad [7:5$$

7:3 RECOMMENDED METHODS

7:3:1 Cation exchange capacity

Using sodium as index ion
A rapid method sufficiently accurate for routine work provided that due care is taken to ensure adequate dispersion and shaking of the sample.

Reagents

Sodium acetate solution, 1·0 M: 136 g dm^{-3} sodium acetate trihydrate and adjusted to pH 8·2.
Ammonium acetate solution, 1·0 M: Add 57 cm^3 glacial acetic acid and 68 cm^3 of strong ammonium hydroxide to 800 cm^3 of water. Dilute to 1 dm^3 and adjust to pH 7·0; *or,*
Magnesium acetate solution, 0·5 M: 107 g dm^{-3} crystalline salt.

Procedure

Weigh 5 g of 2-mm soil into a 50-cm^3 centrifuge tube, add 30 cm^3 of sodium acetate solution and shake for 5 minutes. The tubes should be stoppered with polythene or clean rubber stoppers and not corks which introduce errors. Centrifuge the tubes at 200 rev/s for about 5 minutes until the supernatant liquid is clear. Decant and discard the liquid and repeat the shaking and centrifuging four times more with fresh portions of acetate solution. Shake the soil with 30 cm^3 of 95% ethanol for 5 minutes, centrifuge and discard the liquid. Repeat the ethanol washing three times. Finally extract the soil with three 30-cm^3 portions of ammonium acetate solution and collect the extracts in a 100-cm^3 graduated flask. Occasionally it is necessary to filter the extracts after centrifuging.

Dilute the combined extracts to 100 cm^3 and determine the sodium content, preferably by flame photometry (section 9:3:5).

If the more rapid procedure using the sodium electrode (section 9:3:7) is used to determine the sodium content, make the final extraction with three portions of magnesium acetate instead of the ammonium salt.

Note: Some soils become sticky during the saturation process and it is extremely important to ensure complete re-dispersion in the next added solution. For soils giving difficulty in this respect it may be better to use a leaching technique.

Calculation

$$\text{C.E.C. m.e./100 g soil} = \frac{10 \times \text{Na concentration in m.e. dm}^{-3}}{\text{weight of sample}} \qquad [7:6$$

Using barium as index ion (After Bascomb)

This method can be used for all kinds of soil including calcareous or organic samples; some modification may be necessary if the soil is rich in sulphate. The accuracy is $\pm 5\%$.

Reagents

Triethanolamine solution: Dilute 90 cm³ of triethanolamine to 1 dm³ and adjust to pH 8·1 with 2 M HCl; this will require about 150 cm³ of acid. Dilute to 2 dm³ and protect from CO_2.

Barium chloride solution: 244 g dm⁻³ $BaCl_2.2H_2O$.

Buffered barium chloride solution: Mix equal volumes of barium chloride and triethanolamine solutions.

Magnesium sulphate solution: 6·2 g dm⁻³ $MgSO_4.7H_2O$.

EDTA solution: 3·723 g dm⁻³ disodium salt.

Catechol violet indicator: 0·1 g reagent/100 cm³ of water.

Procedure

Weigh 1 g of 2-mm soil into a 50-cm³ centrifuge tube, stopper the tube and record its weight (w_1). If the soil is calcareous treat it with 20 cm³ of buffered barium chloride solution with gentle shaking for 1 hour; centrifuge at 250 rev/s until the supernatant liquid is clear (about 15 minutes) and discard the liquid. This initial treatment can be omitted for non-calcareous soils.

Treat the soil sample in the tube with 40 cm³ of buffered barium chloride solution overnight. Centrifuge and discard the liquid. Add approximately 40 cm³ of distilled water and thoroughly shake the tube; centrifuge and discard the washing. Re-weigh the stoppered tube (w_2). Pipette 20 cm³ of magnesium sulphate into the tube and shake the mixture for 2 hours. Centrifuge and transfer the liquid to a stoppered flask.

Take 5 cm³ of the final extract, add 6 drops of 2 M ammonium hydroxide solution (or 5 cm³ of 2 M triethanolamine) and titrate with standard EDTA solution using two drops of indicator solution. The end-point is shown by a change in colour from blue to reddish violet. The catechol violet indicator can be replaced with a 0·4% solution of Eriochrome

Black T in methanol and containing 4% hydroxylamine hydrochloride. This solution must be prepared fresh every week. For titration of the 5 cm³ aliquot of extract raise the pH by adding 5 cm³ of 2 M triethanolamine. Let the titre be A_1 cm³. This must be corrected (A_2 cm³) for the effect of volume (not chloride content) of liquid retained by the centrifuged soil.

Titrate 5 cm³ aliquots of the original magnesium sulphate solution under similar conditions (B cm³).

Calculations

$$\text{Corrected titre } (A_2) = A_1(100 + w_2 - w_1)/100 \text{ cm}^3 \qquad [7:7$$

$$\text{C.E.C.} = 8(B - A_2) \text{ m.e.}/100 \text{ g soil.} \qquad [7:8$$

Note: If the C.E.C. exceeds 50 m.e./100 g, repeat the determination using less soil and modify the calculation accordingly.

7:3:2 Total exchangeable bases (after Metson)

Apparatus

The extraction of the soil may be done by the centrifuge technique or, more efficiently, by leaching. For leaching the apparatus of Black and Smith (Figure 7:2:1) or its modification by Legg (Figure 7:2:2) can be used but the most convenient is that of Gillingham (Figure 7:3).

A 0·5-mm inverted capillary U-tube, with its shorter end extending to the bottom of a 125-cm³ conical flask, siphons ammonium acetate solution dropwise on to the soil held in a 19-mm × 50-mm paper thimble. The thimble is placed in the siphon cup with three pieces of glass tubing at its sides to centre it and to prevent air-locks. With the rubber tube on the siphon pinched closed, the cup is filled with extractant and left overnight. The tube is then opened and extractant allowed to drip into the cup; 5 hours are needed to leach 5 g soil with 100 cm³ of extractant.

Reagents

Ammonium acetate solution, neutral, 1·0 M: As prepared in section 7:3:1.
Hydrochloric acid: 0·2 M and standardized.
Ammonium hydroxide solution, 0·1 M and standardized.

Procedure

Extract 5–10 g of 2-mm soil with 100 cm³ of ammonium acetate solution. Make the leachate to standard volume with ammonium acetate solution. If the centrifuge technique is adopted the tubes should be stoppered with polyethylene or rubber stoppers, not corks. It is essential that a blank determination be carried out through all stages including shaking and centrifuging.

The final solution is used not only for total exchangeable bases but for

5 + T.S.C.A.

the determination of individual exchangeable cations. Its pH should be measured (see section 7:2:2).

Evaporate a suitable aliquot of extract to small volume, not to dryness, on an electric hotplate. Transfer the residual liquid to a small (7·5-cm diam.) silica basin, rinsing the original glass dish with hot water. Slowly and carefully evaporate the solution to dryness, avoiding spattering; when the residue is quite dry, heat the basin in an electric muffle furnace at 475 to 500°C for 1 hour.

Cool and add an excess (at least 5 cm³) of 0·2 M hydrochloric acid from a burette, breaking up the residue with a glass rod. If the residue is black or grey it may contain manganese which can be included in the determination by adding 3 drops of 30% hydrogen peroxide solution and boiling (Schollenberger, 1948). Digest on a water-bath for 30 minutes and titrate with 0·1 M ammonium hydroxide solution using methyl red as indicator. The titrated solution may, if necessary, be used for the determination of individual bases, but in general it is best to use a new aliquot of soil extract for this.

Calculation

$$\text{T.E.B. m.e./100 g soil} = \text{cm}^3 \text{ of 0·1 M HCl used to neutral-} \quad [7:9$$
$$\text{ize the residue corresponding to}$$
$$\text{10 g of soil}$$

or if

weight of soil leached	$= w$ g
final leachate	$= V$ cm³
aliquot evaporated	$= v$ cm³
cm³ 0·1 M acid used to neutralize	$= m$ cm³

then   T.E.B. m.e./100 g soil $= \dfrac{10\,mV}{wv}$ [7:10

As a check on the T.E.B. measurement, the individually determined exchangeable bases can be summed or exchangeable hydrogen deducted from the C.E.C.

7:3:3 Individual exchangeable bases

Preparation of leachate

For the determination of individual exchangeable bases the organic matter must first be destroyed. This may be done by evaporation and ignition of the leachate as described in section 7:3:2, or the leachate can be digested with acids.

Evaporate 100 cm³ of soil leachate to dryness in a glass basin on a steam-bath. Treat the residue with 5 cm³ of concentrated hydrochloric acid and 1 cm³ of concentrated nitric acid, the basin being covered with a clock-glass until the initial violent reaction is over. Replace the clock-glass with a 'Speedy-vap' cover (Vogel) and take the solution to dryness.

Add 1 cm³ of concentrated hydrochloric acid to the residue and again take to dryness. Cool and add 5 drops of concentrated hydrochloric acid and 15 cm³ of water; cover the basin and warm on a steam-bath to complete solution. Cool and transfer to a 50-cm³ graduated flask, dilute to the mark and filter through a Whatman No. 40 paper.

The solution will contain the exchangeable bases as chlorides and is suitable for their determination by spectrographic or chemical methods.

It is essential that a reagent blank be prepared and in some cases it is not necessary, or even advisable, to digest the soil leachate for sodium and potassium determinations by flame spectrometry as the blank is often high.

7:3:4 Anion exchange capacity (Mehlich)

Reagents

Triethanolamine solution: Dilute 90 cm³ of triethanolamine to 1 dm³ and adjust to pH 8·1 with HCl. Dilute to 2 dm³ and mix with 2 dm³ of water containing 100 g $BaCl_2.2H_2O$.

Calcium chloride solution: 50 g dm⁻³ $CaCl_2.2H_2O$, adjusted to pH 8·0 with saturated $Ca(OH)_2$ solution.

Phosphoric acid solution: 0·01 M in H_3PO_4.

Reagents for measuring phosphorus by the vanadate method: see section 12:3:1ɪɪɪ.

Procedure

Leach 10 g of 2-mm soil with 100 cm³ of triethanolamine solution and wash six times; Mehlich washed with water but 95% ethanol is preferable. Leach the soil with 100 cm³ of calcium chloride solution and again wash. Dry the calcium-saturated soil at 45°C and weigh into a centrifuge tube sufficient to give 0·2 m.e. C.E.C. Add 20 cm³ of phosphoric acid solution and shake for 30 minutes. Allow to stand for 24 hours and again shake for 30 minutes. Centrifuge and take a 1 cm³ aliquot of liquid to determine phosphorus. In a separate sample of soil, extract the phosphorus with a solution of ammonium fluoride in hydrochloric acid (section 12:2:5) and determine its content.

Calculation

$$\text{A.E.C. m.e./100 g soil} = \text{(extractable P + P adsorbed),}$$
$$\text{expressed as m.e./100 g soil} \qquad [7:11$$

8

Calcium and magnesium

8:1 INTRODUCTION

Calcium, magnesium, potassium and sodium have been grouped in two consecutive chapters—following the discussion of soluble salts and exchangeable cations—as being the four main cations determined. They are all macronutrient elements with the possible exception of sodium which is, however, of importance in saline and alkali soils. Furthermore, many of the analytical techniques described apply to more than one, and sometimes to all, of these four elements.

8:2 CALCIUM

8:2:1 Background and theory

Calcium occurs widely and abundantly in soils as the carbonate, phosphate, silicate, fluoride and sulphate. The carbonate is the most important source of soil calcium, although phosphate and sulphate become predominant in certain types of soil. Calcareous soils as such have been discussed in Chapters 4 and 5 and gypseous soils in Chapter 5.

Calcium is typically deficient in very acid soils, but can be deficient also in sodium-rich alkali soils where it may be precipitated as the carbonate and where it can also become involved in a complementary ion effect, exchangeable calcium being displaced least by sodium compounds. Calcium is also typically deficient in soils derived from serpentines and which have a relatively high magnesium content. In neutral soils calcium can be fixed by phosphorus but otherwise, until such time as it is leached away, it remains readily available to plants, although the presence of montmorillonitic clays reduces the availability of exchangeable calcium (Allaway, 1945).

In plants, calcium is essential for the growth of meristems and root tips and tends to accumulate in leaves as calcium pectate. At least 30% of the adsorption complex of a soil must be saturated with calcium for the average crop to obtain sufficient amounts. Moore (1961) found that the growth of eucalyptus was influenced by the degree of saturation of the soil with calcium rather than by the actual amount of exchangeable calcium. A deficiency of calcium stunts plant roots and gives recognizable leaf symptoms. Albrecht and Smith (1952) considered that calcium deficiency is a prominent feature of the adverse effects of soil acidity upon plant growth. On the other hand, Åslander (1952) stated that calcium deficiency is seldom, if ever, the limiting factor. Colwel and Brady (1945) found that a shortage of calcium was the principal cause of poor yield of ground-nuts grown in an acid soil. These workers showed that if the soil contained more than 1·4 m.e. exchangeable calcium per 100 g, then liming did not increase yield. Brady and Colwel (1945) further showed that calcium sulphate greatly increased the ground-nut yield without raising the pH of the soil; in fact the treatment slightly lowered the pH. Their findings, however, do not apply to all crops; ground-nuts have an exceptionally high calcium requirement.

An insufficiency of calcium permits the accumulation of undesirable ions in plants and, apart from any specific calcium effect, there can be no doubt that liming a soil improves crop growth by preventing this.

To some extent calcium can be replaced in plants with strontium. Myttenaere (1963), for example, showed by water culture tests that partial replacement of available calcium by strontium was beneficial to plant growth and that it causes increased calcium uptake. Analyses of plant tissues showed a strontium–calcium ratio of approximately one, indicating low discrimination by the plants. Complete substitution of calcium by strontium was harmful to plant growth.

For many plants, too much calcium can be detrimental. Wadleigh *et al.* (1951) reported the specific toxicity of calcium salts to orchard grass in saline soil solutions. An excess of calcium causes various disturbances in the organic acid metabolism of calcium-sensitive plants; in particular it increases the concentration of citric and malic acids in plants susceptible to chlorosis (Lattes, 1962). Magnesium and potassium uptake is impaired by excess calcium, and Jacoby (1961) showed that reduced magnesium uptake from soils with low magnesium–calcium ratios is due to excess calcium and not to low magnesium.

8:2:2 Total calcium

The total calcium in a soil can be brought into solution by acid digestion or fusion with sodium carbonate as described in Chapter 16 (section 16:2:4). Göhler and Becker (1961) calcined soil samples at 600°C and extracted the residue with nitric acid in order to determine total calcium.

Once in solution, the calcium can be determined as in section 8:4:3.

8:2:3 Exchangeable calcium

Exchangeable calcium is extracted almost invariably with neutral, 1 M ammonium acetate solution, as described in Chapter 7, unless the soil is calcareous. For soils containing less than about one per cent calcium carbonate Metson (1961) considers that extraction with ammonium acetate is still suitable, provided that a separate carbonate determination is made to correct the calcium results. If the soil contains up to 5% calcium carbonate Metson recommends destroying the carbonate with acetic acid before leaching with ammonium acetate as described by Thomas (1953).

For carbonate contents in excess of 5%, ammonium acetate is an unsuitable extractant. Normal sodium chloride solution was used by Hissink (1923) as this would dissolve relatively little carbonate whilst replacing exchangeable ions. Bower and co-workers (1952) found that leaching with 1 M sodium acetate solution buffered at pH 8·2 gave better results than with sodium chloride solution as the acetate dissolves very little carbonate.

A knowledge of the exchangeable calcium in a calcareous soil, however, is of limited practical use as the soil may be considered to be fully saturated with bases, and such a determination would be made only in connection with the measurement of related quantities such as cation exchange capacity and other exchangeable ions.

Similar difficulties arise in the determination of exchangeable calcium if the soil contains much gypsum. In such cases it is best to include the calcium of the gypsum as exchangeable calcium and to correct the result after separately determining the gypsum.

Blume and Smith (1954) used the isotope ^{45}Ca to investigate the efficiency of ammonium acetate solution as an extractant for exchangeable calcium and they found that with most soils tested the amounts of soil calcium equilibrating with ^{45}Ca were equivalent to the amounts extracted by ammonium acetate solution.

Orlova (1961) modified the electrodialysis method of Purvis and Hanna (1949) to extract exchangeable calcium and magnesium. The exchangeable cations were extracted in 30 minutes by electrolyzing a suspension of soil in boric acid solution with a current strength of 100–200 milliamps and a tension of 300–500 millivolts at 40–50°C.

8:2:4 Calcium in solution

If many calcium determinations are to be made and the equipment is available, the most convenient method of determination is by emission spectrophotometry. For calcium measurement an acetylene flame spectrograph is most suitable, but as other exchangeable or soluble cations will presumably be measured at the same time, some method of reducing the calcium concentration will usually be necessary. Calcium is normally predominant in the test solution and if exposures are made permitting the determination of other cations, the calcium line will be too dense. In many cases the simplest procedure will be to use a specially diluted solution for the separate determination of calcium but methods of reducing the exposure time have been evolved, such as with a stepped sector for example.

The classical routine method for determining calcium is by complexometric titration using ethylenediamine tetra-acetic acid (EDTA) first introduced by Schwarzenbach *et al.* (1946). The method is particularly convenient as it is easily adapted for the simultaneous determination of magnesium. The reagent strongly complexes various polyvalent cations in an order depending upon the dissociation constant of the complex.

$$Na{-}OOC{-}CH_2 \diagdown \atop HOOC{-}CH_2 \diagup {>}N{-}CH_2{-}CH_2{-}N{<} {CH_2{-}COO{-}Na \atop CH_2{-}COOH}$$

M (cation)

$$Na{-}OOC{-}CH_2 \diagdown \atop {CH_2 \atop | \atop O{=}C{-}O} {>}N{-}CH_2{-}CH_2{-}N{<} \cdots M \cdots {CH_2{-}COO{-}Na \atop {CH_2 \atop | \atop O{-}C{=}O}}$$

Cations whose EDTA complex is less dissociated than that of calcium include copper, zinc, iron(III) and manganese and these will be complexed before calcium and thus interfere with the titration.

Several methods of removing interfering ions have been suggested. Jackson (1958) removed iron by ammonium hydroxide separation. Cheng *et al.* (1953) eliminated the effects of iron, manganese and copper with diethyl-dithiocarbamate, which forms brown complexes with these metals that are removed into iso-amyl alcohol leaving calcium (and magnesium) in the aqueous phase. Interference of heavy metals can also be eliminated by addition of hydroxylamine hydrochloride which converts the metals to their lower valency states, or by complexing with sodium cyanide (Vogel, 1962).

Excess phosphate interferes with EDTA titrations but Mattocks and Hernandez (1950) reduced this effect by adding excess reagent and back-titrating with a standard magnesium solution. Göhler and Becker (1961) removed phosphorus by passing a dilute nitric acid solution of calcium through the anion exchange resin Wolfatit C.150.

The procedure generally followed is to titrate first the calcium in solution and then, using a different indicator, to titrate calcium plus magnesium and to obtain the magnesium by difference.

The classical indicator used for calcium titration is ammonium purpurate, commonly known as murexide. Murexide at a pH of 11 changes colour from red to violet at the end-point of the titration; the colour change, however, is not satisfactory and can be extremely difficult to determine. An attempt has been made to improve the end-point by screening with naphthol green when the colour change is from olive-green to blue. The murexide indicator is very susceptible to oxidation and for this reason is added as a powder and not in solution. The indicator is affected by the presence of oxidizing agents and if total calcium is being determined in a soil, perchloric acid digestion is not a suitable procedure.

Two indicators superior to murexide are available for calcium titration with EDTA. One is 2-hydroxy-1-(2-hydroxy-4-sulpho-1-naphthylazo-)-3-naphthoic acid, known as Patton and Reeder's indicator or merely as HHSNNA. At pH 12 to 14 this indicator changes from red to blue at the calcium end-point. The second indicator is Na-1-(2-hydroxy-1-naphthy-lazo-)-2-naphthol-4-sulphonate and is known variously as Calcon, Solochrome Dark Blue and Eriochrome Blue Black R. At a pH of about 12 the indicator changes from pink to blue at the calcium end-point, and if magnesium is present also, it is quantitatively precipitated as the hydroxide. The end-point is improved by adding a few drops of a 1% aqueous solution of polyvinyl alcohol before adjusting the pH; this reduces the absorption of indicator on the precipitate (Vogel, 1962).

For the combined determination of calcium and magnesium, Eriochrome Black T, 1-(1-hydroxy-2-naphthylazo-)-6-nitro-2-naphthol-4-sulphonate, is used as indicator. The optimum pH for formation of the magnesium complex is 10 and this is maintained by a buffer solution. Calcium is complexed before magnesium and the indicator changes from red to blue when the magnesium end-point is reached.

The end-points of EDTA titrations can be estimated more accurately by using a photoelectric titrimeter, and a calcium-free blank prepared in exactly the same way as the test-solution forms a useful end-point comparison when using indicators.

If the volume of test solution is small, the murexide method can be adapted to give calcium and magnesium in the same aliquot. After calcium has been titrated the colour of the murexide is discharged with dilute hydrochloric acid or with bromine water, Eriochrome Black T indicator is added and the titration continued to the calcium plus magnesium end-point.

Barrows and Simpson (1962) described a method convenient for the determination of calcium and magnesium separately, yet in the same aliquot. Their method depends upon the separation of calcium as the sulphate and they use Calcon as the calcium indicator and Superchrome Black TS as the magnesium indicator. Barrows and Simpson investigated soils containing from 5% to 59% clay, 0·1% to 7·5% organic matter, 0·4 m.e. to 42·2 m.e./100 g calcium and 0·8 m.e. to 11·0 m.e./100 g magnesium. The method does not require destruction of organic matter or of ammonium acetate and gave mean coefficients of variation of 2·23% for calcium and 3·35% for magnesium which are within required limits of accuracy. The method has the advantage that ammonium acetate extracts of soil can be analyzed directly without digestion and as sodium and potassium can also be determined directly in acetate solution (section 9:3:2), the digestion step can be omitted altogether. Different methods of separating calcium from mangesium were investigated, as was the effect of ammonium acetate upon the separation. Provided that the solution is evaporated to dryness where indicated, ammonium acetate proved to have no effect.

Barrows and Simpson gave a list of interfering ions together with their maximum permitted concentrations and this list is reproduced in Table 8:1.

Table 8 : 1

     Interfering ions in the EDTA titration of calcium and magnesium (Barrows and Simpson)

Ion	Maximum concentration of ions which when added to untreated solutions did not interfere with the titration of Ca and Mg*	
	Ca titration mg/130 cm^3	Mg titration mg/130 cm^3
P	0·1	4·0
Al	5·0	0·2
Mn	0·15	0·04
Fe	0·10	0·10
Cu	0·02	0·01
Zn	0·08	0·04

* Solutions contained 1 mg Ca and 0·4 mg Mg.

The only serious interference was from manganese and this can be eliminated with hydroxylamine.

Zugravescu (1961) described a method of measuring calcium in the presence of sesquioxides in which heavy metals are complexed by potassium cyanide and excess manganese removed as the sulphide. Magnesium and ammonium are complexed with nitrophenol and the calcium titrated with EDTA.

Gravimetrically calcium can be determined as the oxalate, as the carbonate or oxide via the oxalate and as tungstate. Procedures depending upon precipitation as oxalate are subject to interference from heavy metals and from magnesium; the tungstate procedure is applicable in the presence of considerable amounts of magnesium. The gravimetric methods are fully described in most inorganic quantitative analysis books but are not particularly suitable for soil analysis. However, precipitation of calcium as oxalate, with subsequent estimation of oxalate volumetrically with potassium permanganate or cerate, has often been applied to soil analysis and experimental details are given in standard works. Reitemeier (1943) gave details of a volumetric procedure for estimating from 0·005 to 0·08 m.e. of calcium in centrifuge tubes by titration of precipitated oxalate with ammonium hexanitrate cerate. Jackson (1958) gives details of a macromethod adapted from Ellis (1938) but which is far too tedious to be commonly employed. Jackson also describes a semi-micro procedure which is almost completely free from magnesium interference but which again is tedious.

Several colorimetric procedure have been developed for the determination of calcium. As little as 4 µg of calcium can be determined by the

5*

method of Harrison and Raymond (1953) in which the element is precipitated as the molybdate and the molybdenum determined colorimetrically as the thiocyanate.

After treatment of precipitated calcium oxalate with cerate as in the volumetric method, excess cerate can be measured colorimetrically (Weybrew *et al.*, 1948; Cornfield and Pollard, 1951). Sendroy (1942) treated excess cerate with potassium iodide in acid solution and determined iodine colorimetrically. Pereira (1944) precipitated calcium as the phosphate and used this to form molybdenum blue but the method is subject to too much interference to be of use in soil analysis. In neutral solution picrolinic acid (4-NO_2-3-Me-1, p-NO_2-phenylpyrazalone-5) precipitates calcium and when the precipitate is made alkaline it forms a reddish colour which is proportional to the concentration of picrolinic acid (Alten *et al.*, 1933); heavy metals must be absent but magnesium, sodium, potassium and ammonium do not interfere.

Young and Sweet (1955) developed a procedure for colorimetrically measuring the complexes of calcium and magnesium with Eriochrome Black T. At high pH (11·7) calcium and magnesium are both completely reacted but at pH 9·52 magnesium only is reacted. By reading the transmittance at 630 nm at both pH values, calcium and magnesium can be measured at concentrations from 0·3 to 6 μg cm^{-3}; an accuracy of 0·1 μg cm^{-3} was claimed.

Calcium reacts with glyoxal *bis* (2-hydroxyanil) to give a red-coloured complex which can be extracted into chloroform from an alkaline medium (Goldstein and Stark-Mayer, 1958). The technique was improved by Kerr (1960) who stabilized the complex in ethanol–butanol in the presence of an alkaline buffer and Peaslee (1964) adapted the reaction to produce a microquantitative estimation of calcium in soil. For the quantitative estimation up to 25 μg of calcium is the limit recommended by Peaslee, although Beer's Law is followed for 0–40 μg calcium. Interference from heavy metals is eliminated by addition of sodium diethyl-dithiocarbamate unless their concentration is too great, when sodium cyanide is also added. Phosphorus interferes at concentrations greater than about 0·05 μg cm^{-3} in the solution measured, which at the dilution of soil extract employed permits a soil content of soluble phosphorus of at least 200 μg cm^{-3}. The final pH of solution is important and a value of 12 is optimum. With a mean concentration of 1·61 μg cm^{-3} calcium, Peaslee obtained a relative standard deviation of 0·6%.

Peaslee also proposed a 'quick-test' for calcium as an alternative to flame photometric methods. In the test a drop of clear soil extract containing from 50 to 500 μg cm^{-3} calcium is mixed with reagents in a small vial and the colour measured in a simple colorimeter with a green filter. Using this procedure Peaslee measured calcium with less than 8% error in solutions containing from 120 to 525 μg cm^{-3} Ca, 200 μg cm^{-3} Mg, 125 μg cm^{-3} Al, 75 μg cm^{-3} Fe, 60 μg cm^{-3} Mn and 20 μg cm^{-3} P.

A 'spot-test' was also devised to classify the calcium content of soils at various fertility levels. In the test a small drop of extract is placed on a spot-tile and mixed with a composite reagent; the colour produced is compared with standards.

Peaslee designed the glyoxal *bis* (2-hydroxyanil) procedure for use with Morgan's solution (0·734 M sodium acetate + 0·518 M acetic acid) as soil extractant, but it can be adapted for standard ammonium acetate or water extracts of soil.

8:3 MAGNESIUM

8:3:1 Background and theory

In soils, magnesium occurs principally in the clay minerals, being common in micas, vermiculites and chlorites; it sometimes occurs as the carbonate. Smaller quantities are present as exchangeable ions, water-soluble forms and in organic combination. Magnesium occurs in dioctahedral silicates such as montmorillonite as an isomorphous substitution for aluminium, and in trioctahedral silicates such as biotites, having magnesium and iron as dominant octahedral cations. In chlorites magnesium occurs in the layers alternating with silicate layers and it is a common interlayer cation in vermiculites (Salmon, 1963). Barshad (1949) showed that the greater hydration of magnesium ions compared to that of potassium ions causes wider spacings between silicate layers and hence magnesium is more easily exchanged than is potassium.

Some solonetz soils, particularly in the North American continent, contain relatively little sodium but have nearly 50% of the total exchangeable ions as magnesium. It thus appears as if magnesium can bring about the development of solonetz soils in a manner similar to that exerted by sodium (Russell, 1961). Barshad (1960) found that magnesium ions account for a large percentage of the total exchangeable ions in hydrochloric acid-acidified clays and considered this to explain the high percentage of exchangeable magnesium found in solonetz soils and in near-neutral soils derived from serpentine rocks. Barshad pointed out that due to the exchangeable magnesium present in the acid clays, full cation exchange capacity occurs at pH values between 10·3 and 10·6 when magnesium is precipitated as hydroxide. Adsorbed hydrogen passes into the lattice and displaces aluminium, magnesium and possibly iron, and suggests a mechanism whereby breakdown of the silicate mineral proceeds beyond the first stage of hydrolysis. The release of magnesium from soils and minerals was found by Stahlberg (1960) to be inversely related to particle size.

Magnesium can be lost from soils by leaching and this will be influenced by soil physical conditions as well as by rainfall. Many years ago Voelcker (1871) reported that application of superphosphate or of potash will increase the amount of magnesium lost in drainage water. More recently Hogg (1962) investigated the leaching of magnesium from soils ranging from clays and loams to sands and peats and found that displacement of magnesium following application of potassium chloride and superphosphate was greatest in sandy and pumice soils and accounted for up to 59·5% and 17·5% respectively of the initial exchangeable magnesium within 3 weeks. If potassium carbonate, potassium hydrogen carbonate or potassium dihydrogen phosphate were used, then little displacement of magnesium took place.

Schroeder and Zahiroleslam (1963) classified magnesium in soils as—

i. Available; that is, soluble in 0·0125 M calcium chloride solution,

ii. Moderately-available; that is, that soluble in 0·05 M hydrochloric acid minus that soluble in calcium chloride solution,

iii. Reserve A; that is, that soluble in 10% hydrochloric acid minus that soluble in 0·05 M acid, and

iv. Reserve B; that is, that insoluble in 10% hydrochloric acid but decomposed by hydrofluoric acid.

These workers found that about 50% of 'Reserve' magnesium was in the less than 20 μm fraction of soils and the stability of reserve magnesium as characterized by the ratio $A:B$ was associated with the stage of soil development.

Magnesium is an essential constituent of chlorophyll and also is involved in enzyme reactions. The element affects the translocation of phosphorus (Truog *et al.*, 1947) and has been reported (Semenova, 1962) to increase sugars, vitamins, starches and inulin in root crops. A deficiency of magnesium typically causes chlorosis and has been associated in recent years with certain animal disorders (Reith, 1963). The appearance of magnesium deficiency has become more frequent of late in agricultural land due to greater removal by high-yielding crops, leaching from acid, coarse-textured soils and with less magnesium being applied in fertilizers.

A magnesium deficiency can be caused not only by small concentrations of the nutrient in a soil but by ionic antagonism, particularly in acid and potassium-rich soils. In alkali soils magnesium can be deficient due partly to precipitation and partly through the complementary ion effect.

Sometimes high calcium–magnesium ratios impair the uptake of magnesium but the most commonly encountered antagonistic ion is potassium. Hovland and Caldwell (1960) consider that potassium interferes with magnesium uptake in at least three general ways:

i. Addition of potassium to soils may decrease the ease of displacement of magnesium and result in less available magnesium.

ii. Increased soil potassium may compete with magnesium for exchangeable sites on plant roots.

iii. High concentrations of potassium in the plant may prevent magnesium from functioning properly.

Welte and Werner (1963) investigated the uptake of magnesium by plants as influenced by hydrogen, potassium, ammonium and calcium ions. They found that hydrogen ions suppressed magnesium uptake most and with a strongly acid substrate, magnesium deficiency could be remedied by

applying magnesium and raising the pH. Raising the soil pH without adding magnesium increased magnesium availability more than by adding
magnesium without raising the pH. An excess of lime, however, is to be
avoided as causing calcium antagonism. The depressive effects upon magnesium uptake by the ions tested were additive and thus the effect of potassium antagonism was increased with decreasing pH. Welte and Werner
found that in a magnesium-deficient soil with a high sorption capacity and
in which magnesium ions were strongly bound, application of potassium
by releasing the magnesium can actually increase magnesium uptake, but
in light soils of low sorption capacity, leaching of magnesium was promoted. Welte and Werner concluded that the antagonistic effect of potassium is important only if the magnesium level is below the critical value for
optimum growth. The effect can be overcome by supplying magnesium
and, if necessary, lime and there is no necessity to reduce the amount of
potassium supplied. McColloch *et al.* (1957) considered that a knowledge
of exchangeable magnesium–potassium ratios is more useful than knowing
merely the exchangeable magnesium content.

Bould (1963) reported that the ratio of potassium to magnesium in soil
solutions will be higher in wet than in dry soils (by the Donnan rule) and
hence magnesium deficiency will be more pronounced during wet seasons.

Al-Abbas and Barber (1964) studied the uptake of magnesium, as well
as of calcium, by plant roots in relation to the quantities moving to the
roots by mass flow and the amounts contacted by root growth. Mass flow
depends upon the amount of water used per unit of plant growth and the
content of magnesium (or calcium) in that water. The amount of magnesium moving by mass flow was calculated from the concentration of
magnesium in the saturation extract and the amount of water utilized.
Al-Abbas and Barber found that the level of exchangeable magnesium in a
soil gave the highest correlation with plant uptake but also highly correlated was the exchangeable magnesium in the root volume plus mass flow.
At low magnesium levels there was more magnesium uptake than could be
attributed to mass flow plus root interception and this was attributed to
diffusion.

Control of magnesium deficiency by organic manuring was investigated
by Saalbach and Judel (1961). Apart from the magnesium added in the
manure it was concluded that micro-organisms were stimulated whose
metabolic products mobilized unavailable magnesium.

8:3:2 Total magnesium
The total magnesium in a soil can be brought into solution by fusion or
acid digestion (Chapter 16, section 16:2:4) and the magnesium determined
by one of the methods in section 8:4:7.

8:3:3 Exchangeable magnesium
Exchangeable magnesium is normally extracted with neutral, 1 M ammonium acetate solution and this solution is suitable even for calcareous

soils unless magnesium carbonate is present. If magnesium carbonate is present the extraction procedure of Bower *et al.* (1952) using sodium acetate can be followed, although eliminating the effect of the carbonate in this manner makes the subsequent determination of magnesium difficult; Metson (1961) therefore recommends the use of successive leachings with normal sodium chloride solution (Hissink, 1923; Piper, 1942). The first leachate is presumed to contain all exchangeable magnesium plus some carbonate-magnesium and succeeding leachates to contain carbonate-magnesium only. Values are plotted and extrapolated to find the carbonate-magnesium in the first leachate. For soils known to contain dolomite it would seem preferable to report them as being fully saturated and if necessary the calcium plus magnesium approximated from the difference between total exchangeable bases and exchangeable potassium plus sodium.

The electrodialysis method of Orlova (1961) is claimed to give results in good agreement with those obtained by acetate extraction.

It should be remembered that exchangeable magnesium in a moist soil will increase with time of storage in the wet state (Metzger, 1929).

8:3:4 Available magnesium

Available magnesium is often regarded as exchangeable magnesium but many other extractable forms have been suggested. Schachtschabel (1954) introduced the use of 0·0125 M calcium chloride solution as being less arbitrary than other extractants as calcium is the dominant cation in most soils. Welte *et al.* (1960) found no definite relationship between magnesium extracted in the dilute calcium chloride solution and plant uptake of magnesium or yield. Welte concluded that the mobility of magnesium in soil is better estimated from exchangeable magnesium or from that soluble in 0·05 M hydrochloric acid. Hoffman and Schroeder (1960), however, found that the calcium chloride-soluble magnesium gave the best correlations with magnesium status of potatoes. Graham *et al.* (1956) preferred 0·05 M hydrochloric acid as extractant and found the results more reliable than calculation of calcium–magnesium ratios or percent magnesium saturation. Geissler and Kurnoth (1960) modified the bioassay method using *Aspergillus niger* to an accuracy of less than 1 mg Mg per cent and found the results in close agreement with those using Schachtschabel's calcium chloride solution. Gärtel (1962) extracted available magnesium with Egnér–Riehm's (1955) calcium lactate reagent and obtained higher values than with calcium chloride solution. Schroeder and Zahiroleslam (1963) combined Schachtschabel's solution and the 0·05 M hydrochloric acid solution techniques to distinguish between available and moderately-available magnesium and, together with Hoffman (1963), they carried out pot experiments using rye grass as test plant. Their results indicated that the major part of the magnesium released by 2 M hydrochloric acid minus exchangeable magnesium was non-available and that a close correlation existed between magnesium uptake and calcium chloride-soluble magnesium, exchangeable magnesium or hydrochloric acid-soluble magnesium.

They concluded that 0·0125 M calcium chloride solution is the best general measure of magnesium availability. Reith (1963) compared ammonium acetate and 2·5% acetic acid as extractants of available magnesium and found that both removed about the same amount of magnesium from a soil. Reith concluded that in general, before any yield response can be expected from magnesium fertilization, the readily soluble magnesium would have to be less than 30 µg g^{-1}.

The search for a universally applicable extractant for available magnesium is typical of all attempts to determine available nutrients and demonstrates the absolute necessity to correlate extracted amounts of nutrient with the results of field trials on a specific soil using a specific crop. It is futile to try and apply the results to different soils or crops. As a normal routine it would seem to be satisfactory to assume exchangeable magnesium as being available; other extracted forms would be used only for specific investigation of magnesium availability when, presumably, several extractants would be tried. Recently a more meaningful determination of available magnesium has been introduced by Beckett (private communication) by measuring activity ratios in a manner similar to that for potassium (section 9:3:3ɪɪ).

8:3:5 Magnesium in solution

As in the case of calcium, the flame spectrographic method of determining magnesium is most convenient if many samples are to be analyzed as a routine (Scott and Ure, 1958). Measurement of magnesium using atomic absorption spectrophotometry is a well established practice and the sensitivity of this method has recently been improved by Wacker *et al.* (1964) by directing the flame through a horizontal, tubular absorption cell, thereby elongating the path. Williams and co-workers (1966) investigated the suitability of the atomic absorption technique relative to various chemical methods using 244 soils and a range of extractants. Williams concluded that atomic absorption has advantages over chemical methods and is a decidedly superior procedure when measuring low levels of magnesium.

Magnesium can also be measured with the simple flame photometer. Metson (1961) recommends a wavelength of 382 nm for the determination and Snell and Snell (1959) recommend 371 nm; Rich (1965) gives 285 nm as the preferable atomic line, but many instruments do not permit measurements at such short wavelength and, in which case, Rich gives 371 nm as next best.

Magnesium can be determined volumetrically by titration with EDTA using Eriochrome Black T as indicator. When, as is usual, calcium is also to be determined, the EDTA titration is particularly suitable; the method has been fully discussed in section 8:2:4 and experimental details are given in section 8:4:3.

Another volumetric procedure is that involving precipitation of magnesium with 8-hydroxyquinoline (Berg, 1927). Calcium is first removed by precipitation and the filtrate is evaporated with nitric and hydrochloric acids and heavy metals removed with ammonium hydroxide and bromine water. The magnesium is then precipitated with oxine in acetic acid, dissolved in dilute hydrochloric acid and a bromate–bromide solution added. After adding excess potassium iodide, iodine is titrated with thiosulphate solution (Metson, 1961). The method is subject to a great many errors and can be controlled only if the approximate magnesium concentration is known.

The reaction of magnesium with oxine has been made the basis of a colorimetric procedure for the determination (Willson, 1951) and has been developed by Schouwenburg and Walinga (1962) for micro-determination of magnesium. In this method an aliquot containing up to 0·02 mg of magnesium is placed in a separating funnel and mixed with a tartrate solution, a boric acid buffer and sodium diethyl-dithiocarbamate. The mixture is extracted with chloroform and the aqueous phase discarded. A masking reagent, n-butylamine, and oxine are added and the transmittance of the chloroform extract read at 380 nm.

Magnesium has been determined colorimetrically by precipitating it as the phosphate and converting this to the phosphomolybdate (e.g. Reitemeier, 1943) but this method also, is very susceptible to error.

The classical colorimetric method of determining magnesium is with Titan Yellow, the sodium salt of dehydrothio-*p*-toluidine sulphonic acid, which reacts with precipitated magnesium hydroxide in the presence of sodium hydroxide to give an orange-red coloured complex (Barnes, 1918). No definite compound is formed and Beer's Law is not obeyed. Heavy metals, ammonium and organic matter interfere and must be removed or suppressed. Phosphate can be tolerated up to 100 μg cm^{-3} and calcium to 500 μg cm^{-3}. It has been claimed that provided all precautions are taken, magnesium can be estimated to 0·015 mg with an accuracy of 0·0001 mg; however, taking such precautions with soil extracts is difficult and the method has been much criticized. The addition of a protective colloid improves the stability of the complex and the reproducibility of reaction (Urbach, 1932). Heagy (1948) found polyvinyl alcohol to be the best protective additive. Yien and Chesnin (1952) stabilized Titan Yellow suspensions with starch and also stabilized the colour with sodium hydrogen sulphite which inhibits oxidation of the dye. A common complaint is of the variability in reaction between magnesium and different batches of the commercial reagent. Some improvement can be obtained by preparing standards for each new reagent solution. Hall *et al.* (1966) selected a Titan Yellow reagent by thin layer chromatography; Hall further studied the protective colloid effect and proposed the use of gelatine with a lithium hydroxide–glycine buffer. In spite of its known difficulties and doubtful accuracy the Titan Yellow method is still widely used in soil laboratories. King (1967) described the preparation of a purified Titan Yellow from the commercial product. The dyestuff is reduced with tin and hydrochloric acid, converted to the ammonium salt and acidified with acetic acid. The product is treated with sodium hydroxide, sodium nitrite and hydrochloric acid and the Titan Yellow formed is purified by exhaustive washing with hot acetone in a Soxhlet apparatus.

Chenery (1964) described a modification of the Titan Yellow procedure permitting the estimation of magnesium in the presence of excess manganese. To the test solution is added the reagent and an excess of triethanolamine; this is followed with an excess of hexacyanoferrate(III) solution. On addition of sodium hydroxide, a brown colour due to manganese appears but after 15 minutes fades completely leaving the red magnesium complex colour unimpaired. The brown colour reappears if the solution is left standing too long.

Mikkelsen and Toth (1947) announced the reagent Thiazole Yellow as superior to Titan Yellow for the determination of magnesium. The new reagent was introduced into agricultural analysis by several workers, *inter al.*, Young and Gill (1951) and Yien and Chesnin (1952); the latter compared Titan Yellow, Clayton Yellow and Thiazole Yellow as reagents for magnesium. Mikkelsen and co-workers (1948) removed interfering elements by precipitating them as tungstates. An aliquot of test solution is treated with a solution of sodium tungstate, made alkaline with sodium hydroxide and the precipitate centrifuged off after digestion on a water-bath. Yien and Chesnin preferred to add a compensating solution containing a known excess of interfering cations and of phosphate to the test solutions and to the standards. The colour is stabilized with sodium metahydrogen sulphite. Instead of measuring the transmission of the magnesium complex the optical density of excess reagent can be measured (Hunter, 1950).

Note: A word of warning regarding Titan and Thiazole Yellow. The dyestuff listed in the Colour Index as Direct Yellow 9 has fifteen synonyms which include Titan Yellow, Clayton Yellow and five Thiazole Yellows (each distinguished by its manufacturer by a following capital letter). Thus fifteen manufacturers sell this reagent under their own trade name; as a consequence it can be difficult to obtain the required reagent. An order from one manufacturer for Thiazole Yellow may result in a sample of Titan or Clayton Yellow and which may or may not be suitable. There appears to be no doubt that all the synonyms of Thiazole Yellow are not of the same exact chemical composition and when ordering the reagent it should be specifically stated that it is for magnesium.*

Grad (1962) developed a rapid method for colorimetrically determining magnesium in soil extracts using Brilliant Yellow (2,2'-disulphostilbene-4,4'-diazo-*bis*-phenol). The reaction of magnesium with Eriochrome Black T can be used for its colorimetric determination.

Mann and Yoe (1957) proposed a colorimetric method for determining magnesium based upon its reaction with 1-azo-2-hydroxy-3-(2,4,dimethyl carboxanilido)-naphthalene-1'-(2-hydroxybenzene) to produce a soluble

---

* The author is indebted to Edward Gurr Ltd, London, for information regarding the various grades of Titan Yellow and its synonyms; this company provides a suitable reagent for magnesium determination under the name Titan Yellow, Michrome No. 163.

red-coloured complex. The reaction is tolerant to high amounts of calcium and the complex is stable for 48 hours. Peaslee (1966) improved the procedure to permit the determination of from 1 to 10 μg magnesium in the presence of 1000 μg calcium, 300 μg ammonium-nitrogen and 10 μg of phosphorus. The borax buffer solution proposed by Mann and Yoe was found to be insufficient for masking interference from iron, aluminium and copper and Peaslee replaced it with a special buffer solution. Peaslee found that the highest interference was given by copper but could be eliminated with potassium cyanide.

Magnesium in soil extracts was determined fluorometrically by Swanson *et al.* (1966) who added small portions of extract to 0,0'-dihydroxy-azobenzene and measured the fluorescence. Swanson found the fluorometric procedure far more accurate than the Thiazole Yellow method and once the necessary apparatus has been set up the procedure is simple.

The determination of calcium, magnesium and potassium has in recent years been adapted to auto-analysis; for example an auto-analyzer in conjunction with an atomic absorption spectrophotometer (Lacy, 1965).

8:4 RECOMMENDED METHODS

8:4:1 Determination of total calcium

The total calcium in soil can be brought into solution by acid digestion or by fusion as described in Chapter 16 (section 16:2:4). If perchloric acid is used for digestion, excess should be evaporated off before measuring calcium. The dissolved calcium is then determined by one of the methods in section 8:4:3.

8:4:2 Determination of exchangeable calcium

Exchangeable calcium is extracted from soil as described in Chapter 7 (section 7:3:2) and is usually determined in an ammonium acetate extract. For determination of the dissolved calcium by spectrophotometry the element should preferably be in dilute hydrochloric acid solution but not necessarily so; similarly the ammonium acetate extract can be analyzed directly if the procedure of Barrow and Simpson is followed. For other methods of determining the extracted calcium the extract should be digested with acids to destroy the acetate, as described in section 7:3:3.

8:4:3 Determination of calcium in solution

I / BY EMISSION SPECTROPHOTOMETRY

For spectrographic determination of calcium it should preferably be in dilute hydrochloric acid solution. Usually the dilute acid solution as such is used for spectrographic determination of potassium, manganese, sodium, magnesium and strontium, but for calcium determination the concentrated solution must be diluted unless a stepped sector is used on the instrument.

If much aluminium is present in the solution this will affect the spectrographic readings for calcium (Mitchell and Robertson, 1936). The usual method of overcoming aluminium interference is to add an excess of strontium to the standard solutions. In contrast, the effect of aluminium can be utilized to reduce the intensity of the calcium emission, thus avoiding dilution of the extract; this permits the determination of all the exchangeable cations in the same spectrogram (Lundegardh, 1941).

Detailed practical instructions for spectrographic analysis are outside the scope of this book but can be found in books such as that of Mitchell (1964) which also gives an abundance of references. A simple procedure using a Hilger medium quartz spectrograph is given by Black and Smith (1950).

II / BY TITRATION WITH EDTA

This procedure permits the determination of calcium and magnesium in the same solution. The accuracy is $\pm 3\%$.

Reagents

Standard calcium solution: Weigh 6·2431 g of $CaCO_3$ dried at 150°C into a 500-cm³ volumetric flask. Add 100 cm³ of water and then 150 cm³ of 1 M HCl slowly and with shaking. Dilute the solution to 500 cm³. The solution contains 5 mg cm⁻³ Ca. Dilute an aliquot to give a solution containing 0·5 mg cm⁻³ Ca.

Standard magnesium solution: Dissolve 2·500 g of pure, unoxidized Mg metal in dilute hydrochloric acid. Dilute to 500 cm³ when the solution will contain 5 mg cm⁻³ Mg. Dilute an aliquot to give a solution containing 0·5 mg cm⁻³ Mg.

EDTA solution: Dissolve 2·5 g of ethylenediamine tetra-acetic acid (di-Na salt) in water and dilute to 2 dm³.

Sodium hydroxide solution: 10% w/v, aqueous.

Buffer solution: 67·5 g NH_4Cl/400 cm³ of water; to this add 570 cm³ of 0·880 NH_4OH solution and dilute to 1 dm³.

Hydroxylamine hydrochloride solution: 5% w/v, aqueous. Prepare fresh each week.

Potassium hexacyanoferrate(II) solution: 4% w/v, aqueous.

Potassium cyanide solution: 1% w/v, aqueous.

Triethanolamine: commercial reagent.

Calcon solution: 0·2 g reagent/50 cm³ methanol; prepare fresh every 2 weeks.

Eriochrome Black T solution: 0·2 g reagent/50 cm³ methanol; prepare fresh every 2 weeks.

Procedure

Standardization of EDTA solution. Pipette 5·0 cm³ of standard calcium solution into a graduated, tall-form, 100-cm³ beaker. Dilute to 10 cm³ and add 15 cm³ of buffer solution. Add 10 drops each of potassium cyanide, hydroxylamine hydrochloride, potassium hexacyanoferrate(II), triethanolamine and Eriochrome Black T solutions. Place the beaker on a magnetic stirring plate and commence stirring. Prepare a blank solution in exactly the same manner, taking 5 cm³ of water instead of calcium solution. Usually the blank solution will be blue in colour, but if not, it should be titrated with EDTA solution until blue, and the blank titre noted. Stand the blue, blank solution alongside the standard calcium solution and titrate the standard with EDTA solution, stirring the whole time. The solution is titrated to a permanent blue colour matching the blank; dilute the blank with water now and again to equalize the volumes of the two solutions as the titration proceeds.

Repeat the standardization using Calcon as indicator. In this case add to the diluted calcium solution 10 drops each of potassium cyanide, hydroxylamine hydrochloride, and triethanolamine solutions. Then add 2·5 cm³ of sodium hydroxide solution and 1 cm³ of Calcon solution. Prepare a reagent blank and titrate both solutions with EDTA solution until blue.

From the results calculate the mg of calcium equivalent to 1 cm³ of EDTA solution; alternatively, calculations can be on a molar basis. The titrations can be made with a photoelectric titrimeter using an orange (670-nm) filter.

Determination of calcium in solution. Pipette an aliquot of test solution containing up to 3 mg calcium into a 100-cm³ tall-form beaker and dilute to 10 cm³. For soils containing less than 10 m.e. Ca/100 g or solutions containing less than 1 mg Ca/10 cm³, it will be necessary to take larger aliquots in a 250-cm³ beaker. Add 10 drops each of potassium cyanide, hydroxylamine hydrochloride and triethanolamine solutions. Add 2·5 cm³ of sodium hydroxide solution and 1 cm³ of Calcon solution. Titrate with EDTA and calculate the mg Ca in solution.

Determination of calcium in the presence of excess phosphate. For most soil extracts there will be no interference from phosphorus but interference is likely when analyzing soil digests for total calcium. Phosphorus can be pre-extracted but the interference is best overcome by back-titrating an excess of EDTA solution.

Pipette an aliquot of test solution into a beaker, dilute and add the masking reagents up to and including the triethanolamine. Add a known excess of EDTA solution (determined by a preliminary titration) and raise the pH to 12 with sodium hydroxide solution. Heat the solution to nearly boiling for 3 minutes, cool and add 1 cm³ of Calcon solution. Titrate to a red colour with standard calcium solution, matching the colour against a blank.

Determination of calcium plus magnesium. Pipette an aliquot containing up to about 3 mg (Ca + Mg) into a beaker. Dilute to 10 cm³, add 15 cm³ of buffer solution and then 10 drops each of potassium cyanide, hydroxyl-amine hydrochloride, potassium hexacyanoferrate(II) and triethanolamine solutions while gently warming the solution on the magnetic stirrer. When all reagents have been added continue warming for 3 minutes, cool and add 10 drops of Eriochrome Black T solution. Titrate with EDTA.

Phosphorus interference can be overcome as before, except that after adding the excess EDTA solution, the pH is brought to 10 by adding 15 cm³ of buffer solution. The solution is then heated to nearly boiling, cooled and 10 drops of indicator solution added. Finally titrate the excess EDTA with standard calcium solution.

Determination of magnesium. For most purposes magnesium is calculated from the difference between the (Ca + Mg) and the Ca determinations. If it is desired to specifically measure the magnesium in solution the procedure of Barrow and Simpson is recommended.

Reagents

The reagents are as in section 8:4:3II, plus
Nitric acid solution, 1·0 M: 63 cm³ dm⁻³ conc. HNO_3.
Sulphuric acid, 3 M: 168 cm³ dm⁻³ conc. H_2SO_4.
Ethanol.

Procedure

Evaporate an aliquot of soil extract containing up to 10 mg calcium to dryness. Cool and add 3 cm³ of 1 M nitric acid to dissolve the residue; transfer the solution to a 40-cm³ centrifuge tube using the minimum amount of water. Concentrate to about 5 cm³ in a water-bath and cool. Add 1 cm³ of 3 M sulphuric acid and 34 cm³ of ethanol. Cover the tube with aluminium foil and leave overnight. Centrifuge and decant the liquid into a beaker. If so desired, the calcium can be determined in the precipitate by dispersing it in water, adding 2·5 cm³ of sodium hydroxide solution and titrating with EDTA using Calcon as indicator.

Evaporate the magnesium solution in the beaker to small volume but not to dryness. Cool and dilute to about 100 cm³. Add 5 cm³ of buffer solution and 10 drops each of potassium cyanide, hydroxylamine hydro-chloride, potassium hexacyanoferrate(II) and triethanolamine solutions. Stir and allow to stand for 10 minutes. Add 10 drops of Eriochrome Black T solution and titrate with EDTA solution.

III / BY COLORIMETRY (PEASLEE)

Reagents

Standard calcium solution: As in section 8:4:3II.
Glyoxal *bis* (2-hydroxyanil) solution: Dissolve 0·150 g of reagent in 30 cm³ of absolute methanol.

Stock buffer: Dissolve 5·28 g of $Na_2B_4O_7$ in 800 cm³ of water and add
   10 g NaOH. Cool, dilute to 1 dm³ and store in a polythene bottle protected from CO_2.
Buffer solution: Dissolve 0·20 g sodium diethyl-dithiocarbamate in 100 cm³
   of stock buffer solution. Prepare fresh every 5 days.

Procedure

Pipette an aliquot of soil extract or standard solution containing up to
25 μg of calcium into a tube marked at 10 cm³ and dilute to that mark.
Add 10 cm³ of absolute methanol and exactly 1·0 cm³ of buffer solution.
The pH of the solution is important and should be adjusted if necessary to
12. Mix and add exactly 0·5 cm³ of reagent solution. Allow to stand for
25 minutes and read the optical density at 535 nm. The colour is stable for
at least 20 minutes and Beer's Law is followed for the range 0–40 μg Ca.
   The sodium diethyl-dithiocarbamate eliminates interference from iron,
copper and manganese unless these exceed the concentrations 0·32, 0·15
and 0·30 μg cm⁻³ respectively. If these critical concentrations are exceeded
then 5 mg of sodium cyanide should be added. Phosphorus present to less
than 1 μg per tube permits calcium determination up to 25 μg; phosphorus
present to less than 0·1 μg permits determination of calcium up to 40 μg.
Soil extracts usually do not contain sufficient phosphorus to interfere.
Shake coloured extracts with activated carbon before removing the aliquot.
In Peaslee's experiments 50 g of soil were shaken with 100 cm³ of Morgan's
reagent for 25 minutes and the solution re-shaken with activated charcoal
before filtering through a Whatman No. 5 paper. 0·5 cm³ of extract were
diluted to 100 cm³ for measurement.
   The procedure has a relative standard deviation of 0·6%.

8:4:4 Determination of total magnesium

Digest the soil with acids or fuse with sodium carbonate as described in
section 16:2:4. The residue should be brought into dilute hydrochloric
acid solution and aliquots measured as in section 8:4:7.

8:4:5 Determination of exchangeable magnesium

Exchangeable magnesium is most conveniently extracted in the routine
manner with neutral, 1 M ammonium acetate solution (section 7:3:2); for
soils containing magnesium carbonate Hissink's method of leaching with
sodium chloride solution is recommended (section 7:2:1).

8:4:6 Determination of available magnesium

For general purposes exchangeable magnesium plus water-soluble magnesium should be regarded as available. For more specific purposes, such
as in fertility investigations involving field or pot experiments, Schachtschabel's 0·0125 M calcium chloride solution can be used. In this method
10 g of air-dry soil are shaken for 30 minutes with 100 cm³ of a solution

containing $1\cdot4$ g dm^{-3} CaCl$_2$ which has been standardized and diluted to $0\cdot0125$ M. The magnesium content of the extract is determined as in section $8:4:7$.

8:4:7 Determination of magnesium in solution

I / BY EMISSION SPECTROPHOTOMETRY

As in the case of calcium, magnesium should preferably be in dilute hydrochloric acid solution for its spectrographic determination. Experimental details will vary according to the type of spectrograph available (Mitchell, 1964).

Scott and Ure (1958) developed a small direct-reading spectrograph for measuring magnesium over the range $0\cdot3$ to $24\cdot0$ μg cm^{-3} by porous cup spark excitation. In this method an ammonium acetate extract of soil can be used directly, without further treatment apart from addition of acetic acid, and of strontium chloride as internal standard.

II / BY TITRATION WITH EDTA

The complexometric determination of magnesium with EDTA has been fully described in section $8:4:3$II.

III / COLORIMETRICALLY WITH THIAZOLE YELLOW (YIEN AND CHESNIN)

Reagents

All reagents give better results if stored overnight before use; rubber stoppers should not be used.

Standard magnesium solution: As in section $8:4:3$II. For measuring exchangeable magnesium the standard solution should be made normal in ammonium acetate.

Compensating solution: Dissolve $3\cdot3$ g CaCO$_3$, $0\cdot74$ g Al$_2$(SO$_4$)$_3$, $0\cdot36$ g MnCl$_2$ and $0\cdot60$ g Na$_3$PO$_4$ in 500 cm^3 of water containing 10 cm^3 conc. HCl. Dilute to 1 dm^3.

Polyvinyl alcohol, 1%: Add the solid reagent to water and shake. Warm on a steam-bath until all froth disappears.

Sodium hydrogen sulphite solution: $0\cdot5$% w/v, aqueous. Store in a brown bottle.

Sodium hydroxide solution, 10 M: 400 g dm^{-3} NaOH.

Thiazole Yellow solution: $0\cdot04$% w/v, aqueous and stored in a brown bottle. The reagent should be obtained from the manufacturers specifically for Mg determination (see section $8:3:4$).

Procedures

Determination of magnesium in water extracts. Pipette a suitable aliquot of test solution into a 25-cm^3 volumetric flask. Add 5 cm^3 of a mixture of equal parts of polyvinyl alcohol, sodium hydrogen sulphite solution and compensating solution and swirl the flask. Add 1·00 cm³ of Thiazole Yellow reagent from a microburette, dilute the solution to about 20 cm³ and again swirl. Add 3 cm³ of sodium hydroxide solution and dilute to volume. Read the optical density of the solution at 540 nm (or green filter) after 10 minutes.

The colour is stable for 24 hours and the range of measurement is up to 3 μg cm⁻³ Mg.

Determination of magnesium in ammonium acetate extracts. Pipette a suitable aliquot of test solution into a 25-cm³ flask and dilute to about 15 cm³ with ammonium acetate solution. Add 5 cm³ of a mixture of equal parts of polyvinyl alcohol, sodium hydrogen sulphite and compensating solutions. Mix by swirling as shaking causes frothing. Add 1·00 cm³ of Thiazole Yellow solution from a microburette, 4 cm³ of sodium hydroxide solution and dilute to volume. Measure the optical density at 540 nm. The colour is stable for 1 hour and the range of measurement is up to 3 μg cm⁻³ Mg.

The results are reproducible and the standard error between duplicates is 0·01. It is important to maintain the final concentration of ammonium acetate to within 0·5 to 0·7 M.

9

Potassium and sodium

9:1 POTASSIUM

9:1:1 Background and theory

Forms of potassium in soil

The potassium content of a soil depends primarily upon the parent material and degree of weathering. In weakly weathered soils the feldspars and micas are the most abundant potassium-bearing minerals and with the exception of illite which is found in the clay fraction, the potassium minerals predominate in the sand and silt fractions of soil.

Most of the total potassium in a soil is in non-exchangeable forms; Williams (1962) suggested three categories of potassium status in soil:

i. That immediately available and which is water-soluble and exchangeable;

ii. That which is moderately available and which is known as 'fixed' potassium;

iii. The main reserve of potassium and which is rendered available very slowly. This form of potassium is lattice-bound and is released by weathering.

The quantitative significance of the immediately available potassium may vary for different soils and crops and may be quite insignificant for long term predictions of yield, particularly on heavy soils.

Fixation of soil potassium

That plants can utilize non-exchangeable potassium was first postulated by Fraps (1929) who suggested that some non-exchangeable potassium must be considered as available potassium. Bray and De Turk (1939) showed that removal of exchangeable potassium from a soil upsets the equilibrium, and potassium is released from non-exchangeable forms to replenish the exchangeable form; it is this newly-formed exchangeable potassium that is utilized by plants. Micaceous clay, illite for example, is largely, but not entirely, responsible for this release of exchangeable potassium from non-exchangeable forms (Rouse and Bertramson, 1950). Milford (1962) found that in soils whose potassium availability is under stress from leaching or cropping, the availability is related to the amount of exposed illite edge surface. Bolt *et al.* (1963) found that the rapidly exchangeable potassium in a soil occupied two types of site, a 'planar' and 'edge interlattice'. Analysis of exchange curves showed that 96% of the rapid exchange sites was on the external, planar side of the lattice and

4% was on the edges; the latter site showed high preference for potassium ions and explains the difficulty of removing potassium from that site with calcium or sodium.

Many soils can 'fix' large amounts of potassium in a non-exchangeable form. Herrera (1958), on the basis of montmorillonite being the constituent which fixes potassium most strongly, suggested that the percentage of montmorillonite in a soil can be used as an index of its potassium-fixing capacity. Fieldes and Swindale (1954) suggested that vermiculite may fix potassium ions to form a mineral indistinguishable from illite.

The fixation and release of potassium by soils can be affected by the temperature and moisture régimes. Thus Barber (1960) reported that areas receiving low rainfall had low available potassium and that low available potassium was also associated with areas having excessive rainfall. In his laboratory experiments Barber found that moisture had no effect upon release of non-available potassium, whereas increasing temperature increased the rate of release. Scott and Hanway (1960) found that a moist soil contained 26 $\mu g\ g^{-1}$ extractable potassium while the same soil contained 62 $\mu g\ g^{-1}$ extractable potassium if air-dried, 210 $\mu g\ g^{-1}$ if oven-dried at 110°C and 714 $\mu g\ g^{-1}$ if dried at 500°C. Potassium chloride added to the soil under moist conditions was largely fixed.

Dowdy and Hutcheson (1963) investigated the effect of drying soils (six series) over sulphuric acid upon potassium release and fixation. Differences in behaviour of different soils were related to the clay minerals present. Illite was the main source of potassium released by drying. The fixation of potassium at moisture levels greater than 4% was due to vermiculite and at lower moisture levels to montmorillonite. If the exchangeable potassium levels were greater than 0·45 ± 0·1 m.e./100 g for field-moist samples, potassium was fixed when the soil dried, but if the exchangeable potassium was less than this figure then non-exchangeable potassium was released by drying.

Richards and McLean (1963) examined several soil minerals for their potassium-fixing activities when moist and dry and their findings may be summarized thus: Kaolinite fixed little potassium regardless of moisture conditions, vermiculite fixed little potassium in suspension but an appreciable amount when dry. In both kaolinite and vermiculite the exchangeable potassium remained unchanged when the material was re-wetted after drying. Illite fixed considerable amounts of potassium when wet, more when dried and still more if re-wetted. Bentonite fixed considerable amounts of potassium in suspension, released the potassium again on air-drying, fixed the same amount as when in suspension if oven-dried and remained unchanged when re-wetted. Prochlorite fixed some potassium in suspension, none when dried but considerable amounts if re-wetted.

A considerable amount of work has been done on the mechanism of potassium release and fixation as affected by heat and moisture and Metson (1961) gives a concise review of this. It appears that the potassium ion is a special case when its adsorption on the soil complex is considered.

Its radius in the dehydrated state is just slightly larger than the radius of the cavities in the hexagonal rings of oxygen in the silica–oxygen sheets of micaceous clay and leads to the possibility of fixation of potassium when clays of the 2:1 lattice type dry out.

The role of potassium in plant nutrition

In plants potassium is an essential element associated with metabolism; for example, it may be involved in enzymatically controlled reactions of transphosphorylation. It has been suggested (Webster and Varner, 1954) that potassium is needed for protein synthesis by linking amino-acids to form peptides. However, as yet, no specific organic compounds containing potassium have been found in plants, which makes it distinctive from all other essential plant nutrient elements. When plants are deficient in potassium, soluble nitrogen compounds accumulate in the tissues due to a decrease in protein formation from amino-acids (Black, 1957). Carbohydrate production is also limited by potassium supply due to a decrease in carbon dioxide assimilation and to a decrease in leaf area owing to premature dropping of leaves. It is this reduction in carbohydrate formation that causes weak straw in cereals and weak fibre in fibre crops (Black, 1957). The roots of root crops contain much more potassium than the tops and such crops need ample potassium for development. Other potassium deficiency effects are: poor keeping quality of fruit, increase in disease susceptibility, increase in incidence of low temperature damage and retardation of maturity. Vȳskrebentenseva (1963) in water culture studies with pumpkins found that absence of potassium disturbed the Krebs cycle in roots.

Too high a potassium content in a soil can induce iron chlorosis (Walsh and Clarke, 1942) and magnesium deficiency (Boynton and Burrell, 1944). The antagonism between magnesium and potassium in plant nutrition has been discussed in section 8:3:1.

Potassium levels can affect a plant's water utilization. High concentrations of potassium in a soil contribute to excessive osmotic pressure in the plant and so increase its absorptive capacity for water. On the other hand, high potassium in a soil solution affects also the osmotic pressure of that solution and makes the water less available. Wallace (1958) reported that increasing soil moisture under aerobic conditions increases potassium uptake by plants, but high moisture content under anaerobic conditions does not assist in potassium uptake which will be poor. Humbert (1958) found that potassium in solution sprayed over a sugar cane crop at the rate of 21 kg K_2O per 10^4 m^2 (19 lb/acre) was more effective in increasing the potassium and water contents of the crop than 220 kg hectare (200 lb/acre) applied directly to the soil. This low rate of application stimulated potassium uptake from the soil potassium. The soil in question was a dark, heavy clay with pH 7·5 and high magnesium and calcium content.

A good supply of potassium is essential if saline alkali soil is to be utilized. Toxicity of the sodium ion can be reduced by potassium due to an antagonistic ion effect and thus, if potassium is added to sodium-rich water, that water can be used for irrigation. In sodium soils a high content of available or partially available potassium is important to maintain the sodium–potassium balance and a well-developed microflora will be important in this respect (Heinmann, 1958).

Zazvorka (1959) has pointed out that rainfall should be considered in conjunction with plant analyses as he found that 8 hours of rain removed 22% of the total potassium in barley leaves and 35% from lucerne.

Soileau and co-workers (1964) determined some of the effects of cutans upon uptake of potassium by plants, specifically to find if potassium-free clay coatings on potassium-rich aggregates affects the release of potassium. They prepared illite aggregates of known potassium content (total, ammonium acetate-extractable and dilute sulphuric acid-extractable) and by alternatively leaching with a solution of $Fe(OH)_3 . xH_2O$ and a sodium-saturated kaolinite suspension, obtained clay coatings upon the aggregates of a thickness comparable to that in natural soils. Plant experiments were then made. It was found that potassium uptake from the coated aggregates was less than from the uncoated but the reason for this was not clearly understood. It was suggested that the decreased uptake was due to the iron present or to the coating restricting ionic diffusion.

9:1:2 Determination

Total potassium

One of the earliest described methods for determining the total potassium content of a soil is that of Smith (1881). In Smith's method the soil is mixed with solid ammonium chloride and calcium carbonate and fused. The melt is leached, the leachate precipitated with ammonium carbonate and barium chloride solutions and the filtrate evaporated to dryness. Ammonium chloride is driven off by heating and the potassium in the weighed residue determined. Sodium is determined also from the weight of combined residues. Smith's method has not become obsolete and Chew *et al.* (1962) have modified it only slightly by first mixing the soil with calcium carbonate and moistening it with 15% ammonium chloride solution before fusing.

A simple sodium carbonate fusion can just as well be made provided that sodium is not to be determined simultaneously and if sesquioxides are removed. The most convenient method, however, is the wet digestion of the soil with hydrofluoric acid which permits the determination of sodium as well. Emission spectrography can be used to determine total potassium in soils, when arc or spark procedures are followed (Mitchell, 1964). Skobets *et al.* (1962) have described a polarographic method for determining the total alkali metal content of soils. In this method the sum of the potassium and sodium is measured by differential polarography in lithium chloride and calcium hydroxide buffer; the half-wave potential for potassium is $2 \cdot 128$ volt and for sodium $-2 \cdot 104$ volt.

Kogan *et al.* (1961) determined the potassium content of topsoils directly by measuring energy of gamma-quanta at the soil surface with a scintillator-γ-spectrometer; this could be done on the land surface or from an aeroplane. An investigation of the procedure using ^{40}K was made by Redies and Vimpany (1966) who found a liquid scintillation technique unsatisfactory due to continuum interference. Better results were obtained using a sodium iodide crystal in conjunction with a multi-channel pulse height analyzer.

Exchangeable potassium

For the extraction of exchangeable potassium, neutral 1 M ammonium acetate solution is recommended as the conversion of non-exchangeable potassium to the exchangeable form is retarded (Merwin and Peech, 1950). When dealing with saline soils water-soluble potassium must be determined separately and deducted from exchangeable potassium found experimentally.

Available potassium

Available potassium can be separated into that immediately available which is the water-soluble and exchangeable, and that potentially available or 'fixed'. McLean (1961) examined eleven soils and found that the amount of exchangeable potassium in surface soils is significantly correlated with potassium uptake by plants grown in the greenhouse. The percentage potassium saturation and water-soluble potassium indicated the percentage potassium uptake from the exchangeable potassium in the soil. Semb and Øien (1961) determined readily soluble potassium by extraction with a monoacetate solution; the difference between this acetate-soluble potassium and that extracted in boiling 1 M nitric acid they called 'acid-soluble' potassium. The use of boiling 1 M nitric acid was suggested originally by De Turk *et al.* (1943). Various soils differed little in readily soluble potassium in the surface layers but the acid-soluble potassium was found to be from four to five times more abundant in clays and in mica-rich sands than in sands or silts. In the clays and micaceous sands acid-soluble potassium increased with depth. Semb and Øien reported a good correlation between acid-soluble potassium and yield. McLean (1961) also extracted non-exchangeable potassium with boiling 1 M nitric acid and with 0·01 M hydrochloric acid and by shaking the soil with a hydrogen-saturated exchange resin. The quantities of potassium so extracted were significantly related to each other and with plant uptake of potassium.

Vries (1961) obtained what he termed the 'K-value' by extracting soil with 0·1 M hydrochloric acid and expressing the results in units of 0·9 kg K_2O/1000 kg organic matter. Vries related K-values to available potassium and plotted curves for several different kinds of soil.

Evans and Simon (1949) found that repeatedly drying a hydrogen-saturated soil and then extracting it with 0·5 M hydrochloric acid gave a potassium figure related to that removed by exhaustive cropping of the soil. Potentially available potassium has also been determined by electro-dialysis (Ayres *et al.*, 1946).

Barber and Matthews (1962) extracted soil with ammonium acetate with and without a cation exchange resin and the difference in potassium extracted was considered as moderately available potassium. Warren and Cooke (1962) compared water, citric acid, acetic acid and hydrochloric acid as extractants of available potassium and found citric acid the best. The biological method in which the dry weight of mycelium growth of *Aspergillus niger* is determined, has been improved by Nowosiels (1962). Gouveia and co-workers (1960) compared the *A. niger* method with ex-traction of soil with sodium nitrate solution and found the two methods closely related. Blanchet *et al.* (1962) by pot experiments found that the potassium nutrition of plants was governed by exchangeable potassium, the absorbing surface of the root system and the requirement of the par-ticular plant. Milchera (1962) compared Neubauer's technique for deter-mining available potassium with several methods of extraction and found that the potassium soluble in 2 M hydrochloric acid gave the best correla-tion with plant uptake.

From the great number of different extraction procedures used with varying success it is obvious that no universally satisfactory extractant can be postulated for determining available potassium. As is so often the case in soil analysis the particular soil, crop and environment must all be con-sidered together.

A method which is of universal application, however, is that utilizing the concept of standard free energies. Woodruff (1955) introduced this concept to determine whether a soil has sufficient available potassium to support a crop. In the method the standard free energies of potassium replacement from the soil's exchange complex, at which nourishment of plants becomes critical, are used to define potassium availability. By suitably controlling the procedure it is possible to determine whether the amount of potassium required by a crop can be removed without exceed-ing the energy of replacement necessary.

The 'intensity' of available potassium has thus been defined in terms of free energy of potassium replacement by calcium by measuring the potassium potential relative to calcium; that is $pK - \frac{1}{2}pCa$. This is analo-gous to the lime potential, $pH - \frac{1}{2}pCa$, discussed in Chapter 3 (section 3:3). In practice, as when considering the lime potential, it is more con-venient to regard calcium and magnesium as a single ionic species and the potassium potential is determined by $pK - \frac{1}{2}p(Ca + Mg)$ in 0·01 M calcium chloride solution.

Matthews and Beckett (1962) showed that the relationship between exchangeable potassium and the activity ratio

$$a_K / \sqrt{(a_{Ca} + a_{Mg})}$$

of the solution with which it is in instantaneous equilibrium is a character-istic independent of the ratio of soil to solution or total electrolyte concentration (up to 0·06 M) of the solution. Thus instantaneous equilibrium curves provide a basis against which to measure change in exchangeable potassium due to fixation and release. The activity ratio gives a satisfactory measure of the chemical potential of labile potassium in a soil provided that soils of widely differing calcium or magnesium content are not compared.

Different soils, however, although having the same activity ratio—that is, intensity of labile potassium—may not have the same capacity to maintain the intensity as potassium is removed. Consequently to obtain a clear picture of the potassium status of a soil, we must specify not only the current potential in the labile pool, but also the form of the 'quantity–intensity' relation, known as the 'Q/I' relation (Beckett, 1964). The ability of a soil to maintain its potassium intensity depends partly upon the character of the labile pool, the rate of release of fixed potassium and the rate of diffusion and transport of potassium ions in the soil solution. Beckett determined the immediate Q/I relationships for a number of soils and found them to be linear over the activity ratios commonly encountered. The form of the immediate Q/I relationship regulates the uptake of potassium over short periods and will regulate also the total potassium uptake where each volume of soil is drawn upon only once and for a short time.

The determination of the Q/I relationship is fairly straightforward provided that temperature is controlled. Samples of soil are shaken with aliquots of solutions containing potassium and calcium chloride of known concentration and covering the equilibrium activity ratio of the soil. After equilibrating for 12 hours the solutions are filtered and potassium, sodium, calcium, magnesium and aluminium are determined. For each suspension the amount ($\pm \Delta K$ in m.e./100 g soil) by which the exchangeable potassium content of the soil had changed is calculated from the difference between the potassium concentrations of the initial and resultant solutions. Beckett showed that the values of ΔK thus calculated owed nothing to the release or fixation of potassium or to soluble salts present. The activity ratios, $a_K/\sqrt{(a_{Ca} + a_{Mg})}$ of the resultant solutions were calculated from their compositions by a modified Guggenheim's (1950) expression for activity coefficients. A typical Q/I relationship curve as obtained by Beckett is shown in Figure 9:1.

Tinker (1964) when investigating highly acid soils of the tropics (West Africa) found the slopes of their Q/I curves much lower than those reported by Beckett who used soils from England. This showed that the tropical soils were less well buffered against potassium depletion. The soils examined by Tinker contained small amounts of calcium and magnesium, and appreciable quantities of calcium were absorbed from the equilibrating solutions. An equilibrium was set up between (Ca + Mg) and aluminium and the soil solution always contained aluminium, an ion which did not enter seriously into Beckett's investigation. Tinker thus found it necessary to modify Beckett's procedure slightly by adding aluminium to the equilibrating solutions. He prepared groups of tubes of varying potassium content as usual, but each group also contained a different level of aluminium chloride. The results gave different activity ratio − ΔK curves (Figure 9:2) and the intercept ΔK = 0(A.R.$_0$) was found for each curve

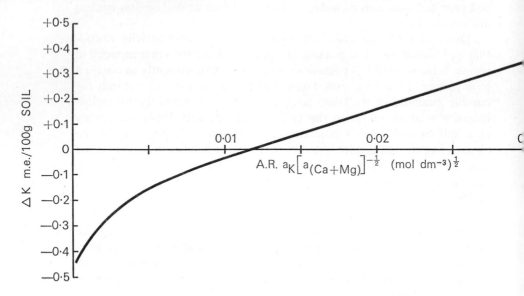

Figure 9:1 Immediate Q/I relation of a soil (Beckett, 1964; *J. Soil Sci.*)

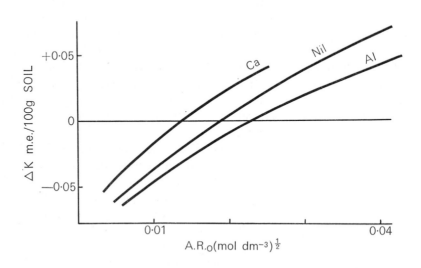

Figure 9:2:1 Relationship between exchangeable potassium and activity ratio in the presence of added Ca(OH)$_2$ or AlCl$_3$ in soil (Tinker, 1964; *J. Soil Sci.*)

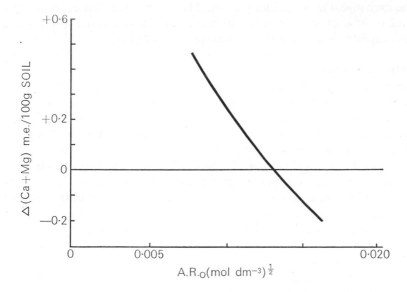

Figure 9:2:2 Relationship between A.R.$_0$ and exchangeable calcium plus magnesium with varying concentrations of CaCl$_2$ and in the presence of either AlCl$_3$ or HCl (Tinker, 1964; *J. Soil Sci.*)

and plotted against change in (Ca + Mg), {Δ(Ca + Mg)}. Then the intercept Δ(Ca + Mg) = 0 gave the true activity ratio A.R.$_{00}$. The method, however, was not quite satisfactory for soils with too low a pH value as the large amounts of aluminium chloride needed produced very steeply-sloped lines making the determination of the ΔK = 0 intercept difficult.

Tinker applied his work to the potassium fertilization of oil palms in Nigeria. He found that the potassium–calcium plus magnesium activity ratio was not related to yield response to potassium in acid soils and this was supported by the failure of liming to affect potassium response. A better relationship was obtained between yield response and what Tinker called a 'unified activity ratio',

$$\frac{K^+}{\sqrt{(Ca^{2+}) + (Mg^{2+}) + P^3/(Al^{3+})}}.$$

Moss (1963) used an ethanol displacement technique to obtain equilibrium soil solutions from different soils in the moisture range of from 25% to ten times field capacity. For one soil the activity ratio pK − ½p(Ca + Mg) was constant over the whole range of moisture, showing an undisturbed equilibrium and defining the characteristic potassium intensity status of the soil. For the other soils examined the ratio was constant only over the field range of moisture. Equilibrium studies with dilute calcium chloride solution showed an increasing release of potassium from soil to solution with increasing soil–solution ratio and increasing strength of calcium chloride solution. The release of potassium considerably exceeded

6 + T.S.C.A.

the exchangeable potassium present. Moss suggested that excessive dilution or low concentration of divalent cations can cause depletion of exchangeable potassium from interlayer sites of illites.

Potassium in solution

Once in solution potassium can be determined spectrographically, volumetrically, colorimetrically and gravimetrically, although nowadays the flame photometric method has superseded all others for speed and convenience. A turbidimetric method depending upon a suspension of potassium cobaltinitrite can be used where great accuracy is not required and, as described by Olson (1953), can be used in the rapid testing of soils.

The potassium compound of tetraphenylboron, $KB(C_6H_5)_4$, is extremely insoluble and thus is a suitable compound for isolating potassium from solution. Sunderman and Sunderman (1958), for example, proposed a turbidimetric method for determining potassium using tetraphenylboron as reagent and the method was adapted by Benett and Reed (1965) for determining extractable soil potassium. Benett and Reed found that the most suitable soil extractant was 0·1 M sodium nitrate solution used at a ratio of 5 of solution to 1 part soil.

The soil (5 g) is shaken for 5 minutes with 25 cm^3 of extracting solution and filtered. 1 cm^3 of filtrate is mixed with a solution of formaldehyde and EDTA in a 10-cm^3 volumetric flask, after which 1 cm^3 of an alkaline 5% solution of sodium tetraphenylboron is forcibly injected with a hypodermic needle. After standing and diluting, the turbidity of the solution is measured using a spectrophotometer set at a wavelength of 525 nm. The EDTA masks the effects of certain ions such as calcium and magnesium and the formaldehyde reacts with ammonium which is otherwise coprecipitated. The purpose of injecting the reagent is to standardize the rate of its addition and hence of reaction.

The flame photometric method for the determination of potassium and sodium is to be preferred to the flame spectrographic method partly because of the direct reading technique but also because the sensitive lines in the visible part of the spectrum can be employed. This is inconvenient when using photographic techniques owing to the extreme sensitivity of the sodium line and the high photographic contrast of the potassium line (Mitchell, 1964).

The use of sodium tetraphenylboron to extract soil potassium has led to some difficulty over the subsequent determination of potassium by flame photometry. Hunzinker (1958) dissolved the potassium precipitate in acetone and Neeb and Gebauhr (1958) used tetrahydrofuran as solvent. In such methods, however, errors and difficulties were found due to volatilization of solvent. Hunzinker eventually decided that oxidation of the potassium precipitate with aqua regia in order to liberate the potassium was the best procedure. Scott and Reed (1963) hydrolyzed the precipitate with boiling, dilute hydrochloric acid and determined the potassium in the filtrate. Schulte and Corey (1963) developed a method whereby the potassium precipitate is dissolved in acetone and the solution evaporated with concentrated sulphuric acid; boron up to 500 μg cm^{-3} did not interfere and sodium interference was overcome by adding sodium to the standards.

In flame photometry the intensity of radiation emitted when an element is introduced into a flame is correlated with the concentration of that element. Two types of instrument can be used for measuring the emission intensity, a simple flame photometer and a flame spectrophotometer. For most purposes of soil analysis the simple photometer is adequate and there are several different models commercially available. Experimental details will vary according to the particular instrument but the basic procedures are the same in each case. The sample in solution is contained in a small cup of an atomizer to which is fed a mixture of air or oxygen and a combustible gas such as petrol gas or acetylene. The solution is then vaporized and at the high temperature of the flame the salts vaporize. The vapours of the metal atoms are then excited by thermal energy to give the emission radiation which is measured photoelectrically after passing through a suitable filter.

Detailed, practical instructions for the operation of the Beckman flame photometer are given by Metson (1961). With this instrument potassium is determined before sodium at a wavelength of 770 nm. Metson also describes briefly the operation of the Beckman flame spectrophotometer and this instrument is recommended by Jackson (1958) who describes its general principles. Mitchell (1964) gives many useful references in connection with flame photometry and mentions that the two instruments in use at the Macaulay Institute are both based on the Lundegardh-type atomizer and air–acetylene burner. For most routine analyses the simple EEL flame photometer is sufficiently accurate and explicit details for its setting up and operation are given by Vogel (1962). The flame photometric procedures can be adapted to auto-analysis (section 8:3:5).

Gravimetrically potassium can be determined as the cobaltinitrite, the chloroplatinate or the tetraphenylborate. Jackson (1958) recommends the cobaltinitrite method as being rapid, economical and adaptable to semi-micro technique. The procedure, however, involves several stages of evaporation, precipitation and filtration, each of which is a source of error, and the method is not recommended except in such cases where flame photometry is impossible and even then the tetraphenylborate procedure would be preferable. As an alternative to weighing the precipitate of potassium cobaltinitrite, it can be determined volumetrically by titration with cerate or permanganate (Jackson, 1958; Metson, 1961), and Munk (1962) has described a 'dead-stop' method of volumetrically estimating potassium tetraphenylborate. In Munk's method the precipitate is formed under alkaline conditions and the excess of reagent back-titrated with thallium nitrate.

Colorimetrically, potassium can be determined as the dipicrylaminate (which is the salt of di-2,4,6-trinitrophenylamine) (Williams, 1941) and details of this method are given in the USDA Handbook No. 60. Jackson (1958) describes the development of a cobalt hydrocarbonate colour from precipitated potassium cobaltinitrite which, as in all methods involving the cobaltinitrite precipitate, involves many steps and sources of error. The chloroplatinate precipitate can similarly be evaluated colorimetrically either by directly measuring the transmission of the yellow colour as such or, more sensitively, by developing potassium iodoplatinate with potassium iodide solution (Shohl and Bennett, 1928).

Glass electrodes responsive to potassium ions have been described by Eiseman *et al.* (1957) and by Portnoy *et al.* (1962). These electrodes were examined by Mattock and Uncles (1964) who found the glass used to be unsuitable. A new electrode known as BH.115 was developed by Mattock and Uncles and which had improved stability and reproducibility and is easily prepared. Such electrodes are now commercially available and have an inner reference system of silver–silver chloride dipping into 0·04 M potassium dihydrogen orthophosphate solution mixed with 0·01 M disodium hydrogen orthophosphate and 0·01 M potassium chloride. The external reference is a 3·8 M potassium chloride–calomel electrode. A special intermediate salt-bridge is incorporated to prevent potassium chloride leaking into the test solution.

The electrode shows selective response to potassium and ammonium ions; the response is logarithmic to concentration and the slope is in good agreement with that predicted from the Nernst equation. Hydrogen ions and alkaline-earth elements do not exert a significant effect upon the electrode but sodium ions interfere up to pK (or pNH_4^+) 3, when the concentration ratio of $Na^+ - K^+$ (or NH_4^+) of 1:5 is exceeded. In practice one can apply a correction based on the sodium content of solution.

9:2 SODIUM

9:2:1 Background and theory

Soils in which sodium is specifically determined are nearly always those of arid or semi-arid regions which have become or are liable to become, alkali or saline alkali. Occasionally sodium has to be determined in coastal soils which have been inundated with sea-water.

The formation of saline and saline alkali soils has been discussed in Chapter 6. An alkali soil contains an excess of exchangeable sodium and may or may not be saline as well, although the two conditions often go together. A saline alkali soil has a pH value of less than 8·5 but if the salts are leached out, excess sodium reacts with carbon dioxide (atmospheric and from plant roots) to form sodium carbonate and the pH of the soil may rise to 9 or 10. The presence of sodium carbonate causes drastic changes in a soil's physical properties; permeability and pore-space are decreased, structure is destroyed and the soil shrinks into hard-surfaced, prismatic units. Clay particles are then likely to be washed down between the soil blocks and may eventually form a pan. Chemically, the sodium carbonate disperses humic material to give a black coating on the soil and hence the term 'black alkali' soils. The so-called 'white alkali' soils occur only amongst saline alkali soils and the name is due to the white crust of sodium-containing salts which forms on the land surface. In the continued absence of calcium a podsolization-like process occurs, giving rise to degraded alkali soils in which a bleached horizon overlays a sesquioxide-rich horizon.

Another mechanism whereby exchangeable sodium increases in a soil, other than by leaching of salts, is by precipitation of calcium and magnesium as carbonates, when the soil solution becomes concentrated by evaporation. The increased proportion of sodium then exchanges with calcium and magnesium on the soil colloidal complex.

Wilding (1963) found that the high percentage of extractable sodium in alkali soils derived from loess originated mainly in situ by weathering of sodium-rich feldspars and other sodium-bearing constituents of the parent material. Hsi and Chao (1962) considered that the sodium in certain alkali soils originated from the transformation of sodium silicates in weakly mineralized groundwater and from igneous rocks. According to Hsi and Chao the first stage of genesis of saline–alkali and alkali soils is salination accompanied by accumulation of chlorides, sulphates and hydrogen carbonates. Later the sodium migrates and accumulates. With improved drainage the desalination process causes soda to dissociate and sodium ions to enter into the exchange complex. However, sodium usually becomes the dominant ion in an arid soil due to irrigation, and the ratio of sodium to calcium plus magnesium is a very important factor when considering the quality of irrigation water.

The degree of alkalinity of a soil is normally assessed by what is called the 'exchangeable sodium percentage' or E.S.P. and which is defined as

$$\frac{\text{ex. Na m.e.}/100 \text{ g soil}}{\text{C.E.C. m.e.}/100 \text{ g soil}} \times 100.$$

It has proved difficult to decide upon a precise upper limit for the E.S.P. to denote toxic levels of alkalinity, but 15% is generally accepted. As calcium and magnesium are preferentially absorbed by a soil, the soluble sodium must be quite high for the soil to have an E.S.P. of 15%. The E.S.P. is thus of most use for classifying non-saline alkali soils because in saline alkali soils the considerable amount of soluble sodium must be allowed for by a correction factor and, according to Black (1957), this correction factor is questionable. Black suggested that the E.S.P. be confined to classification of the physical properties of alkali soils and that when considering other properties a soluble sodium percentage be used. One important reason for distinguishing two classes of saline soils on the basis of their sodium content is that if salts are removed, a sodium-rich soil will become highly dispersed, whereas a sodium-poor soil will not.

In plant nutrition the harmful effects of alkali soils are due partly to high pH but also to the excess of sodium ions (Bower and Wadleigh, 1949). It is not definitely known why an excess of sodium is toxic to plants but one explanation is that it induces deficiency of other elements, for example calcium, magnesium, iron and phosphorus (Black, 1957). Plants vary in their tolerance to sodium and there have been few examples of specific sodium toxicity; where it does occur it appears usually as leaf scorching. The extremely poor physical condition of sodium-rich soils probably accounts for most plant nutrition troubles on alkali land.

In potassium-poor soils sodium has been reported as causing increase in yield but no response is shown to sodium if potassium is in adequate supply (Koter and Wachwola, 1962). Morani and Fortini (1962) conducted pot tests to investigate the role of sodium in nutrition and they found that when potassium was in short supply, addition of sodium to the soil caused increased uptake by the plants of phosphorus. The economic plants found to be most sodium-demanding are sugar-beet and mangolds. Once sufficient potassium is present these crops require a good supply of sodium to maintain their yield (Crowther, 1947). The effect appears to be due to an increase in the amount of water held per unit dry weight of leaf tissue which makes the plant more succulent and drought-resistant. Rathje (1961) investigated the function of sodium in sugar-beet nutrition and found that it is associated with formation of an osmotic energy accumulator in leaves during daylight followed by its removal during darkness. This economizes on sugars by reducing respiration energy during the night.

9:2:2 Determination

Total sodium

The total sodium content of a soil can be brought into solution by an acid digestion as described for total potassium (section 9:1:2). The classical method of Smith (1881) in which the soil is heated with ammonium chloride and calcium carbonate is still sometimes employed but is not recommended. Skobets *et al.* (1962) have described a method of determining total sodium plus total potassium in soil by polarography, and various spectrographic procedures can be used (Mitchell, 1964).

Exchangeable sodium

Exchangeable sodium in soil is usually and conveniently extracted with ammonium acetate solution in the usual manner (see Chapter 7) although Aleshin and Boldȳrev (1962) used ammonium carbonate and barium chloride solutions with apparently good results. Bower and Hatcher (1962) found that with some saline soils, ammonium acetate extraction gave low results for sodium, whereas with other saline soils where the sodium was not readily soluble in water but dissolved in the acetate, a high result was obtained. Bower and Hatcher concluded that for saline soils in general, the sodium adsorption ratio of an equilibrium extract is the best value for relating sodium to plant growth.

Bower (1960) recommended using normal magnesium acetate solution to remove exchangeable sodium from soils as it removes more sodium from vermiculites than does ammonium acetate solution. In most agricultural soils the exchangeable sodium will be low and it may be necessary to use larger quantities of soil than for the determination of other exchangeable ions. In saline alkali soils and soils of coastal regions a large proportion of the sodium present will be water-soluble. This must be determined separately and deducted from that soluble in the extractant in order to obtain the correct value for exchangeable sodium.

Although exchangeable sodium and water-soluble sodium have both been taken as indices of available sodium, Tinker (1966) recommends for this purpose the determination of the equilibrium activity ratio, $Na/\sqrt{(Ca + Mg)}$, in a manner similar to that for potassium. Tinker applied such measurements in a study of sugar-beet fertilization.

Sodium in solution
Soluble sodium can be determined gravimetrically, volumetrically, colorimetrically, electrometrically and by emission spectrophotometry.

As in the case of calcium, magnesium and potassium, for the routine determination of sodium in many samples the flame spectrographic method offers some advantages, but most laboratories nowadays use a flame photometer. Once set up and calibrated, the flame photometric method is rapid and reliable. The ammonium acetate extract of soil can, if so desired, be used in the flame photometer directly with no further treatment other than filtration. Direct use of the acetate extract obviates the digestion of the extract with acid and so eliminates the risk of certain errors. Even the most careful of digestion procedures results in a higher blank than when the extract is not digested; if the extract is used direct it is necessary to make the standards in ammonium acetate solution and the solution must be filtered to avoid blockage of the atomizer.

Gravimetrically sodium can be determined by precipitation as sodium zinc uranyl acetate or as sodium magnesium uranyl acetate; the magnesium complex is preferable as it is less soluble than the zinc complex. Provided that their concentrations are not too high, ammonium, calcium and magnesium do not interfere but phosphate and, if potassium is present, sulphate, must be absent. Temperature control is critical during precipitation as the temperature coefficient of solubility is large (Vogel, 1962). Although the method is still given in most books on soil analysis, there is little to recommend it unless the laboratory has no other means of making the analysis. A volumetric procedure has been developed whereby, after precipitation as sodium zinc uranyl acetate, the excess uranium is determined with EDTA (Vogel, 1962). Jackson (1958) describes a volumetric method in which sodium magnesium uranyl acetate is redissolved and titrated with sodium hydroxide solution.

Precipitation of sodium with zinc or magnesium uranyl acetates forms the basis for several colorimetric methods of determining sodium. The precipitates can be dissolved and the optical density of their aqueous solutions read directly, or they can be treated before measurement with various reagents. Parks *et al.* (1943) treated sodium zinc uranyl acetate with sulphosalicylic acid and determined sodium in the range 0·01–0·36 mg. Other reagents for developing colour are potassium hexacyanoferrate(II), ammonium thiocyanate, hydrogen peroxide and alizarin (Snell and Snell, 1949). When precipitation is made with manganese uranyl acetate the complex can be oxidized with periodate and the manganese determined. A more recent colorimetric method is by reaction of sodium with violuric acid (5-isonitroso-barbituric acid); calcium and potassium interfere but can be corrected for (Muraca and Bonsack, 1955).

Glass electrodes which respond to sodium ions in neutral or alkaline solution have been developed (Lengyel and Blum, 1934; Eisenman *et al.*, 1957) and are now commercially available (for example the Beckman electrode No. 78187V). By using the glass electrode in conjunction with a calomel reference electrode, sodium ions can be determined potentiometrically in the same manner as hydrogen ions are determined (Chapter 3). The glass electrode is filled with 0·1 M sodium chloride solution into which is dipped a silver–silver chloride electrode. In an early form of the electrode the glass was a sodium-alumino-silicate which responded to both sodium and potassium ions, but in later types the glass was a lithium-alumino-silicate which responds selectively to sodium ions provided that the sodium concentration is equal to or greater than that of potassium. As both types of electrode respond to hydrogen ions, the pH of the solution must be at least 6. Use of the sodium electrode has been criticized, but on the whole it appears to be as good a method of determining sodium in non-acid solution as the flame photometric method. Thus Fehrenbacher and co-workers (1963) analyzed one hundred and twenty-six soil–water samples from solonetzic soils by flame photometry and with a sodium sensitive electrode. The results agreed closely, the correlation coefficient being $r = 0·9989$. Orlov and Tsikuria (1962) found the potentiometric method less accurate than the flame photometric but increased its accuracy by using a ballistic circuit with a mirror galvanometer and a high capacity condenser.

When using the glass electrode to determine exchangeable sodium, Bower (1960) recommended leaching the soil with 0·5 M magnesium acetate solution rather than with ammonium acetate as not only does the magnesium salt remove more sodium from vermiculites, but the electrode responds slightly to ammonium ions. Furthermore, magnesium has a buffer action ensuring negligible concentration of hydrogen ions and if it is present in high concentration relative to other elements, it controls the ionic strength and hence activity coefficient of sodium ions.

9:3 RECOMMENDED METHODS

9:3:1 Total potassium

Total potassium in soils is most conveniently determined by a wet digestion of the sample with hydrofluoric and perchloric acids followed by
flame photometry. Digest the finely-ground soil in a platinum crucible
with the acid mixture as described in Chapter 16, section 16:2:4III, the
final solution being made to 100 cm³. Suitable aliquots are then taken for
potassium and or sodium determination, preferably by flame photometry.
The accuracy using a simple photometer is about ±5%.

9:3:2 Exchangeable potassium

Exchangeable potassium is extracted from the soil with neutral, 1 M
ammonium acetate solution as described in Chapter 7, section 7:3:2. The
ammonium acetate leachate is treated with acids as described in section
7:3:3, and aliquots of the final hydrochloric acid solution analyzed
directly for potassium by flame photometry or spectrography. The acetate
extract can be analyzed directly by flame photometry without acid digestion but in which case, the standard solutions must be made in ammonium
acetate solution.

9:3:3 Available potassium

For specific purposes using particular crops, soils and conditions a variety
of extractants can be used, but, as discussed (section 9:1:2), the most
promising method of determining available potassium in a soil is by
measuring activity ratios and quantity–intensity relationships. It should
be remembered, however, that the activity ratio is dependent upon the
difference in chemical potential of at least two types of ions and the potential of the reference ion (i.e. Ca + Mg) should not vary much between the
various soils or during the process of equilibrium. Use of activity ratios
should be restricted to soils where the Ratio Law holds.

I / WOODRUFF'S METHOD

A simple, routine method of determining if a soil has sufficient potassium
available for a crop is to equilibrate definite quantities of soil with definite
quantities of an electrolyte of known concentration. The experimental
details are those of Woodruff and McIntosh (1960).

Reagents

Calcium chloride solution, 0·01 M: Approximately 1·3 g CaCl₂ (anhyd.)
    is dissolved in water and the solution diluted to 1 dm³. Standardize
    with EDTA and dilute to exactly 0·01 M.

6*

Procedure

Weigh 5 g of 2-mm soil into a 100-cm³ screw-capped bottle and add 1·5 cm³ of water; leave overnight. Add either 35 cm³ or 80 cm³* of 0·01 M calcium chloride solution, shake intermittently during 1 hour and filter. Determine the potassium content of the filtrate with a flame photometer calibrated with standard solutions containing 5 μg cm⁻³ and 10 μg cm⁻³ potassium in 0·01 M calcium chloride solution.

If the concentration of K in the 35 cm³ extract is 9 μg cm⁻³ or greater, little response to potassium fertilization can be expected; similarly for a concentration of greater than 4 μg cm⁻³ K in the 80-cm³ extract. At smaller concentrations of K the crop will probably respond to potassium fertilization. Using corn as test crop Woodruff found excellent response to potassium when its concentration in the 35-cm³ extract was 4–5 μg cm⁻³ and in the 80-cm³ extract, 2–2·5 μg cm⁻³. By using 35 cm³ of extracting solution the results will give a margin of safety for crop production, whereas 80 cm³ will give results acceptable for considering the cost of fertilizer.

II / ACTIVITY RATIOS AND QUANTITY–INTENSITY (BECKETT, TINKER)

Reagents

Calcium chloride solution, 0·01 M: As in section 9:3:3i, standard reference solutions being 0, 0·5, 1·0, 2·0, 3·0, etc. × 10⁻³ M potassium chloride made up in 0·01 M calcium chloride solution. A 0·01 M solution of potassium chloride contains 0·7455 g KCl dm⁻³.

EDTA solution, 0·01 M: 3·7226 g dm⁻³.

Potassium cyanide solution: 10% w/v, aqueous.

Eriochrome Black T indicator: 0·1 g/25 cm³ methanol containing 1 g hydroxylamine. Prepare fresh.

Titrated magnesium solution: 0·01 M magnesium chloride solution (2·03 g MgCl₂.6H₂O dm⁻³) titrated with EDTA to the Eriochrome Black T end-point.

Procedure

Weigh 5-g to 10-g portions, depending upon cation exchange capacity, of 2-mm soil into 50-cm³ centrifuge tubes. Add 30 cm³ of standard reference solutions, doing each tube in duplicate. Stopper the tubes and gently shake for 1–2 hours avoiding considerable temperature changes. Centrifuge and pipette aliquots of clear solution into small conical flasks. Add a few drops of potassium cyanide solution, 1 cm³ of titrated magnesium solution, 4 drops of indicator and titrate with EDTA solution. The titrated magnesium solution is added to improve the end-point and is not necessary if the extracts contain normal amounts of magnesium.

---

* 35 cm³ of solution will remove 141 kg K per 10⁴ m² 15-cm (125 lb/acre 6-in) at a concentration of 9 μg cm⁻³, while at 4 μg cm⁻³ 80 cm³ are required. 141 kg K per hectare was chosen as an estimated crop requirement.

Soluble salts, the presence of large amounts of a cation other than potassium, calcium or magnesium and, to a lesser extent, the exchange of potassium for calcium, may cause the result of the titration of $(Ca^{2+} + Mg^{2+})$ in the equilibrium solution $(Ca^{2+} + Mg^{2+})_E$ to vary from the original calcium concentration.

The potassium content is found by 'bracketing' the equilibrium solution from the soil (E) and the original, standard solution from which it was derived (S_1) and the standard solution immediately higher or lower in concentration (S_2), according to whether the equilibrium solution has gained or lost potassium. The solutions are measured for potassium using a flame photometer. Set the instrument arbitrarily to give reasonable readings for all solutions, the range $S_1 - S_2$ being as large as possible on the scale. Read solutions S_1, E, S_2, S_2, E, S_1, S_1, E, S_2 in succession. The potassium content of the equilibrium solution is then found by simple proportion after averaging the readings:

$$K_E = K_{S_1} - \left[(K_{S_1} - K_{S_2}) \times \frac{(S_1 - E)}{(S_1 - S_2)} \right] \quad \text{if} \quad S_1 > S_2 \quad [9:1$$

For this calculation to be valid the instrument should have a linear calibration curve. If the curvature of the calibration graph is very marked then the difference between S_1 and S_2 should be as small as possible and it may be necessary to prepare intermediate standards.

The purpose of this lengthy method of measurement is to obtain the change from the original solution (from which potassium is calculated) very accurately.

Calculations

The concentration ratio (C.R.) is calculated by:

$$\text{C.R.} = \frac{K_E}{\sqrt{(Ca + Mg)_E}}, \quad [9:2$$

all quantities being in mol dm^{-3}. If soluble salts are low then $(Ca^{2+} + Mg^{2+})_E$ may be very close to the original calcium concentration, making the denominator 0·1. Very rarely do changes in potassium level change the $(Ca^{2+} + Mg^{2+})_E$ sufficiently to make it necessary to recalculate the denominator for each solution in the set.

The activity ratio (A.R.) is calculated by multiplying the concentration ratio by the activity coefficient,

$$\frac{f_K}{\{f(Ca^{2+} + Mg^{2+})\}^{-2}}.$$

It may be considered sufficient to use the simple form of the Debye–Hückel relationship to obtain the activity coefficient ratio but more accurate calculated values are given by Beckett (1965). Tinker (1964) found that with 0·002 M solutions the correction factor calculated from the simple Debye–Hückel formula usually approximates to 1·10 and could be ignored for comparative work in solutions of similar concentration.

A graph is prepared of A.R. − ΔK, where ΔK is the change in exchangeable potassium in the soil in equilibrium suspension from its original state and is found from the difference in initial and final potassium concentration in the solution $(K_{S_1} - K_E)$ in m.e. dm^{-3}. Thus,

$$\Delta K = (K_{S_1} - K_E) \times \frac{V}{W} \times \frac{1}{10} \quad \text{m.e./100 g soil,} \qquad [9:3$$

where W is weight of soil taken and V is the volume of standard solution added.

If $K_E > K_{S_1}$, ΔK will be negative, that is, potassium is lost from the soil.

The graph relates the intensity function (A.R.) to the quantity function (ΔK) and the slope of the graph is a capacity function referred to as the 'buffer power' for potassium and which measures the change of potassium potential with loss or gain of potassium. The point of intersection of the ΔK = 0 line and the curve gives the 'activity ratio' in the original solution, A.R.$_0$.

Note, however, that if the soil contains much soluble salt then A.R.$_0$ is not strictly the activity ratio of the original soil, which can be more accurately found by plotting the original activity ratio in the solutions against the final activity ratio after equilibrium; the line drawn at 45° on the graph for original A.R. = final A.R., intersects this graph at the original A.R. value of the soil.

9:3:4 Total sodium

Digest a suitable weight of soil with hydrofluoric and perchloric acids as described in Chapter 16, section 16:2:4III, the final solution being made to 100 cm^3. Aliquots of this solution can be taken without further treatment for determination of sodium by flame photometry. If it is desired to use the lithium glass electrode to determine the sodium, the solution must be neutralized with magnesium hydroxide before making to standard volume.

9:3:5 Exchangeable sodium

For the routine determination of exchangeable sodium in normal soils and when other exchangeable cations are also to be determined, extraction with neutral, 1 M ammonium acetate solution (section 7:3:2) is the best procedure, and the resulting dilute hydrochloric acid solution can be measured directly with the flame photometer or spectrograph. It is essential to make a blank determination right from the beginning of the experiment, especially if the centrifuge procedure is used. If the ammonium acetate extract is used for flame photometric measurements without acid digestion, the standard sodium solutions must be made up in ammonium acetate also.

When exchangeable sodium specifically is to be determined, extraction with 0·5 M, neutral magnesium acetate solution is preferable. The resulting extract can be used directly with a sodium-sensitive electrode for potentiometric determination of the sodium, or it can be worked up for spectrographic or photometric analysis; the presence of large quantities of magnesium must be allowed for by including similar quantities in the standards.

Reagent

Magnesium acetate solution, 0·5 M: 107 g $Mg(C_2H_3O_2)_2.4H_2O$ dm^{-3}.

Procedure

Weigh 5 g of 2-mm soil of known moisture content into a 50-cm^3 centrifuge tube with a narrow neck. Add 30 cm^3 of magnesium acetate solution, stopper and shake the tube for 5 minutes. Centrifuge until the supernatant liquid is clear (about 5 minutes at 400 rev/s) and decant into a 100-cm^3 volumetric flask. It may be necessary to filter the extract for some soils. Repeat the extraction twice more and dilute the combined extracts to 100 cm^3. Measure the sodium content as in section 9:3:7.

9:3:6 Exchangeable sodium percentage

The E.S.P. is calculated from the measured exchangeable sodium and cation exchange capacity using the expression,

$$\text{E.S.P.} = \frac{\text{ex. Na}^+ \text{ m.e./100 g} \times 100}{\text{C.E.C. m.e./100 g}} \qquad [9:4$$

Note that when measuring the exchangeable sodium, due allowance must be made for water-soluble sodium.

9:3:7 Sodium in solution

The two methods recommended for the routine determination of sodium in solution are flame photometry and potentiometric measurement using a lithium glass electrode. Precise experimental details for using the flame photometer will depend upon the type of instrument and are given by the manufacturers; see also section 8:4:3.

Potentiometric measurement (Bower)

Place 10 cm^3 of test solution which should be at pH 6 or more* in a 100-cm^3 beaker and add 10 cm^3 of 1 M magnesium acetate solution. This step may be omitted if the sodium is already in magnesium acetate

---

* The electrode manufacturer's instructions should be considered also. It may be desirable to adjust the pH of test solutions to between 8 and 12 and it has been found satisfactory to add 1 or 2 drops of triethanolamine for this purpose.

solution. Read the e.m.f. of the solution in millivolts at intervals of 15–20 seconds until constant, using lithium and calomel electrodes. The pH meter may need modifying in order to read accurately to within 1 mV; a Beckman Zeromatic meter, for example, requires a 700-ohm resistance across the thermocompensator terminals to expand its millivolt scale ten times.

Prepare a calibration curve by measuring the e.m.f. of solutions containing 0, 1, 10 and 100 m.e. Na^+ dm^{-3} in normal magnesium acetate solution for each set of unknown samples. The e.m.f. results are plotted against lg Na^+ on semi-log paper and should be a straight line at 20–25°C. The slope of the graph is 58 to 59 mV per tenfold change in sodium concentration. If the solution has been diluted with magnesium acetate solution, the results must be multiplied accordingly.

10
Nitrogen

10:1 INTRODUCTION

Plant growth is limited by nitrogen more than by any other element and, for this reason, methods of determining nitrogen in soils have occupied soil chemists for a long time. Probably also, the interpretation of nitrogen analyses has led to more misunderstanding than with other elements for the form and content of nitrogen in soil is profoundly affected by climate (Jenny, 1930). Nearly all the important nitrogen changes that occur in soil are due to microbial activity, itself strongly influenced by micro-climatic conditions and thus the time and nature of sampling, handling and storing of soil samples, pre-treatment and method of analysis all affect the analytical results and render an intelligent interpretation difficult.

Forms of nitrogen in soils

In most soils the bulk of nitrogen is in organic forms and usually near the surface. The inorganic forms of nitrogen include nitrate which is soluble and easily leached or taken up; nitrite which is usually a transitional stage between nitrate and ammonium; ammonium which occurs as the easily-removed exchangeable ion and also in more unavailable, fixed forms; and traces of gaseous forms such as dinitrogen and nitrogen monoxides and elemental nitrogen.

The nature of the organic compounds of nitrogen in soil still presents problems, one of the greatest being the apparent lack of proteins in soil (but see section 11:2:4). Nearly all the organic nitrogen added to soil is proteinaceous and in due course is converted to microbial protein but the fate of this remains a mystery. Nitrogen added to soil as protein is rapidly broken down to inorganic forms, yet soil organic nitrogen is resistant to attack under normal conditions and this has led to hypotheses that soil protein nitrogen must be bound up in some way to give resistant complexes (Ensminger and Gieseking, 1939; Mattson and Koutler-Anderssen, 1943).

The principal source of plant-available nitrogen in unfertilized soils is the organic nitrogen which is microbiologically converted via ammonium and nitrite to nitrate, and the rate of production of nitrate is controlled by the rate of the first process, the breakdown of organic nitrogen to ammonium. Under certain conditions, for example waterlogging, the mineralized nitrogen will accumulate as ammonium and in neutral or alkali soils containing much ammonium, nitrite will tend to accumulate (Chapman and Liebig, 1952). In both cases the accumulations are due to certain micro-organisms being unable to operate under the given conditions.

Thus on occasion, the soil chemist has to analyze a soil for total nitrogen, nitrate, nitrite, exchangeable and fixed ammonium and organic nitrogen. Organic nitrogen is estimated as the difference between total and mineral nitrogen, although certain individual organic compounds can be determined. Modern fertilization practices have made it necessary to measure soils for their capacity to absorb and hold anhydrous ammonia and, in contrast, it is sometimes necessary to measure the capacity of a soil for losing its nitrogen, other than by drainage or plant uptake. Such loss of nitrogen is termed denitrification or sometimes nitrate dissimilation and refers to the microbial conversion of nitrate to gaseous products, chiefly elemental nitrogen, which are lost to the atmosphere. Dentrification is likely to occur when nitrate or nitrite is present together with decomposable organic matter under conditions of oxygen deficiency. In such cases the nitrogen acts as a hydrogen acceptor during the oxidation of organic matter.

Gaseous forms of nitrogen which occur in soil include nitrogen, ammonia, dinitrogen and nitrogen monoxide and nitrogen dioxide. Chemical and physical methods of determining gaseous forms of nitrogen are discussed by Cheng and Bremner (1965).

The latest techniques of isotope-ratio analysis of nitrogen in ^{15}N tracer investigations are presented by Bremner (1965).

10:2 DETERMINATION: BACKGROUND AND THEORY

10:2:1 Total nitrogen

The traditional method of measuring organically bound nitrogen is that of Dumas (1831) in which nitrogenous organic compounds are ignited with copper(I) oxide in a stream of carbon dioxide and the gaseous products decomposed by copper into water, carbon dioxide and nitrogen gas. The latter is measured manometrically after absorption of the carbon dioxide. The method and its micro-modification were adapted to soil analysis but did not remain popular owing to the cumbersome apparatus and tedious experimental procedure. Kjeldahl (1883) developed an alternative method involving conversion of organic nitrogen into ammonia by boiling with sulphuric acid; the ammonia was subsequently liberated from its sulphate by distillation with alkali and estimated. The Kjeldahl process with certain modifications has been used for soil analysis ever since.

The Kjeldahl digestion process

The original process of simple digestion with the acid is slow and subsequent modifications included addition of potassium sulphate to raise the boiling point of the acid (Gunning, 1889) and of catalysts to hasten reaction.

Temperature of digestion

McKenzie and Wallace (1954) investigated the effect of temperature of digestion with and without catalysts by inserting glass-sheathed thermo-couples into the digestion flasks. Bremner (1960) investigated the effects of different quantities of potassium sulphate in the absence of catalysts upon the boiling point of the acid mixture and upon the amount of nitrogen mineralized. Bremner found that the period of digestion necessary to re-cover all the organic nitrogen decreased as the concentration of potassium sulphate increased, unless the temperature was allowed to exceed about 400°C, when nitrogen was lost. During the course of digestion some acid is lost by volatilization, some by reaction with soil minerals and some by oxidation of organic matter; such acid loss causes an increased salt con-centration and hence a higher boiling point and consequently must be con-trolled to avoid loss of nitrogen. Bremner found that the amount of acid lost varied from 0.2 cm^3 g^{-1} of soil with sandy soils to 6.1 cm^3 g^{-1} with peats, and he estimated that one can assume 10 cm^3 of acid lost per gramme of organic carbon present. Sesquioxides also increased loss of acid which suggests that a special treatment involving more acid should be given to lateritic soils. Bremner also investigated the use of sodium sulphate as an alternative to the potassium salt. This was introduced by Poe and Nalder (1935) and recommended by Jackson (1958). Bremner found the sodium salt less effective and more inclined to spatter than potassium sulphate.

Catalysts and time of digestion

The catalysts recommended to hasten the Kjeldahl reaction are numerous and have received a deal of investigation and caused much controversy. The catalysts most commonly used are copper sulphate, mercury(II) oxide (Wilfarth, 1885) and selenium, either separately or in various mixtures. Koch and McMeekin (1924) added hydrogen peroxide to increase the efficiency of digestion but McKenzie and Wallace (1954) found that this results in incomplete recovery of nitrogen. Lauro (1931) first introduced selenium as a catalyst and it has remained one of the most efficient means of hastening the reaction. Ashton (1936) compared the effects of selenium and copper sulphate on time of digestion for three soils and found that, whereas the digestion must be continued for 24 hours after clearing when copper sulphate is used, only 3 hours are necessary if selenium is employed. Thus Ashton's findings cast considerable doubt upon the effectiveness of short digestion procedures which were, and are, all too commonly in practice. Murneek and Heintze (1937) used thirteen soils to investigate the effects of various catalysts and found that selenium greatly decreased the necessary time of digestion, especially in the presence of mercury. Murneek and Heintze finally recommended a selenium, copper sulphate and mer-cury(II) oxide mixture. Chibnall and co-workers (1943) found that to re-cover all protein nitrogen using a mixture of selenium and copper as catalyst, eight hours' digestion was necessary, whereas Hiller *et al.* (1948) considered that 30 minutes' digestion was sufficient if mercury was used. Alves and Alves (1952) compared selenium, copper and mercury singly and together; they obtained the highest nitrogen recoveries with copper sulphate and the lowest with selenium, a finding at variance with most other workers. Alves and Alves finally recommended, however, the use of a mixture of selenium and mercury(II) oxide which they thought gave almost as good a recovery as copper sulphate but with the shorter digestion time of two and a half hours. Alves and Alves further noted that continuation of digestion beyond the time necessary for maximal nitrogen recovery resulted in loss of nitrogen, a fact previously reported by Patel and Sreenivasan (1948).

Periodic reviews of the effects of catalysts have appeared—Kirk (1947, 1950) and Salt (1953) for example—and a fairly comprehensive study of all recommended methods was undertaken by Bremner (1960). Bremner adopted the macro method currently in use in his laboratory as a 'standard' method for means of comparison and which employed a mixture of 10 g potassium sulphate, 1 g copper sulphate ($5H_2O$) and 0·1 g selenium. The mixture was digested with 30 cm³ of concentrated sulphuric acid and heating was continued for 5 hours after clearing. Bremner investigated nineteen different catalysts, these being various quantities and combinations of copper sulphate, selenium and mercury(II) oxide, using three different soils. Bremner concluded that both selenium and mercury are better catalysts than copper sulphate and that mercury is probably the best single catalyst. Mercury, however, gives rise to practical difficulties when subsequently determining the ammonium and so was not favoured. Selenium alone gave good results provided that not too much was used and that the digestion time was not too long. As a result of this work Bremner emphasized that clearing of the digestion mixture is not a criterion of complete digestion as is assumed by, *inter al.*, the Association of Official Agricultural Chemists (1955). Furthermore, the use of an iron(II) sulphate–copper sulphate catalyst, as in the A.O.A.C. method, leads to low results.

Effect of clay

The efficiency of the Kjeldahl method has several times been questioned. Bal (1925) found that the routine procedure failed to recover all the nitrogen from certain clay soils but he solved the problem by pre-treating the soils with water. The effect was explained by Bal as being due to a cementing material, probably containing iron, which is insoluble in concentrated sulphuric acid and thus protects some organic matter from oxidation; it is, however, soluble in dilute sulphuric acid. Walkley (1935) found that ball-milling of clay soils resulted in the recovery of more nitrogen than by merely soaking the soil in water as recommended by Bal, and considered that the low nitrogen recovery from untreated soils was due not to cementing materials but to failure of soil crumbs to disperse in the non-polar sulphuric acid. Bremner and Harada (1959) considered that Walkley's findings were consistent with the presence of fixed ammonium which was not being determined by the Kjeldahl process unless ball-milling broke up the crystal lattice or soaking in water expanded the lattice. However, the results of an experiment did not support this theory as some soils gave no increased nitrogen recovery on wetting and yet yielded fixed ammonium on treatment with hydrofluoric acid. It was concluded therefore that the low nitrogen results found by Bal in untreated soils were due to a short digestion period insufficient to liberate all the nitrogen. Not all soils need pre-treatment with water, but as the operation is so simple, it can well be adopted as a routine.

Inclusion of nitrate and nitrite

The Kjeldahl method sometimes fails to include certain forms of nitrogen, for example, from some heterocyclic compounds and from N—N and N—O linkages. Friedrich *et al.* (1933) pre-treated soil with hydriodic acid and red phosphorus before digestion to permit the determination of N—N and N—O linked nitrogen. Friedrich's method was modified by several workers; Clark (1941), for example, refluxed soil with hydriodic acid and removed excess acid and iodine by acid evaporation before digestion. Steyermark *et al.* (1958) used a zinc–iron reduction method for including compounds containing N—N and N—O linkages.

That the routine Kjeldahl process fails to estimate nitrate and nitrite nitrogen is well known but modifications exist to include these forms of nitrogen. Cope (1916) introduced the salicylic acid method for including nitrate and nitrite in the determination. In this modification the sulphuric acid used for digestion contains salicylic acid and sodium thiosulphate. The sulphuric–salicylic acid mixture converts nitrate and nitrite to nitro compounds which are then reduced to amino compounds on heating with the thiosulphate. For many years it was thought that the salicylic acid method suffered from the drawback that the presence of water inhibited the reaction (e.g. Piper, 1944) and it was recommended that soil samples be dried before analysis. Bremner and Shaw (1958) obtained up to 75% recovery of nitrate and nitrite nitrogen from 5 g soil samples wetted with 10 cm³ of water and more recently it has been shown (Bremner, 1965) that by using a semi-micro procedure nearly quantitative recovery can be obtained in the presence of 1 cm³ of water.

Davisson and Parsons (1919) heated soil samples with Devarda's alloy and alkali in order to reduce nitrite and nitrate to ammonia. The ammonia was collected in an absorption bottle containing sulphuric acid and when reaction was complete, the contents of the absorption bottle were tipped on to the soil in the Kjeldahl flask for digestion.

Olsen (1929a) oxidized nitrites to nitrate with acid potassium permanganate and reduced the nitrate to ammonia with reduced iron and sulphuric acid before digestion. Olsen's method was modified slightly by Bremner and Shaw (1958) who treated the soil in aqueous suspension with acidified potassium permanganate solution added dropwise whilst shaking. Afterwards a few drops of octanol were added with the reduced iron and the mixture gently refluxed before digestion. The method of Bremner and Shaw is applicable to wet, or even waterlogged, soils and may thus be considered as more reliable than the salicylic acid method.

In many, if not most, soils the nitrate and nitrite content is negligible and little harm is done by neglecting a pre-treatment, but if nitrates are likely to be present in any but very small amounts, they should be included in a total nitrogen analysis or determined separately. With plant analyses there is no latitude and it is imperative to include nitrates. This was strikingly revealed by a statement in *Nature* (Chakraborty and Sen Gupta, 1960) that the rice plant could fix atmospheric nitrogen. Investigation of this surprising finding by Hart and Roberts (1961) revealed that routine Kjeldahl analyses of plant material had been assumed to include nitrate, or more probably, the possibility of nitrates being present had been overlooked.

Inclusion of fixed ammonium

Rodrigues (1954) described a method of pretreating soil so as to include in the total nitrogen analysis the indigenous fixed ammonium-nitrogen, that is, the ammonium taken up by clay minerals when interlayer cations in the expanded lattice are replaced by ammonium. In Rodrigues's method the soil is treated in a platinum crucible with a mixture of sulphuric and hydrofluoric acids for 2 hours, and is afterwards rapidly boiled with sulphuric acid to remove hydrogen fluoride and water before digesting with a catalyst. As some subsoils may contain up to 40% or more of their total nitrogen as fixed ammonium (Bremner, 1959) this pre-treatment would appear to be essential. Bremner (1960), however, compared his 'standard' procedure with that of Rodrigues and found that with four soils known to contain fixed ammonium, treatment with hydrofluoric acid had no effect upon the total nitrogen recovery. Bremner thus concluded that the standard method was sufficient. Recovery of fixed ammonium was further studied by Stewart and Porter (1963) who pointed out that of the four soils used by Bremner, only one had more than 6% of its total nitrogen as fixed ammonium. With soils having more than 15% of the total nitrogen as fixed ammonium Stewart and Porter found low recovery by Bremner's standard method, sometimes by as much as 25%. Using suitable soils they determined total nitrogen by a semi-micro Kjeldahl method, ammonium-nitrogen extractable with potassium hydroxide, non-ammonium-nitrogen extractable with potassium hydroxide, fixed ammonium-nitrogen by extraction with a mixture of hydrofluoric, hydrochloric and sulphuric acids, and the residual nitrogen by the Kjeldahl process. A further Kjeldahl analysis was made on a soil sample pre-treated with the acid mixture, 1 g being shaken for 16 hours with a solution 5 M in hydrofluoric acid, 0·75 M in hydrochloric acid and 0·3 M in sulphuric acid, the whole then being digested as usual. The results, some of which are shown in Table 10:1,

Table 10:1

Nitrogen recovered from each fraction and the sum of the fractions compared with the total nitrogen recovered by two Kjeldahl procedures using a clay soil (from Stewart and Porter)

Soil depth	NH_4-N extracted by KOH	Non-NH_4-N extracted by KOH	Fixed NH_4-N	Residue N	Sum of N fractions	Kjeldahl N	Deficit (sum less Kjeldahl N)	Kjeldahl N HF-pre-treated
					μg g^{-1}			
0–18 cm	216	937	262	253	1668	1569	99	1669
75–112 cm	47	139	366	132	684	528	156	651

Table 10:2

Kjeldahl methods of determining total nitrogen in soil as investigated by Bremner

Method	Reagents					Time of digestion after clearing (hours)
	Catalyst (g)	K_2SO_4 (g)	Na_2SO_4 (g)	H_2SO_4 (cm³)	$\dfrac{K_2(Na_2)SO_4}{H_2SO_4}$	
Bremner 'standard' (1960)	$CuSO_4$ (1·0) Se (0·1)	10·0	—	30·0	0·33	5·0
Ashton I (1936)	Se (0·2)	10·0	—	40·0	0·25	3·0
Ashton II (1936)	$CuSO_4$ (1·0)	10·0	—	40·0	0·25	24·0
Murneek and Heintze (1937)	$CuSO_4$ (0·25) HgO (0·7)	—	10·0	30·0	0·33	1·5 × clearing time
Belcher and Godbert (1941)	$HgSO_4$ (0·265) Se (0·53)	1·68	—	4·0	0·42	0·75
Chibnall et al. (1943)	$CuSO_4$ (1·0) Na_2SeO_4 (0·06)	4·0	—	20·0	0·20	8·0–16·0
Piper (1944)	Se (0·2)	10·0	—	30–35	0·29–0·33	1·0–1·5
White et al. (1948)	HgO (0·7)	15·0–18·0	—	25·0	0·60–0·72	2·0
Willits et al. (1949)	HgO (0·6)	15·0	—	25·0	0·60	3·0
Middleton and Stuckey (1951)	HgO (0·3)	—	3·0	6·0	0·50	2·0
Lake (1952)	HgO (1·3)	20·0	—	35·0	0·57	1·0
Alves and Alves (1952)	HgO (0·5) Se (0·2)	—	10·0	30·0	0·33	2·5
McKenzie and Wallace (1954)	$HgSO_4$ (0·069)	1·5	—	1·5	1·0	0·25
A.O.A.C. (1955)	$FeSO_4$ (0·87) $CuSO_4$ (0·43)	8·7	—	30–40	0·22–0·29	0
A.O.A.C. as modified, Bremner	$FeSO_4$ (0·87) $CuSO_4$ (0·43)	8·7	—	30–40	0·22–0·29	5·0
Jackson (1958)	$CuSO_4$ (0·832) HgO (0·125) Se (0·042)	—	20·0	35·0	0·57	0·75–1·25

Beet (1954) Oxidation with potassium permanganate and sulphuric acid

showed considerable differences between the sum of the nitrogen fractions
and total Kjeldahl nitrogen, the deficits being related to fixed ammonium.
The pre-treated soils, on the other hand, gave total Kjeldahl nitrogen contents very close to the sum of the fractions. Using the same soil the
'standard' Kjeldahl procedure was varied by altering the time of digestion
but keeping the salt–acid ratio constant at 0·33. Only a very slight increase in nitrogen was obtained. By making the salt–acid ratio 1·33, a
better recovery was obtained, but only if the time of digestion was increased
to 7 hours. Stewart and Porter concluded that indigenous fixed ammonium
in the clay minerals is not fully recovered by Bremner's standard method.

General findings

In his comprehensive investigation of the accuracy and efficiency of the
Kjeldahl procedure Bremner analyzed seven soils differing in nitrogen
content from 0·03 to 2·69% and refractory nitrogen compounds such as
nicotinic acid and tryptophan by sixteen different published methods; both
macro and micro techniques were examined (Table 10:2).

As a general result Bremner considered that his 'standard' method was
as good as, if not better than any other; it gave consistent results, it successfully coped with refractory compounds and with the Bremner–Shaw
modification it quantitatively included nitrate and nitrite.

The methods of Ashton (1936), Piper (1944), Alves and Alves (1952)
and Jackson (1958) gave results in close agreement with those by Bremner's
method, but those of Murneek and Heintze (1937), Beet (1954; 1955), and
the A.O.A.C. (1955) gave low or inconsistent results.

Bremner prefers the macro-Kjeldahl procedure to avoid having to use
finely-sieved soil, although he obtained almost identical results with the
semi-micro procedure. The choice between macro and semi-micro procedures must remain a matter of individual preference; Jackson also recommends the macro procedure in which the soil sample varies from 1 g
for peats to 20 g for sands and needs 20 g of catalyst and 35 cm³ of sulphuric acid. Even so, Jackson advises fine grinding of the sample. Metson
(1961) on the other hand recommends a semi-micro procedure using from
0·2 g to 0·5 g of soil and 3 cm³ of acid. The semi-micro procedure is
quicker, neater and consumes far smaller quantities of chemicals than the
macro; the apparatus described by Jackson is a major installation in the
laboratory. Considering the results of several investigators as well as those
of personal experience, there seems to be no reason for not adopting the
semi-micro procedure as a routine and this is facilitated by modern equipment. Small electrically-heated digestion units are commercially available
and usually incorporate their own 'fume cupboard' which is exhausted by
a water pump, the fumes being dissolved before reaching the atmosphere
and so enabling the apparatus to be set up in the open laboratory (Plate 10).
A minor but useful adaptation has been published by Mader and Hoyle
(1964) who place a lead sheet collar with a flange around the neck of the

digestion flask (Figure 10:1) which can be pushed up into the exhaust tube
to increase its efficiency.

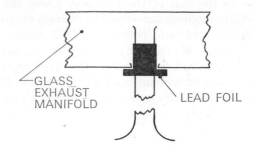

GLASS
EXHAUST
MANIFOLD

LEAD FOIL

Figure 10:1 Lead foil flange for Kjeldahl flasks

Comparison of the Kjeldahl and Dumas methods

Dyck and McKibbin (1935) compared the Kjeldahl method, as it then was,
with the Dumas combustion procedure using twenty-six organic soils.
They found in every case that the Dumas method recovered more nitrogen
than did the Kjeldahl method and concluded that the Dumas must be the
more efficient procedure. Bremner and Shaw (1958) obtained identical
results by both methods when using mineral soils but with organic soils the
Dumas method gave 10–20% higher values, thus confirming the findings of
Dyck and McKibbin. However, Bremner and Shaw did not consider this
sufficient evidence for concluding that the Dumas method was superior,
and after further work, suggested that the higher Dumas nitrogen values
were erroneous and due to atmospheric nitrogen being occluded by the
organic matter. The investigation was followed up by Stewart and co-
workers (1963) who used a modern combustion apparatus to analyze soils
by a micro-Dumas procedure. Stewart obtained results comparable to
those of the Kjeldahl method for soils containing less than 0·3% nitrogen
but obtained higher results when analyzing organic soils. By analyzing the
gas collected, by gas chromatography Stewart demonstrated that the high
Dumas results were due to incomplete combustion of the soil which caused
the formation of methane that was being measured as nitrogen. Later
Stewart *et al.* (1964) modified the experimental technique so as to ensure a
complete combustion by incorporating into the apparatus a platinum reforming catalyst. The modified Dumas method was considered suitable for
analyzing all kinds of soil including those with high organic matter content
and is as convenient to perform as to make a Kjeldahl analysis.

Distillation of ammonia in the Kjeldahl method

Once the Kjeldahl digestion has been completed satisfactorily, it remains
to determine the ammonia produced, and this is usually done by distillation
from an alkaline medium into dilute acid.

In the macro procedure the whole digest is distilled and all that is necessary is to attach the digestion flask to a splash-head and condenser (Figure 10:2). In the more recent apparatus for semi-micro methods a similar distillation of the whole digest is possible (Figure 10:3). Usually for

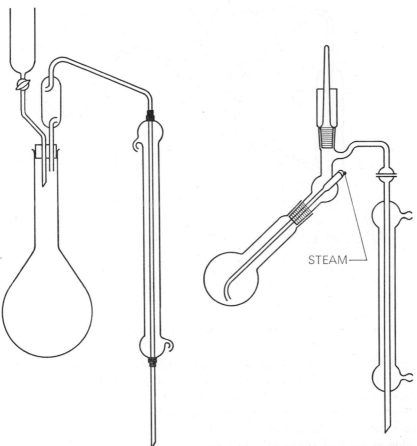

STEAM

Figure 10:2 Apparatus for macro-Kjeldahl distillations

Figure 10:3 Micro-Kjeldahl distillation apparatus permitting distillation directly from digestion flask

the semi-micro procedure the digest is made to standard volume and aliquots are distilled: this has the advantage over the macro procedure in that several replicate distillations are possible. Normally the semi-micro digest is made to 50 cm³ and it is not necessary to filter off the solid matter. Transfer to the volumetric flask is made by decanting, leaving the bulk of silica behind. The small quantities of silica that invariably are included in the volumetric flask can be ignored as even if 25% of the weight of the soil is present in suspension and settles out before aliquots are taken, the total error involved will be only about 0·2% of the nitrogen present (I. A. Black, private communication).

Several designs of still have been used for the semi-micro steam distillation of ammonia. Parnas and Wagner (1931) designed the still shown in Figure 10:4 which was self-emptying by means of pressure differences

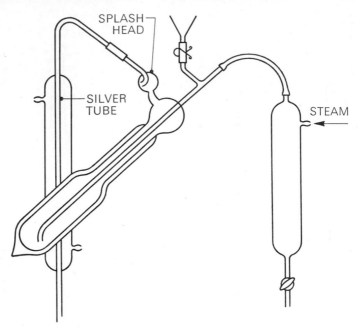

Figure 10:4 The Parnas–Wagner nitrogen still (*Biochem. Ztg.*, 1931)

caused by disconnecting the steam source; the condenser was a silver tube. Kirk (1936) modified the Parnas–Wagner still by using an all-glass condenser and eliminating rubber connections. The main disadvantage of the Parnas–Wagner still, and its modification by Kirk, is that the flame has to be removed at the end of each distillation and that the water must be allowed to cool sufficiently to create the partial vacuum necessary for emptying the reaction chamber. Shepard and Jacobs (1951) and Sheers and Cole (1953) improved the design by including an escape valve on the steam generator. Markham (1942) produced a compact and efficient piece of apparatus (Figure 10:5) which also permits emptying and washing of the reaction chamber without dismantling or stopping the steam generator. A similar still (Figure 10:6) was described by Hoskins (1944) which is really a vertical form of the Markham still but having a larger distillation chamber permitting the safe distillation of up to 80 cm³ of liquid: the Hoskins still has an interchangeable macro and micro reaction chamber. For those who favour the Hoskins still, a useful investigation of its capabilities has been made by Bremner (1960) who used it in the macro form. During this investigation Bremner found that the volume of distillate to be collected for quantitative recovery of ammonia is increased by a decrease in pH of

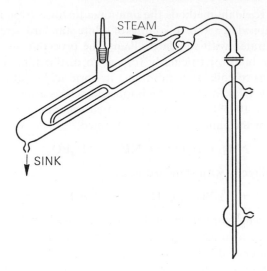

Figure 10:5 The Markham nitrogen still (*Biochem. J.*, 1942)

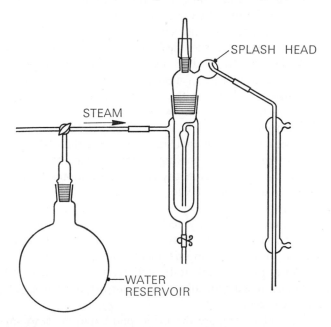

Figure 10:6 The Hoskins nitrogen still (*Analyst*, 1944)

the solution distilled and by increase in the volume of the solution distilled; it was not affected by the rate of distillation from 3 cm³ to 6 cm³ per minute. Bremner found that considerable time was saved by using a water pump to empty the distillation chamber rather than by disconnecting the steam supply. McKenzie and Wallace (1954) recommended a double splash-head for all nitrogen stills (as in Figure 10:2).

In the early Kjeldahl methods the ammonia distilled from ammonium sulphate was trapped in a known volume of dilute standard acid, the excess of which was titrated with standard alkali. The procedure is still recommended in some books, particularly for macro distillation. A more convenient method is to collect the ammonia in boric acid solution. This was first suggested by Winkler (1913). As boric acid acts as a weak, monobasic acid ($K_a = 5 \cdot 8 \times 10^{-10}$), it cannot be titrated accurately with alkali but the borate ions of the ammonium borate formed,

$$NH_3 + H_3BO_3 = NH_4{}^+ + H_2BO_3{}^- \qquad [10:1$$

can be titrated directly with standard acid,

$$H_2BO_3{}^- + H^+ \longrightarrow H_3BO_3 \qquad [10:2$$

Thus the exact strength or volume of the boric acid used need not be known.

At first considerable difficulty was experienced in determining the end-point of the titration and Yuen and Pollard (1953) compared three different indicators, methyl red, methyl red–methylene blue, and methyl red–bromcresol green (Ma and Zuazaga, 1942). With all three indicators high blanks and low sensitivity were experienced. Yuen and Pollard found that the sensitivity of the indicators decreased as the concentration of the boric acid increased. Furthermore, so-called 'pure' boric acid exerted a considerable buffering effect on the alkaline side even at concentrations of 0·1 %. It was concluded that a mixture of methyl red and methylene blue was the most suitable indicator but that the concentration of boric acid must be less than 1 %. Satisfactory recovery with an error less than 0·5 %, of 500 μg ammonium-nitrogen was obtained using 10 cm³ of 1 % Analar boric acid. The ratio of methyl red to methylene blue affects the end-point and the indicator is best prepared by mixing equal volumes of 0·2 % ethanolic methyl red and 0·1 % aqueous methylene blue; ammonia-free water must be used and three drops of indicator per 10 cm³ of boric acid are recommended. The proportions used by Ma and Zuazaga in the preparation of their indicator were 0·5 % methyl red and 0·1 % bromcresol green in 95 % ethanol. Jackson (1958) found that increasing the proportion of bromcresol green five times was an improvement, the indicator then being purple at its mid-point at pH 4·5, pink at 4·2 and green at 4·9. Sher (1955) introduced a different mixed indicator containing bromcresol green, *p*-nitrophenol and new coccine.

It is convenient to prepare the boric acid solution ready mixed with indicator and the sharpness of the end-point will depend in part upon the quality and source of the indicator. Bremner and Edwards (1965) recommend magnetic stirring of the distillate while titrating.

Bremner found that water condensing on the outside of the condenser sometimes drips into the receiver and affects the titration; a small collar of filter paper at the base of the condenser was used to prevent this. In a later form of the apparatus, a special funnel trap was fixed to the condenser to prevent extraneous water entering the receiver (see Figure 10:7). Bremner

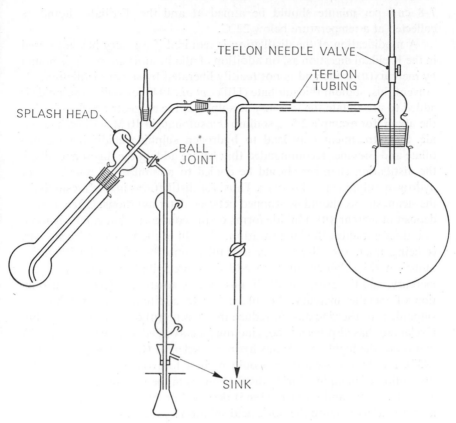

TEFLON NEEDLE VALVE

TEFLON TUBING

SPLASH HEAD

BALL JOINT

SINK

Figure 10:7 Semi-micro nitrogen still of Bremner and Edwards (Soil Science Society of America, *Proceedings*, 1965)

found also that it was not necessary to have the end of the condenser dipping into the boric acid solution, complete recovery of ammonia being obtained with the condenser tube just above the surface of the acid. In a later study Bremner and Edwards (1965) found that quantitative recovery of up to 10 mg nitrogen could be obtained even in an empty flask. Yuen and Pollard found that the weak boric acid solution could not retain ammonia against the bubbling of distillation and they used boiling tubes as receivers to increase the depth of acid and reduce ammonia loss. Thus, by not dipping the tube into the acid at all, more efficient absorption of ammonia may be expected.

Bremner and Edwards finally recommended the apparatus shown in Figure 10:7 which was designed to use standard tapered ground glass jointed Kjeldahl flasks, the size of which can vary from 50 cm³ to 250 cm³. This permits distillation of ammonia for the determination of all forms of soil nitrogen regardless of the quantities involved, eliminates transfer errors and simplifies working procedures. It is recommended to acidify the water in the steam generator with sulphuric acid; a distillation rate of

7–8 cm³ per minute should be aimed at and the distillate should be collected at a temperature below 22°C.

A modification of the distillation is required if mercury has been used in the Kjeldahl digestion as, on addition of alkali, some ammonia is bound by mercury(II) oxide and is not readily liberated by steam distillation. To prevent this, sodium thiosulphate (Hiller *et al.*, 1948) or sodium sulphide is added to precipitate the mercury as sulphide; the reagents are dissolved in the alkali, for example 2·5% sodium thiosulphate in 10 M sodium hydroxide. Such treatment may lead to hydrogen sulphide acidifying the distillate and Bremner recommended that before adding the digest and alkali, the distillation chamber should be cooled to minimize the possibility of hydrogen sulphide passing over. Thus, for distillations involving sulphide, the steam supply should be stopped between determinations. Often a black deposit of mercury(II) sulphide forms on the condenser, but can be ignored as it does not affect the final titration. If a still of the Parnas–Wagner type is being used, the silver condenser tube must be replaced with glass. Sufficient thiosulphate must be added to precipitate completely all the mercury, and the excess of alkali must not be so great as to cause distillation of metallic mercury. An alternative to addition of thiosulphate or sulphide is to use zinc dust to reduce the mercury(II) oxide to mercury, but if selenium has also been used, zinc must be avoided as it leads to the formation of the highly poisonous hydrogen selenide (Campbell and Hanna, 1937); for the same reason it would be best to avoid the use of zinc for prevention of bumping during distillation, as in the method of Jackson for example. Bouat and Crouzet (1965) designed an apparatus in which ammonia is transferred to the boric acid solution by means of a current of air rather than by steam distillation.

Omission of distillation in the Kjeldahl process

If the distillation of ammonia from the Kjeldahl digest could be omitted, this would be an attractive modification. Marcali and Rieman (1946; 1948) described a method whereby ammonia is titrated directly in the digest by neutralizing excess acid with sodium hydroxide after displacing ammonia from its mercury(II) complex with sodium bromide. Carbon dioxide is boiled out, formaldehyde added to complex ammonia as hexamethylenetetramine and the mixture titrated with standard alkali using phenolphthalein as indicator. However, as pointed out by Jackson, the constituents of a soil digest, such as iron, aluminium, phosphorus and silica, interfere with the procedure and it would be difficult to adapt it for soil analysis. One modification of the procedure was to oxidize ammonia to nitrogen in alkaline solution with sodium hypobromite and to determine excess hypobromite iodimetrically. Adams and Spaulding (1955) further modified the method so as to permit the use of selenium as catalyst. The soil digest was boiled with sulphurous acid and made to standard volume; aliquots adjusted to pH 6·8 with sodium hydroxide were titrated with formaldehyde which reduced the pH and were then titrated to pH 9 with sodium hydroxide solution and the nitrogen calculated.

Ashraf *et al.* (1960) published a method in which the sample is digested for 40 minutes only using mercury(II) sulphate as catalyst. After cooling and diluting, 60% sodium hydroxide solution was added dropwise with shaking until a yellow precipitate of mercury(II) oxide was formed. The mixture was then cooled and treated with sodium hydrogen carbonate followed by potassium bromide to dissolve completely the yellow precipitate. An excess of 0·1 M sodium hypochlorite solution was added until the colour appeared pale yellow, and after standing 5 minutes, a known excess of arsenite was added and back-titrated with hypochlorite; 1 cm³ of 0·1 M NaOCl ≡ 0·4670 mg N. This procedure was applied to soil analysis by Qureshi *et al.* (1962) who claimed an accuracy of nitrogen recovery comparable to that of the A.O.A.C. method. This is a rather unfortunate comparison as the latter is one of the least reliable methods.

Bradstreet's (1965) book on the Kjeldahl process gives a wealth of general information but unfortunately deals very summarily with soil analysis.

Other methods for determining total nitrogen in soil

Emmert (1929) measured total nitrogen by oxidizing soil with a mixture of sulphuric acid and sodium chlorate; after distillation, nitrate was determined with phenol disulphonic acid. Later (Emmert, 1934) the distillation step was omitted; the digest was neutralized with sodium hydroxide, precipitated silica and sesquioxides were removed by filtration, excess chloric acid removed with fuming sulphuric acid and nitrate determined.

Grabaraov and Visherskaya (1960) consider that the only possible way to measure total nitrogen in soil with accuracy is the Tiurin method. A small amount of soil, 0·2–1·0 g, is weighed into a conical flask and 2·5 cm³ of a 25% solution of chromic anhydride and 5 cm³ of sulphuric acid are added. After shaking, the mixture is boiled for 10–20 minutes, transferred to a Kjeldahl flask and distilled. In a somewhat similar procedure, Vestervall (1963) takes an aliquot of the dichromate–sulphuric acid digest of soil obtained during carbon determination (section 11:2:3), dilutes it, adds sodium hydroxide and distils the ammonia into boric acid solution.

Nagai and Hiraya (1962) described a method of determining nitrogen in wet soils in which magnesium oxide is added as an absorbent and copper sulphate and sodium sulphate as a catalyst. After adding sulphosalicylic acid and warming for half an hour, the mixture is digested with sodium thiosulphate anhydride. It is then heated in an oven at 300°C to vaporize sulphur and distilled with sodium hydroxide.

Conclusions

For the average laboratory the most suitable procedure for determining the total nitrogen content of a soil is the semi-micro Kjeldahl method using the catalyst, potassium sulphate–sulphuric acid ratio and time of heating recommended by Bremner (1960) and after pre-soaking the soil with water. For soils in which indigenous fixed ammonium is likely to occur the pre-treatment of Stewart and Porter should be given, and if nitrate and nitrite are to be included the method of Bremner and Shaw or the modified salicylic acid method of Bremner should be followed. Subsequent distillation of ammonia is best done from the apparatus of Bremner and Edwards, Hoskins or Markham.

After the recent findings of Stewart and co-workers, the use of the semi-micro Dumas method may well rise in popularity. The main drawback to the method for most existing soil laboratories would be the specialized apparatus necessary (see Chapter 11, section 11:3:1ı).

The student should be aware of the fact that both the Dumas and the Kjeldahl methods have now been adapted for automated soil analysis (e.g. Dabin, 1965; Keeney and Bremner, 1967).

10:2:2 Ammonium-nitrogen

The quantity of ammonium in a soil at any one time is dependent upon prevailing conditions of moisture, temperature, degree of aeration, pH and, *inter al.*, microbial activity. It follows, therefore, that a random measurement of ammonium in a soil is almost meaningless and such a determination should be made always as part of a well-planned experiment which takes into account all the variables. Microbial changes will continue at a measurable rate after the soil has been sampled right up to the time of analysis, and cannot be prevented by adding, say, toluene as is sometimes recommended for stabilizing nitrate contents of a soil extract. As discussed in Chapter 2 (section 2:4), during both moist storage (during transportation, for example) and dry storage, changes in the ammonium content of a soil will occur and although it has been recommended that rapid drying at 50°C may be resorted to if the soil cannot be analyzed immediately (Piper, 1944; Jackson, 1958), there is no acceptable alternative to analyzing the soil as soon as it is sampled. In practice this is feasible as most ammonium determinations are part of a laboratory experiment rather than a field investigation.

Extraction of ammonium-nitrogen

Ammonium is held chiefly in exchangeable form by soil and can be displaced by potassium or sodium. Early methods of analysis consisted of mixing soil with alkali and covering with a bell-jar; a bowl of acid was placed under the jar to absorb the evolved ammonia. Matthews (1920) improved this technique by drawing air through a column of soil mixed with alkali and absorbing the displaced ammonia. Matthews's apparatus is shown in Figure 10:8. The soil was placed in the aerator tube A, C and D were splash-bulbs and an additional optional splash-trap was placed at E. Incoming air at B was first purified by bubbling through dilute sulphuric acid, and the absorbing tube F contained dilute sulphuric acid with methyl red as indicator. The soil was treated with a solution of sodium chloride

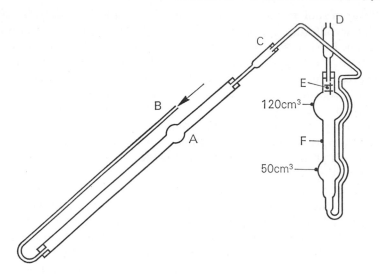

Figure 10:8 Apparatus for removing ammonia from soil by aeration (Matthews, 1920; *J. agric. Sci.*)

and sodium carbonate and finally the acid in F was titrated with standard alkali. The possibility of hydrolysis of organic nitrogen by such procedures was not allowed for by the early investigators.

McLean and Robinson (1924) found the procedure of Matthews to be unsatisfactory due to the fact that the water supply of their laboratory (like many of today) was too poor to maintain suction. McLean and Robinson thus adopted a completely different approach and, assuming that ammonia could be exchanged in the same manner as other ions, they leached the soil with cold 1 M sodium chloride solution. Olsen (1929b) used a mixture of hydrochloric acid and potassium chloride at pH 1·0 for extracting ammonium and this was modified by Richardson (1938) who made the pH 1·0–1·5 with hydrochloric acid after the extractant had been mixed with the soil, thus allowing for the natural pH of the soil. Metson (1961) recommends that the potassium chloride solution should be boiled with magnesium oxide before use, in order to remove any ammonia present, and that all filter papers used should be leached free from ammonia with potassium chloride solution. The average filter paper exhibits a strong tendency to absorb ammonia from the atmosphere and Metson's precaution is well taken. Instead of using a potassium chloride–hydrochloric acid extractant, ammonium can be removed from soil with potassium sulphate–sulphuric acid and this was advocated by Bremner and Shaw (1955) as being less likely to involve frothing during distillation, and also because commercially available potassium sulphate is more free from ammonia than is the chloride. Furthermore, nitrate in the extract can then be determined directly with phenoldisulphonic acid. Jackson uses sodium to replace ammonium and in his procedure 100 g of soil are shaken for half an hour with

7 + T.S.C.A.

200 cm³ of sodium chloride solution acidified to pH 2·5 with hydrochloric acid; after filtration the soil is washed with 250 cm³ aliquots of extracting solution. Lewis (1961) recommended extraction of ammonia with a solution 0·01 M in copper sulphate and 1·0 M in sodium sulphate; copper is then precipitated as the hydroxide and ammonium measured in the filtrate. Lewis introduced copper into the extractant for two reasons, both concerning the extraction of nitrate rather than ammonium. Aqueous extracts always contain some organic nitrogen and biological loss of nitrate occurs; the presence of copper adequately prevents biological loss and the subsequent precipitation of copper hydroxide removes organic compounds by absorption.

Bremner and Shaw (1958), when studying denitrification in soil, were faced with the problem of extracting ammonium-nitrogen in the presence of nitrites. This precluded the use of acid extractants as some nitrite is converted to nitrate which also had to be determined. Addition of sulphamic acid to the acid extracting solution prevents the conversion of nitrite to nitrate, but if it is desired to measure also the nitrite, then extraction of a duplicate soil sample is necessary and this can present difficulties if dealing, as is normally the case, with waterlogged soils. In Bremner and Shaw's final procedure the soil is shaken for two hours with neutral, 1 M potassium chloride solution and filtered. An aliquot of the filtrate is taken for nitrite determination and to another aliquot is added sulphamic acid (10 cm³ of extract + 1 cm³ 2% w/v aqueous sulphamic acid), the mixture is vigorously shaken for 2–3 minutes and the ammonium and nitrate determined. Sulphamic acid decomposes nitrite rapidly and quantitatively to nitrogen:

$$NH_2SO_3H + HNO_2 \longrightarrow N_2 + H_2SO_4 + H_2O \qquad [10:3$$

As a general extractant of ammonium from soil, 2 M, neutral potassium chloride solution was finally decided upon by Bremner (1965) as being the most suitable. The reagent extracts the same quantity of ammonium as does the acidified extractants and gives highly reproducible results. An acidified solution is not only unsuitable for soils containing nitrite or nitrate but hydrolysis of organic nitrogen is liable to occur.

Determination of ammonium-nitrogen in solution

Three methods are in current use for determining the ammonium content of a solution. One depends upon alkaline steam distillation, the liberated ammonia being absorbed in acid and titrated; one depends upon the reaction of ammonia with Nessler's reagent to give a coloured solution; and the third is a micro-diffusion method first introduced by Conway (1947). Sometimes the first two methods are combined, ammonia being distilled from the soil extract, trapped in acid and the distillate analyzed by reaction with Nessler's reagent (e.g. Metson, 1961).

DETERMINATION OF AMMONIA BY DISTILLATION

This method depends upon the simple reaction,

$$NH_4^+ + OH^- = NH_3 + H_2O \qquad\qquad [10:4$$

and the soil extract is treated with a base, sometimes sodium hydroxide but usually magnesium oxide, and the mixture distilled or steam distilled. Ammonia is quantitatively expelled and is absorbed in excess standard acid which is then back-titrated or, more usually, in excess boric acid and titrated directly with standard acid. A suitable apparatus is shown in Figure 10:9. The magnesium oxide (preferably the heavy form as it is

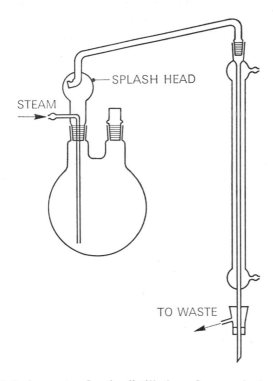

Figure 10:9 Apparatus for the distillation of ammonia during determination of soil ammonium and nitrate

easier to handle and is less likely to creep around the glass joints of the apparatus) must be ammonia-free and the commercial reagent should be freshly ignited in a muffle furnace before use to remove traces of carbonate; it is most conveniently added to the reaction flask from a standard cup through a powder funnel. The procedure must be carefully standardized, particularly with regard to volume of distillate and amount of indicator added. The water used for generating steam should be pre-distilled and made slightly acid. The method has the advantage that nitrate can be measured in the same aliquot of soil extract.

DETERMINATION OF AMMONIA BY NESSLERIZATION

Nessler's reagent is an alkaline solution of mercury(II) iodide in potassium iodide, but several modifications have since been proposed. When the reagent is added to a dilute solution of an ammonium salt ammonia is liberated and reacts to produce a colloidal orange-brown solution which is then compared with standards or measured in a spectrophotometer at 410 nm. As described by Metson (1961) 2 cm³ of reagent are added to an aliquot of soil extract and the transmission read at 425 nm.

If too much ammonium is present in the test solution the reagent produces a cloudiness and the colour of the complex is affected also by soluble salts. To overcome these difficulties a new reagent was developed using, instead of potassium mercury(II) iodide, sodium mercury(II) chloride, which is more tolerant to salts and ammonium content (Snell and Snell, 1949).

Vogel (1962) points out that when determining ammonia by Nesslerization it is essential to use ammonia-free water throughout the experiment. Vogel recommends either using conductivity water or re-distilling 500 cm³ of distilled water from a Pyrex flask containing 1 g of anhydrous sodium carbonate; the first and last 100 cm³ of distillate are rejected.

Instead of using a spectrophotometer a more rapid, if less accurate, procedure is to compare the colour of the test solution with the permanent standards of a Lovibond comparator. Four discs covering a wide range of ammonium-nitrogen are available and special instructions are provided for preparing the reagent as it was used for calibrating the coloured discs.

In alkaline solution ammonia reacts with a phenol–sodium hypochlorite reagent to give an intense blue colour which is proportional to the amount of ammonia involved (Van Slyke and Hiller, 1933). The intensity of the colour increases if the reaction is carried out at 100°C but is not so reproducible. The colour is stable for 1 hour and the optical density of solution can be read at 625 nm. Experimental details are given by Snell and Snell (1949) and the method has been applied to soil analysis by, *inter al.*, Ferry and Blanchere (1957).

DETERMINATION OF AMMONIA BY MICRO-DIFFUSION

If a volatile substance is liberated in one vessel where it exerts a tension it can be absorbed by simple gaseous diffusion into another chamber where its tension is reduced to zero on the surface of the absorbing fluid. This principle was applied to the determination of ammonia by Conway (1947) who needed a micro method for large numbers of biological analyses.

Conway designed what became known as the 'Standard Conway Unit' and which consisted of a small Pyrex glass dish, rather like a small petri dish, having a sealed-in centre well (Figure 10:10) and which was covered with a flat glass lid sealed with Vaseline or gum. If the ambient temperature of the experiment exceeds 38°C Conway recommended adding paraffin wax to the Vaseline; gum solutions such as tragacanth or acacia are more easily cleaned off than Vaseline.

A modified diffusion cell was designed by Öbrink (1955) in which the lid fits over the cell and dips into an annular chamber containing some of the reagent used for liberating the ammonia, thus eliminating the use of fixatives (Figure 10:11).

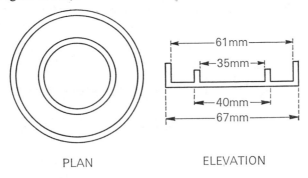

PLAN   ELEVATION

Figure 10:10 Standard Conway unit for microdiffusion (Conway, 1947; *Microdiffusion Analysis and Volumetric Error*, Crosby Lockwood, Lond.)

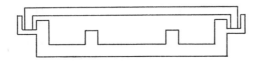

Figure 10:11 Modified Conway unit (Öbrink, 1955; *Biochem. J.*)

1 cm^3 of standard acid is placed in the inner chamber and the ammonium-containing liquid to be analyzed is placed in the outer chamber. With the lid almost covering the unit, the alkali (Conway used 1 cm^3 of saturated potassium carbonate solution) is added, the lid is then closed and the acid titrated after the unit has been standing for some suitable time.

The procedure is fast, distillation or aeration are eliminated and the accuracy is limited only by that of the pipette and burette used. The pipettes should be able to deliver from 0·5 cm^3 to 2·0 cm^3 and for accuracies of the order of 0·1–0·2%, a good tube pipette is ideal (upper Figure 10:12). An

Figure 10:12 Micropipettes for microdiffusion analysis

ordinary bulb-type pipette is not sufficiently accurate and the Ostwald pipette (lower Figure 10:12), although often recommended, is less suitable than the straight tube pipette. Conway designed a special microburette for the titrations and several modifications have since been advocated. Figure 10:13 shows the Conway burette as modified by

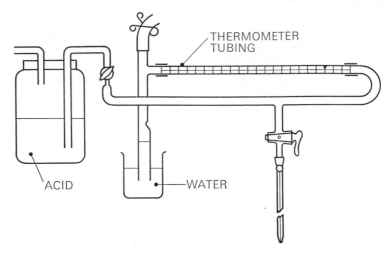

Figure 10:13 Conway microburette as modified by Wilson (Conway, 1947, *Microdiffusion Analysis and Volumetric Error*, Crosby Lockwood, Lond.)

Wilson (Conway, 1947) with a device for diminishing the pressure and slowing down the rate of delivery. Extreme care is necessary when cleaning the burette and pipettes.

Bremner and Shaw (1955) found the standard Conway units too small for routine soil analysis and designed modified units made from perspex (Figure 10:14). Acquaye and Cunningham (1965) modified the Bremner–Shaw diffusion units to rectangular boxes and held down the lids with a stronger glue supplemented by rubber bands.

Full experimental details for the procedure can be found in *Agronomy No. 9* (1965).

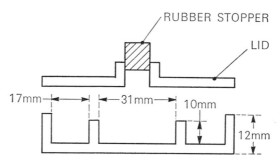

Figure 10:14 Modified Conway unit made in perspex (Bremner and Shaw, 1955; *J. agric. Sci.*)

Freney (1964) carried out diffusion analyses in conical flasks, the ammonia being absorbed on a piece (1 cm²) of muslin dipped in 0·5 M sulphuric acid.

Non-exchangeable or 'fixed' ammonium-nitrogen

Ammonia, unlike nitrate, reacts with soil constituents to give insoluble compounds. Such ammonium fixation appears to be associated with clay minerals such as illite and vermiculite that can also fix exchangeable potassium. Ammonia has been demonstrated to react with lignin to form complexes but there has been no proof that this occurs in soil. Such fixation of ammonia reduces its susceptibility to nitrification and hence its availability. Some soils contain non-exchangeable ammonia which has been fixed during the genesis of the soil and this is sometimes called 'native-fixed' or 'indigenous-fixed' to distinguish it from that fixed when ammonium compounds are added to a soil. Bremner (1965) suggests that the term 'non-exchangeable', that is, not extracted by potassium chloride solution, is more suitable than the term 'fixed'.

It has been shown by Bremner that the strong hydrofluoric–sulphuric acid treatment of Rodrigues (section 10:2:1) decomposes some organic nitrogen, thus accounting for the high 'fixed' ammonium results. Bremner and Bremner and Harada (1959) suggested the milder treatment of 1 M hydrofluoric acid mixed with 1 M hydrochloric acid for 24 hours at room temperature. According to Schachtschabel (1960) the method of Bremner and Harada does not liberate more than 50% of the native fixed ammonia from soils and he suggested that the soil should be extracted with a mixture of hydrofluoric and sulphuric acids whilst heating on a sand-bath, organic matter having been destroyed with hydrogen peroxide; the soil is extracted finally with potassium chloride solution. Dharival and Stevenson (1958) recommended pre-treatment of soil with hot potassium hydroxide solution in order to remove the labile organic nitrogen compounds which would otherwise be decomposed by the strong acids. Clays, however, may contain organic matter protected from extraction by hydroxide and yet which would be removed by hydrofluoric acid, and so Dharival's pre-treatment was not always satisfactory. The procedure was modified by Young and Cattani (1962) by shaking the soil overnight with the acid mixture and eliminating the effect of incomplete disintegration of silicate minerals. Stewart and Porter (1963) modified the procedure of Dharival and Stevenson by continuously shaking the mixture for 16 hours in polythene centrifuge tubes.

To measure the quantity of ammonium that a soil can absorb from solution and render non-exchangeable, Bower (1950) added a known amount of ammonium to a weighed sample of soil and after washing, leached the soil with an extracting solution; the ammonium content of the leachate was compared with that obtained from an untreated soil sample. A similar procedure, but with total nitrogen determinations, was described by Allison and co-workers (1951). These methods were considered by Leggett and Moodie (1962) to be time-consuming and needing too much attention; moreover, with some soils there is a greater loss of organic nitrogen from treated than from untreated samples during extraction.

Barshad (1951) distilled duplicate ammonia-saturated soil samples for 3 hours with sodium hydroxide and potassium hydroxide. Assuming that sodium hydroxide can release non-exchangeable ammonium whereas potassium hydroxide is unable to do so, the difference in ammonium released by the two alkalis was taken as a measure of fixed ammonium. The method, however, may release some indigenous fixed ammonium and was considered by Leggett and Moodie to be too drastic. Nommik (1957) used radioactive $^{15}NH_4$ to saturate a soil sample which was then extracted with potassium chloride solution. After drying, the soil was treated with hydrofluoric acid to liberate the ammonia fixed.

To overcome the inherent disadvantages of previous methods Leggett and Moodie devised what they called the 'Aeration Recovery' method. This method is essentially that of Allison *et al.* except for the experimental procedure. A known amount of ammonium, as acetate, is added to the soil sample; this and another, untreated, sample are treated with potassium carbonate solution and the liberated ammonia removed by aeration using the apparatus shown in Figure 10:15. Liberated ammonia is titrated and the difference between the ammonium added and that recovered gives the amount fixed. The method can be used also to determine the exchangeable ammonium in soil; a macro form of the apparatus is recommended by Chapman (1965) for determining exchangeable ammonium during the measurement of cation exchange capacity of soil.

Bremner (1965) recommends treating the soil with alkaline hypobromite to minimize interference from organic nitrogen compounds before extracting with a mixture of 5 M hydrofluoric acid and 1 M hydrochloric acid. Bremner found that this pre-treatment removed up to 99% of the organic nitrogen in soils and that the residue released only trace amounts of ammonia upon subsequent treatment. After treatment with acid mixture for 24 hours, the soil suspension is added to 10 M potassium hydroxide solution and the ammonia steam-distilled into boric acid solution.

The phenomenon of fixed ammonium was investigated by Freney (1964). Ammonia was fixed in a soil by shaking overnight with 1 M ammonium chloride solution, centrifuging, decanting and washing with ethanol. After drying for 24 hours at 105°C, exchangeable ammonium was removed with 1 M potassium chloride solution. The 'fixed' ammonium was then determined by four methods; that of Bremner (1959) in which 5 g of soil are shaken with 100 cm^3 of 1 M hydrofluoric acid and 1 M hydrochloric acid for 24 hours in polythene bottles and the extract distilled with magnesium oxide; that of Dharival and Stevenson where 2 g of soil are treated with 50 cm^3 of 1 M potassium hydroxide at 120°C under a pressure of 100 kN m^{-2} (15 lb/in^2) for 8 hours and after centrifuging and washing with 0·5 M potassium chloride solution are extracted for 16 hours with a mixture of 5 M hydrofluoric acid, 0·75 M hydrochloric acid and 0·3 M sulphuric acid; that in which duplicate 2-g samples of soil are leached with up to 2 dm³ of boiling 1 M potassium chloride solution and boiling 1 M sodium chloride solution, the ammonium in the leachates being deter-

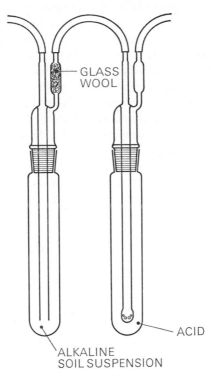

Figure 10:15 Apparatus for determining fixed ammonium by the aeration recovery method of Leggett and Moodie (Soil Science Society of America, *Proceedings*, 1962)

mined by distillation with magnesium oxide; and that of Allison and Roller (1955) where duplicate 5-g samples of soil are distilled with 0·2 M sodium hydroxide solution and 0·2 M potassium hydroxide solution (after Barshad), the sodium hydroxide-extracted ammonium less the potassium hydroxide-extracted ammonium being taken as fixed ammonium. Freney modified the original Allison and Roller method by adding an excess of sodium tetraphenylboron to the sodium hydroxide solution if potassium ions were present.

As a general result of his investigation Freney concluded that ammonium fixation is largely a laboratory phenomenon brought about by extractants containing potassium ions which contract the lattice, thus trapping some ammonium ions. He considered that the so-called 'indigenous fixed' ammonium results from degradation of labile organic nitrogen compounds by the action of the extracting hydrofluoric acid or alkali. The more recent

7*

work reported by Bremner (1965), however, would appear to negative this conclusion as virtually all the labile organic nitrogen is removed by alkaline hypobromite before acid extraction.

Determination of the capacity of a soil to retain anhydrous ammonia

Nowadays anhydrous ammonia is used extensively as a fertilizer and the capacity of a soil to retain the ammonia can be important. Methods of studying the sorption of ammonia by soil have been complicated by the various mechanisms of that sorption. Jackson and Chang (1947) and Stanley and Smith (1956) described apparatus for injecting ammonia into laboratory closed soil systems, and retention was estimated by subjecting the system to specific aeration procedures. Papendick and Parr (1965) considered that in both these procedures the method of ammonia injection was cumbersome, potentially dangerous and of limited accuracy.

Blue and Eno (1952) passed ammonia gas through columns of soil and assumed maximum retention when free ammonia could be detected at the far end of the column; sorbed ammonia was then extracted by leaching the soil with a salt solution. Nommik and Nilsson (1963) described an injection technique simulating field conditions and the soil after saturation was leached with a salt solution. Methods involving extraction with a salt solution, however, do not recover fully the sorbed ammonia, only that held in exchangeable form.

Sohn and Peech (1958) ammoniated an air-dry soil and then aerated it for one week at room temperature to remove the physically sorbed ammonia; ammonia retention was then reported as the difference between the total nitrogen of the ammoniated soil after aeration and the total nitrogen of an untreated control sample. A similar procedure was followed by Young and Cattani (1962), except that instead of aerating for 1 week, they removed the physically sorbed ammonia by 12 hours' evacuation. The technique of Young and Cattani was to expose 15 g of 0·5-mm soil in a conical flask to a steady flow of ammonia for 10 minutes. The flask was then stoppered for 1 hour and finally again flushed with ammonia for 1 minute; the flask was shaken continuously. The soil was then placed in a shallow paper cup and evacuation, using a water aspirator, continued overnight. Ammonia retention was measured by total nitrogen analyses on treated and untreated soil samples.

Papendick and Parr evaluated all published methods of determining ammonia retention and pointed out the importance of considering the initial concentration of ammonia, the time necessary for reaction and for reaching equilibrium and the method of desorption of the physically held ammonia. As a result of their investigation Papendick and Parr evolved an improved technique; they decided to include the ammonia absorbed by all mechanisms, that is, chemically sorbed or exchangeable ammonium, fixed ammonium, ammonium reacting with organic matter and physically sorbed ammonia. An ammonium injection apparatus (Figure 10:16) was designed on a principle of positive displacement in which gaseous ammonia

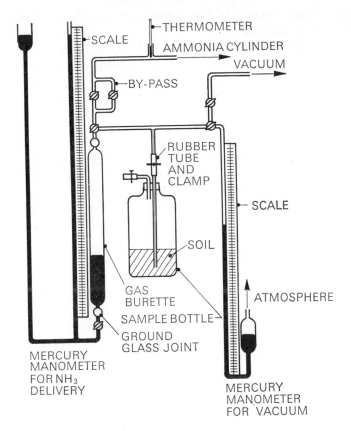

Figure 10:16 Apparatus for injection of predetermined quantities of anhydrous ammonia into laboratory soil systems (Papendick and Parr, *Soil Science* © 1965)

from a cylinder displaces a retaining liquid (mercury) in a gas burette. The apparatus was calibrated by substituting standard acid for soil in the reaction vessel. After delivery of ammonia the bottles were closed and swirled and the acid titrated; standard curves were obtained for several different-sized burettes to allow a wide range of delivery. The standard curves permitted the accurate determination of the quantity of ammonia injected into the soil system without having to employ directly correction factors for temperature and pressure. The apparatus must be used in a constant temperature room.

Papendick and Parr injected ammonia at concentrations of 1000, 2500, 5000 and 7500 $\mu g\ g^{-1}$ N on an oven-dry basis of soil and found the retention capacities of the soils they studied to vary from 400 to 7000 $\mu g\ g^{-1}$ N. Time of reaction was varied from 15 minutes to 40 hours. Unreacted and weakly-sorbed ammonia was removed at a constant, known rate of aeration using the apparatus of Parr and Reuszer (1959) in which air is bubbled through sodium hydroxide and sulphuric acid solutions before being drawn through the soil.

As explosive mercury(II) ammonium oxide can form in trace amounts at high pressure in the presence of water vapour, extreme caution is advocated in the use of the apparatus; the explosions are catalyzed by metals. Papendick and Parr recommended using only pure, dry ammonia, triple-distilled mercury frequently replaced, and not leaving mercury and ammonia in contact when the apparatus is not in use.

The procedure proved accurate for injection of even very small amounts of ammonia and full-scale greenhouse experiments could be simulated needing only a few minutes to inject the ammonia. It was found difficult, however, to evaluate ammonia retention as a discrete, fixed value as it is a dynamic process depending largely upon experimental conditions such as rate of aeration.

10:2:3 Nitrite-nitrogen

Extraction

Nitrites are generally extracted from soil with the same reagents as used for ammonium and nitrate extraction except that the solution must be neutral and not acid. Bremner and Shaw (1958) found neutral, 1 M potassium chloride solution to be adequate although it is often difficult to obtain a clear extract without decomposing some of the nitrite. In most cases nitrite will be found in measurable content only in waterlogged soils. Bremner (1965) subsequently found neutral 2 M potassium chloride to be the best general extractant.

Determination

The standard method of determining the nitrite content of a solution is to react it with sulphanilic acid to give a diazo compound which is then coupled with α-naphthylamine to give the red azo-dye α-naphthylamine-p-azobenzene-p-sulphonic acid (Griess, 1879).

Ilosvay (1889) modified Griess's procedure by making the diazotization in acetic acid solution and the method became known as the Griess–Ilosvay method. Rider and Mellon (1946) considered that the Griess–Ilosvay method did not satisfy the conditions necessary for accuracy which they listed as,

   i. Diazotization must be in strongly acid solution.
   ii. Diazotization should be made in as cool a solution as possible.
   iii. Coupling should not be attempted before diazotization is complete.
   iv. Coupling should be at as low a degree of acidity as possible.

Thus Rider and Mellon modified the procedure accordingly. In the new procedure hydrochloric acid-acidified 4-aminobenzene-sulphonic acid was allowed to cool to room temperature and then to stand in diffuse light with the added sample. α-Naphthylamine hydrochloride was added and the mixture buffered to pH 2·0–2·5 with sodium acetate. After dilution to standard volume the transmittance was read at 520 nm.

Wallace and Neave (1927) found that the colour produced with α-dimethylamino-naphthylamine is superior to that obtained with α-naphthylamine. The reagent is used in a similar manner except that it is dissolved in 95% methanol made 4 M with respect to acetic acid.

Bratton and Marshall (1939) recommended the use of sulphanilamide and n-(1-naphthyl)-ethylenediamine hydrochloride as giving better reproducibility and more rapid coupling than dimethylamino-naphthylamine; moreover, there is no need of a buffer solution and the azo dye produced is more acid-soluble.

Shinn (1941) developed the sulphanilamide method and as nitrite standards are difficult to prepare, a standardized solution of sulphanilamide was substituted. Dried sodium nitrite is dissolved in water and made to standard volume and is then assayed by titration with permanganate; this standard solution is then used to determine the nitrite equivalent value of a sulphanilamide solution. The sample to be analyzed must be neutral or acid and should contain about 1 mg dm^{-3} NO_2.

Montgomery and Dymock (1961) reverted to the use of sulphanilic acid as they found it gave solutions with more stable a colour.

For less accurate, routine determinations, the Lovibond comparator can be used in conjunction with the Griess–Ilosvay method. In this method the sulphanilic acid (0·5 g) is dissolved in 125 cm^3 of boiling water, and after cooling, is acidified with 30 cm of acetic acid and filtered. 0·1 g of α-naphthylamine is dissolved in 120 cm^3 of water and mixed with 30 cm^3 of acetic acid. Either reagent can be decolorized if necessary by shaking with zinc dust and filtering.

Vogel (1962) uses sodium nitrite to prepare standard solutions and Jackson (1958) recommends precipitation of silver nitrite from the nitrate with sodium nitrite solution which is then used to prepare standard nitrite solutions. Snell and Snell (1949) advocate addition of 0·1 g dm^{-3} of sodium hydroxide to the nitrite standard to prevent liberation of dinitrogen mon-oxide by any carbon dioxide present.

10:2:4 Nitrate-nitrogen

Extraction

The same reagents used for ammonium and nitrite extraction are used for extracting nitrate and these have been discussed in section 10:2:2. It is usual to determine ammonium and nitrate in the same extract and, with some methods, in the same aliquot of extract.

As the nitrate content of a soil is much influenced by microbial activity it should be determined as soon as possible after sampling the soil. It has been thought that the nitrate content can be temporarily stabilized by addi-tion of a little toluene (Davies *et al.*, 1940) or by rapid drying at low tem-perature (50–55°C). An investigation by Robinson (1967a), however, showed that these treatments are unreliable. There is no reason why ex-periments should not be arranged so as to permit the immediate analysis of nitrate when a soil is sampled, and this will undoubtedly give the most reliable results.

The commonly employed extractant is potassium chloride solution but others have been used. Robinson and Gacoka (1962) used dilute calcium sulphate solution, Bremner (1965) uses a saturated calcium sulphate solu-tion and Metson (1961) recommends shaking the soil with water containing a flocculating agent (copper sulphate) which was erroneously thought to inhibit biological changes, and a base (calcium hydroxide and magne-sium carbonate) to prevent loss of nitrate by acidity when the extract is concentrated for analysis. If nitrites are present an acid potassium chloride solution will convert some nitrite to nitrate and this was prevented by Bremner and Shaw (1955) by first destroying the nitrite with sulphamic acid (section 10:2:2). If it also is desired to measure the nitrite the best extractant is neutral, 2 M potassium chloride solution, sulphamic acid being added to an aliquot of extract for nitrate determination.

Determination

Nitrate in soil extracts can be determined either by reduction to ammonia which is then liberated by alkaline steam distillation and titrated, or colori-metrically, either directly or after reduction to nitrite or ammonia.

The distillation method has the advantages of permitting the estimation of nitrate and ammonia in the same aliquot of extract, in that, once filtered from the soil, the extract needs no further preparatory treatment and that coloured extracts make no difficulty. The method depends upon reduction of nitrate to ammonia by iron and sulphuric acid or, more usually, by Devarda's alloy (50 Cu : 5 Al : 5 Zn) and alkali,

$$3NO_3^{2-} + 8Al + 5OH^- + 2H_2O = 8AlO_2^- + 3NH_3 \qquad [10:5$$

and it should be rememberd that any nitrite present will also be reduced and must be allowed for. The apparatus has been shown in Figure 10:9

and the experimental technique is given in section 10:3:3 for determination of ammonium. After distillation of ammonia a new portion of boric acid is placed below the condenser, approximately 1 g of Devarda's alloy added to the reaction flask, and heating recommenced to distil out the ammonia formed by nitrate reduction. The subsequent titration of ammonia is carried out as previously described.

Instead of distillation, the micro-diffusion technique of Conway (1947) as adapted by Bremner and Shaw can be followed. In this procedure (section 10:3:3) a 15% solution of titanium(III) sulphate is used as reductant and is added to the soil extract in the outer chamber just before mixing in the magnesium oxide suspension. Bremner and Shaw considered the micro-diffusion method particularly useful when studying denitrification in soils which involves the addition of nitrate and organic matter to the soil; colorimetric methods of determining nitrate are affected by the presence of soluble organic matter (Panganiban, 1925). Bremner and Shaw obtained theoretical recovery of 100 μg NO_3^--N in the presence of up to 25 mg of glucose with the diffusion method. Conway claimed to eliminate interference from nitrite by acidifying the solution to pH 1·0 and exposing it to the atmosphere for 1 hour but Bremner and Shaw found that, although this treatment destroyed up to 98% of the nitrite, from 10 to 12% is likely to be converted to nitrate.

Some methods, although depending finally upon colorimetric measurements, are basically methods of nitrate reduction in the same way as distillation and diffusion methods. Haag and Dalphin (1943), for example, reduced nitrate to nitrite with zinc and determined it with sulphanilic acid (section 10:3:5). Wolf (1947) reduced nitrate to ammonia with titanium(III) chloride and sodium hydroxide and estimated the ammonia by Nesslerization.

The traditional direct colorimetric method for determining nitrate is with phenoldisulphonic acid (Sprengels, 1864) and depends upon nitration of the 2,4 phenoldisulphonic acid in fuming sulphuric acid,

The product, although colourless in acid solution, becomes yellow when made alkaline with ammonium hydroxide due to formation of the tri-ammonium salt. Water must be absent during the reaction and thus the soil extract must be evaporated to dryness before adding the reagent. Chloride in excess of about 10 μg cm^{-3} interferes with the reaction and must be removed; this is usually accomplished by adding silver sulphate to the original extracting solution. Coloured soil extracts are troublesome when a colorimetric procedure is used and must be decolorized with activated carbon, or a blank must be prepared from the extract and used to set the zero of the colorimeter. Roller and McKaig (1939), while investigating the phenoldisulphonic acid method, found that the best way of obtaining a clear soil extract was to add calcium sulphate to the extractant; as mentioned, the most recent methods of extracting nitrate from soil utilize calcium sulphate solution. Before evaporating the soil extracts to dryness calcium carbonate should be added to prevent loss of nitrate by acidity, and Roller and McKaig decolorized soil extracts with hydrogen peroxide freshly purified by vacuum distillation. The transmission of the yellow solution is measured with a spectrophotometer at 410 nm and a new standard curve should be prepared for each new batch of reagent. A Lovibond comparator can be used to estimate nitrate using phenoldisulphonic acid. The accuracy of the method using a spectrophotometer is about 5%.

Alternative direct colorimetric procedures to the phenoldisulphonic acid method have at times been introduced. Mäkitie (1963) evaporated the decolorized soil extract to dryness and dissolved the residue in a sulphuric acid solution of 1-naphthol-4-sulphonic acid. After dilution the solution was buffered at pH 8·5 with a borax–ammonium hydroxide mixture and the extinction measured at 435 nm. Swan and Adams (1956) developed the method depending upon reaction of nitrate with iron(II) sulphate in sulphuric acid solution to give red iron(II) nitrosyl sulphate ($FeSO_4.NO$). This procedure, first described by English (1947), is free from interference from chloride, iron and organic acids. Clarke and Jennings (1965) adapted the reaction between nitrate and chromotropic acid (4,5-dihydroxy-2,7-naphthalene-disulphonic acid) in order to measure small quantities (0·02 μg cm^{-3}) of nitrate in small volumes (2 cm^3) of soil extract; iron(III) and chloride interfere with the method. Lewis (1961) measured nitrate in soil extracts by the xylenol method in which the sample is first treated with silver sulphate to remove chlorides, and then sulphamic acid to remove nitrites. After acidifying with dilute sulphuric acid, 3:4 xylenol dissolved in acetone is added and the volatile nitroxylenols distilled from sodium hydroxide solution. The transmittance is read at 430 nm.

Skyring and co-workers (1961) used a polarographic method for determining nitrate and found that, besides being independent of coloured solutions, organic matter and chlorides, it is faster than colorimetric procedures. Sulphate, however, does interfere and if much sulphate is present the method is unreliable. A series of standard potassium nitrate solutions are prepared in 25-cm^3 volumetric flasks each containing 2 cm^3 of a zirconyl chloride solution (8·58 g ZrCl$_2$/100 cm^3); the solutions are diluted with de-ionized water. Each standard solution is de-oxygenated with nitrogen gas for 5 minutes before being placed in the polarograph cell and for 30 seconds after; the diffusion current is measured at $E_a - 1·16$ volts versus a saturated calomel electrode. A linear relationship exists between nitrate and E_a over the range 0–12 μg NO$_3^-$-N cm^{-3}. The accuracy of the determination varies from $\pm 0·3\%$ with 8 μg N cm^{-3} to $\pm 15\%$ with 0·04 μg N cm^{-3}.

Skyring found that the soil suspension itself cannot be directly analyzed polarographically and he recommended extraction of the soil by shaking 50 g with 200 cm³ of 0·01 M barium chloride solution for 15 minutes and taking from 5 cm³ to 10 cm³ aliquots of the filtered extracts for analysis in the same manner as for the standards. A typical curve obtained by Skyring is shown in Figure 10:17.

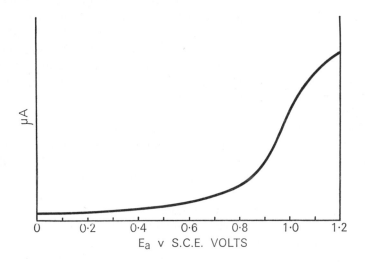

Figure 10:17 A typical curve obtained during polarographic determination of nitrate (Skyring, *Soil Science* © 1961)

10:2:5 Available nitrogen

Although the bulk of plant-available nitrogen in soil is that present in mineral forms, the simple determination of nitrate and ammonia is of limited use for estimating available nitrogen.

The procedures which have been used to find the nitrogen available in a soil include the determination of the total nitrogen content as correlated with plant response, measurement of the amount of nitrogen mineralized during a definite period of time or by some arbitrary treatment, chemical extraction of inorganic forms of nitrogen present, determination of the carbon dioxide produced when the soil is incubated with nitrogen-free energy material, and measurement of microbial growth.

The mineralization method depending upon microbiological oxidation has proved the most popular and reliable, although Richard and co-workers (1960) considered the chemical analysis devised by them to be a better index of nitrogen availability than any other method. Richard's procedure, which measures the ammonium and nitrate in a soil, is to boil the soil rapidly with iron, sulphuric acid and salicylic acid, to distil for a few minutes and discard the distillate and to continue the distillation after adding sodium hydroxide. Displaced ammonia is determined by Nesslerization and expressed as available nitrogen in kg per hectare or lb/acre.

Other chemical methods of estimating available nitrogen include that of Truog and others (1951) which involves the distillation of ammonia from soils in the presence of alkaline permanganate, and that of Purvis and Leo (1961) involving hydrolysis and mild oxidation with dilute sulphuric acid and subsequent determination of the ammonia produced. McClean (1964) on the basis that organic nitrogen accumulates in acid soils and that mineralization of the nitrogen is increased when the pH is raised, assumed a relationship between solubility and availability of nitrogen compounds. He selected sodium hydrogen carbonate as a suitable mildly alkaline extractant which would give conditions similar to those of liming the soil. After shaking the soil with 0·01 M sodium hydrogen carbonate solution, an aliquot of filtrate is digested with sulphuric acid and potassium sulphate, and Nesslerization of the ammonia produced is carried out directly in the digestion tube.

Cornfield (1960) found that the amount of ammonia released by treating a soil with 1 M sodium hydroxide solution was positively correlated with the amount of nitrogen mineralized during incubation. Cornfield thus proposed a rapid, approximate method of estimating available nitrogen in which the soil is mixed with 1 M sodium hydroxide solution in the outer chamber of a Bremner–Shaw micro-diffusion unit (section 10:3:3); 2% boric acid solution is placed in the centre well and the soil incubated at 28°C for 40 hours before titrating the ammonia displaced.

Prasad (1965) estimated potentially available nitrogen by alkaline hydrolysis using calcium hydroxide and found his results correlated highly with nitrogen uptake by millet under greenhouse conditions.

It must be concluded, however, that chemical extraction procedures are too empirical for general use. It may be that, for soils of known characteristics and behaviour, a chemical extraction can be found suitable to estimate the nitrogen available to a specific crop, and this applies to all available nutrient elements. For an unknown soil, or when the crop and land management are new factors, it would be well to avoid an arbitrary chemical extraction as an estimation of available nitrogen. A chemical procedure for estimating available nitrogen which may prove of wide application was suggested by Keeney and Bremner (1966) and involves the extraction of hot-water-soluble forms of nitrogen. After refluxing the soil with water, the extract is subjected to a Kjeldahl digestion and ammonium determined. The results appear to be unaffected by previous drying and storage of the soils.

Biological methods depending upon the growth of *Aspergillus niger* have been used to estimate nitrogen availability. Boswell and co-workers (1962), whilst adopting the more reliable incubation method as a standard procedure, found it too time-consuming for routine. They compared such an incubation where the soil was kept at 0·75% of field moisture capacity at 30°C for 4 weeks with the chemical methods of Truog and of Purvis and found the results correlated. Boswell then investigated a new microbiological technique using a strain of the proteinaceous bacterium *Pseudo-*

monas aeruginosa. This organism produces a pigment, pyrocyanin, during growth in a medium where soil serves as the source of nitrogen. A highly significant negative correlation was obtained between pigment production and nitrifying capacity in thirty soils. The method requires 4 days for completion.

Hunter and Carter (1965) utilized isotopic ^{15}N to determine available soil nitrogen. Air-dry soils were extracted with water and the soluble ammonium and nitrate determined. Nitrifiable nitrogen was then measured by incubation. Pot tests were carried out using Sudan grass as test crop, fertilizers being added; nitrogen was added as $^{15}NH_4^+$ or as $^{15}NO_3^-$. After harvesting, the grass was analyzed for total nitrogen and for ^{15}N (using a mass spectrometer). Nitrogen values were then calculated as follows:

$$\frac{\text{Amount of indigenous soil N}}{\text{Amount of added fertilizer N}}$$

$$= \frac{\text{Amount of N in plant from indigenous soil N}}{\text{Amount of N in plant from fertilizer N}} = x \quad [10:6$$

Proportion of total N in plant derived from fertilizer is

$$\frac{\text{Amount of N in plant from fertilizer N}}{\text{Amount of N in plant from soil} + \text{N from fertilizer}} = y \quad [10:7$$

N-value (the indigenous soil N available to plant during the growing period as the applied fertilizer N) is

$$\frac{\text{Amount of fertilizer N added } (x - y)}{y} \quad [10:8$$

Mineralization of soil nitrogen

Measurement of the amount and rate of nitrogen mineralization in a soil, under specific conditions and for specific lengths of time, remains the most reliable index of nitrogen availability. Most of the criticisms of the method are based upon the fact that it is too time-consuming. This is a finicky argument. The time involved is a few weeks and well within the scope of a fertility investigation. It is difficult to envisage a situation in which the nitrogen available in a soil must be known within a few hours. Furthermore, the number of replicates or of separate incubation experiments involving different soils or treatments is limited only by the number of flasks and incubator space available.

Mineralization of soil nitrogen refers to the conversion of organic nitrogen to inorganic nitrogen and, in practice, refers to production of ammonium and nitrate as other inorganic forms are usually transitional, although for the most accurate results nitrite should be considered also. Mineralization is largely a microbiological process, the organic nitrogen being first changed to ammonium-nitrogen, this being referred to as ammonification, and then via nitrite to nitrate-nitrogen, the process of nitrification. Thus, in normal well-drained soils mineralization is synonymous with nitrification, whereas in waterlogged soils the process is more likely to stop at the ammonification stage. The quantity of ammonium and/or nitrate produced in a soil is an indicator of biological activity and taking into account the total nitrogen present, is a measure of the soil nitrogen that is potentially available to plants.

As the quantity of mineral nitrogen is so dependent upon other soil variables it is meaningless to determine nitrate or ammonium in a soil at random. On a practical basis it is almost as meaningless to measure the rate of nitrification or ammonification unless the possible fate of the released nitrogen is considered also—leaching losses for example—and unless the technique is standardized.

Nitrification and ammonification are affected by factors such as temperature, moisture, aeration, pH, C/N ratio and nature of substrate and, as a consequence, are very variable processes. For analytical results to be meaningful, strict standardization of procedure is absolutely essential and factors to be taken into account include time and method of sampling, time and method of preparation, drying and storing and time and method of incubation.

The increase in mineralizable nitrogen caused by drying a soil before incubation was noted by Budin (1914). It was found by Birch (1959) that the longer a soil remains dry before re-moistening and incubating, the greater the amount of nitrogen mineralized, and that this amount is a significant, linear function of the logarithm of the time the soil remains dry. The magnitude of the nitrogen mineralization after drying a soil was also a function of the organic carbon content of the soil (Figure 10:18). Similar findings were reported by Harpstead and Brage (1958). The effect of dry storage upon nitrification in a soil is not a negligible one of purely academic interest. With one soil having 6·9% carbon, for instance, 9 weeks' dry storage before incubation resulted in extra nitrogen mineralized (that is, in addition to that produced if the soil had not been stored before incubation) equivalent to over 2600 kg of ammonium sulphate per hectare 15-cm (more than 1 ton/acre 6-in). The effect is repetitive, that is, the soil on re-drying after incubation will again show increased nitrification on re-wetting and incubating, and, up to a point (about 60°C), is a function of the temperature of drying as well as of the time.

For his experiments Birch incubated soil at field capacity moisture content, which is the optimum moisture condition for mineralization (Birch, 1958). When the original flush of decomposition had ceased (after about 14 days), moisture, ammonium and nitrate were determined; one of six replicate samples was replaced in the incubator and the other five dried for different lengths of time before reincubating. The mineral nitrogen determined after the initial flush was taken as that present at the start of the experiment and was deducted from subsequent determinations, in order to establish the effect of drying.

Birch's findings emphasize the necessity for standardization of technique when determining the nitrifying capacity of a soil. Quite ob-

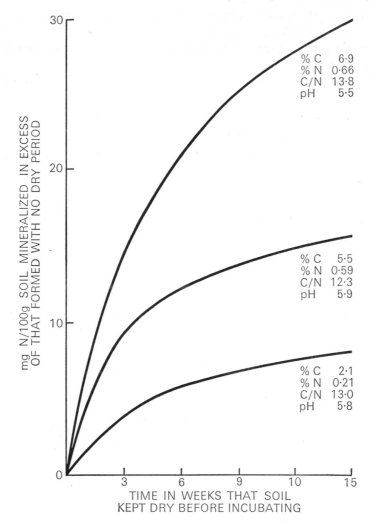

Figure 10:18 Mineralization of soil nitrogen as affected by dry storage and carbon content (from the data of Birch)

viously, the result of an incubation will be meaningless unless the period that a soil has been dry before moistening and the intensity of drying are known (see also Chapter 2, section 2:2). For field experiments, such factors as length of dry and rainy seasons and time of planting are essential prerequisites for interpretation of results.

Gasser (1958) suggested that nitrogen changes in a soil could be prevented before analysis by storing at sub-zero temperatures. Robinson (1964) stored soils at temperatures from −4°C to −10°C for up to 8 days and found that significant nitrogen changes were not prevented. Furthermore, on subsequent incubation, nitrification was found to have been stimulated by the cold storage, a fact noted also by Ross (1964). Clement and Williams (1962) stored soils at $-15°C$ in polythene bags until required for incubation and found that this treatment, whilst depressing microbial activity, did not affect the re-activation of microflora, even after two years' storage at the low temperature.

Nitrification is depressed by puddling and occurs hardly at all in water-logged soil. Waring and Bremner (1964a) utilized the cessation of nitrification when a soil is waterlogged to improve the incubation method of determining available nitrogen. As aerobic incubation involving nitrification is such a capricious process and is very difficult to control or standardize, Waring and Bremner suggested incubation under waterlogged conditions with subsequent determination of ammonium. Such a procedure is easy to control, is precise and is more rapid than aerobic incubation. Either freshly sampled or air-dried soil can be used and special moisture contents do not have to be obtained. The one factor that must be standardized is soil crumb size (Waring and Bremner, 1964b; 1964c) and Maraghan and Pesek (1963) caution against the use of metal apparatus for grinding soil as traces of metallic iron result in the production of nascent hydrogen on waterlogging, with consequent reduction and loss of nitrate. Robinson (1967b) investigated the effect of crumb size upon mineralization of nitrogen using aerobic and anaerobic incubation methods. He concluded that there is little point in using soil finer than the routine 2-mm sample. In the soils he examined, Robinson found that the effect of crumb size is really one of simple mechanical sorting of particles leading to sand fractions acting as an inert diluent.

Incubating a pre-dried soil after waterlogging it does not eliminate the effect of drying upon mineralization. When interpreting the results, the drying effect must be considered in the same way as for aerobic incubations.

Techniques and apparatus for measuring nitrogen mineralization

The common procedure for measuring nitrification is to incubate the soil in vessels under controlled conditions for a definite period of time and then to determine the quantities of ammonium and nitrate produced. Jackson (1958) recommends incubation in pots; the soil and pot are weighed and water is added to bring the soil to $0·7$–$0·8$ of the field moisture capacity. The pots are covered with paper and kept at 28–$30°C$, water being added periodically to maintain constant weight. Every few days the surface crust of soil is broken up and once a week the entire contents of the pots are thoroughly mixed for aeration.

Fraps and Sterges (1947) incubated soil in beakers at 50% of the water-holding capacity and at $35°C$. Nitrification was stimulated by adding an inoculum prepared by grinding a small quantity of an actively nitrifying soil in water. Saunder and co-workers (1957) sampled soils in Africa near the end of the dry season and incubated for 2 weeks at $35°C$ after moistening with $0·1\%$ monocalcium phosphate solution to field capacity. At first, nitrate was determined as an index of nitrogen availability, but as some

soils yielded no nitrate at all, it was realized that ammonium must be determined as well. Saunder found that the method gave a good index of the
nitrogen available during the next growing season.

Clement and Williams (1962) used the apparatus shown in Figure 10:19

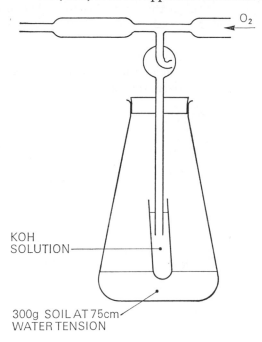

Figure 10:19 Incubation apparatus of Clement and Williams (*J. Soil Sci.*,
1962)

for incubation, and the soil was brought to 75 cm moisture tension as
Stanford and Hanway (1955) had reported a lower coefficient of variation
for nitrification when soils are incubated at known moisture tension (dependent upon soil texture) than at standard moisture content. The soil
samples were placed in the flask, together with a tube containing potassium
hydroxide solution. After keeping at 30°C for 1 hour to allow the air to
expand, the T-tube was inserted with its stem 1 cm below the surface of the
hydroxide solution. Flasks could be connected in series to an oxygen
supply at atmospheric pressure. Incubation was continued for 20 days before ammonium and nitrate were determined. Investigation showed that a
depth of soil up to 15 cm permitted adequate gaseous diffusion for nitrification and that size of soil crumbs is an important factor. Very large soil
crumbs resulted in ammonia accumulation due to creation of reducing
conditions; 4-mm samples were chosen for routine work.

Kresge and Merkle (1957) mixed soil with an equal volume of quartz
sand and placed the mixture in a Gooch crucible. Nitrates already present
were leached out with lime-water and the soil was then leached with a
nutrient solution containing potassium, calcium, magnesium, phosphate

and sulphate at pH 5·8. The crucible was then incubated at 35°C in a desiccator maintained at 99% relative humidity, and mineralized nitrogen was leached out as before at 14-day intervals.

Stanford and co-workers (1965) mixed soil with an equal weight of silica sand and placed the mixture in a filter tube. Ammonium and nitrate were removed by leaching with potassium sulphate solution and the soil was then leached with a nitrogen-free solution containing phosphorus, potassium, calcium, magnesium and sulphur. Excess solution was removed by suction and the soil incubated in the tube for 2 weeks at 29°C, after which the mineralized nitrogen was leached out and determined. A similar

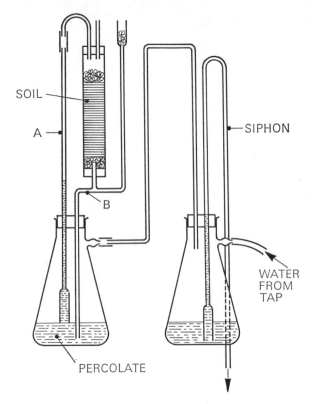

Figure 10:20 Soil percolation apparatus designed by Lees and Quastel (*Biochem. J.*, 1946)

method had been used by Olsen and co-workers (1960) who incubated soil mixed with sand in percolation tubes after leaching with water.

All methods involving pre-leaching of the soil sample to remove soluble and exchangeable inorganic nitrogen are subject to errors arising from removal of soluble organic nitrogen, which is the most mineralizable form.

Eagle and Matthews (1958) sandwiched soil between layers of vermiculite in tubes, and incubated after adding water sufficient to moisten the top two layers and part of the third. The use of vermiculite was rejected by

Bremner and Harada (1959) as it is capable of fixing ammonia in non-exchangeable form. To facilitate leaching, purified sand is superior to materials such as vermiculite, but even this procedure produces artificial conditions which will affect mineralization.

Bremner (1965) recommends incubation of soil samples in containers closed with polythene film or some form of semipermeable membrane, to permit aeration whilst restricting water loss. For simple incubation of a known quantity of soil, as is the case in all the methods so far described, it is refreshing to find van Schreven (1963) unashamedly using jam jars.

A more complicated procedure is that of continuous percolation of a soil as originally described by Lees and Quastel (1944; 1946) whose apparatus is shown in Figure 10:20. The left-hand part of the apparatus is the percolation section and the right-hand part alternately forces in and sucks out air by means of a siphon. As air enters the percolation flask it forces some liquid up the tube A, through a capillary and on to the soil. The amount of liquid so injected is controlled by the depth of the base of A below the surface of the liquid in the flask. When air is sucked out by the siphon operating, air is pulled through tube B and mixes and aerates the percolating liquid. The process is continuous and as percolation is

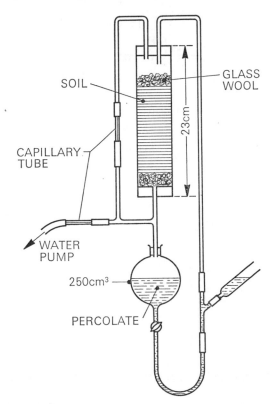

Figure 10:21 Modification of the Lees–Quastel percolation apparatus (Audus, 1946; *Nature, Lond.*)

intermittent the soil does not become waterlogged, although its moisture content remains unchanged. At various periods of time, aliquots of the percolate can be analyzed, the soil remaining undisturbed until the end of the experiment when it, too, can be analyzed. The technique is applicable to investigations other than nitrification, as any desired liquid can be used. Extra substances, such as biological poisons, can easily be introduced at any time.

Audus (1946) modified the apparatus of Lees and Quastel to the form shown in Figure 10:21. About 30 g to 50 g of soil are placed in the tube and 200 to 250 cm³ of percolating liquid placed in the funnel. All joints are sealed with collodion and the water pump is adjusted so as to gently suck a mixture of solution and air up the side-arm. At intervals of the time the pump is disconnected and aliquots of liquid pipetted from the side tube for analysis. Chase (1948) dispensed with the by-pass tube in the Audus apparatus, thus passing all the air through the soil.

Sperber and Sykes (1964) for their investigations needed a percolation system with variable aeration, and they modified the Audus apparatus by inserting a time-regulating valve in the air line (Figure 10:22). When the

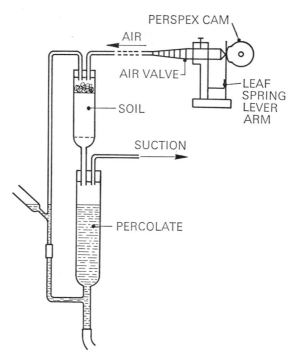

Figure 10:22 Percolation apparatus of Sperber and Sykes (*Pl. Soil*, 1964)

air supply is cut off by the valve closing, a mixture of air and liquid passes up the side arm and percolates; when the valve is open, air alone passes.

Morrill and Dawson (1964) described a percolation system using a simple air pressure siphon; their apparatus is shown in Figure 10:23. Circulating air is scrubbed with dilute sulphuric acid and water and is controlled at 205 kg cm^{-2} by a diaphragm regulator. Morrill and Dawson

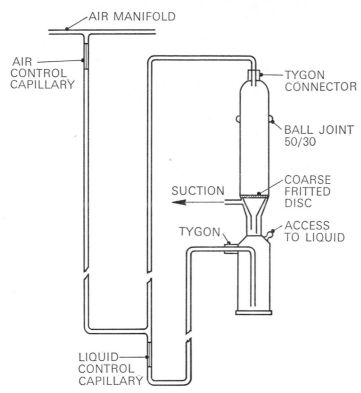

AIR MANIFOLD

AIR CONTROL CAPILLARY

TYGON CONNECTOR

BALL JOINT 50/30

SUCTION

COARSE FRITTED DISC

TYGON

ACCESS TO LIQUID

LIQUID CONTROL CAPILLARY

Figure 10:23 Percolation apparatus of Morrill and Dawson (Soil Science Society of America, *Proceedings*, 1964)

recommended pre-treating the soil sample with 0·1% by weight of krilium to prevent slaking; no effect of krilium additions upon microbial activity was noted.

Macura and Málek (1958) and Macura (1960) devised the apparatus shown in Figure 10:24 for the continuous flow percolation of soil. Soil is placed in the tube and water is allowed to flow for 5 days before starting measurements in order to remove any effects due merely to wetting and drying. Chosen amounts of substrate can be added at will and at a controlled rate. An electromagnetic valve and electronic time switch make possible the continuous regulation of the number of times the valve opens and the time interval between openings. Carbon dioxide evolved is collected and measured in sodium hydroxide solution and the percolating solution is collected for analysis. The solution is collected at low temperature in order to prevent further biological changes. The method is one of selective culture, when, due to the continuously added substrate, an accumulation of microbial association, utilizing that substrate under the given conditions, takes place.

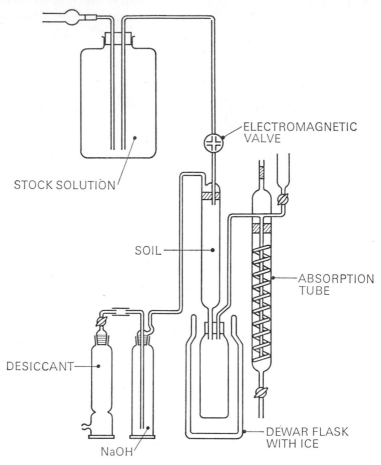

Figure 10:24 Apparatus for continuous flow percolation of soil (Macura, 1960)

Greenwood and Lees (1956) used the apparatus shown in Figure 10:25 and which was at first driven by suction, but later by air pressure. The incoming air was water-saturated to reduce evaporation losses and was sterile; ammonia was removed by scrubbing with 5% copper sulphate solution in 2·5% v/v sulphuric acid. Whenever it was desired to measure carbon dioxide evolved, a second apparatus, containing glass chips in place of soil and barium hydroxide solution in its reservoir, was connected.

The technique of percolation is really of more use in studying the process of nitrogen mineralization rather than for simple determination of the quantity of nitrogen mineralized. It was using their original apparatus,

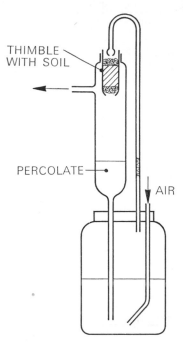

THIMBLE
WITH SOIL

PERCOLATE

AIR

Figure 10:25 Percolation apparatus of Greenwood and Lees (*Pl. Soil*, 1956)

for example, that Lees and Quastel showed that ammonification is a biological rather than a chemical process. Greenwood and Lees used their apparatus to study the breakdown of amino-acids and Hesse (1957) used the Audus modification to study the effects of sulphur compounds and plant extracts upon nitrogen mineralization. A typical nitrification curve showing the nitrate content of a water percolate of soil against time is shown in Figure 10:26.

If it is desired to study the breakdown of organic matter in conjunction with nitrification then the most convenient apparatus is the macrorespirometer of Birch and Friend (Chapter 11, section 11:3:7).

10:2:6 Denitrification in soil

The term denitrification, or nitrate dissimilation, refers to the biological process whereby nitrate is reduced to gaseous compounds such as dinitrogen monoxide and elemental nitrogen and thus lost from the soil. Arnold (1954), by infra-red spectrographic analysis of soil air, showed that dinitrogen monoxide evolution is a factor of importance to the nitrogen economy of soils.

Provided that an oxidizable substrate, which may be organic or inorganic, is present, the denitrifying bacteria can live under anaerobic conditions in the presence of nitrate, and under aerobic conditions in the presence of any suitable nitrogen source. The confusion that exists over the

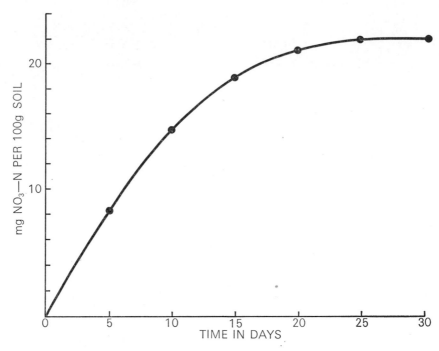

Figure 10:26 Nitrification in a forest soil as shown by continuous water percolation

importance of denitrification in aerobic soils is due, according to Bremner and Shaw (1958), to methodology.

Bremner and Shaw followed denitrification by total nitrogen analyses, as it is difficult to analyze quantitatively, or even qualitatively, the gaseous products. If it is desired to identify and determine gaseous nitrogen compounds it is best to use mass spectrometry. Bremner and Shaw used their modification (section 10:2:1) of Olsen's (1929a) technique to include nitrate and nitrite in the total nitrogen analyses. Soil samples were incubated at 25°C for 80 days at two moisture levels, 60% and 120%, of water-holding capacity. 100 µg g^{-1} of nitrate-nitrogen were added to the soils as potassium nitrate and the incubation was done in stoppered Kjeldahl flasks, with aeration every 4 days; controls had no added nitrate. By incubating in Kjeldahl flasks the sample does not have to be transferred or sub-sampled for analysis, which is important when dealing with water-logged soils. Bremner and Shaw included the determination of hydroxylamine by the method of Csaky (1948) which is a colorimetric method based on the oxidation of hydroxylamine to nitrite with iodine in acetic acid.

Greenwood (1962) studied denitrification using a percolation technique, his experiments being specifically designed to study the effect of size of soil aggregates and the oxygen partial pressure around them. Greenwood used an electrolytic rocking percolator (Greenwood and Lees, 1959; 1960) shown in Figure 11:13, and a Warburg manometer. For gas output the percolator was connected to the side arm of the manometer enclosing the copper cathode and the compensating percolator was joined to the side arm enclosing the platinum electrode. Water was added to the compensating percolator so that the gas space in the two percolators was equal. Any gas evolved from the soil caused an equal volume of oxygen to be evolved at the anode, its volume being obtained by calculation from the measured hydrogen at the cathode.

Percolation was carried out at 25°C using 10 g of soil and 50 cm³ of a solution containing glucose, ammonium-nitrogen and nitrate-nitrogen. Aliquots of percolate were withdrawn and analyzed after 3 hours; an aliquot from the percolator measuring gas evolution was also analyzed, but for nitrite only. Anaerobic conditions could be imposed by an atmosphere of oxygen-free nitrogen. One of Greenwood's conclusions was that oxygen uptake by soil can be used as an approximate measure of denitrification.

Wheeler (1963) percolated amino-acids through soil and found that denitrification depended upon the presence of nitrite and nitrate and a readily available supply of carbon metabolites.

Hauk *et al.* (1958) followed denitrification by means of the heavy isotope ^{15}N. Nitrate-nitrogen enriched with ^{15}N was added to soil which was then incubated and the evolved nitrogen gas collected, purified and analyzed in a mass spectrometer.

10:3 RECOMMENDED METHODS

10:3:1 Standard nitrogen solutions for calibration and checking analytical procedures (Bremner)

Standard ammonium-nitrogen solution: 0.236 g $(NH_4)_2SO_4$ dm^{-3} and store in a refrigerator.

$$1 \text{ cm}^3 \equiv 50 \text{ } \mu g \text{ } NH_4^+\text{-N}$$

Standard (ammonium + nitrate)-nitrogen solution: 0.236 g $(NH_4)_2SO_4$ and 0.361 g KNO_3 dm^{-3}.

$$1 \text{ cm}^3 \equiv 50 \text{ } \mu g \text{ } NH_4^+\text{-N and } 50 \text{ } \mu g \text{ } NO_3^-\text{-N}$$

Standard (ammonium + nitrate + nitrite)-nitrogen solution: 0.236 g $(NH_4)_2SO_4$ and 0.361 g KNO_3 and 0.123 g $NaNO_2$ dm^{-3}.

$$1 \text{ cm}^3 \equiv 50 \text{ } \mu g \text{ } NH_4^+\text{-N, } 50 \text{ } \mu g \text{ } NO_3^-\text{-N and } 25 \text{ } \mu g \text{ } NO_2^-\text{-N}$$

10:3:2 Determination of total nitrogen (Kjeldahl; and after Bremner)

Apparatus

Micro-Kjeldahl flasks of 30-cm³ or 50-cm³ capacity. Electrically heated semi-micro Kjeldahl stand as shown in Plate 10. Modified semi-micro Hoskins still as shown in Figure 10:7.

Reagents

Sulphuric acid, concentrated and nitrogen-free.

Catalyst mixture: 100 g K_2SO_4, 10 g $CuSO_4.5H_2O$ and 1 g selenium powder separately ground, mixed together and the mixture again finely ground in an agate mortar.

Sodium hydroxide solution, 10 M: 420 g dm^{-3} NaOH. Allow to stand for several days for any carbonate to settle.

Boric acid–indicator solution: 20 g Analar boric acid dissolved in about 900 cm^3 of hot water. Cool and add 20 cm^3 of a mixed indicator solution (prepared by dissolving 0·1 g bromcresol green and 0·07 g methyl red in 100 cm^3 of ethanol). Add 0·1 M NaOH solution dropwise until the colour is reddish-purple, and dilute to 1 dm^3.

Hydrochloric acid, M/70: Dilute a standard solution of HCl to give a solution exactly M/70.

Procedures

i. Routine. Accurately weigh a sample (about 0·5 g to 1·0 g) of 0·15-mm (100-mesh) soil containing about 1 mg of nitrogen into a micro-Kjeldahl flask and moisten with 1 cm^3 of water. Allow the moist soil to stand for at least 30 minutes and, as a routine, it will be found convenient to weigh the samples at the end of a working day and leave them, protected from the atmosphere, overnight.

  1 g of catalyst mixture is added from a standard cup and follow with 3 cm^3 of concentrated sulphuric acid. Gently heat the flask on the digestion stand until all the water has boiled off and frothing has ceased. Then increase the heat and maintain until the digest clears; gently boil the clear digest for $2\frac{1}{2}$ hours. Regulate the heat so that the acid condenses about one-third of the way up the neck of the flask.

  After cooling, cautiously add approximately 20 cm^3 of water, shake the solution and allow it to stand until all the sediment has settled. Decant the liquid into a 50-cm^3 volumetric flask and wash the sediment several times by decantation. Although the bulk of the silica should be left in the digestion flask it does not matter if some is transferred to the volumetric flask (section 10:2:1). Dilute the digest to volume, mix and allow to stand until any silica present has settled.

  Experimental details for completing the determination depend largely upon the type of still used; the procedure using a semi-micro Hoskins still is given. The accuracy of the distillation is ±0·5%.

  The apparatus (Figure 10:7) has a 5-dm^3 flask as steam generator and is best heated in an electrothermal mantle. The water in the flask should be double-distilled and slightly acid with sulphuric acid; a few boiling rings or glass beads prevent bumping. A funnel attached to the condenser prevents water from dripping into the receiver. Before use the apparatus should be steamed out and the rate of steam generation adjusted to produce about

6 cm³ of distillate per minute. The condenser should be such as to keep the temperature of the distillate below 25°C.

Place 5 cm³ of boric acid–indicator solution in a 50-cm³ conical flask marked at the 35 cm³ level; the solution is best added from an automatic pipette. Place the flask beneath the condenser with the tip of the condenser about 2 cm above the liquid.

With the by-pass tube of the steam jacket open, pipette an aliquot, usually 10 to 20 cm³, of the soil digest into the reaction vessel. Place about 15 cm³ of sodium hydroxide solution in the funnel. Allow the alkali to run slowly into the reaction chamber until about 1 cm³ remains in the funnel. Add water to the funnel and allow the dilute alkali to run into the vessel, again leaving about 1 cm³ behind. Fill the funnel with water. Close the steam by-pass and commence the distillation. When the distillate reaches 35 cm³, open the by-pass tube, remove the receiver and allow the water in the funnel to enter the flask which then automatically empties. Rinse the vessel in this manner with several portions of cold water before commencing the next distillation.

Titrate the distillate with M/70 hydrochloric acid using a semi-micro burette graduated at 0·01-cm³ intervals, until the green colour changes to pink.

A blank experiment should be made right from the start using 0·03 g of pure starch in place of soil.

$$1 \text{ cm}^3 \text{ M/70 HCl} \equiv 0\cdot2 \text{ mg N} \hspace{3cm} [10:9$$

ii. Modification to include nitrate and nitrite (Bremner)

Reagents

As in section 10:3:2i, except that 3·4 g salicylic acid are dissolved per 100 cm³ of the sulphuric acid.
Sodium thiosulphate, solid, 0·25-mm (60-mesh).

Procedure

Proceed as described in section 10:3:2i, except that 3 cm³ of the sulphuric–salicylic acid mixture is added to the wet soil before it is left overnight. Add 0·5 g sodium thiosulphate just before heating the mixture until frothing ceases; then add the catalyst and continue the digestion.

iii. Modification to include 'fixed' ammonium (Stewart and Porter). This method should be used for soils having more than 15% fixed ammonium nitrogen.

Reagents

As in section 10:3:2i.
Acid mixture: Separately dilute 68 cm³ conc. HCl, 17 cm³ conc. H_2SO_4 and 190 cm³ 40% HF; mix the diluted acids and make to 1 dm³ in a polythene bottle.

8+T.S.C.A.

Procedure

Shake 1 g of soil for 16 hours with 15 cm³ of acid mixture in a polythene bottle. Transfer the mixture to a semi-micro Kjeldahl flask and heat gently to drive off the water. Add 3 cm³ of concentrated sulphuric acid and 1 g of catalyst; continue the digestion as in section 10:3:2*i*.

10:3:3 Determination of exchangeable ammonium-nitrogen

Extraction

Reagents

Potassium chloride solution, 2 M: 150 g dm⁻³ KCl. Boil the solution for about 15 minutes with solid magnesium oxide to remove any ammonia, and cool and filter before making to final volume.

Procedure

Shake 5 g of 2-mm soil for 1 hour with 50 cm³ of potassium chloride solution and allow to settle. There is normally no need to filter the extract and aliquots can be taken from the supernatant liquid.

Measurement by distillation

Apparatus

A suitable apparatus is shown in Figure 10:9 but an alternative set-up is given by Bremner (1965). The steam generator should be as described for the determination of total nitrogen (section 10:3:2*i*) and the apparatus should be steamed out before use.

Reagents

Magnesium oxide: Heavy magnesium oxide is heated at 650°C for 2 hours in an electric muffle furnace, allowed to cool in a desiccator over solid potassium hydroxide and stored in a tightly stoppered bottle.
Boric acid–indicator solution: As in section 10:3:2*i*.
Hydrochloric acid, M/70: As in section 10:3:2*i*.

Procedure

Pipette a suitable aliquot of potassium chloride extract of soil into the distillation flask. Place 5 cm³ of boric acid solution in a 50-cm³ conical flask beneath the condenser.

Add 0·5 g of magnesium oxide to the solution through a long-stemmed powder funnel placed through the second neck of the flask, stopper the flask and steam distil the ammonia into the boric acid solution. The final volume of distillate should be 35 cm³ and it is titrated with standard hydrochloric acid as in section 10:3:2*i*.

10:3:4 Determination of non-exchangeable ammonium-nitrogen (after Silva and Bremner)

Reagents

Potassium hypobromite solution: Add 6 cm³ of bromine slowly (1 cm³ per 2 minutes) to 200 cm³ of 2 M KOH solution which is constantly stirred and cooled in ice. Prepare as required.

Hydrofluoric–hydrochloric acid: Add 84 cm³ conc. HCl to about 700 cm³ of water in a polythene bottle marked at 1 dm³. To this add 163 cm³ of 52% HF solution and dilute to 1 dm³.

Potassium chloride solution, 0·5 M: 37 g dm⁻³ KCl.

Potassium hydroxide solution, 10 M: 560 g dm⁻³ KOH kept in a refrigerator.

Procedure

Weigh 1 g of 100-mesh soil into a tall-form beaker and add 20 cm³ of potassium hypobromite solution. Allow the mixture to stand for 2 hours, add 60 cm³ of water, cover the beaker with a clock-glass and vigorously boil the mixture on a hot-plate for 5 minutes. Allow the boiled mixture to stand overnight at room temperature and discard the supernatant liquid by decantation. Transfer the residue with 0·5 M potassium chloride solution (contained in a plastic wash-bottle) to a 100-cm³ polythene centrifuge tube marked at 80 cm³ and add more potassium chloride solution to the mark. Stopper the tube, shake by hand and centrifuge at 400 rev/s for about 10 minutes. Discard the liquid, add more potassium chloride solution to 80 cm³ and repeat the process. Add 20 cm³ of hydrofluoric–hydrochloric acid mixture and shake the stoppered tube for 24 hours.

Place 15 cm³ of cold 10 M potassium hydroxide solution in the ammonia distillation flask (Figure 10:9) which should be marked to indicate 60 cm³. Transfer the contents of the centrifuge tube to the flask through a long-stemmed, polythene funnel dipping below the level of the hydroxide solution. Rinse the tube well with water, adding the washings to the flask and add sufficient water to bring the total volume to 60 cm³. Stopper the flask, swirl and allow to stand for a few minutes; steam distil the ammonia into 10 cm³ of boric acid solution until the distillate amounts to 100 cm³. Titrate the ammonia with standard acid. A blank distillation of reagents is necessary.

10:3:5 Determination of nitrite-nitrogen (Bratton and Marshall; Barnes and Folkard)

Extraction

Nitrites are conveniently and quantitatively extracted from soil with 2 M potassium chloride solution and thus can be determined in the same extract as used for determining exchangeable ammonium-nitrogen.

Measurement

Reagents

Diazotizing reagent: Dissolve 0·5 g sulphanilamide in 100 cm³ of 2·4 M
HCl and store in a refrigerator.

Coupling reagent: Dissolve 0·3 g of N-(1-naphthyl)-ethylenediamine
hydrochloride in 100 cm³ of 0·1 M HCl and store in a bottle of low
actinic glass in a refrigerator.

Standard nitrite solution: 4·93 g dm⁻³ $NaNO_2$. Dilute 10 cm³ of this solution to 1 dm³ and store in a refrigerator. 1 cm³ ≡ 0·01 mg N. The solution can be standardized if necessary against potassium permanganate
solution.

Procedure

Place a 2 cm³ aliquot of soil extract, neutralized if necessary with sodium
hydroxide or hydrochloric acid, in a 50-cm³ volumetric flask and dilute to
about 40 cm³. Add 1 cm³ of diazotizing reagent, mix and after 5 minutes
add 1 cm³ of coupling reagent. Allow the mixed solutions to stand for
20 minutes and dilute to volume. Measure the transmission at 520 nm.
A reagent blank must be measured and the standard curve should cover
the range 0–0·2 µg cm⁻³ NO_2^--N.

10:3:6 Determination of nitrate-nitrogen

Extraction

Nitrates are quantitatively extracted from soils by 2 M potassium chloride
solution and thus can be measured in the same extract as is made for determining ammonium- or nitrite-nitrogen in soil.

Measurement by distillation

The apparatus and reagents are those described for the determination
of exchangeable ammonium-nitrogen by distillation (section 10:3:3).
In addition Devarda's alloy is required and should be finely ground to
pass a 0·15-mm sieve. For the modification of the method a 2% w/v solution of sulphamic acid is needed and should be prepared from freshly recrystallized (from hot water) reagent and be stored in a refrigerator.

Procedure

After distillation of the ammonium-nitrogen as described in section
10:3:3, remove the stopper of the second neck of the flask and add 0·2 g
of Devarda's alloy. Rapidly replace the stopper and steam distil the ammonia into a fresh portion of boric acid solution. Collect 30 cm³ of distillate and titrate with standard hydrochloric acid.

Note: The procedure will recover nitrite-nitrogen as well as nitrate and if it
is desired to distinguish the two forms add 1 cm³ of 2% aqueous sulphamic
acid to the flask and swirl for a few seconds to destroy the nitrite before
adding the Devarda's alloy.

10:3:7 Determination of available nitrogen (Waring and Bremner)

The only reliable procedure for estimating available nitrogen in an unknown soil is determination of mineralizable nitrogen on incubation. For the reasons discussed in section 10:2:5, the method of Waring and Bremner is to be preferred to aerobic measurement of nitrification unless the soil contains high amounts of nitrate. The crumb size of the sample is important and must be standardized; 2-mm soil has here been adopted as a routine.

Procedure

Place 12·5 ± 1·0 cm³ of water in a 16-mm × 150-mm test-tube and add 5 g of air-dry 2-mm soil. Stopper the tube and incubate at 40°C for 7 days. Transfer the contents of the tube to the flask of the steam distillation apparatus (Figure 10:11) and rinse the tube into the flask with three 4-cm³ portions of 4 M potassium chloride solution. Alternatively, transfer to a conical flask in the same manner and filter the suspension after shaking; aliquots are then distilled. Add 0·25 g of heavy magnesium oxide and distil the ammonia into 5 cm³ of boric acid–indicator solution (section 10:3:3) and titrate with standard acid.

The mineral nitrogen present at the start of the experiment must be separately determined and deducted to give the mineralizable nitrogen. A separate experiment must be made to determine how much nitrogen is mineralized as a sole consequence of air-drying and re-wetting the soil; this is then taken into account. It may be found necessary to incubate the sample for much longer than 7 days as the drying effect could affect mineralization for up to 14 days.

10:3:8 Determination of amino nitrogen

The determination of amino forms of nitrogen is discussed and described in Chapter 11 as part of the subject of organic matter fractionation.

11

Carbon and organic matter

11:1 INTRODUCTION

The term 'soil organic matter' embraces the whole non-mineral fraction of soil and any vegetable or animal matter forming part of the sample analyzed will be included in the determination. Thus analytical results depend in part upon the mean size of the sieve used in the preparatory stage although, in practice, macro-organic matter is largely excluded by fine grinding and sieving (0·5 mm) of the sample.

Organic matter contributes to the physical condition of a soil by holding moisture and by affecting structure. It is a direct source of plant nutrient elements, the release of which depends upon microbial activity and, by affecting the cation exchange capacity, organic matter is directly involved in the availability of nutrient elements.

Total soil organic matter is estimated as a routine from measurement of its organic carbon content, but to determine the organic material, which is intimately incorporated into soil and in equilibrium with the soil environment, special techniques are required. Such material is loosely referred to as 'humus' and is a relatively stable, dark-coloured substance distinct from the immediate decomposition products of animal or vegetable residues. The chemical composition of humus is variable and fluctuates with time, facts which complicate its extraction and determination.

11:2 BACKGROUND AND THEORY

11:2:1 Total carbon

Carbon occurs in soils in the elemental form (coal, graphite and so on), in the inorganic forms of carbonate, hydrogen carbonate and carbon dioxide, and organically as plant and animal matter, their immediate decomposition products and as the more resistant humus. The principle of total carbon analysis is to convert the element completely into carbon dioxide which is then determined gravimetrically or volumetrically. Oxidation is achieved by combustion, either wet or dry, and all forms of carbon can be included. As inclusion of the inorganically bound carbon is seldom, if ever, desired, it is usual to determine carbonate and hydrogen carbonate separately or to destroy them with acids before combustion of the sample; thus the quantity normally determined is 'total organic carbon'.

11:2:2 Total organic matter

The total organic matter content of a soil is estimated either from a knowledge of the total organic carbon content or from the loss in weight when the organic matter is destroyed. As mentioned, results will depend upon the degree of exclusion of macro-organic matter from the sample analyzed; as a routine the 2-mm soil is ground to pass a 0·5-mm sieve. To reduce sampling errors some workers prefer to use the finer 0·15-mm sample.

Total organic carbon

BY DRY COMBUSTION

For the dry combustion of organic matter is needed a furnace fitted with some means of collecting the carbon dioxide evolved and which is usually an absorption train as used in the elemental analysis of pure organic chemicals (Figure 11:1). Owing to sampling errors, micro combustion

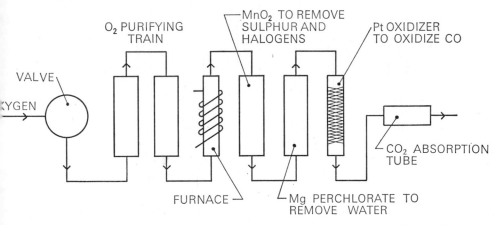

Figure 11:1 Schematic diagram of combustion train for determining total carbon in soil

units are not suitable, although semi-micro procedures have recently been introduced. Jackson (1958) recommends using an induction furnace in which the heat comes from high frequency electromagnetic radiation. Carbonates are destroyed by passing sulphur dioxide for 12 hours through a warm aqueous suspension of the soil containing iron(II) chloride to prevent oxidation of organic matter by manganese dioxide; chloride in excess of 1% (as sodium chloride) is leached out and the soil dried at 100°C. A weighed quantity of the dried soil is mixed with electrolytic iron and tin in an alandum boat and placed in the furnace. Combustion is carried out in a stream of pure dry oxygen. Sulphur dioxide is trapped by manganese dioxide, water vapour is removed by anhydrous magnesium perchlorate and any carbon monoxide produced is oxidized over platinum metal. The issuing carbon dioxide is collected in soda-lime and weighed; alternatively it is absorbed in standard alkali and titrated. Jackson suggests also the possibility of measuring the carbon dioxide gasometrically.

Young and Lindbeck (1964) used a high frequency induction furnace and obtained complete recovery by applying an internal, quartz-enclosed graphite igniter or by using an external, auxiliary resistance furnace. Stewart and co-workers (1964) described the use of a commercial Dumas apparatus for measurement of both carbon and nitrogen (see section 10:2:1). In this method the soil is mixed with copper(II) oxide and the post-heater tube (Figure 11:2) at 600°C is filled half with a 3:1 mixture

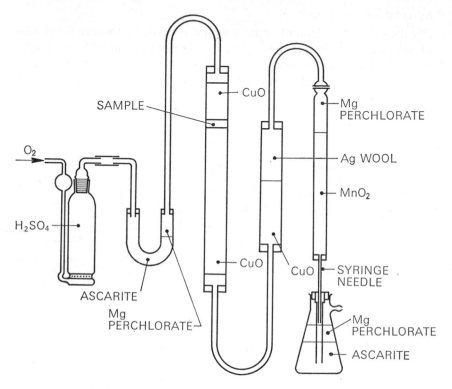

Figure 11:2 Apparatus for determining total carbon in soil by dry combustion (Stewart *et al.*, 1964; Soil Science Society of America, *Proceedings*)

of copper(II) oxide and platinum and half with silver wool to remove halogens. The carrier gas is oxygen and combustion is made at 875°C slowly increasing to 925°C.

BY WET COMBUSTION

For the wet combustion of organic matter the soil is heated with an oxidizing agent and strong acid to convert all forms of carbon into carbon dioxide which is then determined as in the dry combustion methods. Carbonates must be pre-determined or destroyed. Van Slyke and Folch (1940) described a method for analyzing organic compounds depending upon combustion of organic carbon in a mixture of boiling chromic, iodic, sulphuric and phosphoric acids. Hydrazine was added to the digestion mixture to reduce any halogens formed and the evolved carbon dioxide was absorbed in an alkali in a Van Slyke–Neill manometric apparatus. McCready and Hassid (1942) used the Van Slyke–Folch procedure to determine organic carbon in soil.

Clark and Ogg (1942) simplified the oxidizing agent to a mixture of sulphuric acid, phosphoric acid and potassium dichromate and their procedure was modified further by Allison (1960) who included successive traps of potassium iodide and silver sulphate solutions to remove chlorides. Cornfield (1952) discarded the expensive manometric apparatus of the Van Slyke–Neill method in favour of a digestion–purification train using a column of heated silver to eliminate chloride interference and then determining the carbon dioxide gravimetrically. Jackson used the apparatus of Heck (1929) shown in Figure 11:3 and a mixture of sulphuric,

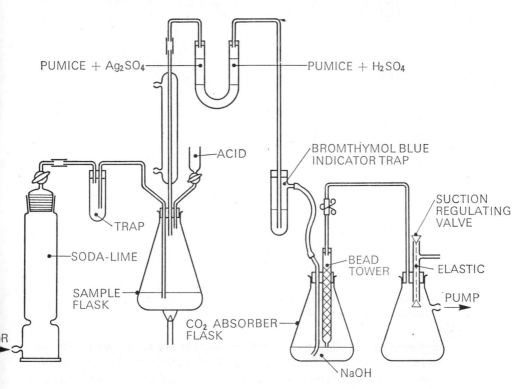

Figure 11:3 Apparatus for organic matter determination by wet combustion (Heck, *Soil Science* © 1929)

phosphoric and chromic acids; the carbon dioxide was determined volumetrically after absorbing it in standard alkali. Shaw (1959) found the absorption procedure of Clark and Ogg inconvenient and that only small quantities of soil could be analyzed by this method. Shaw consequently

8*

mixed soil in a flask with potassium dichromate and water and added the digestion acids after all the carbon dioxide had been swept out of the apparatus; carbon dioxide produced was then absorbed in U-tubes and determined gravimetrically.

Enwezor and Cornfield (1965) further simplified the apparatus (Figure 11:4) and omitted the column used to absorb volatile mineral vapours;

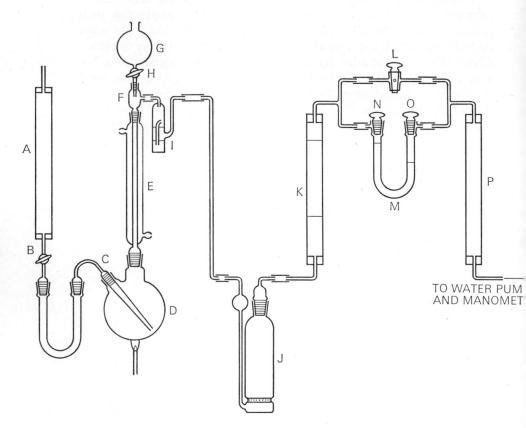

Figure 11:4 Apparatus for determining total carbon in soil by wet combustion (Enwezor and Cornfield, 1965; *J. Sci. Fd. Agric.*)

they employed a singe trap of sodium acetate–acetic acid and potassium iodide to fix volatile mineral vapours, chlorine and chromium(III) chloride. Enwezor and Cornfield reported good reproducibility and agreement of results with those from dry combustion. The proposed single trap was tested by adding sodium chloride to the test sample and it was found that in the absence of the trap chloride greater than 0·004 g, equivalent to 0·2% Cl in the soil, led to high results for total carbon but with the trap in use there was no interference, even up to 4% Cl in the soil. Part of the chloride produced on heating reacts to form chromium(III) chloride but on further heating this largely decomposes again and the chlorine evolved reacts in the trap to liberate iodine. Most of the iodine is retained in the liquid and the remainder is caught in the solid potassium iodide. The main volatile mineral acid formed is nitric acid and this is removed by condensation. Carbonates are removed with sulphurous acid and if significant amounts of elemental carbon are present the time of heating has to be extended.

Expression of results

Total organic carbon can be reported as such or it can be expressed as total organic matter by use of a multiplication factor. The conventional assumption is that organic matter contains 58% carbon making the factor 1·724, although, in fact, different soils contain different sorts of organic matter and the factor varies. Read and Ridgell (1922) examined thirty-seven soils and concluded that it is preferable to assume 51–52% carbon in organic matter. Lunt (1931) found that the carbon content of organic matter in very humus soils was less than 58% and that the factor 1·724 was too low. For example, in peats a factor of 1·86 was more accurate and for well-decomposed humus the factor should be 1·80, for decomposing organic matter 1·85, and for freshly fallen leaf litter, 1·89. Wilson and Staker (1932) examined fifty-five peat soils and found an average factor of 1·736 to apply, which approximates to the conventional factor of 1·724 and thus Lunt's factor does not apply to all peat soils. Jackson points out that the estimation of organic matter by the total nitrogen content by assuming 5% nitrogen in organic matter and multiplying by twenty may be as accurate as using the carbon content. This calculation had been proposed by Leighty and Shorey (1930) but with reservations. Anderson and Byers (1934), on the other hand, considered that as the C/N ratio of soils is so variable, a determination of nitrogen cannot be used to derive an estimate of organic matter.

Determination of total organic matter by weight loss

The principle of determining total organic matter by weight loss is simple; the organic matter in a weighed quantity of soil is destroyed completely, the sample re-weighed and the loss in weight represents the organic matter. The practice of such methods is not simple.

Three methods of destroying organic matter have been advocated, ignition of the soil at low temperatures (350–400°C), ignition at high temperatures (800–900°C) and treatment with hydrogen peroxide.

IGNITION OF SOIL AT LOW TEMPERATURE

Determination of organic matter by low temperature ignition has been described by, *inter al.*, Mitchell (1932). Water is first driven off at 110°C and a weighed quantity of dried soil is heated in a furnace for 8 hours at a temperature of 350–400°C. Such procedure was considered by many not adequate to discriminate between loss of organic matter and loss of mineral matter such as carbon dioxide from carbonates and water and hydroxyl groups from clay. Rather (1917) had recommended the prior destruction of hydrous silicates by treatment with hydrochloric and hydro-fluoric acids in order to eliminate loss of hydroxyl groups, but invariably some organic matter is decomposed by this treatment. Ball (1964) pointed out that the greatest part of the weight loss due to the clay mineral water occurs in the temperature range of 450–600°C and considered the errors of the method to have been overrated provided that the temperature was kept below 450°C. Ball's contention is supported by the observations of Keeling (1962) who, by heating clays at 375°C for 16 hours, destroyed over 90% of the carbonaceous material without loss of structural water. Ball in his experiments used forty-five upland soils having carbon contents varying from 1% to 40%, twenty-two lowland soils and eleven highly organic soils. By heating at 375°C ± 5°C for 16 hours he obtained the regression equation,

$$y = 0.458x - 0.4 \qquad [11:1$$

where y is organic carbon and x is loss on ignition. Ball concluded that the method gives an estimate of organic matter (in non-calcareous soils) sufficiently accurate for most purposes.

IGNITION OF SOIL AT HIGH TEMPERATURES

The usual procedure for high temperature ignition of soil is to remove moisture by heating at 105–110°C and to ignite a weighed amount of dry soil at 800–900°C for 30 minutes; the weight loss is expressed as per cent organic matter. The method has generally been considered as only very approximate due to loss in weight caused by volatilization of clay mineral structural water. Ball (1964) examined the procedure for over one hundred soils having a wide range of organic matter content. The regression equation obtained by correlating the results with those obtained by Tinsley's method (section 11:2:3) was,

$$y = 0.476x - 1.87 \qquad [11:2$$

where y is organic carbon and x is loss on ignition. Ball concluded that the traditional errors of the ignition method have been exaggerated. The high temperature procedure, however, was less satisfactory than low temperature ignition. Ball recognized that his derived regressions may not be applicable to all soils but suggested that a similar one may be. Experiments with soils from East Pakistan (Hesse, unpublished) have shown that different regression equations must be derived from those of Ball.

Richer and Masson (1964) critically discussed the loss on ignition method and introduced as a preferable procedure the use of a thermo-balance. The thermogravimetric method takes into account loss of carbon dioxide from calcareous soils and loss of structural water.

DESTRUCTION OF ORGANIC MATTER WITH HYDROGEN PEROXIDE

The estimation of organic matter from the loss in weight when a soil is treated with hydrogen peroxide was introduced by Robinson (1927). Judd and Weldon (1939) described a modification of Robinson's procedure and this method is given by Jackson. Thirty per cent hydrogen peroxide is used and the temperature is kept below 110°C to retain hydroxyls and clay mineral water. As elemental carbon and the more resistant organic compounds are unattacked, the method is comparable with the chromic acid methods for estimating oxidizable carbon rather than a method for total organic carbon. Carbonates are destroyed with hydrochloric acid, the pH is brought to 5·8 and, after digesting with the peroxide, excess liquid is evaporated off. The hydrogen peroxide treatment is repeated and the oxidized mixture is finally centrifuged with excess ammonium carbonate to re-convert calcium, magnesium and other ions to the carbonates; the soil is then washed, dried and weighed. For more accurate results the supernatant liquid and washings are evaporated to dryness, ignited at 550°C and the residue weighed; this gives the weight of soluble salts removed from the sample.

11:2:3 Oxidizable organic matter

Of all the forms of carbon in soil it is the chemically active form that is of most interest and this includes the immediate decomposition products of raw organic material and soil humus; that is, the inorganically combined and elemental forms of carbon are to be excluded from the analysis.

A method of determining the oxidizable, or active, organic carbon was introduced by Schollenberger (1927) and involves the partial oxidation of organic matter with chromic acid. Although the method is somewhat empirical, Schollenberger obtained reproducible results and found that the fraction of organic matter oxidized was approximately constant for different soils. Schollenberger obtained the chromic acid by mixing potassium dichromate and sulphuric acid with the soil and heating the mixture to 175°C for about 2 minutes; after cooling, the unused dichromate was determined by titration with ammonium iron(II) sulphate solution. Carbonates do not interfere with the method and do not have to be destroyed first. The method was calibrated for organic carbon by correlation with results from dry combustion. Later, Schollenberger (1931) recommended the addition of sodium fluoride or phosphoric acid before titration to eliminate interference from iron(III).

Metson (1961) found that the temperature of 175°C was unnecessarily high and that satisfactory results can be obtained by heating to 150°C, when thermal decomposition of the dichromate is less and evolution of acid fumes is reduced. As the soils investigated by Metson contained high amounts of organic matter, he slightly modified the Schollenberger method by heating the soil with sulphuric acid and excess solid potassium dichromate under controlled conditions of temperature (150°C ± 5°C); excess chromic acid was then titrated with iron(II) sulphate. The necessary amount of potassium dichromate was calculated from the approximate carbon content estimated from the nitrogen percentage. Metson averaged an 87% recovery of organic carbon by this method.

Tinsley (1950) investigated the factor of thermal decomposition of the dichromate. He found that on boiling in an open flask with sulphuric acid alone or with a mixture of sulphuric and phosphoric acids, considerable decomposition occurred. The dichromate was less decomposed when boiled with sulphuric and perchloric acids. Tinsley reduced thermal decomposition to negligible amount by using a cold finger in the neck of the flask and he recommended using this apparatus with a mixture of perchloric and sulphuric acids with a boiling time of 2–3 hours.

Jackson (1958) heated the soil and oxidant in a bath of phosphoric acid maintained at 155°C with frequent mixing, and Purvis and Higson (1939) used an electric oven.

Several substances interfere with the determination of oxidizable carbon, notably chloride, nitrate, iron(II) and higher oxides of manganese. Chloride competes with organic matter in reducing the chromic acid, one milliequivalent of Cl^- being equal in reducing power to 3·5 mg of carbon. Schollenberger (1945) suggested leaching out chlorides before digesting the soil. This, however, would remove soluble organic matter and a better and more simple method is to add silver sulphate or mercury(II) oxide to the digestion mixture. Quin and Salomom (1964) found that the addition of 2 g of mercury(II) oxide to the sulphuric acid was sufficient to immobilize the maximum chloride concentration that can be expected in neutralized hydrochloric acid hydrolysates of soil. Walkley (1935) suggested that the effect of chloride ions could be allowed for by calculation. From the equations,

$$Cr_2O_7{}^{2-} + 6Cl^- + 14H^+ = 2Cr^{3+} + 3Cl_2 + 7H_2O \qquad [11:3$$

and

$$4Cr^{6+} + 3C \longrightarrow 4Cr^{3+} + 3C^{4+} \qquad [11:4$$

a chloride factor of 1/12 is calculated, that is,

$$\% \text{ C} = (\text{uncorrected } \% \text{ C} - \% \text{ Cl}/12). \qquad [11:5$$

Metson pointed out that if sufficient chloride is present to interfere seriously, it would almost certainly be determined in its own right and thus the calculation method can most easily be used. The error involved is normally low enough to ignore as 1% chloride in the soil would give an error of about 0·1% in the carbon result.

Nitrate interferes only if present in amounts greater than $\frac{1}{20}$ of the carbon present. Iron(II) interferes giving high results but, as Jackson points out, air-dry soils seldom contain appreciable amounts of iron(II). Water-logged soils, on the other hand, often contain considerable quantities of iron(II); in most cases this can be oxidized by drying the soil before analysis, but if the soil is to be analyzed wet then the effects of iron(II) must be allowed for. Similarly most soils do not contain sufficient manganese oxides to interfere seriously with the method, but if necessary the interference can be eliminated by treating the soil with iron(II) sulphate in amount calculated from the results of a separate titration (Jackson, 1958).

Degtjareff (1930) recommended the addition of hydrogen peroxide to the soil sample followed by chromic acid in concentrated sulphuric acid, no external heat being applied. Tiurin (1931) reported Degtjareff's modification as being ineffectual and recommended boiling the soil for 5 minutes in chromic acid made in 1:1 sulphuric acid.

Walkley and Black (1934) and Walkley (1935; 1947) modified the Schollenberger technique by using the heat of reaction between dilute dichromate solution and concentrated sulphuric acid to effect oxidation. Such procedure gave 76% recovery of total organic carbon as opposed to 90.4% recovery by Schollenberger (1945). The Walkley–Black procedure recovers only the more active organic matter and over 90% of elemental carbon is excluded; thus the method is more selective than others giving higher recoveries. The method of Tiurin (1931), for example, recovers up to 84% of any elemental carbon present.

Allison (1935) critically examined Schollenberger's method and its subsequent modifications and concluded that all the modifications were inferior to the original method, as they decreased the simplicity, rapidity or accuracy. Tiurin's modification of boiling for 5 minutes was found to be no better than Schollenberger's method in which boiling is for $1\frac{1}{2}$ minutes and, furthermore, if a hotplate was used, evaporation occurred. Allison criticized the Walkley–Black procedure as the temperature reaches only 124°C resulting in a low recovery. Nevertheless, the Walkley–Black method, owing to its simplicity, is the most widely used procedure for obtaining a good estimate of oxidizable organic matter in a large number of samples. Several minor modifications of the published method exist but basically a weighed quantity of finely ground soil is treated with a known volume of M/6 potassium dichromate solution followed by the rapid addition of an excess of concentrated sulphuric acid. As the degree of oxidation depends upon the heat generated, all samples must be allowed to cool uniformly. After cooling, the mixture is diluted, an excess of phosphoric acid or sodium fluoride added to suppress iron(III) and the excess chromic acid titrated with ammonium iron(II) sulphate solution using diphenylamine as indicator. One modification is to use an aqueous solution of barium diphenylamine sulphonate as indicator, this being more stable than the sulphuric acid solution of diphenylamine (Peech *et al.*, 1947). Alternatively *o*-phenanthroline can be used as indicator and this is theoretically superior to diphenylamine as the colour change occurs at a higher oxidation–reduction potential; the end-point however, is less easily determined. Markert (1962) used a 0.2% solution of phenylanthranilic acid as indicator.

With a little practice the diphenylamine end-point can be established very accurately although the colour of some soils can make difficulties.

Normally an intense dark blue colour is formed on addition of the indicator and as the titration proceeds this changes to a dirty green. Towards the end of the titration the colour passes through violet to a clear, deep blue and at this point the titration should be continued dropwise, the end-point being a sharp change to a bright, clear green. If the end-point is overshot, a small measured quantity of dichromate solution can be added and the titration continued dropwise. Some workers recommend such additions (of about 0·5 cm³) of extra dichromate as a routine but provided that the deep blue stage of the titration is recognized and that further additions of iron(II) solution are made dropwise, the procedure is unnecessary. Inexperienced analysts can confuse the dirty green colour at the beginning of a titration with the green colour of the end-point but this is merely due to lack of practice. Sometimes too much soil has been weighed out, considering its organic matter content, and all the dichromate is reduced and the indicator turns green at once. It is not unknown for a beginner to continue titrating such a mixture and, to his frustration, never reaching the deep blue stage. For soils containing high amounts of organic carbon it is preferable to use correspondingly more dichromate rather than to reduce the amount of soil analyzed.

The same interferences are found with the Walkley–Black method as with the Schollenberger method and their elimination is effected in similar ways.

Instead of relying entirely upon colour change, the excess dichromate can be titrated potentiometrically, in which case no phosphoric acid is necessary. As an alternative to titration of dichromate Smith and Weldon (1941) added an excess of iron(II) sulphate to the reaction mixture and back-titrated with potassium permanganate.

As the colour of dichromate itself changes by reduction from yellow to green, it should be possible to measure the degree of reduction by direct colorimetry and several workers have proposed methods of doing this. Two variations are possible, to measure the yellow colour of the unreacted dichromate or to measure the green colour of reduced dichromate. If the yellow colour is to be measured the amount of dichromate added to the soil must be accurately known, but if the green colour is to be measured, standardization of the dichromate solution is unnecessary and thus, in principle, it is easier to measure the green colour. Graham (1948) treated 1 g of soil with potassium dichromate solution and sulphuric acid, rapidly added from a large burette. After standing 5 minutes the mixture was diluted and allowed to stand for 5 hours before the supernatant liquid was poured off and its extinction measured at 645 nm.

Carolan (1948) investigated both alternatives of colour measurement and measured yellow colours using a violet filter and green colours using a red filter. Samples of soil were treated with M/6 potassium dichromate solution and concentrated sulphuric acid. Experiment showed that removing turbidity by standing, as in Graham's method, was not practical and that it is better to filter the solution after standing for half an hour and diluting. Carolan oxidized different weights of soil and plotted weight against extinction using both filters. Three completely reduced (with potassium metabisulphite) solutions of dichromate were treated with sulphuric acid and water in the same manner as soils, and the average value of their extinctions appeared on the graphs showing that the yellow and the green colours can be measured without interference from other soil colours. Carolan corrected all extinction readings for colour absorption by addition factors and calibrated the readings for determination of carbon, assuming organic matter to contain 58% carbon and the method to give a 77% recovery. It was concluded that measurement of the green colour is, on the whole, to be preferred. Metson (1961) gives details of measuring the green colour of reduced dichromate with an orange filter (about 600 nm). The method was calibrated against carbon values as obtained by Schollenberger's method. Metson recommends the use of sodium, rather than potassium, dichromate as it is more soluble. After the diluted reaction mixture has stood for about 4 hours, some of the supernatant liquid is centrifuged off as Metson found attack by chromic acid of the filter paper caused high results. Datta and co-workers (1962) measured the green colour with a red (660 nm) filter and prepared a standard curve using known solutions of glucose.

Expression of results

Methods of determining oxidizable organic matter which depend upon partial oxidation make the expression of results difficult. To express the results as total organic carbon, an arbitrary conversion factor must be used. Allison (1935) calculated the apparent percentage recovery of carbon from five soils by the Schollenberger method, as compared with the dry combustion method, and obtained an average of 87%. Thus to convert uncorrected carbon percentages to total organic carbon he recommended multiplying by the factor 100/87 or 1·15. Schollenberger (1945) used nine soils in which the carbon by dry combustion varied from 0·5 to 17·8% and obtained an average of $90.4 \pm 1.3\%$ recovery by his method. Using the Walkley–Black procedure Walkley (1947) obtained 76% recovery making the conversion factor 1·31, although the conventional factor for the Walkley–Black method is 1·33 (75% recovery). Davies (1950), working with New Zealand soils, found Allison's factor too high and, obtaining a recovery of 93·8%, Davies used the factor 1·07. Metson (1961), also working with New Zealand soils, found Allison's factor suitable as long as the soils contained less than about 12% carbon. With soils having more than 12% carbon and using his modified Schollenberger technique, Metson compensated by subtracting from the carbon values calculated before employing the moisture factor, a quantity read from a graph and which increased from zero to about six as the carbon increased from 10% to 50%. Little *et al.* (1962), using the Walkley–Black procedure, found a factor of 1·302 best for the soils they analyzed.

It would seem a mistaken idea to attempt to find a generally applicable factor by averaging the recoveries of several soils with widely differing carbon contents. The most advisable procedure is to determine the average percentage recovery for each general type of soil, using the chosen method if it is desired to express results as total organic carbon or as organic matter. Alternatively, the results as obtained can be quoted, provided that the method is given as representing a particular fraction of organic matter.

11:2:4 Fractionation of soil organic matter

The true soil organic matter or humus, as opposed to plant and animal residues and their immediate decomposition products, is of variable and varying constitution. Baumann (1887) considered that soil organic matter contained amino compounds as considerable quantities of ammonia could be released by heating a soil with hydrochloric acid or by adding cold alkali. Berthelot and André (1887) found that even boiling humus with water produced ammonia and deduced the presence of amides. Dojarenko (1902) found up to 70% of the nitrogen in an alkaline extract of soil to be in amino form. For many years it was thought that free amino-acids do not occur in soils but Dodd and co-workers (1953), Putnam and Schmidt (1959) and Paul and Schmidt (1960) demonstrated the presence in soil of free amino-acids by chromatographic techniques.

The manner in which amino-acids are bound in the soil is still largely unknown. Until quite recently no humus material had been shown to contain protein and, although Bremner (1949) found up to 36% of total soil nitrogen in α-amino forms, he could not show that it came from protein. Simonart *et. al.* (1967) were the first to isolate a protein from humus. The separation was made by chromatographic analysis of humic acid in phenol solution and the protein contained 14·8% nitrogen and gave rise to twenty different amino-acids.

As carbohydrates constitute a large proportion of plant material they play an important part in soil organic matter composition but are easily decomposed. Shorey and Lanthrop (1910) obtained evidence of pentosans in soil by isolating furfural after hydrolyzing the organic matter with hydrochloric acid, and Thomas and Lynch (1961) separated pentosans from soil hydrolysates with exchange resins. The complexity of soil organic matter composition has been discussed in a review by Whitehead and Tinsley (1964).

Extraction of organic matter from the mineral fraction of soil is not easy. Chemical extractions which are sufficiently drastic to dissolve the organic compounds are liable to alter the nature of those compounds and milder extractants remove correspondingly less material.

Very early work on the nature of organic matter led first to the extraction of 'humus' from the bulk of soil and then to its separation into two main fractions named 'humic' and 'fulvic' acids. Achard (1786) extracted soils with sodium hydroxide and Vaquelin (1797) treated soil with acid before extracting humus with alkali. Sprengel (1826) extracted humus from

peat with ammonium hydroxide after an acid treatment and further purified the extract by precipitating with hydrochloric acid, re-dissolving the
precipitate in sodium carbonate solution and again precipitating with acid.
The principle of Sprengel's method, that is, extraction of the soil with
alkali and treatment of the extract with acid to obtain two different
fractions of humus, is still used in organic matter fractionation. Soil
organic matter is first separated into humic and non-humic material, the
latter comprising undecomposed plant remains, carbohydrates and so on.
The humic material, on treatment with alkali, gives an insoluble 'humin'
and a soluble 'humus'. The initial alkaline extraction is liable to cause
hydrolysis and for this reason alternative extractants for humus have been
sought. Bremner and Lees (1949) showed that neutral sodium pyrophosphate formed a stable complex with humus-bonding cations and removed
up to about 40% of the organic carbon from soils. Other reagents investigated included ethylenediamine tetra-acetic acid and oxalate, but these
were found less effective than pyrophosphate. Bremner and Harada (1959)
attempted to isolate organic matter from mineral matter and thought that
one of the difficulties was due to organic matter reacting with clay minerals
to give complexes. Bremner and Harada thus investigated the use of
hydrogen fluoride for extracting organic matter; being a weak acid this
should not have much effect on the organic matter itself. It was found that
hydrogen fluoride had little effect upon organic matter extraction in the
presence of neutral or alkaline reagents, but with dilute hydrochloric acid
considerable amounts of ammonium and organic matter were released.
Coffin and De Long (1960) extracted 75% of the organic matter from a
podsol (B-horizon) with benzene-8-quinolinol and water. Tinsley and
Salam (1961) found sodium pyrophosphate, citrate, oxalate and sulphite
all good reagents for the partial extraction of humus. Boiling for 1 hour
with M/8 sodium pyrophosphate at pH 8·4 dissolved more organic matter
than shaking for 24 hours at room temperature. Dubach *et al.* (1961)
removed 90% of the organic matter from a podsol B-horizon with neutral
ethylenediamine tetra-acetic acid. Monnier and others (1962) used a
series of organic solvents and Ferrari and Dell'Agnola (1963) extracted
with 0·1 M sodium fluoride solution. Whereas fluoride is effective in removing humus closely linked with clay, sodium dithionite has been used to
separate humus from sesquioxides. Dubach and Mehta (1963) defined the
objectives of extraction procedures as being,

    i. Universal applicability to obtain comparable fractions from all
soils.
    ii. Isolation of unaltered material.
    iii. Selective extraction free from contamination.
    iv. Completeness of extraction to ensure representation of fractions of
the whole molecular weight range.

The most universally applicable reagents are the alkali hydroxides but
these do not result in extraction of unaltered material. The effectiveness of

complexing reagents like pyrophosphate, can be increase by pre-treatment of the soil with hydrochloric and hydrofluoric acids to destroy carbonates and silicates, and Dubach and Mehta pointed out that the pre-treatment itself will remove some humic material which must be isolated.

Posner (1966) was particularly concerned with obtaining an extract as free as possible from inorganic material and he found extraction with pyrophosphate gave material with the lowest ash content (about 0·2%) although the yield of humus was much lower than with sodium hydroxide extraction.

Whitehead and Tinsley (1964) investigated dimethyl formamide as an extractant of organic matter from soil and their recommended procedure is to make a double extraction (for 15 minutes and 45 minutes) with boiling dimethyl formamide made 0·4 M with respect to boric or oxalic acid and 0·2 M with fluoboric acid. A single extraction gave a lower yield although the product was less contaminated with inorganic material.

As a suitable starting material for the extraction of humus Roulet (1963) recommended dispersion of the soil aggregates by water turbulence and separation of a root-free material by flotation sieving. The less than 0·06-mm fraction is collected from the 2-mm soil for extraction and con- tains 50–70% of the total organic carbon.

Once the initial extraction of humus has been effected it is customary to separate the acid-soluble (fulvic) from the acid-insoluble (humic) frac- tion. This is done by acidifying the soil extract with hydrochloric or sul- phuric acid when humic acid is precipitated, although exchange resins can be used instead (Dubach *et al.*, 1961).

Yuan (1964) pointed out that insufficient attention has been paid to the degree of acidity when precipitating humic acid from humus solutions. A few workers have adopted particular pH values which have varied from 1 to 3 but most analysts merely add an 'excess' of acid. As their solubility in mineral acids is the chief way of distinguishing the two main fractions of humus, Yuan rightly considered that the pH of the precipitation should be more closely controlled. For his investigation Yuan used three different extractants, being those most commonly employed, namely 0·5 M sodium hydroxide at pH 12, M/40 sodium pyrophosphate at pH 10·2 and 0·5 M sodium fluoride at pH 7·1. A fourth extractant was a synthetic chelating resin, Dowex A-1, having iminodiacetic acid groups. This resin had been advocated by Bremner and Ho (1961) as being not only effective in the extraction of organic matter but eliminating risk of artefaction and con- tamination. Yuan shook 100-g samples of soil with 500 cm³ of extractant (or 480 cm³ of water and the equivalent of 10 g oven-dry resin) in polythene bottles. After standing 1 week to allow very fine material to settle, the supernatant liquid was siphoned off and centrifuged. The soil remaining in the bottle was washed three times with water and the washings added to the extract. Having been made to volume (1 dm³) aliquots of the extracts were brought to various pH values with dilute sulphuric acid, and after standing 5 minutes the mixtures were centrifuged and any precipitate washed once with water. pH values varied from that of the original extractant down to 1·0 in steps of 1·5, with the exception of the sodium fluoride extract which was taken to pH 2·0. Organic carbon and nitrogen analyses were made of the products.

Yuan found that sodium hydroxide removed most humus from the soil but that the different extractants seemed to extract different kinds of organic matter. The exchange resin was found unsatisfactory for highly organic soils. The pH of precipitation had a considerable effect which varied with reagent. In general the humic acid precipitated increased as pH decreased; an exception was in the sodium hydroxide extract of an organic pan layer of soil when maximum precipitation occurred between pH 5·4 and 4·0. The increased solubility of the precipitate below pH 4 of this humic acid was found to be related to its aluminium content. For completeness of separation of humic from fulvic acid it is not desirable to include aluminium with the precipitate and Yuan recommended that a pH of 1·0 be adopted as a standard value to be attained when acidifying humus extracts.

Dubach and Mehta (1963) considered that the fulvic acid fraction is best removed from the acidified humus extract by adsorption on charcoal. Alternatively, the fulvic acids can be precipitated from the filtrate as barium, copper or iron salts provided that the extractant itself is not also precipitated. The salts are then decomposed with acidic cation exchangers. Morrison (1963), after precipitating humic acid with sulphuric acid, coagulated it by freezing for ease of filtration.

Having made the broad distinction between humic and fulvic acids, both fractions can be further broken down. For example, treatment of humic acid with alcohol gives two fractions, one soluble and one insoluble, and treatment of fulvic acid with sodium hydroxide to raise the pH to 4·8 gives a soluble fraction and a precipitate of so-called β-humus.

Fractionation of humic acid

The isolated humic acid should be purified before further fractionation. Waxes and fats can be extracted with organic solvents and inorganic compounds can be removed by electrodialysis (Forsyth and Fraser, 1947); with complexing reagents (Hori and Okuda, 1961) or with ion exchange resins (Sowden and Deuel, 1961). The removal of carbohydrates and lignins is more difficult as they may be linked with the humic substances. Dubach and Mehta (1963) suggested that these compounds could be determined and their removal followed by periodic analysis. Chemical procedures such as hydrolysis for removing carbohydrates must be avoided in case the humic material is also affected; in certain cases, for example with B-horizons of podsols, separation has been effected by use of exchange resins (Wright and Schnitzer, 1959).

Removal of aromatics such as lignins and tannins has proved almost impossible and Dubach and Mehta considered that such purification merges with the process of fractionation.

The partially purified humic fraction of soil can be fractionated by a whole variety of methods, although in every case the separates are still complex and as yet no discrete fraction has been obtained. Fractionation methods include differences in solubility at various pH values, extraction with organic solvents, precipitation with salts, electrophoresis and chromatography. MacKenzie and Dawson (1962) carried out column electrophoresis on soil extracts; the column was prepared with cellulose powder mixed with a phosphate buffer and sucrose was used as a reference substance as it does not exhibit electrophoresis, thus permitting measurements to be corrected for electro-osmosis. After the zone electrophoresis was complete, the column was placed on an automatic fraction collector and various fractions obtained for analysis. MacKenzie and Dawson, although obtaining some fractionation of humic acid, could not get any clear-cut differences between fractions. Kononova and Titova (1961) and Jacquin (1961) attempted humic acid fractionation by paper electrophoresis, and Waldron and Mortensen (1961) used continuous flow electrophoresis followed by paper chromatography. According to Burges (1960) electrophoretic separation of humic acid into fractions is due to separation into particle size groups rather than chemical entities.

Pauli and Grobler (1961) found that the acriflavine adsorbing capacity, as measured with a fluorimeter, of humic material precipitated by acids is directly proportional to the humic acid content of the soil. Certain organic fractions can be characterized by measuring their extinction coefficients from the absorption curves obtained with the extracts at different wavelengths. Yuan (1964), for example, used wavelengths of up to 650 nm and found that extracts containing the same amounts of organic carbon had different extinction coefficients with different extractants.

Fractionation of fulvic acid

By partial adsorption on activated charcoal and serial elution with 90% acetone, water and 0·5 M sodium hydroxide solution, Forsyth (1947) separated fulvic acid into four broad fractions. In each fraction carbon, and nitrogen were determined and Forsyth identified sugars, amino-acids. polysaccharides, organic phosphorus compounds and pigmented materials.

The various fractions of organic matter can all be broken down into more simple compounds by hydrolysis. Such treatment yields amino-acids and shows that a considerable quantity of polysaccharides also occurs. The remainder of the humus is considered to be aromatic and has been partially characterized by ultra-violet absorption spectra, by oxygen uptake and by determination of methoxy groups; polyphenols would appear to be present among other compounds.

A different technique for fractionating soil organic matter depends upon extraction with various solvents of compounds characteristic of plant tissue. Thus Shewan (1938) treated soil successively with ether to isolate fats and waxes, ethanol to isolate resins, water to remove polysaccharides, boiling 2% hydrochloric acid to remove hemicelluloses and 80% sulphuric acid to decompose cellulose; the residue was finally analyzed for carbon and nitrogen to estimate lignin-humus.

Carbohydrates

Waksman and Stevens (1930) brought soil polysaccharides into solution with cold, 12 M sulphuric acid and Sowden and Ivarson (1962) used the same reagent to extract carbohydrates from soil. The acid dissolves organic matter and begins to hydrolyze polysaccharides to monosaccharides; hydrolysis can then be increased by diluting and boiling. By this procedure Ivarson and Sowden (1962) extracted hexoses, pentoses and uronic acids from soil.

Cheshire and Mundie (1966) further examined the use of 12 M sulphuric acid in order to find the conditions for maximum yield of carbohydrates. They found that after shaking the soil with the acid at 20°C a further extraction with 0·5 M sulphuric acid at 100°C was necessary for maximum hydrolysis. Cheshire and Mundie concluded that the best procedure is to extract the soil with cold, 12 M sulphuric acid for 16 hours followed by hydrolysis with 0·5 M acid at 100°C for 5 hours. Such treatment gave maximum yield of pentoses and methyl pentoses from a sandy loam.

Forsyth (1950) isolated polysaccharides from fulvic acid by using a charcoal column and selective elution, and Parsons and Tinsley (1961) extracted polysaccharides with anhydrous formic acid. Mortensen (1960) extracted soil with hot water, concentrated the extract by evaporation under reduced pressure, precipitated humic matter with hydrochloric acid and then the polysaccharides with acetone. Ivarson and Gupta (1967) found that freezing soil samples prior to analysis increased the quantity of free sugars extracted. For example, storage of soil for 3 weeks at -14°C resulted in up to a fivefold increase in glucose and a threefold increase in pentoses.

Pentoses in the purified hydrolysates can be estimated with aniline acetate using the method of Tracey (1950). An aliquot of hydrolysate is added to a solution of acetic acid, oxalic acid and aniline in a test-tube and the extinction of the resulting coloured solution measured (after standing in the dark for 24 hours), using an Ilford filter No. 622. Glucose and galactose do not interfere with the determination to more than 2%, galacturonic acid more than 5% and other sugars more than 1%.

Hexoses can conveniently be determined with anthrone as described by Brink *et al.* (1960). In this procedure a portion of hydrolysate in a test-tube of prescribed size is treated with a 0·2% solution of anthrone in 95% w/v sulphuric acid; after 15 minutes the absorbance of the green solution is measured at 625 nm. A calibration curve is prepared using glucose solution over the range 0–20 μg cm^{-3}. The standardized test-tube was recommended by Morris (1948) who had previously described an anthrone method for carbohydrates, because the reaction is partly controlled by the heat of mixing the solutions.

Determination of cellulose in soil was originally based on the procedure of Waksman and Stevens (1928) in which the soil is given a preliminary extraction with a mixture of benzene and ethanol to remove fats, waxes and resins and is then boiled under reflux for 5 hours with 2% hydrochloric acid. The filtered residue is washed free from chloride and treated with 80% sulphuric acid at room temperature for 2 hours; after dilution with water the mixture is refluxed for 5 hours. The hydrolysate is neutralized and glucose measured by reduction. Daji (1932) extracted cellulose directly from soil with Schweitzer's solution (ammoniacal copper(II) hydroxide) after pretreating the soil sample with dilute sodium hydroxide solution, hot dilute hydrochloric acid, sodium hypochlorite and hydrogen peroxide. To the Schweitzer extract was added an excess of ethanol and the precipitate filtered off and washed with ethanol–hydrochloric acid to dissolve the copper hydroxide; the residue was then washed free from copper with water. The crude cellulose was weighed and ignited and the loss in weight taken as a measure of pure cellulose. Daji's procedure was slightly modified by Gupta and Sowden (1964) who used a centrifuge for separating and washing the precipitates and who made a double extraction of the soil with the reagent. The final precipitate of crude cellulose was hydrolyzed to glucose and the percentage of cellulose calculated.

Dische (1947) discussed the fact that when sugars are treated with concentrated mineral acids such as sulphuric or hydrochloric, they give mixtures of products which react with various organic substances to give strongly coloured solutions. Different groups of sugars, pentoses, hexoses, and so on, show differences in speed of reaction, and by choosing the appropriate temperature, time and acid concentration, some reagents can be made specific for a particular sugar. Dische developed in this way the reaction between carbazole and uronic acids, for example galacturonic acid, $HOOC(CHOH)_4CHO$, in sulphuric acid at 100°C, obtaining a pink colour which increased in intensity over a period of 2 hours; addition of water caused the pink colour to disappear within 3 minutes but resulted in formation of a stable violet colour.

In this reaction glucuronic acid and galacturonic acid, in the free state as well as combined in polyuronides, react with equal intensity. By decreasing the temperature of reaction to 60°C, Dische (1950) found that the intensity of colour depends not only upon the structure of individual hexuronic acids but also upon the specific linkage of the polyuronides. Briefly, Dische's procedure is to add the sugar solution to diluted sulphuric acid cooled in ice and then to warm the solution to 60°C for $1\frac{1}{2}$ minutes, again cool and add the carbazole reagent. The extinction of the purple-coloured solution produced is measured after 1 hour at 527 nm. Protein present in excess of 0·02% interferes and such solutions should be diluted first; if dilution is not practical, a known excess of galacturonic acid is added. Dische gives a table of colour intensities against various hexuronic acids and polyuronides.

Lynch and others (1957) determined uronic acids in the fulvic acid fraction of soil organic matter using a modification of Dische's method in which the acids are reacted with carbazole in sulphuric acid after reduction of iron(III) with tin(II) chloride solution.

Dormaar and Lynch (1962) extracted uronic acids from soil by shaking with 0·5 M sodium hydroxide solution and treating the residue with a mixture of 1 M hydrofluoric and hydrochloric acids before extracting nine times with hot (70°C) 0·5 M sodium hydroxide solution. The alkaline extracts were combined and humic acid precipitated with 3 M sulphuric acid; the filtrate was analyzed for uronic acids by the method of Lynch *et al.* (1957). The acid (HF/HCl) extract of soil was also analyzed for uronic acids after removal of iron with an exchange resin.

Lignin

Swaby and Ladd (1962) estimated lignin from the amounts of vanillin, syringaldehyde and *p*-hydroxybenzaldehyde obtained when humic acid is oxidized with alkaline nitrobenzene. Lignin can be determined directly in organic material by the method of Moon and Abou-Raya (1952). In this procedure the finely ground material is extracted with a mixture of benzene and ethanol to remove fats, waxes and resins and is then refluxed for 1 hour with 5% hydrochloric acid to hydrolyze polysaccharides and hemicellulose. The residue is treated with sodium carbonate solution and trypsin and maintained at 38°C for 18 hours to destroy protein, and then treated with sulphuric acid to remove cellulose. The final residue is filtered through a Gooch crucible, the lignin destroyed by ignition and its amount found from the loss in weight. A duplicate sample is not ignited but used for nitrogen determination in order to correct (approximately) for any remaining protein.

Amino compounds

Early methods of investigating amino compounds in soils were based on van Slyke's (1911) method of protein analysis by acid hydrolysis followed by reacting the amino-nitrogen with nitrous acid to yield gaseous nitrogen,

$$R.NH_2 \xrightarrow{\text{HONO}} R.OH + N_2 + H_2O \qquad [11:6$$

Later, van Slyke and co-workers (1941) introduced the ninhydrin reaction for determining α-amino-acids and which is highly specific. Amino-acids are decarboxylated and deaminated by ninhydrin, the full name of which is triketohydrindene hydrate;

Under certain conditions the liberated ammonia condenses with the reagent but in the van Slyke method the amino-acid is estimated by determining the liberated carbon dioxide. Alternative procedures depend upon the blue colour formed during the reaction.

Acid hydrolysis methods showed (Kojima, 1947; Bremner, 1949) that up to about 40% of the total nitrogen in soil can be present as combined amino-acids and up to about 10% as 2-amino-sugars (hexosamines). Bremner (1958) and Sowden (1959) further found that most of the hexosamine nitrogen is present as glucosamine and galactosamine. Using paper and ion-exchange chromatography, many individual amino-acids have been identified in soil hydrolysates (Bremner, 1965).

Hydrolysis of the soil is carried out with boiling, dilute mineral acids, for example 6 M hydrochloric acid or 3 M sulphuric acid and in the classic procedures excess acid was removed after hydrolysis by vacuum distillation before further analysis. This evaporation requires time and attention and the most recent procedures involve neutralization of excess acid rather than removal.

Aliquots of neutralized hydrolysate are then analyzed for different classes of nitrogen compound and Bremner determines total nitrogen, ammonium-nitrogen, hexosamine-nitrogen, serine and threonine-nitrogen and amino-acid-nitrogen. In each case the nitrogen compound is converted to ammonia which is steam distilled using the apparatus shown in Figure 10:9.

The total nitrogen in the hydrolysate is measured by the usual Kjeldahl procedure and ammonium-nitrogen is separated from amino-nitrogen by a short-time distillation from magnesium oxide. Hexosamines and ammonium are determined together by steam distilling in the presence of a phosphate–borate buffer at pH 11·2. Acid hydrolysis causes a slight loss of hexosamines but this can be allowed for by a correction factor. Serine and threonine are determined after removal of hexosamines by adding periodate to convert the nitrogen to ammonium and meta-arsenite to reduce excess periodate. The amino-acid nitrogen is recovered by steam distilling with the phosphate–borate buffer after treatments with sodium hydroxide at 100°C to decompose hexosamines and remove ammonium, and with ninhydrin at pH 2·5 to convert α-amino-nitrogen to ammonium.

The glucosamine and galactosamine in soil hydrolysates can be estimated directly by the colorimetric method of Elson and Morgan (1933) in which the amino-sugar is heated with alkaline acetylacetone and then treated with an acid alcoholic solution of p-dimethylamino-benzaldehyde when a characteristic red colour is produced. Interfering substances are removed from the hydrolysate by passing it through an anion exchange resin and eluting with a dilute sodium hydrogen carbonate solution; the leachate is then passed through a cation exchange resin and the amino-sugars recovered by elution with 2 M hydrochloric acid.

11:2:5 Acidity in organic matter

The acidity of the humic and fulvic fractions of humus is due to dissociable hydrogen in aliphatic and aromatic carboxyl groups, in phenols and in alcoholic hydroxyl groups. As this dissociable hydrogen is responsible for the cation exchange capacity of organic matter, there is sometimes need to measure it.

Schnitzer and Gupta (1965) developed the calcium acetate method of Blom *et al.* (1957) in order to measure the carboxyl groups by ion exchange. In their procedure the soil organic matter is shaken with 0·5 M calcium acetate solution, made with carbon dioxide-free water, for 24 hours at room temperature. The filtrate and washings are then titrated potentiometrically with standard sodium hydroxide solution to pH 9·8 and the carboxyl groups calculated as milliequivalents per gramme organic matter.

Total acidity was measured by Schnitzer and Gupta by an adaptation of the method of Brooks and Sternhell (1957) by equilibrating with barium hydroxide solution. The organic matter is shaken with 0·125 M barium hydroxide solution in an atmosphere of nitrogen and the filtrate titrated to pH 8·4 with standard hydrochloric acid. The total acidity, that is, carboxyl plus hydroxyl hydrogen, is calculated as m.e./g organic matter and the phenolic hydroxyls obtained by difference between total acidity and carboxyl groups. Schnitzer and Gupta compared the calcium acetate method for carboxyls with the standard decarboxylation procedure (e.g. Hubacher, 1949) where the organic matter is treated with basic copper carbonate in quinoline, and they found the results to agree sufficiently for the more simple method to be adopted. The method for total acidity was checked against the method of discontinuous titration (Pommer and Breger, 1960; Schnitzer and Desjardins, 1962). In this procedure organic matter as an aqueous suspension is mixed with 1 M potassium hydroxide solution and allowed to stand overnight in an atmosphere of nitrogen. 1 M hydrochloric acid is added and the mixture diluted; the potassium chloride formed sharpens the end-point and depresses activity changes. Aliquots are placed in glass-stoppered flasks and multiples of 0·2 cm³ hydrochloric acid added. The pH of each mixture is determined after 30 minutes and then after 1, 2, 7, 14, . . . days. The amount of acid used is plotted against pH on a semi-logarithmic scale and two straight lines intersect at equivalence point. The barium hydroxide equilibrium method gave results in sufficiently good agreement with the more tedious discontinuous titration method to warrant its adoption as a routine.

11:2:6 Carbonyl groups in organic matter

As the carbonyl and quinone groups in organic matter probably affect the oxidation–reduction potential of soils Schnitzer and Skinner (1965) developed procedures for determining these groups. The fulvic acid fraction of soil organic matter was separated and 2,4-dinitrophenylhydrazones, phenylhydrazones, semicarbazones and oxime derivatives were prepared. The four derivatives were then analyzed by infra-red, ultra-violet and visible spectroscopy, X-ray and ultimate analysis. Fritz and others (1959) proposed a method based on the reaction between organic matter in methanol–2-propanol and excess hydroxylamine, followed by potentiometric back-titration of unreacted hydroxylamine with perchloric acid. In the method humic material is mixed in a glass-stoppered conical flask with 2-diamino-ethanol solution and hydroxylamine hydrochloride solution. The mixture is heated on a steam bath, cooled and titrated with perchloric acid using a glass–calomel electrode system. A blank titration is necessary and milli-volts are plotted against cm^3 of acid to determine the end-point. Carbonyl groups are then calculated from

$$\text{(blank titre − actual titre)} \times \text{concentration of acid} \times 1000/\text{wt. in mg} = \text{m.e. CO/g.} \quad [11:7$$

Schnitzer and Skinner (1966) used the method of Fritz *et al.* and compared the results with those obtained by preparing the 2,4-dinitro-phenylhydrazone; the hydrazone was determined by measuring the excess of reagent polarographically.

11:2:7 Organic matter decomposition

Decomposition of organic matter in soil is a biological process and thus soil biology must be considered when investigating the subject of humus and soil fertility. The mode and rate of decomposition of soil organic matter are subjects too vast to be considered here in any detail but the practical laboratory procedures for examining these processes are dealt with at some length. This is because most soil laboratories are now includ-ing such measurements in their routine, usually as part of fertility investiga-tions, and so many different techniques have been advocated that a thorough knowledge of the factors involved is deemed necessary for the modern student of soil analysis.

The fundamental change occurring during the decomposition of organic matter is uptake of oxygen and liberation of carbon dioxide, the hydrogen attached to carbon atoms providing the necessary energy for the micro-flora. Part of the decomposed organic matter is re-assimilated by the micro-organisms to form protoplasm and a certain fraction of the organic matter, being resistant to microbial attack, remains undecomposed.

Early methods of measuring the rate of organic matter decomposition ignored the uptake of oxygen and relied upon periodic determination of the amount of carbon dioxide evolved. Thus Appelman (1927) covered a definite area of soil with a container of known volume for a specific period of time and measured the carbon dioxide produced; Boynton and Reuther (1938) pumped out the soil atmosphere for analysis. In laboratory studies the usual basic procedure was to keep a sample of soil in a vessel which also contained a vial of carbon dioxide absorbent. After a definite period of time the carbon dioxide absorbed was determined. In its simplest form the apparatus consisted merely of a bell-jar covering the soil sample and a dish of absorbent—sodium hydroxide solution for example (Neller,

1918)—and such a procedure was used by Meyer *et al.* (1959) to study carbon dioxide evolution in the field. In the laboratory any desired conditions of temperature, moisture, substrate and so on can be imposed. Lees (1949) used the more elaborate apparatus, one half of which is shown in Figure 11:5. Two reaction vessels with glass taps communicating with

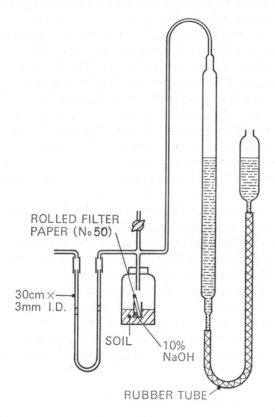

ROLLED FILTER
PAPER (No 50)

30cm × 3mm I.D.

SOIL

10% NaOH

RUBBER TUBE

Figure 11:5 Apparatus for studying organic matter decomposition (Lees, 1949; *Pl. Soil*)

the atmosphere are connected by a manometer; to one vessel is attached a graduated side arm containing water or mercury. One vessel is left empty and the other contains the soil and a vial of sodium hydroxide solution; a rolled piece of filter paper serves to increase the area of hydroxide solution to improve absorption. At the start of the experiment both taps are open, the manometer fluid level and the liquid in the graduated side-arm set to some chosen mark. The taps are then closed, and any gas volume change in one vessel not paralleled by a similar change in the other is reflected by the manometer. The apparatus had to be periodically dismantled to renew the oxygen supply.

Elkan and Moore (1962) incubated samples of freshly sieved soil at 30°C in stoppered conical flasks. The flasks had a built-in centre well in which was placed barium hydroxide solution that was subsequently titrated with hydrochloric acid. Atmospheric carbon dioxide in the flasks was corrected for by a blank.

The main drawback to incubation procedures for following organic matter decomposition is the gradual depletion of oxygen which affects microbial activity. Cornfield (1961) utilized barium peroxide to renew the diminishing oxygen supply during respiration. A solution of barium peroxide, being slightly alkaline, absorbs carbon dioxide and liberates oxygen; at the conclusion of the experiment the vial of peroxide was placed in a Collins's calcimeter (section 5:2:1) and the absorbed carbon dioxide determined. For incubation vessels Cornfield used large test-tubes which were arranged in racks inclining at 20 degrees from the vertical so as to increase the surface area of the peroxide solution. 0·2 g of barium peroxide was found to cope with 12 mg of carbon dioxide per 24 hours without oxygen becoming a limiting factor for respiration, and the solution could absorb up to 50 mg of carbon dioxide.

Clement and Williams (1962) incubated samples of soil at 75 cm water tension in flasks which also carried test-tubes containing potassium hydroxide solution to absorb the carbon dioxide (Figure 10:19). The flasks were connected in series to pure oxygen at atmospheric pressure which was drawn into the flasks as carbon dioxide was absorbed.

A different technique to allow for oxygen depletion was to keep the soil in a vessel and draw over its surface a slow stream of carbon dioxide-free air. The air was then passed through a carbon dioxide absorbent and the carbon dioxide determined at known intervals of time. Gainey (1919) used the apparatus shown in Figure 11:6 and which could be joined in series to similar units; the carbon dioxide was determined by titration with standard acid. Waksman and Starkey (1924) used the more simple set-up

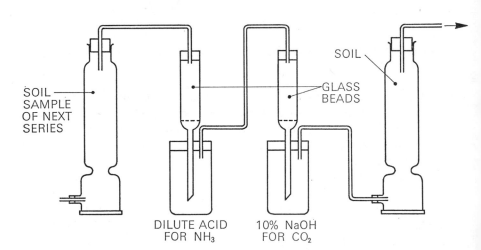

Figure 11:6 Apparatus for measuring carbon dioxide evolution from soil (Gainey; *Soil Science* © 1919)

shown in Figure 11:7 and Heck (1929) used the apparatus shown in Figure 11:8 which included a solution of barium hydroxide in the train to saturate the incoming air with moisture and simultaneously to indicate when the soda-lime in A had lost its efficiency. Vandecaveye and Katznel-

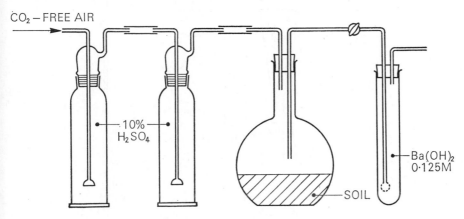

Figure 11:7 Apparatus for measuring carbon dioxide evolution from soil (Waksman and Starkey; *Soil Science* © 1924)

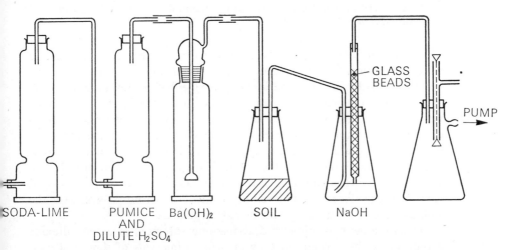

Figure 11:8 Apparatus for measuring carbon dioxide evolution from soil (Heck; *Soil Science*, © 1929)

son (1938) used Heck's apparatus but kept the soil in brown glass bottles and at field capacity. Marsh (1928) used two soil reservoirs in each unit of his apparatus (Figure 11:9) in order to combine the principles of drawing air over the surface of the soil and also through it, so as to remove accumulated carbon dioxide. Marsh set up his apparatus in sets

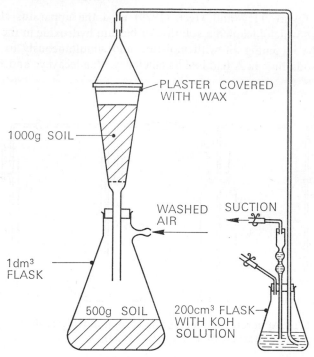

PLASTER COVERED WITH WAX

1000g SOIL

WASHED AIR   SUCTION

1dm³ FLASK

500g SOIL   200cm³ FLASK WITH KOH SOLUTION

Figure 11:9 Apparatus for measuring carbon dioxide evolution from soil (Marsh; *Soil Science* © 1928)

of about twenty units in series and air was drawn through by means of a water respirator, being purified by passing through water, soda-lime, potassium hydroxide and barium hydroxide before reaching the soil. Johnson and Guenzi (1963) incubated soil in a chamber through which water-saturated air was passed and Bartholomew and Broadbent (1950) controlled the moisture, temperature and composition of the air being passed over the soil sample. For carbon dioxide absorption Bartholomew and Broadbent used pettenkoffer tubes. The apparatus is shown in Figure 11:10 from which it can be seen that air of any desired composition and carbon dioxide-free, has its humidity controlled before passing over the soil; the rate of air flow is controlled by capillaries.

In all the methods hitherto described measurement of carbon dioxide evolution can be carried out only at definite intervals of time, not continuously, and they do not give a complete picture of organic matter decomposition. For many reasons it is desirable to know the quantity and rate of oxygen uptake during decomposition; this in conjunction with periodic determination of carbon dioxide evolved permits the calculation of the 'respiratory quotient', CO_2/O_2, which gives some indication of the type of organic material decomposing.

Several workers have used the Warburg (1926) technique for measuring oxygen uptake by soil. The main drawback to this procedure, apart from

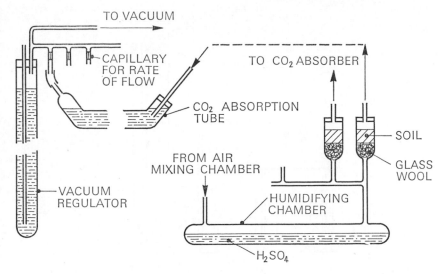

TO VACUUM

CAPILLARY
FOR RATE
OF FLOW

TO CO₂ ABSORBER

CO₂ ABSORPTION
TUBE

SOIL

GLASS
WOOL

FROM AIR
MIXING CHAMBER

VACUUM
REGULATOR

HUMIDIFYING
CHAMBER

H₂SO₄

Figure 11:10 Respirometric apparatus with controlled moisture and air composition (Bartholomew and Broadbent, 1950; *J. agric. Sci.*)

the expensive equipment, is the small size of sample that can be examined. A quick exchange with the air in the flask takes place and after a short time the conditions under which the micro-organisms are working are quite different from those in the field. Thus the Warburg apparatus is limited to small quantities of soil for short periods of time. Rovira (1953) used the Warburg technique for a short-term study of soil respiration and Ross (1964) investigated the effect of freezing soils upon their subsequent decomposition by incubating in a Warburg apparatus for up to 2 hours. Pauli (1965) used a Warburg-type apparatus adapted for larger flasks than usual.

Swaby and Passey (1953) introduced an electrolytic soil respirometer for measuring oxygen uptake utilizing the principle of Helvey (1951) for continuous replacement of oxygen by electrolysis as decomposition proceeds. The apparatus, Figure 11:11, consists of a beaker holding 50 g or more of soil, in a container with a lid held in place with bolts. Sodium hydroxide solution is placed in the bottom of the container to absorb the evolved carbon dioxide and a tube with a platinum electrode connects the container to a solution of sulphuric acid. Another electrode is immersed in the acid, being placed beneath an acid-filled, inverted, graduated cylinder. As the carbon dioxide is absorbed it creates a partial vacuum in the container and acid is drawn up the tube until it makes contact with the electrode. Electrolysis occurs and oxygen is liberated into the container until the pressure is restored when the acid sinks to its original level and electrolysis ceases. Hydrogen liberated at the same time at the cathode can be measured at any time by reading its volume in the cylinder. Swaby and Passey subsequently modified the soil container so as to sample the

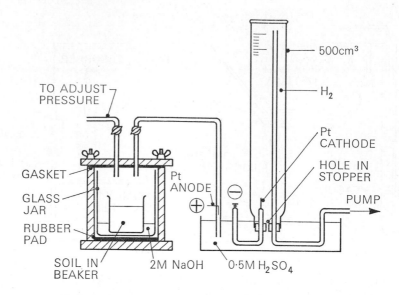

Figure 11:11 Soil respirometer of Swaby and Passey (*Aust. J. agric. Res.*, 1953)

alkali more easily (Figure 11:12). With the right-hand tap closed and the left-hand tap open, sodium hydroxide solution can be sucked out into a titration flask.

Birch and Friend found the apparatus of Swaby and Passey unmanageable. The blank readings obtained when the large container expanded or contracted with changes in temperature were too high for readings to be reliable and bolting down the lid each time made the apparatus unsuitable

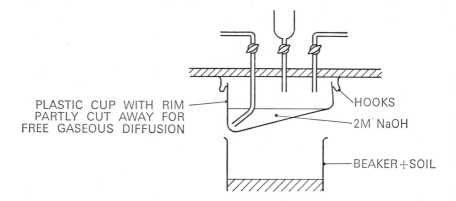

Figure 11:12 Modified CO_2 absorption apparatus to permit easy access to the sodium hydroxide solution in the apparatus of Swaby and Passey (*Aust. J. agric. Res.*, 1953)

for a large-scale experiment. Birch and Friend (1956) thus devised the apparatus shown in Figure 11:13 which employed the electrolytic device of Helvey and yet was suitable for routine laboratory investigations. The apparatus is easily and quickly set up from common laboratory glassware and no expensive, specially made parts are required. As originally described the apparatus had home-made platinum electrodes sealed in plastic, and

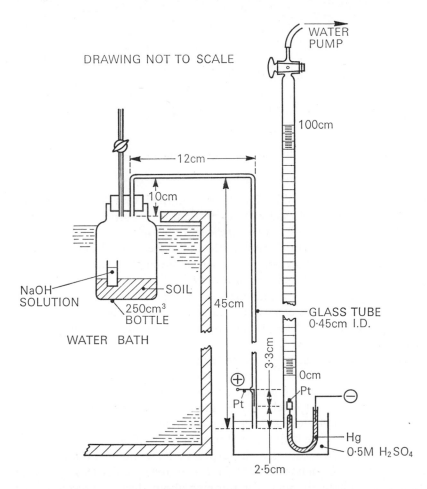

Figure 11:13 The macro-respirometer of Birch and Friend

this was because Birch and Friend had to make do with old platinum crucible lids as a source of platinum. Although the home-made electrodes work satisfactorily, they need constant repair to prevent gas leaks and are as difficult to make as sealed-in glass electrodes; platinum-in-glass electrodes are thus shown in the figure.

The soil sample was brought to the desired moisture content in a wide-mouth bottle and a specimen tube containing sodium hydroxide

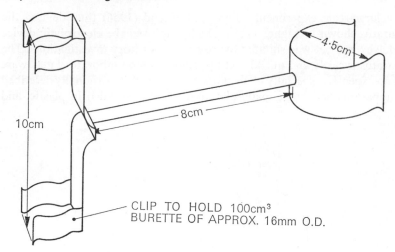

Figure 11:14 Dual clip for holding burette and sample bottle in the Birch–Friend respirometer

solution was placed inside. Experiment showed that the compression effect of the tube on the soil was insignificant. The bottle was closed with a rubber cork carrying the oxygen delivery tube and placed up to the neck in a constant temperature water bath. Full experimental details for use of the Birch–Friend macro-respirometer are given in section 11:3:7. The apparatus, which could still be improved in details, is ideally suited for measuring oxygen uptake and carbon dioxide evolution and for continuously following the course of organic matter decomposition over any length of time, with the exception of very short periods when the Warburg technique can be followed.

The soil sample can be placed in the apparatus and maintained under conditions in which it was collected, the nature of the apparatus automatically maintains the moisture content of the soil, the oxygen content is kept at a constant level, carbon dioxide does not accumulate, and accurate, continuous measurement of oxygen uptake is possible by taking burette readings of the hydrogen volume. In point of fact, one reading per day is sufficient for most research purposes and, as pointed out by Birch, it takes about 10 minutes to obtain twenty-four results leaving, in an average day, about $6\frac{3}{4}$ hours for thinking about them. By virtue of the thermostat any desired temperature can be maintained, light can be admitted or not, and at any desired time portions of soil can be removed for analysis.

One modification made to the Birch–Friend macro-respirometer is to include a stopcock in the rubber cork closing the sample bottle. This permits the introduction of gases other than oxygen and enables the initial acid levels to be adjusted before starting an experiment to ensure uniformity of conditions; otherwise, differences due to the depth to which the tube dips into the acid affect the results.

Lees (1950) devised a percolating respirometer which he called a 'rocking percolator' and Greenwood and Lees (1956) used a modification of the apparatus to study the breakdown of amino-acids. The apparatus is depicted in Figure 11:15 and was made to gently rock between the

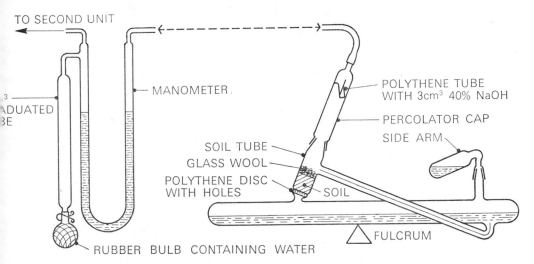

Figure 11:15 The rocking percolator of Greenwood and Lees (*Pl. Soil*, 1956)

horizontal and 60 degrees. The polythene alkali cups had hinged lids which, by bending backwards, jammed the tubes into position. The manometer was permanently attached to the percolator cap and the apparatus was set up with a pair of units being connected via the manometer. Samples of soil were placed in the tubes of both units and water was placed in each main tube. The apparatus was then put on a rocking table, the percolator caps fixed and rocking commenced. When the apparatus had settled down and the manometer reading was steady, the side arms were removed and the material under analysis, for example amino-acid, was placed in one and pure water in the other. Most of the water in the graduated tube was run into the bulb, the side arms were replaced and the apparatus rocked for 1 hour to even up temperature differences. The bulb clip was then released and the manometer levelled. By turning the side arms through 180 degrees the amino-acid solution and water were tipped into the apparatus and rocking commenced. The volume of water needed to again level the manometer was a measure of oxygen uptake and aliquots of the percolate could be analyzed at any time.

Subsequently Greenwood and Lees (1959) improved the design of the apparatus so as to permit electrolytic replacement of oxygen. The side arm was replaced by a rubber turnover closure, as used for vaccines, enabling the percolate to be sampled with a hypodermic syringe; the manometer was redesigned as shown in Figure 11:16. Oxygen is evolved at electrode A, copper being precipitated on the copper wire, and hydrogen is measured in the burette. The oxygen evolved at electrode B is allowed to escape into the atmosphere.

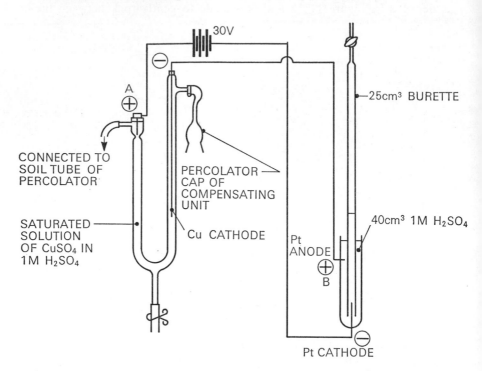

Figure 11:16 Redesigned manometer for the rocking percolator to permit electrolytic replacement of oxygen (Greenwood and Lees, 1959; *Pl. Soil*)

Stotzky (1960) discounted all previously described respirometers as useless for his particular line of research. He required rigid moisture control and periodic measurement of carbon dioxide for which he considered the electrolytic type of apparatus unsuitable. Stotzky also suspected that bases such as methane, hydrogen and so on were involved in his experiments. The apparatus designed by Stotzky to meet his requirements is shown in Figure 11:17 and was composed of two sections connected by a ground glass joint, a bottom section with a side arm and a top section to join on to the manometer. The apparatus was autoclaved before use and kept plugged with sterile cotton until joined to the manometer. Additions of substrate and water were made to field capacity and the vessels were maintained for 24 hours at 3°C; the vessels were then incubated in a water bath at 25°C. Carbon dioxide collectors were attached to the vessels and consisted of towers containing glass beads and filled with sodium hydroxide solution. After connecting with the manometer manifold, carbon dioxide-free air saturated with water was drawn slowly through the apparatus.

For respiratory quotient measurement the carbon dioxide collectors were removed and the alkali washed out into (according to Stotzky) tumblers with carbon dioxide-free water, and the carbon dioxide was determined by titration. The bottom sections of the vessels were then removed, a dish of potassium hydroxide solution placed on the plastic shelf and the re-assembled apparatus returned to the water bath. Oxygen

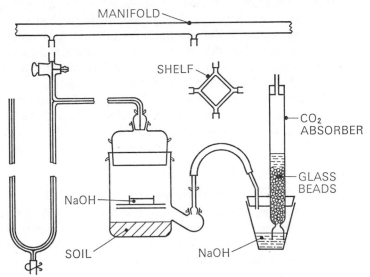

Figure 11:17 The soil respirometer of Stotzky (Reproduced by permission of the National Research Council of Canada from *Can. J. Microboil.*, 6,441, 1960)

uptake could be measured by taking manometer readings every hour, the tap being opened after each reading to replenish the oxygen. After 8 hours the potassium hydroxide solution was emptied into tumblers of carbon dioxide-free water and the carbon dioxide determined. Stotzky found that the passage of carbon dioxide-free, water-saturated air counteracted any tendency of water loss into the alkali, but this would seem to be the only real advantage over the very much simpler apparatus of Birch and Friend who, in any case, found no effect of water loss upon their results.

Freytag and Igel (1964) used a respirometric technique to follow microbial decomposition of organic matter added to soil. However, they used ^{14}C-labelled organic matter and in their apparatus measured radioactive carbon dioxide by intermittent determination of the activity of the carbon dioxide-absorbing chamber or by continuous measurement of the radioactivity of the absorbing solution (sodium hydroxide) or of the carbon dioxide in a system of steady flow.

Carbon dioxide evolved from decomposing organic matter in soil was also measured by Macura (1960) in his continuous flow percolation apparatus (Figure 10:24) by titration of the sodium hydroxide solution in the wash bottle.

Pauli (1965) considered the study of the intensity of microflora taking part in the decomposition process. He recognized that biological activity has often been studied by enzymatic tests but pointed out that this is not an estimation of total biological activity. Thus Pauli investigated the usefulness of the dehydrogenase test (formation of formazan), as opposed to oxygen consumption, and found that, although the test gave a quick overall enzyme indication of biological potentialities, the oxygen uptake procedure was better, particularly if done in conjunction with measurement of humus turnover by acriflavine absorption (Pauli and Grobler, 1961).

To determine the carbon dioxide absorbed in sodium hydroxide solution, the best technique is rapidly to make the alkali to standard volume by washing into a flask with carbon dioxide-free water and to pipette two equal aliquots into small flasks. One aliquot is titrated directly with standard hydrochloric acid using methyl orange as indicator and the other is titrated after precipitating carbonate with barium chloride solution and using phenolphthalein as indicator. The difference in titres is a measure of carbon dioxide.

Factors affecting measurement of organic matter decomposition

The decomposition of organic matter is affected most importantly by oxygen tension, temperature, moisture, pH, texture and presence of salts. In general, decomposition is accelerated by a reasonable increase in temperature but in most methods of determining organic matter breakdown temperature can easily be controlled.

An increase in acidity normally reduces organic matter decomposition owing to adverse effect upon the micro-organisms and this must be taken into account if the soil receives any pre-treatment which may alter its pH value. For example, a soil which becomes strongly acid when dried (section 13:1) will not show a natural decomposition pattern upon subsequent incubation.

Organic matter decomposes over a wide range of moisture content from less than wilting point to saturation. In very wet soils the rate of decomposition falls off due to exclusion of oxygen, and in waterlogged soils decomposition is quite slow as the micro-organisms are limited to the anaerobes. Ekpete and Cornfield (1965) incubated soils at 28°C for 6–8 weeks at various moisture contents from air-dry to waterlogged. They found that carbon dioxide evolution increased with water content up to about 60–80% and then decreased again. Birch (1958) in a similar experiment concluded that field capacity is the optimum moisture regime for organic matter decomposition. Stanford and Hanway (1955; section 10:2:5) strongly recommended studying organic matter decomposition at a constant moisture tension rather than constant moisture content. Bhaumik and Clark (1947) found maximum carbon dioxide evolution over a wide range of moisture tension, 0·1–310 kN m^{-2} (1–3160 cm water) according to the texture of the soil and Clement and Williams (1962) found a tension of 8 kN m^{-2} (76 cm water) optimum for a sandy loam.

To avoid the effect of the flush of decomposition caused by drying, Clement and Williams studied low moisture tension rates by freeze-drying the soil; it is open to question, however, if this is any more suitable than oven-drying.

Birch (1959a) found that the presence of salts in a soil depressed the solubility of organic matter. Birch moistened soil with water and with solutions of sodium, potassium, calcium and magnesium chlorides and compared their rates of decomposition using the macro-respirometer. Most carbon dioxide was evolved from the sample moistened with water, although Birch found no evidence of a concomitant reduction in nitrogen mineralization due to salts. Johnson and Guenzi (1963) made a more thorough study of the salt effect upon carbon mineralization. Soils were first leached with water until residual salts contributed less than 1/30 MN m^{-2} or 1/3 bar (as osmotic tension) to the total soil moisture stress at the moisture content of the soil at which the experiment was conducted. (It should be remembered that this treatment would remove most of the easily-soluble and oxidizable organic matter from the soils.) The soils were then dried and sieved. Salt solutions were made to cover a range of osmotic tension from 0 to 4 MN m^{-2}; sodium chloride, sodium sulphate and mixtures of approximately equivalent amounts of potassium, sodium, magnesium, chloride and sulphate were used. The salt solutions were added to the soils as a fine spray from an atomizer and the soils were incubated at 30°C with a current of water-saturated air passing. Air, having passed over the soil, was bubbled through standard sodium hydroxide solution and the absorbed carbon dioxide determined. Johnson and Guenzi concluded that the salt effect is most probably due to the effect upon moisture tension and salt concentrations should be related to osmotic values. Generally an increase in osmotic tension reduced carbon dioxide evolution and particularly reduced nitrification (c.f. Birch, who found no effect upon nitrification) in a linear manner as salt concentration increased.

The effect on organic matter decomposition of drying and re-wetting a soil has been partly discussed in section 10:2:5. The treatments stimulate decomposition by partial sterilization, rejuvenation of the population of micro-organisms and by an effect upon organic matter solubility. Saunder and Grant (1962) observed very little flush in decomposition when the soils they were studying were dried and re-wetted; only if they were crushed. Saunder and Grant suggested that the dramatic effects found by Birch were due to the allophanic nature of the soils rendering the clay mineral effect important, and they considered that the flush of decomposition may be due to comminution rather than to wetting and drying. Stevenson (1951) found that in a re-moistened soil the soluble organic matter produced contained considerable quantities of amino-acids and this led Griffiths and Birch (1961) to expect the development of a specialized zymogenous population. Griffiths and Birch followed the course of carbon dioxide evolution from a re-wetted soil for 36 hours and made

9*

bacterial counts every 3 hours. The two main cell types, cocci and rods, showed a marked difference in their distribution during the 36 hours (Figure 11:18) and the rods, which were the zymogenous population,

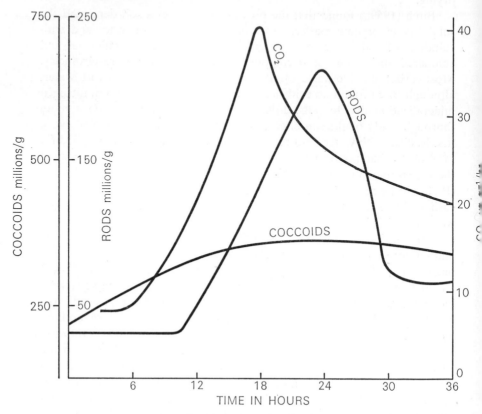

Figure 11:18 Carbon dioxide production and development of bacterial population in moistened soil (Griffiths and Birch, 1961; *Nature, Lond.*)

appeared to be responsible for most of the increased carbon dioxide production.

Skyring and Thompson (1966) studied the availability of organic matter in dried and non-dried soils which had been irradiated with gamma rays to eliminate the effect of the indigenous soil population. They concluded that release of available organic matter during drying was the major factor responsible for increased rates of decomposition in air-dried soils and that changes in the microbial population make no significant contribution.

Although the precise explanation of the drying effect has yet to be agreed upon, there is no doubt that the effect exists and must be taken into account when interpreting the results of organic matter decomposition experiments. As found by Birch (1959b), the longer a soil is stored in a dry state before re-moistening, the larger will be the effect upon subsequent decomposition. Clement and Williams (1962) stored soil at −15°C for up to 2 years and found that this low temperature storage did not affect subsequent carbon dioxide production, and Ross (1964) stored soil at $-20°C$ for 42 weeks with similar results. Ross admitted that it was possible that the cold storage merely reduced the numbers of viable organisms which then counterbalanced the effect of increased decomposable substrate, but thought it more likely to be a simple matter of no effect at all. Hayes and Mortensen (1963) recommended removing a whole column of soil from the field and incubating it immediately to avoid the effects of drying.

Fortunately the drying effect becomes manifest quickly, within about 24 hours of incubation, and usually the decomposition pattern returns to normal after about 15 days. This is demonstrated in Figure 11:19

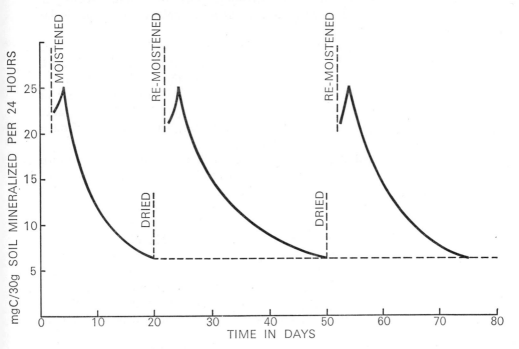

Figure 11:19 The effect of repeated drying and wetting cycles upon mineralization of soil carbon

showing the effect of repeated cycles of wetting and drying a soil upon its decomposition. Thus it is a simple matter to test whether a soil is exhibiting the effect or not by repeating the drying and wetting once more, preferably on a duplicate sample. New experiments can then be commenced at the end of the artificial flush; furthermore the effects of pre-treating a soil can be quantitatively measured with a respirometer and allowed for. If fertility investigations are being undertaken the importance of alternating dry and wet seasons must not be neglected. The behaviour of a soil as regards its mineralization may be very different at the start of a rainy period than later on and this dynamic aspect must be considered in conjunction with laboratory analyses.

11:2:8 Carbon–nitrogen ratios

Vegetable and animal residues added to soil contain more carbon than nitrogen and as long as this carbonaceous material persists in the soil, very little nitrogen will be released. As the raw organic matter decomposes, carbon is lost as carbon dioxide and the C/N ratio gradually narrows. It is generally considered that for most soils the ratio eventually stabilizes at about 10:1. If soil temperature and microbiological activity are high, the ratio may narrow even more. As mineralization of organic matter and immobilization by micro-organisms occur simultaneously in soil, the C/N ratio is one way of expressing the balance at any particular time.

Leighty and Shorey (1930) questioned the ratio of 10/1 as an average value; they analyzed 172 samples and found that the C/N ratios varied from 35/1 to about 3/1. Leighty and Shorey further pointed out that the C/N ratio by itself is not sufficient as an indication of the amount of available organic matter. They quote the example of a high, 16/1, ratio, indicating the presence of decomposable matter, being obtainable from a soil with a very low carbon content, 0·3%.

Anderson and Byers (1934) concluded that variations in C/N ratios are so wide as to make the determination of nitrogen useless as a means of estimating carbon or organic matter (see section 11:2:2), the reason being the difference in kind of organic matter. Young (1962) pointed out that when calculating C/N ratios, allowance must be made for inorganic nitrogen present, which in some cases may make a big difference to the result. Young found that in the soils he examined the organic-C/organic-N ratios varied from 5 to 21 and more than half the values were outside the range 8·5–11·4 and thus he did not accept the principle of all soil humus having more or less constant composition and being stable at a C/N ratio of 10. Fixed ammonium should be allowed for when determining C/N ratios.

When calculating the ratio the carbon figure should represent total organic carbon and thus the values obtained by partial oxidation techniques must be multiplied by the appropriate factor (11:2:3).

11:3 RECOMMENDED METHODS

11:3:1 Total carbon

The total carbon content of a soil is seldom determined but, when necessary, the analysis is best made by the dry combustion method; if the special apparatus required for dry combustion is not available the wet combustion procedure can be followed.

I / BY DRY COMBUSTION (STEWART *et al.*)

Apparatus

The combustion and post-heating apparatus is shown diagrammatically in Figure 11:2. Incoming oxygen is scrubbed with sulphuric acid and is then passed through ascarite and magnesium perchlorate to remove any carbon dioxide and hydrocarbons. The post-heater tube is half filled with silver wool to remove halogens and half with a 3·1 v/v mixture of copper oxide and a platinum reforming catalyst. The platinum catalyst is commercially available (see Stewart *et al.*, 1964). The contents of the post-heater tube must be renewed after every thirty determinations and the platinum catalyst can be regenerated by washing with 0·1 M acetic acid and heating at 850°C for 30 minutes in a current of oxygen. The post-heater tube is connected to a tube containing a layer of magnesium perchlorate over manganese dioxide to remove water and oxides of nitrogen and sulphur. An absorption weighing tube containing magnesium perchlorate over ascarite is attached by means of a hypodermic syringe needle.

The combustion tube is heated by two furnaces which can move round the tube. The upper furnace heats the tube above the sample and can be lowered to heat the sample area when necessary; the lower furnace heats the area beneath the sample.

Procedure

0·15-mm soil is dried overnight at 70°C. Weigh out 0·05–0·5 g of dry soil, depending upon organic matter content, and mix with 3 g of finely ground copper(II) oxide. Fill the combustion tube with wire-form copper(II) oxide to within 13·5 cm from the top, add the sample and fill the tube with more copper(II) oxide. It is important to position the sample correctly. Pack the ends of the tube with Pyrex glass wool.

Connect up the apparatus and purge with pure oxygen for 2 minutes at a rate of 5 cm³ per second. Heat the combustion tube above and below the sample for 120 seconds at 875°C maintaining the post-heater tube at 600°C. Lower the upper furnace to the sample area and stop the oxygen current for 100 seconds. Re-start the oxygen and bring the temperature to a maximum of 1000°C for 67 seconds. Stop the heating and pass oxygen for 2 minutes at a rate of 30 cm³ per second to sweep all carbon dioxide into the absorption tube. Determine the evolved carbon dioxide by weighing the tube to 0·01 mg.

Have the next sample and a weighed absorption tube ready. The procedure includes the determination of carbonates and the average relative standard deviation is 2·24%.

II / BY WET COMBUSTION (ENWEZOR AND CORNFIELD)

Apparatus

The apparatus is shown in Figure 11:4. The tube A, 2 cm × 25 cm, is filled with 1·5-mm (14–20 mesh) Carbosorb soda-lime to remove carbon dioxide from the incoming air. The tap B regulates the air flow and the flask D is of 100 cm³ capacity. The condenser E is 20 cm × 1·5 cm and the dropping funnel G is 100 cm³ capacity. The bubble counter I is filled with syrupy phosphoric acid and the absorbing solution is contained in the trap J. The tube K contains a 5-cm depth of solid 2·5-mm potassium iodide at its base, an 18-cm depth of 1·5-mm anhydrous magnesium perchlorate, and at its top, 2 cm more of potassium iodide. The U-tube M, 1·5 cm × 25 cm effective length, contains 1·5-mm Carbosorb with a 1-cm layer of Anhydrone at the inlet and an 8-cm layer at the outlet. The guard-tube P contains 2·5-mm Carbosorb.

All tubes have glass wool plugs at the ends and to separate the various reagents. All joints and tap H are lubricated with syrupy phosphoric acid and the other taps with silicone grease.

Reagents

Potassium dichromate, powdered.

Digestion acid: conc. H_2SO_4 mixed with 85% o-phosphoric acid in the ratio of 3:2 v/v.

Absorption solution: 100 g KI, 30 g sodium acetate trihydrate and 2 cm³ glacial acetic acid dissolved in 100 cm³ water.

Procedure

Weigh the U-tube M with both its taps N and O closed, as accurately as possible before fitting on to the apparatus. Open tap L and mix 0·25-mm soil containing less than 0·1 g carbon with 3–4 g of potassium dichromate and 3 cm³ of water in flask D. Lubricate the joint of the flask with phosphoric acid and fit the flask on to the apparatus. Suck air through at the rate of 3–4 bubbles per second for 5 minutes to remove carbon dioxide from the apparatus; then close tap L and open taps N and O. Reduce the air flow to 1–2 bubbles per second and slowly add 25 cm³ of digestion acid from the funnel. Heat the flask with a micro-burner until vigorous reaction commences and remove the flame. Once the initial reaction has subsided again heat the flask bringing the acid to the boil within 5 minutes. Boil for 10 minutes, discontinue the heating and bubble air through at 8 bubbles per second for 10 minutes. Stop the air flow, close taps N and O, remove the U-tube and re-weigh.

If necessary the next sample can immediately be analyzed by fitting another, pre-weighed, U-tube and using another reaction flask; there is no need to again flush out the apparatus.

Make a blank digestion using reagents only. Care must be taken to renew the absorbing reagent in good time; the colour of the potassium iodide at the top of the tube K indicates when the trap and lower portion of potassium iodide are exhausted. If significant amounts of elemental carbon are present the time of heating must be increased to 30 minutes. Carbonates, if present, can be removed if desired by treating the soil in the flask with sulphur dioxide solution and drying in vacuo over solid sodium hydroxide.

11:3:2 Total organic carbon

In non-calcareous soils the total carbon content as measured in procedures 11:3:1 can be assumed to be total organic carbon; in calcareous soils the carbonates can be determined separately and allowed for or destroyed with sulphurous acid before analysis.

11:3:3 Oxidizable organic carbon (Walkley–Black)

Reagents

Potassium dichromate solution, M/6: Pure (Analar) potassium dichromate
    is dried in an oven at 105°C and cooled in a desiccator. 49·04 g are
    accurately weighed, dissolved in water and the solution diluted to
    1 dm^3.
Concentrated sulphuric acid.
Concentrated *o*-phosphoric acid.
Barium diphenylamine sulphonate: 0·16% w/v, aqueous.
Ammonium iron(II) sulphate solution, 0·5 M (approx.): 196 g ammonium
    iron(II) sulphate are dissolved in water. Filter if necessary, add 5 cm^3
    conc. H_2SO_4 and dilute to 1 dm^3. In practice it will be found convenient
    to prepare this reagent in 5-dm^3 amounts.

Procedure

Weigh a quantity of 0·15-mm soil containing up to about 25 mg C (usually about 0·5 g to 1·0 g) into a 500-cm^3 conical flask. Add 10 cm^3 of potassium dichromate solution, using a safety pipette, and gently swirl the flask to mix the reagent with the soil. Rapidly add 20 cm^3 of concentrated sulphuric acid from a measuring cylinder and again swirl the flask for about 30 seconds. Allow the flask to stand on an asbestos pad for 30 minutes. Add about 200 cm^3 of water and 10 cm^3 of phosphoric acid and allow the solution to cool (the end-point of the titration is more easily determined with a cold mixture). Add 0·5 cm^3 of barium diphenylamine sulphonate solution and titrate the excess dichromate with ammonium iron(II) sulphate solution. Near the end-point the colour becomes deep violet-blue and at this stage add the iron(II) solution dropwise with shaking. At the end-point the colour changes sharply to green. If the end-point is overshot add 0·5 cm^3 of dichromate solution and continue the titration dropwise. (See section 11:2:3 for discussion of the titration and its difficulties.)

    If more than 8 cm^3 of the dichromate solution has been used the experiment should be repeated using less soil, unless as little as 0·5 g had been originally taken, when the experiment should be repeated using more dichromate solution.

    A blank determination made exactly as above but without soil serves to standardize the iron(II) solution.

Calculations

Per cent oxidizable, organic carbon (uncorrected)

$$= \frac{(\text{blank titre} - \text{actual titre}) \times 0{\cdot}3 \times M}{\text{weight of oven-dry soil in g}} \qquad [11{:}8$$

where M is the concentration of the ammonium iron(II) sulphate solution.

The result can be expressed as per cent organic carbon (Walkley–Black, uncorrected) or converted to total organic carbon by multiplying by 1·33 (or by an experimentally found, more applicable factor, see 11:2:3), or as per cent organic matter by multiplying by 2. The factor 2 is an approximation of 2·29 and incorporates the two assumed factors of 1·33 and 1·724 (see 11:2:2).

If a significant amount of chloride is present and is to be determined it can be allowed for by the calculation given in section 11:2:3. If the chloride is not to be determined, silver sulphate should be added to the sulphuric acid reagent at 15 g dm^{-3}.

11:3:4 Fractionation of soil organic matter (after Stevenson)

Reagents

Hydrochloric acid, 0·1 M: 9 cm^3 dm^{-3} conc. HCl.
Sodium hydroxide solution, 0·5 M: 20 g dm^{-3} NaOH.

Procedure

Place a suitable sample, say 100 g, of fine mesh soil in a sintered glass funnel and wash with 200 cm^3 of dilute hydrochloric acid in several aliquots; drain the soil by suction and dry. The filtrate will contain some soluble organic matter.

Weigh 50 g of the acid-washed soil (or less if very organic) into a polythene centrifuge bottle and shake for 12 hours with 200 cm^3 of sodium hydroxide solution. Centrifuge and decant the liquid into a beaker through a funnel plugged with glass wool. Add a further 200 cm^3 of alkali to the soil and shake the mixture for 1 hour. Meanwhile, add concentrated hydrochloric acid to the first extract until the pH is lowered to 1·0 (using a pH meter and glass electrode). Add the second alkaline extract to the acidified extract and wash the soil by centrifuging with 200 cm^3 of water, the washings being added to the extracts. Adjust the pH to 1·0 and allow the humic acid to settle.

Siphon or pipette off the bulk of the fulvic acid and transfer the residue to a centrifuge tube. After centrifuging, add the liquid to the fulvic acid and purify the solid humic by solution in 0·5 M sodium hydroxide and re-precipitation with hydrochloric acid to pH 1·0. Finally wash the humic acid free of chloride with water, freeze-dry it and powder it.

The humic and fulvic acid fractions can be further fractionated by various methods (11:2:4) depending upon purpose.

11:3:5 Determination of carbohydrates

I / PREPARATION OF HYDROLYSATE (CHESHIRE AND MUNDIE)

Reagents

Sulphuric acid, 12 M: 672 cm^3 dm^{-3} conc. H_2SO_4.
Sodium hydroxide solution, 6 M: 240 g dm^{-3} NaOH.

Procedure

Shake 10 g of 2-mm soil with 50 cm^3 of 12 M sulphuric acid for 16 hours. Dilute the suspension with 1150 cm^3 of water to make the acid 0·5 M and heat the mixture on a water-bath at 100°C for 5 hours. Filter through a No. 3 porosity sintered glass funnel and neutralize the filtrate to pH 6·8 with 6 M sodium hydroxide solution. Remove the dark precipitate by centrifuging and dilute the hydrolysate to standard volume.

II / DETERMINATION OF HEXOSES IN HYDROLYSATE (BRINK *et al.*)

Reagents

Anthrone solution: 0·2% w/v in 95% v/v H_2SO_4.
Standard sugar solution: 0·250 g dm^{-3} glucose. This contains 25 μg sugar per cm^3 and should be prepared 1 hour before use.

Procedure

Place 5 cm^3 aliquots of hydrolysate in test-tubes 150 mm × 25 mm (the size of the tubes is important and must be standardized). Add 10 cm^3 of anthrone solution and shake. After 15 minutes measure the absorption of the green solution at 625 nm. A reagent blank must be prepared and the calibration curve should cover the range 0–25 μg cm^{-3}.

III / DETERMINATION OF PENTOSES IN HYDROLYSATE (TRACEY)

Reagents

Aniline acetate solution: 100 cm^3 glacial acetic acid, 10 cm^3 of 5% w/v aqueous oxalic acid solution, 24 cm^3 of water and 16 cm^3 of colourless aniline are mixed and stored in a dark glass bottle. The reagent is stable for 1 week.
Standard pentose solution: 0·050 g dm^{-3} xylose in saturated benzoic acid solution.

Procedure

Place 6 cm^3 of aniline acetate solution in each of a series of matched spectrophotometer tubes. Add aliquots of hydrolysates containing from 10 μg to 80 μg of pentoses and sufficient water to bring the volume to 8 cm^3. Prepare a blank from 6 cm^3 of reagent and 2 cm^3 of water. Mix and keep in the dark for 24 hours; measure the absorbance of the coloured solutions at 622 nm. The calibration curve should cover the range 0–50 μg dm^{-3} pentose.

IV / DETERMINATION OF URONIC ACIDS IN HYDROLYSATE (LYNCH *et al.*)*

Reagents

Ethanol: Reflux 1 dm^3 of ethanol with 4 g of zinc dust and 4 cm^3 of 1:1 H$_2$SO$_4$ for 24 hours. Distil the ethanol and then redistil from zinc dust and KOH, 4 g of each being added per dm^3.

Carbazole: Purify the reagent by recrystallizing from toluene and dissolve 150 mg in 100 cm^3 of redistilled ethanol.

Hydrochloric acid, 0·5 M: 44·5 cm^3 dm^{-3} conc. HCl.

Sodium hydroxide solution, 0·5 M: 20 g dm^{-3} NaOH.

Sulphuric acid, concentrated.

Tin(II) chloride solution: 2·5 g SnCl$_2$ dissolved in 10 cm^3 of conc. HCl and diluted to 100 cm^3 when required.

Standard galacturonic acid solutions: Galacturonic acid monohydrate is vacuum dried over P$_2$O$_5$. Dissolve 0·1 g of the dry acid in 500 cm^3 of 0·5 M HCl and 500 cm^3 of 0·5 M NaOH solution. Prepare standard solutions from this to cover the range 10–100 μg cm^{-3} by diluting aliquots with the HCl–NaOH mixture.

Procedure

Take an aliquot of the fulvic acid fraction of the soil hydrolysate corresponding to about 0·5 g soil and evaporate if necessary to 2 cm^3. Cool the aliquot in a test-tube in an ice-bath, add 12·5 cm^3 of concentrated sulphuric acid and shake. Place the tube in a boiling water-bath for 30 minutes, cool again in ice and add 0·5 cm^3 of tin(II) chloride solution to reduce the iron(III). Add 1 cm^3 of carbazole reagent to the test solutions and 1 cm^3 of pure alcohol to the blank. Shake the mixtures and leave for 2 hours before reading the absorbance at 520 nm.

2-cm^3 aliquots of standard solutions are treated in the same manner. The standard error is $\pm 4\cdot2$ μg cm^{-3}. The method assumes that all the uronic acids are in the fulvic acid fraction of soil organic matter.

V / DETERMINATION OF CELLULOSE (DAJI; GUPTA AND SOWDEN)

Reagents

Schweitzer's solution: Dissolve 10 g copper(II) sulphate in 100 cm^3 of water and mix with a solution of 5 g KOH in 50 cm^3 of water. Collect the precipitate, wash with cold water, partially dry and macerate with 20 cm^3 of a 20% ammonium hydroxide solution for 1 day with occasional shaking.

---

* See Lowe and Turnbull (1968), *Soil Sci.*, 106, 312, for more recent modifications including removal of interference.

Ethanol, 80%.
Sodium hydroxide solution, 1% w/v, aqueous.
Hydrogen peroxide, 20 volume.
Hydrochloric acid: 30 cm^3 dm^{-3} conc. HCl.
Sodium hypochlorite solution.

Procedure

Heat 10 g of 2-mm soil in a beaker with 100 cm^3 of 1% sodium hydroxide solution; boil the suspension for 30 minutes maintaining the volume. Just acidify the mixture with hydrochloric acid and allow the soil to settle; filter and wash the residue twice with hot water. Replace the soil in the beaker and boil for 20 minutes with 100 cm^3 of the dilute hydrochloric acid; the volume should be maintained. Allow the soil to settle, filter and wash free of acid with hot water. Replace the soil in the beaker, add 5 cm^3 of sodium hypochlorite solution and bring the volume to 100 cm^3 with water. Allow to stand for 30 minutes with occasional shaking and maintain the liquid alkaline. Filter and treat the soil in a beaker with a further 5 cm^3 of sodium hypochlorite solution diluted to 100 cm^3. After standing for 30 minutes decant off the liquid and add 10 cm^3 of hydrogen peroxide. After settling, filter and wash the soil by decantation with hot water; dry the washed soil on a steam-bath.

Crush the dry soil and mix it in a 300-cm^3 centrifuge bottle with 100 cm^3 of freshly prepared Schweitzer's solution. Shake for 6 hours and centrifuge. Add 250 cm^3 of 80% ethanol to 50 cm^3 of the extract and allow to stand overnight. Decant the liquid and wash the precipitate successively with hot water, 1 M hydrochloric acid and water.

Wash the soil residue with water and extract once more with a fresh lot of Schweitzer's solution.

Combine the precipitates of crude cellulose, dry and weigh. Ignite the cellulose in a muffle furnace and re-weigh to obtain pure cellulose. Alternatively, hydrolyze a weighed portion of crude cellulose by refluxing for 5 hours with 2% hydrochloric acid and determine the glucose.

11 :3:6 Determination of amino compounds (Bremner *et al.*)

Reagents

n-Octanol.
Hydrochloric acid, 6 M: 490 cm^3 dm^{-3} conc. HCl.
Sodium hydroxide solution, 5 M: 200 g dm^{-3} NaOH.
Sodium hydroxide solution, 0·5 M: by dilution of 5 M NaOH.

I / PREPARATION OF HYDROLYSATE

Weigh 5 g of 100-mesh soil into a 125-cm^3 conical flask having a ground glass joint. Add 2 drops of octanol to prevent frothing and 20 cm^3 of 6 M hydrochloric acid. Swirl the flask, attach a condenser and reflux the mixture gently for 12 hours.

After cooling filter the hydrolysate through a sintered glass funnel into a 200-cm³ tall-form beaker marked at 60 cm³. Wash the residue with small portions of water until the filtrate reaches 60 cm³. Cool the beaker in ice and neutralize the hydrolysate to pH 5·5 with 5 M sodium hydroxide solution using a pH-meter and glass electrode and with constant stirring. Continue the neutralization to pH 6·5 with 0·5 M alkali. Transfer the neutralized hydrolysate to a 100-cm³ volumetric flask and make to volume.

II / DETERMINATION OF AMMONIUM-NITROGEN IN HYDROLYSATE

Steam distil 10 cm³ of hydrolysate with 0·07 ± 0·01 g of magnesium oxide and collect the ammonia in 5 cm³ of boric acid-indicator solution; the volume of distillate collected should be 15 cm³. Determine the ammonium nitrogen by titration with standard hydrochloric acid (section 10:3:3).

III / DETERMINATION OF (AMMONIUM + HEXOSAMINE)-NITROGEN IN HYDROLYSATE

Reagents

Phosphate–borate buffer solution, pH 11·2: 100 g sodium tribasic phosphate and 25 g sodium borate dissolved in water and diluted to 1 dm³.

Procedure

Mix 10 cm³ of hydrolysate with 10 cm³ of buffer solution and steam distil into 5 cm³ of boric acid solution. Collect 30 cm³ of distillate and titrate the ammonia with standard acid. Calculate hexosamine-nitrogen by difference.

IV / DETERMINATION OF AMINO-ACID NITROGEN IN HYDROLYSATE

Mix 5 cm³ of hydrolysate with 1 cm³ of 0·5 M sodium hydroxide solution and heat in boiling water until the volume is reduced to about 2 cm³. Cool and add 0·5 g citric acid and 0·1 g of finely ground ninhydrin. Return the flask to the boiling water-bath, swirl a few moments and leave in the bath for 10 minutes. Cool and add 10 cm³ of the phosphate–borate buffer solution (section 11:3:6III) and 0·1 cm³ of 5 M sodium hydroxide solution and determine the ammonia by steam distilling into 5 cm³ of boric acid solution.

11:3:7 Measurement of organic matter decomposition

Apparatus and method

Macro-respirometer of Birch and Friend

The macro-respirometer is depicted in Figures 11:13 and 11:14. The incubation vessel is a 250-g wide-mouth bottle closed with a 2-hole rubber bung. One hole carries a capillary stopcock and the other the short arm of the platinum anode tube. The far end of the anode tube dips into a 100-cm³ crystallizing dish filled to a definite depth with 0·5 M sulphuric acid solution. The cathode is prepared from a J-shaped piece of glass tubing, mercury filled and having a spade-shaped platinum electrode sealed into its short arm. A piece of wire, preferably but not necessarily platinum, dips into the open end of the tube for connection to the electric supply. The cathode is covered with an inverted 100-cm³ burette (preferably of the Schellbach type for ease of reading) which is also filled to the upper mark with 0·5 M sulphuric acid. The electric supply is best provided by a 6- or 12-volt accumulator.

The incubation bottle must be kept at constant temperature; ideally a constant temperature room would keep the whole apparatus free from temperature effects but as most laboratories do not run to such luxury, a thermostat must be used. This can be of metal or glass with a metal frame (Birch and Friend used a domestic fish tank) filled with water, and its temperature can be regulated by, for example, a toluene regulator in conjunction with a small 'sausage-type' heating bulb. Birch and Friend actually used the cheap and easily replaceable heating elements in conjunction with a regulator which went with the fish tank for the benefit of tropical fish. This set-up proved very satisfactory; it controlled the temperature of the tank to ± 1°C and took up very little space, no stirrer was found necessary provided that the two heating tubes were situated at the bottom of each end of the tank. A little *p*-nitrophenol or copper sulphate kept the water free from algae and a few pellets of cetyl alcohol, which forms a monomolecular surface layer, reduced evaporation and kept the water free from mosquito larvae; the latter precaution will not, it is hoped, be necessary in all laboratories.

The original macro-respirometer was a forest of retort stands and clamps and it is worth while to prepare, or have prepared, special clips (Figure 11:14) for supporting the bottles in the tank and the corresponding burettes in the acid dishes. The rod of the clip lies across the flat edge of the tank and is secured by means of a Trigrip bosshead.

For adjusting the acid in the burettes it is convenient to connect them to a water pump with rubber tubing; pinching the tubing with the fingers provides good control of the suction.

Incubation bottles can be arranged all round the perimeter of the thermostat and the number of units is controlled merely by the size of the tank. All the positive electrodes can be joined together, similarly all the negative electrodes, the two end wires being connected to the battery.

Reagents

Sulphuric acid, 0·5 M: 28 cm³ dm⁻³ conc. H_2SO_4.

Sodium hydroxide solution: 40 g dm⁻³ NaOH. The solution should be carbonate free; solid sodium hydroxide in a beaker is covered with distilled water, swirled for a few seconds and the solution decanted and discarded. More water (CO_2-free) is then added to prepare the solution.

Hydrochloric acid, 0·1 M: Exactly 0·1 M HCl is prepared by dilution of a standardized, approximate 1·0 M solution (89 cm^3 dm^{-3} conc. HCl).

Methyl orange solution: 0·05% w/v, aqueous.

Barium chloride solution, 0·2 M, with phenolphthalein: 21 g BaCl$_2$ dissolved in 250 cm^3 of water, 2 cm^3 of a 0·5% w/v ethanolic solution of phenolphthalein added and the mixture diluted to 500 cm^3.

Procedure

Place 50 g of 2-mm soil in the bottle and bring to the desired moisture content by adding water dropwise from a burette with frequent shaking and tapping of the bottle. Place a small specimen tube in the bottle, the soil preferably being loosely packed around the tube and compressed as little as possible. By means of a pipette, place 10 cm^3 of 1 M sodium hydroxide solution in the tube, fit the anode tube and place the bottle in the thermostat. The far end of the anode tube should dip about 1 cm below the level of 0·5 M sulphuric acid in the dish. When all bottles are in place, leave them for about 30 minutes to come to constant temperature; during this time bubbles of air usually are displaced through the sulphuric acid solution.

If necessary, adjust the acid level in the burette to the upper mark, and with the capillary stopcock of the bottle open, adjust the acid in the anode tube by gentle suction to a definite, marked level beneath the platinum electrode; the levels should be the same for all the anode tubes in use. The stopcock is closed and the acid in the dish made to a definite, marked level. Connect the end-terminals to the battery.

As respiration proceeds, hydrogen will collect in the burette and its volume can be read at any desired intervals of time giving an immediate, approximate measure of the rate of decomposition. For many purposes one reading per 24 hours is sufficient, say at 9.00 a.m. every day. After taking the readings of all burettes, note the air temperature near the burettes and restore their acid levels by gentle suction. Make good evaporation losses in the acid dishes daily, with distilled water.

A blank unit must be included in which pure, sterile sand takes the place of soil. This will give a small daily reading of hydrogen due to minor fluctuations of temperature, particularly overnight, and this is deducted from the other readings.

At any desired time the experiment can be stopped and the soil analyzed. For many experiments there will be several replicates (there must always be at least duplicates) of soil samples and thus one sample can be removed or treated in some way, while the others remain undisturbed. When it is decided to stop an experiment, disconnect the electrodes, carefully remove the bottle and tube from the thermostat and quickly dry the bottle and rubber stopper with a cloth. Remove the stopper, take out the specimen tube and immediately pour the alkali through a funnel into a 100-cm^3 volumetric flask; rapidly rinse the tube into the flask several times with carbon dioxide-free water (freshly boiled out). Dilute the sodium hydroxide solution to the mark with CO_2-free water and stopper the flask. Treat a 10-cm³ aliquot of the original sodium hydroxide solution in the same way, diluting it to 100 cm³. Just before titration invert the flask several times to mix the contents and pipette a 5-cm³ aliquot into a small conical flask. Add a few drops of methyl orange solution and titrate with 0·1 M hydrochloric acid. Pipette a second 5-cm³ aliquot of alkali into a clean flask and quickly add 5 cm³ of the barium chloride–phenolphthalein solution. This precipitates the carbonate; titrate the excess hydroxide with the standard acid. If the solution fails to turn pink when the barium chloride reagent is added, the experiment must be abandoned as the alkali has been insufficient to absorb all the carbon dioxide evolved.

Calculations

I / DETERMINATION OF EVOLVED CARBON DIOXIDE

If a = cm³ of 0·1 M HCl used in the methyl orange titration, and
  b = cm³ of 0·1 M HCl used in the phenolphthalein titration, then

$(a - b) \times 22\cdot4 =$ cm³ CO_2 evolved from the 50 g soil at s.t.p., or [11:9

  $(a - b) \times 12 =$ mg carbon mineralized. [11:10

II / DETERMINATION OF OXYGEN UPTAKE

The total hydrogen evolved over the experimental period is calculated, and its average daily value worked out. Daily air temperature and barometric pressures are also averaged.

Let t = average temperature °C
  H = total hydrogen volume in cm³
  h = average daily (per 24 hours) hydrogen volume in cm³.

An average of $(100 - h)$ cm³ of acid remained in the burette daily. The height of this volume of acid above the acid level in the dish is measured with a ruler; let this height be A cm.

  The uncorrected total volume of oxygen consumed is given by $H/2$ cm³. Normal atmospheric pressure is 101·3 kN m⁻² which is equivalent to 1000 cm³ of 0·5 M sulphuric acid (hence the use of this strength acid in the dish). Using this fact, the height of 0·5 M sulphuric acid corresponding to the measured atmospheric pressure is calculated; let this value be B cm.

  Then the gas pressure, P, is given by $B - A$. Correcting to s.t.p., the corrected volume of oxygen utilized by the soil during decomposition is given by:

$$V = \frac{273}{(273 + t)} \times \frac{P}{1000} \times \frac{H}{2} \text{ cm}^3 \qquad [11:11$$

If required, the 'respiratory quotient' is given by:

$$\text{R.Q.} = \frac{22\cdot4\,(a - b)}{V} \qquad [11:12$$

Results are conveniently plotted on a graph showing time against mg carbon mineralized or cm^3 oxygen taken up, per gramme of soil per 24 hours.

The soil remaining in the bottle can be sampled for analysis as required, or be given special treatments before being returned to the apparatus with a fresh tube of alkali.

Notes: The J-shaped electrode needs to be examined periodically for leaks and breakage of contact between the mercury and the platinum.

Results will be affected by soil crumb size and, to a certain extent, by the volume of soil; these factors should thus be standardized.

If the soil as originally placed in the apparatus was dry, a flush of decomposition, due entirely to the moistening of the dry soil, will occur (section 11:2:7). This initial flush will be over after about 14 days of incubation, during which time it can be followed on a graph showing time against hydrogen volume, and after which the experiment proper can be commenced.

The titration of the original sodium hydroxide solution serves to show any carbonate present not due to soil respiration.

12

Phosphorus

12:1 INTRODUCTION

As in the case of nitrogen and organic matter the chemistry of soil phosphorus has been so extensively studied that its discussion, even as a précis, is beyond the scope of this book. This chapter, therefore, will be devoted mainly to the determination of the element in soils and to associated analytical problems.

The determination of phosphorus in solution can nowadays be made quickly, easily and with a high degree of precision, and modern research in methodology has been mainly towards the determination of specific phosphorus compounds in soil. Earlier work was concentrated on fractionation of organic phosphorus compounds but, more recently, attempts to isolate and measure the various inorganic forms of phosphorus have been made with increasing success. Much research remains to be done before we can specifically analyze for any particular inorganic phosphorus compound with confidence but the procedures already devised have opened up a vast new field of associated investigation. The interrelationship between various inorganic phosphorus radicals and plant-available phosphorus is one example of obvious interest, and there are many others.

12:2 BACKGROUND AND THEORY

12:2:1 Total phosphorus

The total quantity of phosphorus in most soils is relatively small. For example, Lipman and Conybeare (1936) found an average of 0·06% in the top 15 cm of cropland soils throughout the United States, and Metson (1961) states that the total phosphorus in soils rarely exceeds 0·2%.

Phosphorus occurs in soil in inorganic and organic forms, the relative proportions of which vary with organic matter content but usually the organic forms predominate. The element tends to accumulate in the finer fractions of soil and thus increases as the clay content increases.

Three basic methods have been employed to bring the total phosphorus in a soil into solution: digestion with strong acids, fusion with alkalis and ignition followed by acid extraction. The acids recommended for digestion are perchloric or hydrofluoric and, of the two, perchloric acid is to be preferred as silica is removed by filtration and methods exist for the determination of the resultant soluble phosphate in which perchloric acid does not interfere; furthermore, the use of expensive platinum ware is avoided.

Metson considers that the digestion method is not always quantitative and prefers a method of prolonged fusion or else a double acid digestion. In his final recommendation, Metson advocates the destruction of organic matter with nitric acid before igniting the soil and to follow the ignition with an acid digestion. The acid used is a mixture of nitric and hydrofluoric and the phosphorus is finally measured by the gravimetric procedure of Lorenz (1901).

For fusion methods of bringing phosphorus into solution, sodium carbonate is the usual reagent and typical experimental details are given by Jackson (1958) who, however, points out that, whereas the method is suitable for siliceous soils, it is not satisfactory for soils in which the silica content is less than the calcium plus fluoride, for example calcareous soils. Jackson also describes a modified fusion method in which the melt is taken up in sulphuric acid and iron(III) eliminated by passing the solution through a Jones reductor. Fusion with magnesium nitrate is recommended by the Association of Official Agricultural Chemists (1950) but according to Jackson this does not extract all the phosphorus. It may be that if the magnesium nitrate was carefully purified a more satisfactory extraction would result, as was found in the determination of total sulphur (Chapter 13, section 13:2:1).

Stewart (1910) introduced an ignition method whereby the soil is ignited and then extracted with cold 12% hydrochloric acid. In later modifications the soil was ignited at 550°C for 1 hour and then extracted for 16 hours with 1 M sulphuric acid. Beckwith and Little (1963), after a similar ignition, extracted the soil for 4 hours with dilute (2:1) hydrochloric acid at 100°C. Beckwith and Little considered that one result of the ignition is to render titanium oxide less soluble, thus reducing the absorption of phosphorus.

For the simple determination of total phosphorus in a soil, the perchloric acid digestion is here recommended but if it is also desired to make a broad distinction between inorganic and organic phosphorus, then the ignition procedure is more suitable.

12:2:2 Inorganic phosphorus compounds

Total inorganic phosphorus

An approximation of the total inorganic phosphorus in a soil can be obtained by extraction with strong acid, for example 1 M sulphuric acid.

Parton (1963) suggested that as the solubility of inorganic phosphorus compounds is largely dependent upon pH, the total amount can be brought into solution provided that the pH is made sufficiently low. At pH 1·0 all the inorganic phosphorus, Parton considered, will be mobilized, but not necessarily in solution as some re-sorption of phosphorus will occur. To overcome this difficulty Parton added radioactive phosphorus, ^{32}P, in known amount to the soil and any re-sorption of phosphorus was reflected in a decrease of radioactivity in solution. Parton's procedure was thus to mobilize the inorganic phosphorus using a hydrochloric acid solution at pH 1·0 and containing potassium dihydrogen orthophosphate labelled with ^{32}P. When equilibrium was obtained the inorganic phosphorus in solution was determined and the radioactivity of the solution compared with that of the original solution. The inorganic phosphorus mobilized will then be given by the expression,

$$\left[\frac{\text{Specific activity of applied P}}{\text{Specific activity of P in solution}} - 1 \right] \times \text{quantity of P added.}$$

Experiment showed that using 0·15-mm soil the best extracting procedure was at room temperature and with a 1:1000 soil/solution ratio. The solvent, 0·1 M hydrochloric acid, contained 310 μg of phosphorus labelled with 60 μCi ^{32}P. Time of equilibrium proved difficult to determine and for his experiments Parton says merely that equilibrium was reached in less than 3 weeks.

Fractionation of inorganic phosphorus compounds

The individual forms of inorganic phosphorus in soil are not, as yet, completely identified. Chang and Jackson (1957) postulated three divisions of soil inorganic phosphorus:

i. Discrete phosphates precipitated upon surfaces and which are the most readily available. These include calcium phosphate precipitated upon calcium carbonate, aluminium phosphate precipitated upon aluminosilicates and iron phosphate precipitated upon iron(III) oxides.

ii. Discrete phosphate particles which are slightly available and which include calcium phosphates such as apatite and dicalcium phosphate, aluminium phosphates such as variscite, iron phosphates such as strengite and aluminium–iron phosphates such as barrandite.

iii. Occluded phosphates which are but little available and consist of calcium phosphate occluded in calcium carbonate, aluminium phosphate occluded in iron(III) oxides and reductant-soluble iron phosphate also occluded in iron(III) oxides.

Two main methods of identifying the inorganic forms of phosphorus have been used, chemical fractionation, depending upon the solubility of different forms in different solvents, and a more indirect method based upon the concept of solubility products. If in a system we have a sparingly soluble solid-phase crystalline compound in equilibrium with its ions, the activities of those ions are controlled by the nature of the compound. For example if crystals of $CaHPO_4$ are in equilibrium with Ca^{2+} and HPO_4^{2-} ions in solution, then the product of the activities of those ions is constant and characteristic. Åslyng (1954) attempted to use this method to find whether dibasic calcium phosphate ($CaHPO_4.2H_2O$) or hydroxyapatite ($Ca_5OH(PO_4)_3$) was present in a soil and his paper may be consulted for practical details. Characteristic and constant solubility products, however, are theoretically obtained only with a well crystallized solid phase which is in true equilibrium with its ions and as this is probably seldom the case in soil suspensions, the method is limited in application.

Fraps (1906) was among the first to chemically fractionate inorganic soil phosphates. Fisher and Thomas (1935) extracted soil with a mixture of acetic acid and sodium acetate at pH 5 for different periods of time and with dilute sulphuric acid at pH 3. This procedure resulted in three main groups of phosphorus compounds: calcium, magnesium and manganese phosphates; iron and aluminium phosphates; and adsorbed phosphates and apatites. Holman (1936) plotted pH-solubility curves; Williams (1937) extracted with sodium hydroxide obtaining organic, exchangeable and soluble phosphorus leaving an alkali-insoluble fraction, and Dean (1937) followed the alkali extraction with dilute sulphuric acid extraction to separate calcium phosphates. Ghani (1943a) modified Dean's procedure by extracting first with acetic acid to remove mono-, di- and tri-calcium phosphates, then with sodium hydroxide to remove organic, aluminium and iron phosphates and finally with dilute sulphuric acid to dissolve apatites. The disadvantage of this procedure was that the acetic acid-soluble phosphorus was re-adsorbed by the soil and appeared in the alkali fraction. This led to a modification (Ghani, 1943b) in which 8-hydroxy-quinoline was added to the acetic acid. 8-Hydroxyquinoline forms compounds in which the hydrogen of the hydroxy group is replaced to form substances such as $Al(C_9H_6NO)_3$ which are precipitated in dilute acetic acid. Williams (1950) further modified Ghani's method showing that successive single extractions with 2.5% acetic acid-8-hydroxyquinoline and 0.1 M sodium hydroxide give a better soil phosphorus fractionation. Williams separated the soil fractions by centrifuging instead of using a filter candle, as recommended by Ghani, and he investigated the relative efficiency of 8-hydroxyquinoline and cupferron as reagents for blocking the re-adsorption of phosphate by iron and aluminium. He found that cupferron precipitated aluminium more completely than did 8-hydroxyquino-line at the pH employed, but is less stable and may possibly attack organic phosphorus compounds. As 8-hydroxyquinoline extracts are coloured it is necessary to evaporate them to dryness with a little calcium acetate and measure the phosphorus content of the ash after ignition. Cooke (1951) found that addition of selenious acid to 0.5% was more efficient in preventing re-adsorption than 8-hydroxyquinoline, whereupon Ghani and Islam (1957) reported that a combination of selenious acid with 8-hydroxy-quinoline was even better.

Wiklander (1950) fractionated soil phosphorus as that present in the free soil solution, that present in micellar soil solution, that fixed in the interphase between micellar solution and the solid phase, and precipitated, insoluble forms.

Talibudeen (1957; 1958) suggested four main groups of soil phosphorus based upon the speed with which phosphate groups are exchanged with *ortho*-phosphate in the soil solution thus,

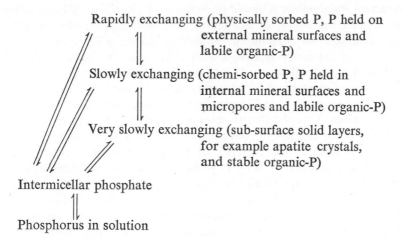

Rapidly exchanging (physically sorbed P, P held on
        external mineral surfaces and
        labile organic-P)

Slowly exchanging (chemi-sorbed P, P held in
        internal mineral surfaces and
        micropores and labile organic-P)

Very slowly exchanging (sub-surface solid layers,
        for example apatite crystals,
        and stable organic-P)

Intermicellar phosphate

Phosphorus in solution

A major advance in the field of phosphorus fractionation was made by Chang and Jackson (1957) who pointed out that in nearly all of the early fractionation methods iron and aluminium phosphates were determined together. Turner and Rice (1954) had reported that neutral ammonium fluoride solution can dissolve aluminium phosphate but not iron phosphate and this was utilized by Chang and Jackson to separate the two phosphates. Acid extractions made in the early stages of fractionation, as in the procedures of Ghani and Williams for example, dissolve not only calcium phosphate but some aluminium and iron phosphate as well. Chang and Jackson arranged the order of the various extractants so as to more discretely separate the fractions; they used synthetics of known composition as test samples.

Water-soluble, saloid-bound (section 12:2:5) phosphates are first removed with dilute ammonium chloride solution and then the aluminium phosphate is removed with neutral ammonium fluoride solution. Iron phosphate is extracted next with sodium hydroxide and finally, calcium phosphate is dissolved in dilute acid. The phosphate not extracted by these four reagents is completely dissolved by a dithionite–citrate treatment and was called 'reductant-soluble' iron phosphate. Chang and Jackson explained the reductant-soluble phosphate as being due to an iron oxide precipitate formed on the surface of iron and aluminium phosphate particles during weathering by hydrolysis of iron(III) salts. Some soils rich in iron oxides contain significant amounts of iron and aluminium phosphate occluded within the oxides and thus not removed by the dithionite treatment; the occluded phosphate, however, can be extracted after the dithionite treatment with fluoride or alkali.

Chang and Jackson's procedure for extracting aluminium phosphate was criticized (e.g. Yuan *et al.*, 1960) as not adequately separating aluminium from iron phosphates due to the pH of the ammonium fluoride solution. Fife (1959; 1962) carefully evaluated ammonium fluoride as a selective extractant for aluminium-bound phosphorus. Using three different soils he investigated the effects of extraction time, soil–liquid ratio and pH of extractant (0·5 M ammonium fluoride solution). Fife found that the efficiency of extraction increased with dilution and more markedly with neutral than with alkaline (pH 8·5) conditions. With alkaline extraction, adsorption of phosphorus by free iron oxides was not influenced by time of extraction between 24 and 72 hours but a linear increase in phosphorus release occurred. This phosphorus release was ascribed to slow hydrolysis of iron phosphate and not to continued attack upon aluminium phosphate. Fife concluded that the presence of high amounts of allophane or of amorphous alumina is unlikely to give problems when determining aluminium phosphate with alkaline ammonium fluoride solution. Difficulties would arise, however, if much gibbsite is present, when aluminium phosphates appear to be released continuously with time. One suggested explanation for this was that the aluminium phosphate may be protected by coatings of iron(III) oxide, but Fife thought it more likely to be due to slow attack of gibbsite by the fluoride ion. A phosphorus release curve extrapolated to zero time and correction of values to an adsorption-free basis was taken as a measure of aluminium phosphate. For routine work Fife recommended a direct alkaline extraction at high dilution, the initial alkalinity of the solution being such as to give a final pH of soil suspension of 8·5.

Askinazi *et al.* (1963) also found that for selective extraction of aluminium phosphate a 0·5 M solution of ammonium fluoride at pH 8·5 was best and that under these conditions removal of iron and calcium phosphates was minimal. Similarly Pratt and Garber (1964) modified the original Chang and Jackson ammonium fluoride extractant by making it to pH 8·5 but reported that whereas the procedure is satisfactory for virgin soils, there is uncertainty about the separation of aluminium phosphate from recently fertilized soils.

Chang and Liaw (1962) examined Fife's procedure for extracting aluminium phosphate and considered that the phosphate was more discretely separated from iron phosphate by extracting with neutral ammonium fluoride solution as in the original Chang and Jackson method, for one hour rather than with the alkaline reagent for longer periods. However, as re-precipitation of phosphate from aluminium by iron occurs, they modified the procedure as being a measurement of total aluminium phosphate, thus dispensing with the need for correcting the result for dissolved iron phosphate.

Smith (1965a) stated that for the accurate distinction between aluminium and iron phosphates a separate correction factor has to be calculated for each level of standard phosphate added to the ammonium fluoride solution. The pre-existing distribution of phosphate between the soil and fluoride solution can be calculated by plotting recoveries from a series of additions against concentration of added phosphate.

Chang and Jackson subsequently modified their extraction procedure by making the ammonium fluoride extractant to pH 8·5 and by extracting reductant and occluded forms of phosphorus before extracting calcium phosphates. The procedure was further studied by Bromfield (1967) who concluded that ammonium fluoride is not a reliable extractant for aluminium phosphate unless the soil calcium and iron phosphates have solubilities similar to those of the controls used by Chang and Jackson and unless reliable corrections are made for re-sorption of phosphorus on sesquioxides. Bromfield found that appreciable amounts of dicalcium phosphate are dissolved in the ammonium fluoride reagent, particularly in soils fertilized with superphosphate, and that some calcium phosphate is dissolved during the preliminary extraction with ammonium chloride. Part of this dissolved calcium phosphate is then resorbed on to sesquioxides, leading to under-estimation of calcium phosphate and over-estimation of iron phosphate.

Before determining the phosphorus extracted by dithionite the excess dithionite and citrate present must be destroyed. This was originally done by oxidation with hydrogen peroxide which is not an easy operation and can lead to errors. Chang *et al.* (1966) thus suggested a procedure whereby the dithionite is oxidized with iron(III) chloride in alkaline solution, citrate being precipitated. After filtration, aliquots are taken for analysis. Petersen and Corey (1966) used potassium permanganate for oxidizing the dithionite.

Avnimelech and Hagin (1965) fractionated soil phosphorus by means of a gradient elution technique in which samples were leached with a linearly increasing acidity. Thus soil phosphorus was extracted according to its solubility in acid. The method involves a non-equilibrium extraction as opposed to conventional methods which employ equilibrium extractions. As the experimental technique may be of use for investigations other than phosphorus fractionation, the apparatus is shown in Figure 12:1 and the experimental details are briefly given. The soil is mixed with pure sand if necessary, to facilitate leaching and is leached with the acid solution at 150 cm^3 per hour; the leachate can be analyzed periodically. If L is the acid concentration, V is the total volume of solution in the two containers and v is the volume removed then the acid content C of leaching solution at a certain moment will be given by

$$C/L = v/V \qquad\qquad [12:1$$

Some inorganic phosphorus compounds, apatite for example, can be identified directly in soil by microscopic examination of the silt and sand fractions.

12:2:3 Organic phosphorus compounds

Total organic phosphorus

Hopkins and Pettit (1908) estimated organic phosphorus in soils by a calculation procedure. Soils were chosen having the same amounts of potassium in the subsoil as in the topsoil and thus assumed to have a uniform

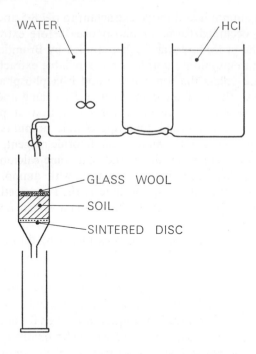

Figure 12:1 Apparatus for the gradient elution of soil phosphorus (Avnimelech and Hagin, 1965; Soil Science Society of America, *Proceedings*)

mineral composition. Phosphorus present in the subsoil was then sub-tracted from that in the topsoil and the difference taken as organic phosphorus. The N/P ratio was then calculated and organic phosphorus determined in other kinds of soil by multiplying the nitrogen content by this ratio. The method was later taken up by Stewart (1910) but was much criticized by Fraps (1911).

The practical determination of total organic phosphorus in soils has been attempted by two main methods, direct chemical extraction and in-directly by the difference between total and inorganic phosphorus. In most cases even the so-called direct method depends eventually upon a distinc-tion between total and inorganic material. According to Russell (1961) neither procedure is acceptable but at present it would be difficult to suggest an alternative.

For the mere determination of total organic phosphorus the indirect method by difference is preferable to extraction procedures and is cer-tainly more simple and expedient. The principle of the method is to deter-mine the total inorganic phosphorus in the soil before and after oxidation of the organic matter, the difference being taken as a measure of organically bound phosphorus. Sources of error lie in the quantitative oxidation of organic matter and, perhaps more importantly, in the extraction of the in-organic phosphorus.

Schmoeger (1897) suggested hydrolyzing soil under pressure at 140–160°C and then extracting with cold 12% hydrochloric acid; the difference in phosphorus between the extracts of hydrolyzed and non-hydrolyzed soil samples was taken as organic phosphorus. Potter and Benton (1916) based a method on the solubility of organic matter in ammonium hydroxide and their procedure was improved by Schollenberger (1918) although it remained tedious.

Destruction of organic matter by ignition was used from early times (e.g. Stewart, 1910) and several different temperatures and times of heating have been advocated. The procedure most commonly followed is that of Legg and Black (1955) who ignited the soil for 1 hour at 240°C and made the acid extractions with concentrated hydrochloric acid. Saunders and Williams (1955) used 1 M sulphuric acid for the extractions and ignited the soil at 550°C for 1 hour.

The chief error in destroying the organic matter by ignition is that the solubility of the inorganic phosphorus already present in the soil is affected, thus rendering the two acid extractions non-comparable. During the initial acid extraction of the non-ignited soil it is possible also that some organic phosphorus will be mineralized by acid hydrolysis (Anderson, 1960) which will cause low results for organic phosphorus.

Dahnke and co-workers (1964) first extracted soil with 0·1 M hydrochloric acid and 0·06 M ammonium fluoride solution for 5 minutes to remove readily extractable phosphorus. The soil was then ignited at 500°C for 1 hour and re-extracted; the phosphorus in the second extract was assumed to be organic.

Dickman and De Turk (1938) suggested a method which was later slightly modified by Bray and Kurtz (1945) in which phosphorus, extracted as acid-soluble plus adsorbed, was measured before and after oxidation with hydrogen peroxide. 1 g of soil in a tube graduated at 50 cm³ was treated with 15 cm³ of 30% hydrogen peroxide and 10 cm³ of water. The suspension was warmed in a steam bath for 30 minutes, diluted, acidified and shaken with ammonium fluoride solution; after filtering, the phosphorus was determined in the filtrate. Jackson gives a similar procedure. The objection to destroying the organic matter with hydrogen peroxide is the difficulty of obtaining a phosphorus-free reagent. The commercial product must be mixed with a little calcium hydroxide and distilled under reduced pressure below 60°C; the distillate will be less concentrated than the original peroxide (20% instead of 30%) due to decomposition.

Most chemical extraction procedures for estimating organic phosphorus differ from those just described only in that oxidation of organic matter is carried out after extracting the organic phosphorus compounds from the soil and not on the soil as a whole.

Wrenshall and Dyer (1939) extracted organic phosphorus from soil with 4 M hydrochloric acid followed by 2·9 M ammonium hydroxide solution, and Pearson (1940) recommended an eighteen-hour extraction with

10 + T.S.C.A.

0·5 M ammonium hydroxide at 90°C. Bower (1945) suggested hot 0·5 M sodium hydroxide solution as a more efficient extractant than ammonium hydroxide and his view was confirmed by Black *et al.* (1948) who added, however, that organic phosphorus may be lost by hydrolysis. Saunders and Williams (1955) compared Pearson's ammonium hydroxide extraction with that in which the soil is pre-treated with 0·1 M hydrochloric acid and then extracted with cold 0·1 M sodium hydroxide. Saunders and Williams found that the results of Pearson's method were influenced by the initial acid extraction and that although the extraction with 0·1 M sodium hydroxide gave better results, it was still affected by hydrolysis losses.

Mehta (1951) investigated the methods of Wrenshall and Dyer, Pearson and Bower and found none of them satisfactory. Mehta *et al.* (1954) later recommended a pre-extraction of soil with concentrated hydrochloric acid followed by two extractions with 0·5 M sodium hydroxide, one cold and the other at 90°C. Mehta's procedure is given by Jackson (1958) who presents it as a flow-sheet and it is here summarized as Table 12:1. Kaila (1962) discovered that the results given by Mehta's method were the same

Table 12:1

Determination of organic phosphorus by the method of Mehta

Soil: Add 10 cm³ of conc. HCl to 1g of 0·15-mm soil and heat on a steam-bath for 10 minutes. Add a further 10 cm³ of conc. HCl and leave for 1 hour at room temperature. Dilute with 50 cm³ of water and centrifuge

---

Soil: Add 30 cm³ of 0·5 M NaOH solution, leave for 1 hour at room temperature and centrifuge		*Solution:* (A) collect in a 250 cm³ graduated flask containing 50 cm³ of water
Soil: Add 60 cm³ of 0·5 M NaOH solution, warm at 90°C for 8 hours and centrifuge	*Solution:* Combine with Solution A	
Soil: Discard	*Solution:* Combine with Solution A	

Combined Solution A: Dilute to mark and mix well. Remove duplicate 15 cm³ aliquots (I) into small beakers. Allow bulk of solution to stand until sediment settles; remove aliquot (II) from supernatant liquid

Aliquot I: Add 1 cm³ of $HClO_4$, evaporate to fuming (not dryness) and determine Organic + Mineral-P	*Aliquot II:* Determine Mineral-P

Organic-P: Calculate from (I − II)

as those obtained by the much simpler procedure of extracting first with 2 M sulphuric acid and then once with 0·5 M sodium hydroxide at room temperature.

Kaila and Virtanen (1955) had attempted an extraction involving organic solvents but difficulty was experienced in choosing a solvent which did not interfere with the subsequent determination of phosphorus. Dormaar and Webster (1963) found that acetone and 0·2 M acetyl acetone did not interfere with the molybdenum blue reaction and modified the Kaila-Virtanen procedure accordingly.

Anderson (1962) found the procedure of Saunders and Williams to be superior to that of Mehta, and Hance and Anderson (1962) developed a modification of Mehta's method in which the soil is extracted with alkali before acid treatment as well as after and which reduces loss by hydrolysis.

Martin (1964) found that 0·3 M potassium hydroxide extracted 83% of the organic phosphorus from soils previously treated with acid. Martin fractionated the extracts by passing through an anion exchange resin and obtaining humic-associated phosphorus, inorganic phosphorus and acid-soluble organic phosphorus.

Boswell and De Long (1959) considered the acid and alkali treatments too harsh and liable to cause hydrolysis and they presented a modified oxine extraction procedure. Unless an acid treatment is given, the soils are first converted to the ammonium form and the pH of their water suspension adjusted to 9·2. A 2·5% solution of oxine in benzene is added and the suspension shaken, with periodic pH adjustment. The benzene phase is separated and washed with water and the aqueous phase is analyzed after centrifuging. A pre-treatment with 0·5 M hydrochloric acid followed by an 18-hour extraction increased the yield of organic phosphorus. The benzene modification permits isolation of the oxine which otherwise interferes with the subsequent phosphorus determination.

Pre-extraction of a soil with acid removes bases, which precipitate dissolved organic phosphorus, and all polyvalent ions which interfere with the subsequent treatments and may disintegrate clay complexes, thus releasing adsorbed organic phosphorus. Saxena (1964) investigated the strength of acid used for pre-treatment and found concentrated hydrochloric acid most effective. However, concentrated hydrochloric acid causes hydrolysis of organic phosphorus and Saxena finally recommended making very short time (2 minutes) extractions with the acid and to neutralize the extracts immediately with ammonium hydroxide. Saxena pointed out that ammonium hydroxide will not dephosphorylate organic phosphorus compounds, whereas sodium hydroxide will, and he found that one extraction with ammonium hydroxide at 75°C removes all the organic phosphorus which would be hydrolyzed by sodium hydroxide. To compensate for the poor extracting power of ammonium compared to that of sodium hydroxide, the ammonia extraction is followed by another using 0·5 M sodium hydroxide at 90°C. Saxena's final procedure for extracting organic phosphorus is as follows:

10 g of soil are shaken for 2 minutes with 50 cm³ of concentrated hydrochloric acid, diluted, centrifuged and washed. The extract is neutralized with ammonium hydroxide. The soil residue is then extracted with 100 cm³ of 1·5 M ammonium hydroxide at 75°C for 2 hours and the extract added to the first, neutralized extract. The soil is finally extracted with 100 cm³ of 0·5 M sodium hydroxide solution at 90°C for 6 hours, the extract neutralized with acid and mixed with the previous extracts. Organic phosphorus is then determined in the usual manner as the difference between total and inorganic phosphorus in the extracts. Saxena claimed that hydrolysis is almost completely eliminated by this procedure.

Many workers have compared the ignition and chemical extraction procedures for estimating organic phosphorus in soil and a wide range of soils has been examined. On the whole, the ignition method has been found to give the highest organic phosphorus recovery (Saunders and Williams, 1955; Hance and Anderson, 1962; Walker and Adams, 1958) although Harrap (1963), using seventeen soils, found little significant difference between the methods. Dormaar (1963) compared several techniques and came to no conclusion other than that organic phosphorus cannot yet be determined quantitatively. Kaila (1962) found that the ignition procedure gave higher values than extraction procedures and that the results were not related to pH, carbon, clay or iron(III); he concluded, however, that for the most accurate assessment of organic phosphorus the average result of the two methods should be taken.

Anderson and Black (1965) developed a procedure for measuring small amounts of organic phosphorus in the presence of large amounts of inorganic phosphorus, this being part of a fertilization investigation and in which the difference methods were unreliable.

0·5 g of soil was extracted with 10 cm³ of 1·0 M hydrochloric acid for 10 minutes at 70°C, cooled and centrifuged. The residue was allowed to stand with 35 cm³ of 1 M hydrochloric acid for 1 hour at room temperature and then with 35 cm³ of 1 M ammonium hydroxide for 4 hours at 90°C. After each treatment the soil was centrifuged and the extract filtered under pressure through a membrane filter of 0·23 μm pore diameter. The extracts were combined and diluted with water and hydrochloric acid to give 140 cm³ of a solution 0·5 M in hydrochloric acid. Norit-A carbon filter columns were prepared with the carbon purified by treating with concentrated hydrochloric acid on a steam-bath overnight, washing with water, drying and then washing successively with ethanol, 6 M ammonium hydroxide and concentrated hydrochloric acid. The bottom of the filter tube (40 cm × 1·6 cm) was closed with a pad of glass wool and a suspension of 0·25 g of diatomite filter-aid in 0·5 M hydrochloric acid added; this was followed by 4 g of the purified carbon in 0·5 M hydrochloric acid and suction applied to compact the solids. 10 to 25 cm³ aliquots of soil extract were placed on top of the carbon and were followed immediately with a suspension of 0·25 g of filter-aid in acid suspension. Suction was applied and the column washed with small amounts of 0·25 M hydrochloric acid before being leached with 0·5 M hydrochloric acid from a reservoir; the volume of leachate was noted. After disconnecting the apparatus the carbon was placed in a 50-cm³ beaker and mixed with 5 cm³ of 6 M ammonium hydroxide solution and 1 cm³ of a 10% magnesium nitrate solution. The mixture was evaporated to dryness and ignited at 500°C. When cold, 20 cm³ of 1 M hydrochloric acid was added and the solution heated on a steam-bath for 10 minutes. Organic phosphorus was measured in the solution as in Mehta's method.

Anderson and Black found that organic phosphorus was retained almost quantitatively by the carbon and that inorganic phosphorus was quantitatively removed; no competition between inorganic and organic phosphorus for sorption on the carbon was noted. Some of the organic phosphorus is retained on the carbon chromatographically, some is precipitated on acidification and retained mechanically and a little, less than 5%, may be washed out during leaching of the inorganic phosphorus. If extremely large amounts of inorganic phosphorus are present, it is feasible that some of it may be retained against leaching, and to discover the extent of this a control is necessary.

As compared with other methods of determining organic phosphorus, the method of Anderson and Black has the advantage of requiring only a small correction for inorganic phosphorus even when the quantities of organic phosphorus are of the order of 10 μg accompanied by as much as 10^4 μg of inorganic phosphorus. The column method of separating the two forms of phosphorus is better than shaking the extracts with carbon as suggested by Goring (1955).

Specific organic compounds of phosphorus

Organic phosphorus compounds in soil are generally considered to fall into three main groups, the phospholipids, the nucleic acids and the inositol phosphates. Nucleic acids have been identified in soil hydrolysates but account for only about 2–3% of the total organic phosphorus. Adams and co-workers (1954) were unable to trace any nucleic acids in soil itself.

The most abundant organic phosphorus compounds yet found in soil are the inositol hexaphosphates. Inositol hexaphosphate is a six-membered carbon ring with a phosphate group on each carbon,

$$H_2O_3PO \qquad OPO_3H_2$$
$$H_2O_3PO \qquad OPO_3H_2$$
$$H_2O_3PO \qquad OPO_3H_2$$

All the phosphate groups are linked as esters having two replaceable hydrogens; in soils it most probably occurs as salts of iron, aluminium, calcium and magnesium (Caldwell and Black, 1958). Caldwell (1955) found that inositol hexaphosphate averaged about 12% of the total organic phosphorus in forty-nine soils and that another 5% was present as an isomer. Smith and Clark (1951) using chromatography found that more than two-thirds of the organic phosphorus-containing material from soil behaved chemically as phytin but did not behave chromatographically as inositol hexaphosphate. Caldwell and Black simplified the chromatographic procedure of Smith and Clark by extracting the soil with cold, concentrated hydrochloric acid, then with boiling hydrochloric acid and finally with 0·5 M sodium hydroxide solution. The acid extracts were passed through the column directly; iron(III) phytate was precipitated from the alkaline fulvic acid fraction with acid before this too was passed through the column.

Anderson (1956) described a method for determining inositol phosphates in which the esters were extracted from soil with hot sodium hydroxide solution. Humic acid and sesquioxides were removed and the esters precipitated with barium. After removal of the barium salts, ethanol was added to precipitate more inositol phosphate from the extract; later Anderson and Hance (1963) discovered that the precipitate formed in this manner contains organic phosphorus not behaving as inositol hexaphosphate. In a further investigation a fulvic acid fraction yielded a substance containing phosphate ester groups and which, on alkaline hydrolysis, gave inositol hexaphosphate. The precise nature of combination of this substance was not determined but it did not behave chromatographically like a free ester. Anderson and Hance pointed out that soil inositol hexaphosphate bound in such a way will not be determined by procedures such as that of Smith and Clark.

Cosgrove (1963) shook air-dry soil for 1 hour with 0·2 M hydrochloric acid, filtered and washed the residue with ethanol. The dry soil was then extracted with 1 M sodium hydroxide solution at 60°C for 4 hours and the centrifuged extract acidified with hydrochloric acid. Humic acid was centrifuged off and the fulvic acid fraction made alkaline with solid sodium hydroxide. After cooling to below 5°C the fulvic acid was treated with bromine for 18 hours. The mixture was heated to 60°C, acidified and excess bromine removed by ether extraction. The pH was brought to 2·0 with sodium hydroxide and iron(III) chloride solution added; the coagulated precipitate was centrifuged off, washed and suspended in water. Phytin, as the iron(III) salt, was purified by solution in hydrochloric acid and re-precipitation with iron(III) chloride and finally was converted to the sodium salt. The solution was made alkaline and fractionated by resin chromatography, paper chromatography and paper electrophoresis. By this means Cosgrove demonstrated the presence of *myo*-inositol hexaphosphate and corresponding derivatives of *d, l*-inositol and *scyllo*-inositol. Of the lower inositol phosphates, pentaphosphates were the major constituent.

The extraction and determination of phospholipids which are found normally in only small concentration in soil (up to 3–4 mg P/100 g) has been investigated by Hance and Anderson (1963a). Phospholipids are indicated by the presence of phosphorus in alcohol and ether extracts of soil and can be determined by the isolation of choline. Wrenshall and McKibbin (1937) found approximately 0·3% of the total extractable organic phosphorus in a soil to be present as phospholipids and Sokolov (1948), using the same extractants as Wrenshall, ethanol and ether, extracted nearly 1·5% of the total organic phosphorus as phospholipids from peat, 0·4% from chernozems and none at all from podsols. Vincent (1937) used acetone, chloroform and ethanol as extractants. Hance and Anderson considered that the extraction procedures previously used were all subject to errors. As the precise nature of the soil phospholipids is not known, it is impossible to choose the perfect extractant and a variety must be employed. For example when Sokolov's ethanol and ether method was used which extracted 0·1 to 0·26 mg P/100 g soil, only very little of this phosphorus could be re-extracted from the residue into a more specific lipid solvent, indicating that some non-lipid material had also been extracted; unless, of course, its solubility had been affected during evaporation of the solvent. The method finally adopted by Hance and Anderson for extraction of phospholipids involved pre-treatment of the air-dry soil with hydrochloric and hydrofluoric acids to overcome the effect of air-drying the soil which causes retention of some lipid phosphorus. The soil was then successively extracted with acetone, petroleum ether, an ethanol–benzene mixture and finally with 1:1 methanol–chloroform mixture. The combined extracts were evaporated under reduced pressure and at low temperature and the residue extracted successively with cold ether–petroleum ether mixture, cold chloroform, boiling ether–petroleum ether mixture and boiling chloroform. After evaporating off the mixed solvents the residue was evaporated with ethanol and magnesium nitrate solution, ignited and then evaporated with hydrochloric acid. Phosphorus was determined in the final residue.

Hance and Anderson (1963b) subsequently identified some of the hydrolysis products of soil phospholipids. After an alkaline hydrolysis glycerophosphate, $C_3H_5(OH)_2OPO_3H_2$, was identified by paper chromatography and was quantitatively estimated by peroxide oxidation. Acid hydrolysis gave choline,

$$(CH_3)_3N \diagup CH_2CH_2OH \diagdown OH$$

provided that non-lipid nitrogen compounds were first removed. It appeared as if phosphatidyl-choline is a predominant lipid component.

12:2:4 Mineralization of organic phosphorus

Although organic forms of phosphorus are not themselves directly available for plant nutrition, they do form the main source of available phosphorus. On the other hand, mineralization of organic matter does not necessarily result in release of plant-available phosphorus and in some cases the reverse is true and immobilization of soluble inorganic phosphorus already present will occur. Thus Black and Goring (1953) showed that during the decomposition of organic matter containing less than 0·2% organic phosphorus, or if the carbon–organic phosphorus ratio was greater than 300, inorganic phosphorus is immobilized. Birch (1961) found that during the decomposition of mature organic matter any available inorganic phosphorus was converted to organic, microbial phosphorus which was not again dephosphorylated. Hesse (1963) found that during decomposition of fibrous matter from a mangrove swamp soil, inorganic phosphorus was converted to organic in spite of oxidation of 19·2 g carbon per 100 g of soil. McCall (1956) found that mineral phosphorus added to a soil increases the rate of organic phosphorus decomposition and Thompson *et al.* (1954) found that the amount of organic phosphorus mineralized during 25 days' incubation at 40°C ranged from −4 to +45 µg g^{-1}, the mineralization depending largely upon total organic phosphorus.

Birch (1961) studied the transformation of phosphorus during decomposition of organic matter using the macro-respirometer (section 11:3:7). Birch pointed out that a high percentage of the phosphorus in plants is inorganic and that this is adequate for both metabolic and synthetic processes of the microbial population throughout decomposition. Thus, in contrast to nitrogen studies, it is mostly changes in microbial phosphorus that are involved and not in plant organic phosphorus. Birch used dry, ground plants mixed with phosphorus-free sand and re-moistened; samples were incubated in the respirometer with sodium hydroxide vials for absorbing carbon dioxide and periodically replicates were withdrawn for analysis. The carbon oxidized was determined by titration of the hydroxide (section 11:3:7) and portions of the sample were extracted with water, 0·5 M sulphuric acid, hot 0·5 M sulphuric acid (under reflux) and with 0·5 M sulphuric acid at 100°C for 5 days in a sealed tube. Phosphorus was determined in each extract.

An initial water extraction gave the original soluble inorganic phosphorus content. The cold acid extracted inorganic and adsorbed inorganic phosphorus from microbial cells, the boiling-acid-extracted phosphorus less the water-soluble phosphorus indicated the acid labile phosphorus (phosphorylated metabolic intermediates) and heating in a sealed tube for 5 days completely hydrolyzed the microbial organic phosphorus to inorganic; thus the difference between organic phosphorus and acid labile phosphorus was found and termed acid-stable phosphorus. By extracting part of a sample at once with water and part after 24 hours' exposure to chloroform vapour, Birch obtained the amount of microbial organic phosphorus enzymatically hydrolyzed to inorganic phosphorus after the cells were killed.

Birch found that during decomposition organic microbial phosphorus was formed from plant inorganic phosphorus and that young plant material forming the substrate was quickly exhausted leaving behind a large microbial population to undergo autolysis with release of inorganic phosphorus, again through enzymatic dephosphorylation. With mature organic matter, the initial build-up of microbial population was much smaller and the supply of substrate was more prolonged and less available. This resulted in a uniform cycle of microbial growth and decay and the mineralized phosphorus was re-utilized. Birch considered that the high proportion of phytin in soils (Bower, 1949) is of plant origin, not a microbial constituent, and this supports the observation that plant organic phosphorus compounds are little mineralized during decomposition. The overall result of decomposition was that residual compounds of largely unchanged plant organic phosphorus were left, together with acid-stable and acid-labile microbial organic phosphorus, in amounts dependent upon the material decomposed and the influence of the soil.

Drying a soil can increase the amount of phosphorus mineralized and mineralization increases as the pH approaches neutrality. Thus liming an acid soil results in increased inorganic phosphorus. An increase in pH affects the capacity of hydrous oxides for sorbing phosphorus, thus leaving more available for mineralization; it also affects the activity of the micro-organisms. That a decrease of pH in alkali soils increases phosphorus mineralization was shown by Gardner and Kelley (1940).

12:2:5 Available phosphorus

Nature, source and determination

Soil phosphorus can be considered as non-available, potentially available and immediately available. Immediately available phosphorus is the inorganic form occurring in the soil solution and which is almost exclusively orthophosphate. Plants are unable to absorb phosphorus directly from solid compounds or from organic phosphorus compounds, even though the latter may be in soil solution. Hannapel and co-workers (1964) considered that the large fraction of organic phosphorus in soil solution (Pierre and Parker, 1927, had found 96% of the phosphorus in soil solution to be organic) must be colloidal as plant roots are unable to utilize it. On mixing sucrose with a soil, Hannapel found that organic phosphorus movement on leaching was increased, whereas treatment of the soil with formaldehyde decreased organic phosphorus movement. This, together with millipore (0·45 μm) filtration, showed that much of the organic phosphorus in soil solution is associated with microbial cells.

Plants do directly take up orthophosphate from solution and have been shown to do so quantitatively. Up to a point, plant growth increases with orthophosphate in the nutrient solution. As the concentration of water-soluble phosphate in soil solution is very small at any one time, it follows that it must be replenished during plant growth and the rate of this replenishment is more important than the concentration.

The most important form of potentially available phosphorus is the organic; inorganic forms other than orthophosphate are largely unavail-
10*

able. Relatively available inorganic phosphorus tends to accumulate in its most stable state under prevailing conditions; thus in calcareous soils the available inorganic phosphorus would be acid-soluble whereas in acid soils the adsorbed phosphorus would be more available. In flooded soils with low oxidation potentials, certain forms of phosphorus normally considered as unavailable, for example iron phosphate, can be regarded as available (Gasser, 1956).

In most soils the main source of orthophosphate is the organic matter unless, of course, direct fertilization with soluble phosphate has been made. Friend and Birch (1960) found that the organic phosphorus in a soil was of significant value in predicting phosphorus availability. Eid and co-workers (1951) found that the significance of organic phosphorus content was of importance only in warm soils and did not apply in cold climates. Semb and Uhlen (1954) more precisely stated that the significance of organic phosphorus content for predicting phosphorus availability was valid only if the pH exceeded 5·5 as more acid conditions adversely affected mineralization. It would seem, therefore, that a knowledge of both the amount of organic phosphorus in a soil and of its rate of decomposition would be most useful for predicting availability. As will be discussed, Friend and Birch go further and include the reassimilation and fixation of the mineralized phosphorus.

The measurement of total phosphorus in a soil is of no use in determining available phosphorus except that a very low total content will indicate a probable phosphorus deficiency.

The amount of phosphorus in a soil that is available to growing plants is determined by one of two methods, biological or chemical, but for the most reliable results both methods should be employed in conjunction. As in the determination of all 'available' nutrients, laboratory methods are of use only as a routine with soils of known characteristics. For the intelligent and practical appraisal of a soil's capacity to provide nutrient phosphate it is essential to first conduct field and greenhouse experiments embodying all probable variables such as crop, land management and so on. Subsequent chemical or biological experiments can then be correlated so as to obtain the best method for that particular soil under particular conditions. There is yet no chemical method of measuring available phosphorus in an unknown soil.

Chemical methods of determination

The chemical technique that has received, and is still receiving, the most attention is that of extraction with one or more solutions. The different extracting solutions that have been recommended are legion and cover the whole gamut between concentrated sulphuric acid and boiling caustic alkali; the number of publications in which these solutions and their applications are discussed probably forms the most voluminous and verbose contribution to soil science.

The underlying principle of the method is to shake a sample of soil with a solution designed to dissolve the fractions of phosphorus that are available to plant roots. The extract is then analyzed for soluble phosphorus and the results correlated with actual uptake of phosphorus from the soil by the plant. It is noticeable that all published methods in which good correlation is obtained between crop response and amount of phosphorus extracted from the soil are applicable to a specific crop and a particular soil only. Many attempts have been made to formulate a 'universal' extractant for available phosphorus but with no success and it is difficult to envisage such a solution ever being found.

Early attempts made were to find an extractant that would remove phosphorus in a manner similar to that of plant roots, it not being realized that different plants can extract different amounts of phosphorus from the same soil. Thus, one of the first procedures was to extract soil with carbon dioxide-saturated water to simulate the action of roots. Puri and Asghar (1936) passed carbon dioxide through a 2% aqueous soil suspension for 15 minutes and determined phosphorus in the filtrate. Dyer (1894) extracted soil with 1% citric acid solution at constant temperature for 7 days and with modifications his method is still used today. Metson (1961), for example, recommends mechanically shaking the soil sample with 1% citric acid solution for 24 hours. Truog (1930) used 0·001 M sulphuric acid containing 3 g dm^{-3} ammonium sulphate and extracted for 30 minutes using a 1:200 soil–solution ratio; Truog's method was modified only slightly by Peech *et al.* (1947) by employing a 1:100 soil–solution ratio. Fraps (1911) used 0·2 M nitric acid. The so-called 'Carolina' procedure (Nelson *et al.*, 1953) is to extract the soil with a mixture of 0·05 M hydrochloric acid and 0·0125 M sulphuric acid, and Warren and Cooke (1962) favoured dilute hydrochloric acid as being the best extractant for acid soils. Williams (1950) used 2·5% acetic acid containing 8-hydroxyquinoline to prevent re-adsorption of phosphorus by iron and aluminium; Joret and Herbert (1955) used 0·2 M ammonium oxalate solution.

Acidic solutions are unsuitable extractants for calcareous soils which are more reliably extracted with alkaline solutions. Das (1930) suggested 1% potassium carbonate solution and Olsen *et al.* (1954) used 0·5 M sodium hydrogen carbonate solution at pH 8·5, an extractant which controls the removal of calcium phosphate. Saunder (1956) found previously described alkaline extractants not applicable to tropical red earths and proposed a much stronger alkaline solution using hot 0·1 M sodium hydroxide. Saunder first leached the soil with acidic sodium chloride solution to remove exchangeable calcium and so increase the phosphate recovery; the hot sodium hydroxide reagent also extracts the less available iron and aluminium phosphates.

A third type of extractant is a weakly acid buffered salt solution such as Morgan's acetic acid–sodium acetate solution or Egnér–Riehm's (1955) reagent which is 0·04 M in acetic acid and 0·1 M in ammonium lactate. If the pH of such solutions is low they are unsuitable for calcareous soils and

Behrens (1962) found Egnér's solution unreliable as it dissolves the relatively unavailable fluor and other apatites; van Diest (1963) on the other hand, found that Egnér's solution gave results which correlated well with yield figures.

Rapidly acid-soluble and adsorbed phosphorus has been extracted with fluoride-containing solutions. Thus Bray and Kurtz (1945), attempting to extract available phosphorus from different soils by different methods, removed acid-soluble phosphorus with 0·1 M hydrochloric acid, adsorbed phosphorus with neutral 0·5 M ammonium fluoride solution, and acid-soluble plus adsorbed by first shaking with the dilute acid and then adding solid ammonium fluoride and shaking again. Dupuis (1950) modified the Bray and Kurtz procedure by shaking 1 g of soil with 8 cm^3 of 0·03 M ammonium fluoride solution in 0·025 M hydrochloric acid, and the method was modified further by Smith and co-workers (1957) by using a 1:50 soil-solution ratio. Datta and Kamarth (1959) extracted soil with a solution 0·03 M in ammonium fluoride and containing 0·1% EDTA.

A highly available form of phosphorus is the 'saloid-bound' phosphorus, defined by Mattson and Karlsson (1938) as that held as an H_2PO_4–Ca–micelle linkage over a limited pH range. Saloid-bound phosphorus is removed by dilute neutral salt solutions such as 5% ammonium chloride or 0·01 M calcium chloride. In general, neutral salt solutions are better extractants of phosphorus than water, although Burd (1948) favoured water-soluble phosphorus as being the best estimate of available phosphorus and Thompson et. al. (1960) extracted available phosphorus with hot water.

The different extracting solutions for available phosphorus have often been compared; Commen et al. (1959), for example found that the results of Das's method (1% K_2CO_3) were correlated significantly with those of Dyer's (1% citric acid) and Olsen's (0·5 M $NaHCO_3$) when dealing with red and lateritic soils but not with alluvial soils. Significant correlations were obtained between Das's and Truog's methods with soils having high cation exchange capacities. Stelly and Ricaud (1960) found that the amount of available phosphorus measured depended not only on the extractant but on the soil–solution ratio. The best general extractant was thought to be 0·03 M ammonium fluoride in 0·1 M hydrochloric acid using a 1:20 soil–solution ratio. Thompson et al. (1960) correlated the results of plant uptake of phosphorus with those obtained by various extractants using twenty-two soils and concluded that the efficiency of the various extractants was as follows: lactic acid–calcium lactate ≤ 2% citric acid ≤ 0·03 M NH_4F–0·025 M HCl < water. Mattingly and Pinkerton (1961) used fifteen soils and correlated the results of several different extraction procedures with yield of rye-grass. The soils were mixed with ^{32}P-labelled superphosphate and kept for 30 days at field moisture capacity before extracting. Mattingly and Pinkerton found that the amounts of ^{32}P that exchanged with soil phosphorus during extraction was comparable to that exchanged in similarly treated soils which were then cropped. Phosphorus extracted with

0·05 M sodium hydrogen carbonate, Morgan's reagent, 0·3 M hydrochloric acid and 0·001 M sulphuric acid correlated well with yields. Weir (1962) used soils whose pH varied from 5·7 to 8·2 and found that the extraction of phosphorus decreased with various reagents in the following order, $0·001 M H_2SO_4 > 0·1 M HCl–0·05 M NH_4F > 0·5 M NaHCO_3 >$ Na acetate–acetic acid > hot 1% NaOH. Warren and Cooke (1962) found that the best general extractant was dilute hydrochloric acid, except for calcareous soils when water or calcium hydrogen carbonate was best. Warren and Cooke considered that methods involving narrow soil–solution ratios are better than those in which large amounts of solvent are used. Breland and Sierra (1962) listed a variety of extractants in decreasing order of efficiency, finding Bray and Kurtz's acid fluoride solution most efficient and carbon dioxide water least efficient. Tombesi and Calé (1962) compared extraction methods using volcanic soils and concluded that dilute solutions of potassium salts of organic acids should be used to bring phosphorus into solution by ion exchange processes, thus reducing fixation by iron, aluminium and calcium. Tombesi and Calé further suggested that extraction of soil by percolation is preferable to shaking. Williams and Knight (1963) considered that the usefulness of an extraction procedure depends largely upon avoidance of attack upon non-available phosphorus; thus the highest correlations with yield were with mild extractants for short times at intermediate pH values, for example the calcium lactate or hydrogen carbonate procedures.

Available phosphorus determined by chemical extraction procedures includes all forms of phosphorus, but predominantly that associated with iron, aluminium and calcium; the relative amounts brought into solution will depend partly upon the amount present in the soil and partly on the solubilities in the various solvents. Thus, from calcium-rich soils, acid extractants such as Truog's reagent will extract more calcium phosphate than other forms. As each different soil will have one or more inorganic forms of phosphorus as the main source of available phosphorus, an extractant should be chosen to best reflect the phosphorus status of that soil; that is, calcium phosphate soils are best extracted with acids, iron and aluminium phosphate soils with alkalis and so on. With these points in mind Chang and Juo (1963) examined different soils for available phosphorus with respect to the forms of phosphorus present, which were identified by the fractionation procedure of Chang and Jackson (section 12:2:2). In one group of eleven soils, iron phosphate was most available to plants and high correlations were obtained with seven different extractants; a second group of eight soils predominated in calcium and iron phosphates and gave high correlations with Olsen's hydrogen carbonate extraction, Peech's acetate extraction and Bray's acid fluoride extraction. For a group of seven soils predominating in calcium phosphate, high correlations were obtained with all the acid extractants and especially those which also contained fluoride.

Several workers have attempted to associate a specific inorganic form of phosphorus with the phosphorus available in that soil. Thus Smith

(1965b), using wheat as a test crop, found that aluminium phosphate was the chief source of phosphorus from a slightly acid soil. Hanley (1962) tested six different soils derived from calcareous and non-calcareous parent materials for plant-available phosphorus by pot tests and fractionated the soil inorganic phosphorus. He found aluminium phosphate to be taken up preferentially, iron phosphate to be important and that calcium phosphate contributed little to plant nutrition; the occluded forms of phosphorus were unavailable.

Further discussion of plant-available phosphorus as determined by chemical extractants can be found in the literature. Numerous reviews have appeared and several articles have been published in Technical Bulletin No. 13 of the Ministry of Agriculture, Fisheries and Food (1965).

As chemical extraction procedures are liable to interfere with the colloid chemical reactions in a soil it has been proposed to use anion exchange resins as a means of extracting available phosphorus. Sheard and Caldwell (1955) extracted phosphate from soil with water and Amberlite IRA-400. Saunder and Metelerkamp (1962), working with Rhodesian soils, claimed a high correlation by this method between phosphorus removed and plant response. Saunder and Metelerkamp used Dowex 21.K resin and the soil was finely ground so that it would pass through a sieve (0·5-mm) on which the resin (1·5-mm) would be retained. Equal weights (4 g) of soil and resin were shaken overnight with 200 cm³ of water and the mixture decanted into the sieve to separate the resin. The extracted phosphorus was leached from the resin with sodium hydroxide and hydrochloric acid solutions. In some cases it is not possible to grind the soil finely and in order to separate the resin, a method other than sieving must be found. Encouraging, but not satisfactory, results were obtained (Hesse, unpublished) by enclosing the resin in small plastic capsules made from Tenoplast tubing by sealing the ends and pricking holes through the sides. Several such capsules were shaken with the soil suspension and subsequently picked out for further treatment.

Birch (1953) found that with acid- or base-unsaturated soils of East Africa no correlation could be obtained between crop response and solvent-extracted phosphorus. Birch found that the lower the pH or, more exactly, the lower the per cent saturation of the exchange complex, the greater was the response to phosphorus, the effect probably being due to iron and aluminium. He showed that much better correlation could be obtained with water-soluble silica which was closely related to base saturation.

Friend and Birch (1960) examined ten different extracting methods for available phosphorus and found that of these only total organic phosphorus and the inorganic phosphorus removed by hot 0·1 M sodium hydroxide were significantly related to crop response. When the amount of organic phosphorus was considered in conjunction with the phosphate retention capacity of the soil to give a measure of available mineralized phosphorus, the response relationships were more significant than with organic phosphorus alone. As mineralized phosphorus is partly re-fixed, Friend and Birch developed an equation for assessing available phosphorus. The approximate amount of mineral phosphorus produced equals $K \times$ total organic phosphorus where K is the fraction mineralized. Of this amount a certain proportion, indicated relatively by phosphorus retention data, will be retained; that is an amount equal to

$$(\text{total organic-P}) \times K \times (\text{P retention})/100,$$

and hence the plant available phosphorus is given by,

$$K \left[(\text{total org.-P}) - \frac{(\text{total org.-P} \times \text{P retention})}{100} \right]$$

and as the value of K does not influence the relation it can be ignored. Friend and Birch applied this formula to various soils and related the results to plant response; regressions were made and correlation coefficients obtained. The approach was recommended as being better than chemical extraction procedures, which take no account of the magnitude of the source of the available phosphorus or of conditions under which the plant has to compete in order to obtain the phosphorus, as for example, phosphorus retention or microbial assimilation.

Phosphate potentials

In an attempt to overcome the arbitrary nature of determining available phosphorus by chemical extraction procedures, Schofield (1955) introduced the concept of 'Phosphate Potential'. When introducing the subject, Schofield likened phosphate potential to water potential, pointing out that it is possible to measure the amount by which the potential of water in a soil is lower than that of pure water. Such decrease in water potential had already (1935) been expressed by Schofield as a vertical height, the logarithm of which was denoted by the term 'pF'. Schofield had found further that water in a soil at pF 4·2 or more is not available to plants and thus available water is that which can be removed before the pF value rises to 4·2.

Most of the easily removed water in a soil (that removed by heating to 100°C, for example) constitutes a labile pool at a common potential. If some water is removed there will be a rise in potential of all the water remaining in the pool. Similarly for phosphorus, and experiments with radioactive phosphorus showed the existence of a 'pool' of labile—that is, isotopically exchangeable—phosphorus from which plants obtain their needs. The larger the phosphate potential, the lower will be the phosphate activity and hence concentration. Thus the opposite effect is obtained from that found with lime potentials (section 3:3) where the higher the potential the higher the calcium activity; this difference is due to the definitions of the two potentials being different.

Schofield suggested that the availability of phosphorus be determined by its chemical potential and by the rate of decrease in potential as phosphorus is removed. The chemical potential cannot with certainty be measured, the only sure determination being that of the phosphorus concentration in the soil solution; the chemical potential of the ions in solution is not necessarily the same as in the solid phase. Hence Schofield's Phosphate Potential is measured as an approximation of the chemical potential and depends upon the satisfaction of his ratio law requiring the activity product,

$$(a_{Ca}{}^{\frac{1}{2}} \times a_{H_2PO_4}),$$

in the soil solution being independent of other ions in solution. Experiment has shown this law to be obeyed in most soils over the range of concentrations normally found.

The potential is obtained by measuring pH and total phosphorus in a dilute calcium chloride solution which has been brought to equilibrium with the soil. Calcium chloride solution at a concentration of 0·01 M is a suitable electrolyte causing little disturbance to the system. The negative potential of monocalcium phosphate in soils is calculated from pH and the concentration of calcium and phosphorus,

$$\tfrac{1}{2}pCa = -\tfrac{1}{2}(\lg \text{concn. Ca} + \lg f), \qquad [12:2$$

where
$$\lg f = \frac{-AZ^2\sqrt{\mu}}{1 + aB\sqrt{\mu}}$$

and where in this case concn. Ca $= 0·01$ M and $A = 0·5$, $aB = 1·5$, Z is the valency and μ the ionic strength.

$$pH_2PO_4 = -\lg(\text{concn. } H_2PO_4 + \lg f) \qquad [12:3$$

where
$$-\lg \text{concn. } H_2PO_4 = p(P) + p\left\{\frac{H}{K'' + H}\right\}$$

and where P is the total concentration of phosphorus in solution and

$$p\left\{\frac{H}{K'' + H}\right\}$$

is a correction factor (Åslyng, 1954) relating pH to H_2PO_4/P (White and Beckett, 1964).

The phosphate potential,

$$(\tfrac{1}{2}pCa + pH_2PO_4),$$

serves as an index of phosphorus availability in the same manner as pF indicates water availability. Schofield recognized that the method is not quite accurate and pointed out that a universally applicable limiting value of phosphate potential corresponding to the pF 4·2 value cannot be expected. The expression

$$(\tfrac{1}{2}pCa + pH_2PO_4)$$

is the logarithmic equivalent of the activity products

$$\sqrt{a_{Ca^{2+}}} \times a_{H_2PO_4^-}$$

and theory predicts that it is independent of calcium concentration, thus being a characteristic of soil in the field.

Russell (1961) considers that Schofield's phosphate potential is a useful measure of the intensity with which a soil can hold its most readily available phosphorus against extraction (the lower the potential the more weakly the phosphorus is held) but points out that there is no evidence to show whether the potential controls the phosphorus uptake or the phosphorus concentration. Wild (1964) found the potential useful to describe the solubility of soil phosphorus but not for predicting availability. Wild experimented to discover whether a low uptake by plants of phosphorus from soil solution was due to a low phosphorus concentration or to too low a potential, and concluded that a low concentration *per se* was the limiting factor.

Le Mare (1960) found that for the red clay loams of Uganda, phosphate potentials gave a good indication of phosphorus uptake by cotton; the potential was linearly and significantly related to pH and often to organic matter as well.

Moorthy and Subramanian (1960) considered that for finding the availability of phosphorus to rice a phosphate 'equilibrium potential' gave better correlations than Schofield's phosphate potential. The equilibrium potential was measured by equilibrating the soil with small graded amounts of $CaH_4(PO_4)_2$ in amounts roughly corresponding to applied fertilizer phosphorus. The initial calcium phosphate potential in solution was then found which would not undergo any change after such equilibration. After shaking the soil with calcium phosphate solution (1:10) for 7 days, a curve was plotted connecting the initial and final potentials, from which the desired potential was found.

White and Beckett (1964) equilibrated 2·5-g samples of soil with 25 cm^3 of 0·01 M calcium chloride solution containing known amounts of calcium dihydrogen phosphate, the phosphate concentration ranging from 0 to 5×10^{-5} M. Phosphorus, calcium and pH analyses were then made and the difference between the initial phosphate concentration of each solution before and after shaking gave the amount, ΔP, of phosphorus gained or lost by the soil. $\pm \Delta P$, expressed as mol P $\times 10^{-8}$ per gramme of soil, was then plotted against phosphate potential giving curves similar to that shown in Figure 12:2, and values of the equilibrium phosphate potential were obtained by interpolating at $\Delta P = 0$. The procedure is valid only if soils of comparable calcium status are examined and if the concentration of the soil solution is not too high. White and Beckett pointed out that although air-drying a soil can cause changes in phosphate potential, the changes are not sufficient to warrant losing the convenience of using air-dried samples.

Åslyng (1964) and Olsen *et al.* (1960) have reported that phosphate

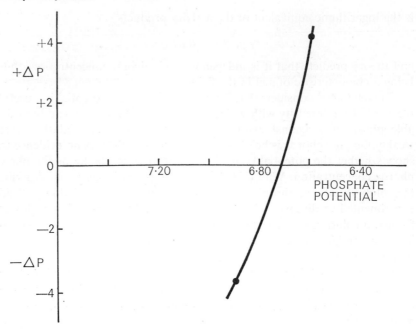

Figure 12:2 Typical curve obtained during the determination of phosphate equilibrium potentials (White and Beckett, 1964; *Pl. Soil*)

potentials vary with the ratio of soil to solution, decreasing as the proportion of soil increases. Åslyng, and Weir and Soper (1963) attempted to correct for the concentration effect by extrapolating to zero concentration, but there is no approach to a limiting value towards dilution as shown by Sutherland and Black (1959) and by Larsen and Court (1960). Olsen *et al.* equilibrated soil with calcium chloride solution for 4 days at constant carbon dioxide pressure and then measured the hydrogen carbonate in solution. This was found to increase with amount of soil and they suggested that the observed decrease in potential, that is, increase in phosphorus concentration, was due to the ability of carbonate ions to displace HPO_4^- ions. Thus they concluded that the effect was due to an accumulation of carbon dioxide from microbial activity in the stoppered flasks. To test this hypothesis Larsen and Widdowson (1964) shook one set of samples in stoppered bottles when it was found that the pH decreased, phosphate concentration increased and phosphate potential decreased with increasing weight of soil. A second set of samples were shaken in open bottles when the same results were obtained but to a lesser extent. A third set had air bubbled through the suspension after passing through 0·01 M calcium chloride solution, the flasks being gently shaken; in this case all values were independent of the soil–solution ratio. Larsen and Widdowson point out that this effect must be considered when interpreting phosphate (and lime) potentials as the carbon dioxide of soil air varies in the field. Because some soils give soil solutions which are greater than 0·01 M in calcium,

Larsen (1965) determined lime and phosphate potentials on four different soils using calcium chloride solutions from 0·002 M to 0·05 M. 10 g of soil were shaken for 16 hours at room temperature with 50 cm³ of solution and with air bubbling through the suspension. The pH of the suspension was then measured and phosphate determined in the filtrates; activities were then calculated. Larsen found the lime potential independent of calcium chloride concentration but the phosphate potential of neutral and alkali soils decreased as the concentration of calcium chloride increased. However, if allowance is made for the formation of soluble $CaHPO_4$, then the phosphate potential is also independent.

It was pointed out by Russell (1961) that in certain acid soils the potentials of aluminium or iron phosphates may be more important than that of calcium. This was found experimentally to apply to potassium potentials, by Tinker (section 9:1:2) who consequently included an aluminium salt in the equilibrating solutions.

In the field, phosphate potential depends upon the volume of soil utilized by the plant roots and therefore upon the physical properties of the soil. The definition of the phosphorus status of a soil requires a 'capacity' factor, potentially available phosphorus, and an 'intensity' factor, strength of retention by soil. The intensity factor (I) is given by the phosphate potential and White and Beckett (1964) taking the symbol Q as the quantity parameter, worked out Q/I relationships as giving a measure of immediately available phosphorus. The capacity factor is the pool of labile, isotopically dilutable inorganic phosphorus and can be determined directly as the 'A' or 'L' value using radioactive phosphorus. A known amount of soluble ^{32}P is added to the soil, and after cropping, the ratio $^{31}P/^{32}P$ is determined in the plant. If the added phosphorus has been adsorbed by the plant in the same form as the readily available phosphorus in the soil, it will be in isotopic equilibrium with that soil phosphorus, and thus we can calculate the 'A' value or amount of phosphate in the reservoir of readily available phosphorus. The term 'A' value is an abbreviation of 'available' phosphorus, as the labile phosphorus is considered as available. Strictly speaking, 'A' values are not the same as 'L' values (labile phosphorus) and cannot be calculated as soil phosphorus and fertilizer phosphorus are considered as distinct sources of phosphorus, and 'A' values are measured from the relative uptake from soil and fertilizer. When soil and fertilizer are mixed, however, there is an isotopic balance of phosphorus, and hence the term 'L' value has been introduced. It can be calculated from the general equation,

$$\text{Labile-P} = b\left[\frac{(1-y)}{y}\right] \qquad [12:4$$

where b is the amount of labelled fertilizer phosphorus added to the soil and y is the fraction of phosphorus in the crop derived from the fertilizer (Fried and Dean, 1952; Mattingly, 1957). Sheard and Caldwell (1955) determined 'L' values with the assistance of exchange resins. The resin was prepared by adding 210 g of air-dry Amberlite IR-4B to 1 dm³ of water containing 10 cm³ of 85% H_3PO_4 and approximately 375 μCi of $^{32}P/g$ ^{31}P were added. After 6 hours the liquid was poured off and the resin well washed with water. Plants were grown in the soil which was mixed with the resin and eventually analyzed for total phosphorus and ^{32}P. Then,

$$\text{'}L\text{'-value} = \frac{\%\,P \text{ from soil}}{\%\,P \text{ from fertilizer}} \times P \text{ added.} \qquad [12:5$$

The resin procedure was found to be superior to and simpler than adding radioactive calcium phosphate to the soil as fertilizer.

Williams (1937) considered that for the determination of phosphorus available to crops with extensive roots and for slow-growing crops, it is best to measure 'L' values, whereas phosphate potentials will give better results for quick-growing, responsive crops with a limited root range.

The 'L' value is related to the 'E' value which is the amount of isotopically exchangeable phosphate and which is measured by including $H_2{}^{32}PO_4$ in each phosphate standard and determining ^{32}P after the solutions reach equilibrium with the soil. Thus if y is the initial radioactivity, y_t the radioactivity after time t and x_t the quantity of phosphate at time t, then

$$\frac{y_t}{x_t} = \frac{y - y_t}{E_t} \qquad [12:6$$

and E_t is measured in mol P $\times$ 10^{-8} per gramme of soil (White and Beckett, 1964). Measurement of 'E' values overcomes the difficulty of measuring small quantities of phosphorus in the soil solution.

Larsen and Sutton (1963) indicated that when determining labile soil phosphorus by isotopic dilution, no phosphorus should be removed, so as not to disturb any equilibrium between labile and non-labile forms. 'L' value measurements, which involve a growing plant, do remove phosphorus, whereas 'E' value measurements do not.

Biological methods of determination

As chemical methods of estimating available phosphorus are limited in their usefulness it is preferable to measure actual plant response. Field tests of this nature, however, are expensive and time-consuming. Mitscherlich (1923) was a pioneer of the plant response method and, assuming that plant growth is proportional to the concentration of phosphorus, he grew plants to maturity in the soil before analysis.

Neubauer (1932) reduced the scale of plant response experiments by using small quantities of soil to grow a large number of seedlings. It was assumed that the plants would exhaust the available nutrient, and after 17 days' growth the seedlings were analyzed. In order to accommodate sufficient plants in the small quantity of soil, sand was used as a diluent. Temperature and moisture conditions must be controlled and seeds of uniform germination are required. A modification of Neubauer's technique using tomato plants grown for 3 weeks was presented by McDonald (1933).

The fungus *Aspergillus niger* has been used to estimate many available nutrients in soil, including phosphorus, and was introduced by Niklas *et al.* (1927). The technique is to add a small quantity of soil to a solution containing sugar, peptone and certain nutrients other than phosphorus, and citric acid to inhibit bacteria. A standard inoculum of mould spores is added and the flask kept in an incubator. After a certain number of days the mycelium is removed, dried and weighed; the weight determines the phosphorus status of the soil. Niklas's technique was criticized as several factors affecting the results were not considered. Smith *et al.* (1932) adapted the method by treating soil with different amounts and different combinations of lime and phosphate and estimating *A. niger*-soluble phosphorus using 1 % citric acid as culture medium. Várallay (1934) eliminated most of the variables by adding known amounts of phosphorus and finding the increase in yield of mycelium over that obtained in untreated soil. Mehlich and co-workers (1935) described a method that depended upon the growth of the fungus *Cunninghamella* sp. The fungus is grown in a soil treated with phosphorus-free nutrients on a special clay plaque, and the best species were found to be *C. elegans* and *C. blakesleeana*. Available phosphorus was estimated by linear measurement of the fungal growth. Arriaga e Cunha (1960) found that the *Aspergillus niger* method gave erroneous results due to the fungus producing acid and lowering the pH of the medium; he found that a similar technique but using *Curvulria* sp. gave better results.

Winogradsky and Ziemiecka (1928) noted that *Azotobacter* produces characteristic colonies upon a soil plaque if the soil has enough energy material and contains certain elements including phosphorus; if any of these elements are absent no colonies are produced. Niklas adapted this fact for measuring available nutrients by covering a soil–sand mixture with a 2 % mannitol nutrient solution and incubating for 7 days. Four such units were incubated, one in which the solution contained potassium sulphate, one with sodium phosphate, one with potassium phosphate, and the fourth with no addition, being used as a control. The amount of bacterial growth was measured and phosphorus (or potassium) deficiencies detected; the method is qualitative rather than quantitative.

12:2:6 Phosphorus fixation

When phosphorus in soil is unavailable to plants it is said to be 'fixed'. A certain proportion of the soluble phosphorus added to soils as fertilizer is always fixed in one way or another depending upon the kind of soil. If phosphate added to a soil increases crop yield, the soil is said to be deficient in phosphorus, even though the amount added may be very small compared with the total phosphorus present in the soil; such soils contain large amounts of fixed phosphorus.

The cause of phosphorus fixation varies with the kind of soil and prevailing soil conditions. In acid soils phosphorus is fixed mainly by aluminium and iron and in alkaline soils by calcium. The calcium phosphate is released if the pH drops below 5·5, and at this pH, aluminium and iron phosphates are not released. In order to remove aluminium or iron phosphates the pH must be greater than 9, or else fluoride ions must be present. In many soils the organic matter is partly responsible for phosphorus fixation.

Williams (1950), studying acid surface soils, found that their phosphate retention capacities ranged from 0·5% to 3·0% P_2O_5 and depended mainly upon soluble aluminium, but were also significantly correlated with the amounts of oxalate-soluble iron, loss on ignition and organic carbon. Volk and McLean (1963) examined four acid soils by adding water-soluble phosphorus labelled with radioactive ^{32}P. They found that nearly all the phosphorus could be recovered as aluminium and iron phosphates, iron phosphate predominating in soils of high fixation capacity and aluminium phosphate in soils with low fixation capacity. Bromfield (1964) assessed the importance of iron and aluminium in phosphorus fixation by pre-treating soil to reduce the amount of free oxides. Bromfield concluded that for the soils studied, phosphorus fixation was dominated by the effect of aluminium and not by the relatively large amounts of reducible iron.

The fixation of phosphorus can be an extremely rapid process, especially in acid soils containing free iron or aluminium. In a mud from West Africa, for example, added phosphorus was completely fixed within 30 minutes and very largely fixed within 5 minutes; the difference in amount of phosphorus fixed in 30 minutes and 30 days was negligible (Hesse, 1962).

In recent years it has been noted by several workers that phosphorus initially fixed by aluminium is slowly converted into iron phosphate; this phenomenon was particularly noticed after the fractionation procedure of Chang and Jackson (section 12:2:2) became known. Yuan and co-workers (1960), for example, found that aluminium was particularly active in phosphorus fixation but that in time the phosphorus appeared as iron phosphate; the effect was enhanced by wetting and drying the soil and by increasing the temperature of drying. Chang and Chu (1961) added soluble phosphorus to six soils varying in pH value from 5 to 7. At first the added phosphorus was fixed mainly by aluminium, probably on the surface of the solid phase, but with increasing time (100 days) the aluminium and calcium phosphates decreased in amount while the iron-bound phosphorus increased. Chiang (1963) fractionated the inorganic phosphorus in paddy soils and found that the amounts of the different forms varied according to pH, *Eh* and parent material. Added phosphorus was fixed initially by aluminium, but as waterlogged conditions were prolonged, iron phosphate increased as aluminium phosphate decreased.

During an investigation of the phosphorus in a mangrove swamp mud (Hesse, 1962), it was found that in spite of high extractable aluminium content (5%) negligible amounts of aluminium phosphate were present, even in soils known to have been fertilized with excess soluble phosphate. From a mud containing 553 μg g^{-1} phosphorus, only 6 μg g^{-1} of phosphorus were associated with aluminium. Subsequent work (Hesse, 1963) showed that although aluminium was in fact responsible for the initial fixation of phosphorus, within 3 months all the aluminium-fixed phosphorus had become associated with iron (Figure 12:3). These findings

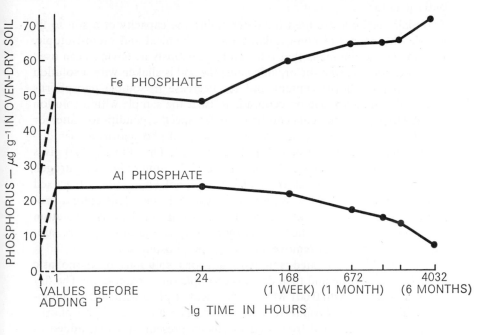

Figure 12:3 Change with time in forms of inorganic phosphorus in phosphate-treated mangrove swamp soil

were of particular interest when considering the use of phosphate as a means of reducing aluminium toxicity. Yuan and co-workers (1960) considered that the explanation for the transfer with time of phosphorus from aluminium to iron lies in the lower solubility product of iron phosphate. Re-sorption on sesquioxides of phosphorus dissolved in ammonium chloride solution during fractionation (Bromfield, 1967) may, however, account for a dominant presence of iron phosphate.

Harris and Warren (1962) investigated the phosphorus fixing powers of highly organic soils by shaking with an aqueous solution of phosphorus (20 µg cm^{-3}) using a soil–solution ratio of 1:20. After 2 hours' shaking the phosphorus still in solution was compared with the original concentration. No correlation of phosphorus fixation with total iron, aluminium or calcium was found.

Birch (1964), in a study of phosphorus transformations in soil, found that one broad group of micro-organisms used only inorganic phosphorus and another group obtained phosphorus by mineralization of organic compounds. The first group may lead to immobilization of inorganic phosphorus, particularly when a dry soil is wetted, and may absorb all the inorganic phosphorus produced by mineralization of organic matter. The phosphorus would again be liberated only on the death and lysis of the microbial cells.

Determination

Two main methods are used for determining the capacity of a soil to fix soluble phosphorus in non-available forms; chemical and radio-isotopic. The simplest technique is that referred to previously as having been used by Harris and Warren for organic soils; the soil is shaken with a solution of known phosphorus content and the decrease in soluble phosphorus measured. An alternative procedure is to treat the sample with a solution of known phosphorus concentration under specific conditions, filter or centrifuge and wash out excess reagent. The fixed phosphorus is then displaced with a suitable extractant and determined. Thus Piper (1942) gives the soil sample a prolonged treatment with 1 M ammonium dihydrogen phosphate solution, washes the soil with ethanol and extracts the fixed phosphorus with hot, M/8 sodium hydroxide solution. This second procedure is not satisfactory as it is difficult to wash out the excess solution without causing loss of phosphorus by hydrolysis and also because subsequent extraction may remove some organic phosphorus.

Doughty (1935) treated soil with aqueous solutions of monobasic potassium phosphate and then dried the sample on a steam-bath. The dry soil was extracted with 0·001 M sulphuric acid at pH 3 with added ammonium sulphate and phosphorus determined. The amount of phosphorus extracted was subtracted from that originally present plus that added to give the amount fixed.

Hibbard (1937) used two variations of the method depending upon reduction in concentration of a solution of phosphate shaken with soil; in one variation the same amount of phosphorus was added to all soils and in the second variation the amount of added phosphorus was varied according to the kind of soil so as to give approximately 1 μg cm^{-3} phosphate in all filtrates. The second variation, which determines the amount of phosphorus necessary to yield a definite, small amount in the extract, was thought to be the better. Hibbard defined the 'phosphorus fixing power' of a soil as the number of mg P/kg soil that must be added so that a 1:1 water–soil extract will contain 1 μg cm^{-3} P.

Bass and Sieling (1950) considered it best to accept the predominant role of iron and aluminium in phosphorus fixation by acid soils and to determine these elements as a measure of the fixation capacity. Bass and Sieling thus extracted samples with 0·5 M citric acid and measured iron plus aluminium gravimetrically together as the phosphate. The phosphate figure obtained gave the relative fixation capacity of the soil for phosphorus.

Phosphorus fixation is influenced by pH, calcium ions, silica–sesquioxide ratios, soil–solution ratio and the physical properties of a soil. Doughty (1935) found that destruction of organic matter with hydrogen peroxide decreased the fixation capacity of a soil and that heating the sample to 800°C completely destroyed the capacity. Phosphorus fixation is also influenced by a time factor and thus two soils having the same fixation capacity may have very different rates of fixation; consequently all methods of determining fixation capacity are empirical.

Olsen and Watanabe (1957) determined the phosphorus adsorption maximum of soils by use of the Langmuir isotherm which gave better results than did the Freundlich isotherm. Olsen and Watanabe found that adsorption of phosphorus was closely related to surface areas as measured by ethylene glycol retention. In the experiments 5 g of soil were shaken with 100 cm^3 of dipotassium hydrogen phosphate solutions of various strengths for 24 hours at 24°C. All solutions were initially neutral and aliquots of filtrates were analyzed for phosphorus. Adsorbed phosphorus was calculated from the change in phosphorus concentration in solution. A second 5-g sample was shaken with water containing a known amount of ^{32}P for 48 hours and after centrifuging the solution was analyzed for ^{32}P and ^{31}P. Surface adsorbed phosphorus was then calculated from the expression,

$$^{31}P \text{ (surface)} = \frac{^{32}P \text{ (surface)} \times \text{ }^{31}P \text{ (solution)}}{^{32}P \text{ (solution)}} \qquad [12:7$$

Larsen and co-workers (1959) measured the fixation capacity of organic soils by first shaking 0·75 g oven-dry equivalent of soil for 14 hours with 24 cm^3 of water, centrifuging aliquots and analyzing the extract for phosphorus. 1 cm^3 of water containing 2·25 mg P as $NH_4(H_2{}^{32}PO_4)$ was then added to the tube which was again shaken. At various time intervals aliquots of suspension were centrifuged and analyzed for ^{31}P and ^{32}P.

Datta and Srivastava (1959) used autoradiography to study phosphorus fixation and movement in soil columns. Split plastic tubes were used for containing soil columns which were leached with a solution containing ^{32}P. After moisture equilibrium had been attained, the columns were cut in half by separating the tubes and the half-columns inserted in a camera loaded with X-ray film.

As phosphorus is fixed in different soils by different mechanisms it is inadvisable to specify one particular method for determination of phosphorus fixation capacity. The definition of the term 'fixed' indicates that the phosphorus becomes unavailable to plants, and if we consider that phosphorus soluble in very dilute sulphuric acid is a measure of available phosphorus, then Doughty's method, for example, is not tenable. The method depending upon reduction in phosphorus concentration of a solution shaken with soil appears to be the best generally applicable chemical procedure. For acid soils the method of Bass and Sieling to obtain the relative fixation capacity gives useful additional information, while for alkaline soils a similar figure based upon calcium phosphate would be relevant. For a meaningful analysis there should be a specific reason for determining the fixation capacity; for example, it may be desired to predict the fate of a phosphorus fertilizer applied to a certain soil in relation to wheat growth. This enables us to define more closely what we mean by 'fixed', and in this case would mean unavailable to wheat under certain prescribed conditions and during a known length of time. A field trial or greenhouse experiment in conjunction with a selected group of chemical experiments will indicate the best procedure to adopt as a routine.

12:2:7 Phosphorus in solution

Once in solution, phosphorus can be estimated volumetrically, complexometrically, gravimetrically or colorimetrically.

Volumetric methods

The classic methods of determining soluble orthophosphate are described, *inter al.*, by Vogel (1962). In one method the phosphate is reacted with ammonium molybdate in the presence of nitric acid and the precipitate of ammonium phosphomolybdate is washed with dilute potassium nitrate solution to convert it to $(NH_4)_3PO_4 . 12MoO_3$. The precipitate is then dissolved in excess standard sodium hydroxide solution, and the excess alkali titrated with standard hydrochloric acid using phenolphthalein as indicator. In another procedure the precipitated ammonium phosphomolybdate is dissolved in ammonium hydroxide, sulphuric acid added, and the molybdenum reduced to the tervalent state by passing the solution through a Jones reductor. Excess iron(III) alum solution is added and the excess iron(II) titrated with potassium permanganate solution. Neither of these two methods has been much used in soil analysis.

Wilson (1951, 1954) developed a volumetric method of determining phosphorus in soil extracts dependent upon precipitation of quinolinium phosphate,

$$(C_9H_7N)_3H_3PO_4 . 12MoO_3 + 26NaOH$$
$$= Na_2HPO_4 + 12Na_2MoO_4 . 3C_9H_7N + 14H_2O \quad [12:8]$$

The quinolinium phosphate is very insoluble and the reaction is quantitative. Interfering silicon, which is also precipitated as silicomolybdate, is suppressed by addition of citric acid which forms a complex with molybdic acid, permitting it to react with phosphorus but not with silica.

For complexometric determination of phosphorus the element is precipitated as magnesium ammonium phosphate which is then dissolved in hydrochloric acid. Excess EDTA solution is added, the pH adjusted to 10 and the excess reagent back-titrated with standard magnesium solution. Other complexing ions must be absent and the method is not recommended for soil extracts.

Gravimetric method

The gravimetric determination of phosphate involves precipitation as phosphomolybdate which is then weighed. The method was applied to soil analysis by Lorenz (1901) and more recently has been described by Strzemienski (1955). The procedure is both lengthy and tedious and is considered unsuitable for routine work.

Colorimetric methods

The most commonly employed methods of determining phosphorus in solution are those dependent upon the formation of coloured complexes. Nowadays in soil laboratories the element is determined in this manner almost exclusively, and with modern techniques the method is by far the simplest and quickest and is capable of a high degree of accuracy.

There are two main colorimetric procedures, each of which has advantages over the other in certain circumstances; one depends upon the combination of phosphorus with the molybdate ion and the other upon its reaction with vanadomolybdate. The most extensively used method is the first and was introduced by Osmund (1887). Phosphorus reacts with molybdate to form a heteropolymolybdic complex with phosphorus as the central coordinating atom,

$$H_3PO_4 + 12H_2MoO_4 \longrightarrow H_3P(Mo_3O_{10})_x + 12H_2O \qquad \text{(Boltz } et\ al., 1949)$$

$$[12 \cdot 9$$

Phosphorus, however, is not unique in this manner of combination, and other ions, for example As^{5+}, Si^{4+}, Ge^{4+} and B^{3+}, can react with molybdate in the same way and suitable precautions have to be taken against this. The complex is yellow-coloured but on partial reduction some of the Mo^{6+} is converted to Mo^{3+} and/or Mo^{5+} and the complex assumes a characteristic blue colour.

Many different reducing agents have been advocated to form molybdenum blue—Snell and Snell (1949) for example, list over thirty—and selection of the appropriate reagent depends upon the required sensitivity and freedom from interference. Jackson (1958) lists four methods of reduction, each most suitable under specific conditions:

    i. Reduction by chlorostannous acid in sulphuric acid.
    ii. Reduction by chlorostannous acid in hydrochloric acid.
    iii. Reduction by molybdenum in sulphuric acid.
    iv. Reduction by 1,2,4-amino-naphtholsulphonic acid in sulphuric or perchloric acid.

I / WITH CHLOROSTANNOUS ACID IN SULPHURIC ACID

Reduction of the molybdate complex with chlorostannous acid in sulphuric acid is the most sensitive of the four methods, is suitable for determining phosphorus in the range $0 \cdot 02$–$0 \cdot 60\ \mu g\ cm^{-3}$ P and was recommended for determining phosphorus in extracts of soils of low fertility. As described by Jackson, the method involves a somewhat tedious standardization of tin(II) chloride solution and the reducing solution requires a complicated system of storage to prevent oxidation. Metson (1961) recommends keeping a stock solution of tin(II) chloride and diluting before use, but even the stock solution can lose its power unless metallic tin is present and would give rise to green solutions instead of blue.

The method can be greatly simplified by preparing as required a fresh solution of tin(II) chloride from metallic tin and hydrochloric acid; this also avoids turbidity of solution. The solution requires to be 0·1 M with respect to tin and is made by dissolving a definite weight of metal in acid and making to a definite volume. An exact quantity of the reducing agent must be added to the test solution and an excess causes turbidity. Woods and Mellon (1941) recommended no more than 0·25 cm³ of reducing agent per 50 cm³ of test solution and Jackson recommends 0·15 cm³ per 50 cm³ as a standard procedure.

In the method a solution of ammonium molybdate is made up in 0·175 M sulphuric acid, and 2 cm³ are added to an aliquot of test solution which is then neutralized, diluted and reduced.

II / WITH CHLOROSTANNOUS ACID IN HYDROCHLORIC ACID

This method is slightly less sensitive than the first and is suitable for determining concentrations of 0·05–1·00 μg cm⁻³ P. Up to 15 μg cm⁻³ iron(III) can be tolerated, but if this concentration is exceeded, the iron effect must be suppressed. Jackson recommends the use of a Jones reductor for this, or the addition of hydrazine sulphate when up to 150 μg cm⁻³ iron(III) do not interfere. For best results the solution in which iron has been reduced should be used as a blank. Fluoride in greater concentration than 5 μg cm⁻³ must be suppressed by addition of boric acid, and Robinson (1941) goes so far as to eliminate fluoride by evaporation with perchloric acid.

In the method 30 g of ammonium molybdate are dissolved in 700 cm³ of water and mixed with 700 cm³ of hydrochloric acid before diluting to 2 dm³. 10 cm³ of molybdate solution are added to an aliquot of test solution which is then diluted to standard volume, neutralized to *p*-nitrophenol and reduced.

III / FACTORS AFFECTING CHLOROSTANNOUS ACID REDUCTIONS

The intensity of blue colour formed when phosphomolybdates are reduced with chlorostannous acid depends upon the amount of phosophomolybdate present, time of formation and pH. The effect of pH has been extensively studied and it has been found that for any particular method there is a range of acidity over which the blue colour is stable. This range is known as the 'acid stability plateau' (Cotton, 1945) and becomes smaller as phosphorus concentration increases. The term 'plateau' was used due to the flat, high portion of the curve showing stability against pH values. At relatively low acid concentration (for example up to 0·175 M sulphuric acid) the molybdate reagent itself will give a blue colour on reduction which interferes with that formed by the heteropoly complex. On the other hand, in strongly acid solution the blue colour of the reduced complex is less intense and may not form at all.

It is usual to eliminate the interference due to molybdomolybdic acid and molybdosilicic complexes by adjusting the acidity of the solution. Thus in the method where the reagent is made up in 0·195 M sulphuric acid, the silica complex does not interfere. At acid concentrations in excess of 0·2 M the silicic, arsenic and phosphoric complexes are reduced under certain conditions, but this normally does not give rise to difficulties in soil analysis (Boltz and Mellon, 1947). If the molybdenum concentration is increased then the concentration of acid must also be increased to prevent the reduction of molybdate. The normal practice is to bring the test solution to pH 2·7–3·0 using 2,4- or 2,6-dinitrophenol as indicator. A buffer solution is often used to control acidity and sodium sulphite is commonly employed as the buffer.

Although molybdoarsenic acid gives a blue colour on reduction, this is seldom a source of error in soil analysis. When appreciable amounts of arsenic are present the interference can be eliminated by pre-reduction to arsenious acid which does not react in the same manner. Similarly germanium is not often likely to interfere; Levine *et al.* (1954) precipitated phosphorus from solution with aluminium hydroxide and treated the precipitate with a strong acid to volatilize germanium, arsenic and silicon. Normally fluoride and iron(III) are the elements interfering with the methods and routine procedures are adapted to take care of this.

Snell and Snell (1949) eliminated iron(III) by using more tin(II) chloride if the iron did not exceed 30 μg cm^{-3}, by heating for 1 hour at 100°C with 5 cm^3 of a 16% solution of sodium metabisulphite for iron contents up to 200 μg cm^{-3}, or by extracting into butanol before reduction. Bacon (1950) utilized iron(II) sulphate to reduce the molybdo complex. The test solution was first reduced with sulphur dioxide and then sulphuric acid; ammonium molybdate reagent was added, followed by iron(II) sulphate solution.

The blue colour produced on reduction of the molybdate complex with chlorostannous acid increases in intensity with time, reaching a maximum after a period depending upon the acidity; thereafter the colour fades slowly. It follows that for reproducible results the colorimetric readings must be taken after a definite period of time from addition of reducing agent. This is a drawback to the method but is not insurmountable. The usual procedure is to add the reducing solution from a burette to the series of test solutions at some definite time interval, say every 2 minutes, and then to commence photoelectric readings after the first solution has stood for 15 minutes. Subsequent readings are then taken every 2 minutes which gives ample time for reading the instrument, preparing the next solution for reading and, if necessary, adding reductant to the next test solution in the series.

Beer's Law is obeyed up to approximately 0·4 μg cm^{-3} P in solution. It is most important to note that too strong a blue colour cannot be diluted for measurement; if the colour is too intense the experiment must be repeated using smaller aliquots of test solution. The reason for this is that the colour is dependent upon several factors such as pH, molybdate–acid ratio and so on, which would be changed by diluting the solution.

Pons and Guthrie (1946) modified the general procedure by extracting the phosphomolybdate complex into butan-2-ol before reduction. This method is sensitive and of particular use for determining inorganic phosphorus in the presence of organic phosphorus. Watanabe and Olsen (1962) demonstrated that the Pons–Guthrie procedure is superior to others which may be affected by the presence of organic matter.

If the soil extract is coloured, as from organic soils for instance, a separate aliquot should be treated in the same way as the test solution, except for reduction, and used as a blank when adjusting the instrument.

IV / WITH MOLYBDENUM IN SULPHURIC ACID

This is a reasonably sensitive method, about a third as sensitive as the chlorostannous acid reduction, and the effects of arsenic are eliminated. The colour produced is stable for about 24 hours. One of the early procedures described was that of Deniges (1920) and was later improved by Truog and Meyer (1929) who increased the acidity of the reagent and doubled its molybdenum content. The reaction was thus made more sensitive and the interference of silicon eliminated. Truog and Meyer's reagent was prepared by dissolving 25 g of ammonium molybdate in 200 cm^3 of water at 60°C and filtering; 280 cm^3 of sulphuric acid were diluted to 800 cm^3 and when cold added to the molybdate with stirring. After cooling, the reagent was diluted to 1 dm^3. As described by Jackson the reagent is made by heating very pure MoO_3 with sulphuric acid, molybdenum powder is added and heating continued to dissolve the powder. The concentration, found by a permanganate titration, is adjusted and iron and arsenic are reduced with sodium hydrogen sulphite. The reagent is added to the test solution at 100°C and heating must be prolonged for 30 minutes to effect complete reduction.

V / WITH 1,2,4-AMINO-NAPHTHOLSULPHONIC ACID

This is a less sensitive method than those previously described and is suitable for determining phosphorus in the range 0·4–2·4 μg cm^{-3} P. It has the advantages of being non-affected by the presence of up to 200 μg cm^{-3} iron(III), there is no arsenic interference and it is less sensitive to acidity. The method is thus particularly useful for the determination of total phosphorus in a soil when considerable amounts of iron may be present in solution. Furthermore, the acid medium may be perchloric acid, thus enabling perchloric acid digests of soil to be analyzed.

The reagent is first purified by recrystallization from a solution of sodium sulphite and sodium hydrogen sulphite, washed and dried and finally dissolved in sodium sulphite–hydrogen sulphite solution. A suitable aliquot of the test solution is adjusted to pH 3, 5 cm^3 of perchloric acid added and the solution diluted to known volume. The reducing agent is added and 15 minutes before reading, 4 cm^3 of ammonium molybdate solution (5% aqueous) are added. The extinction is read at 660 nm after 15 minutes. As many samples as required may be prepared in advance as far as adding the reductant. The standard graph should cover the range 0–2·4 μg cm^{-3} P although Snell and Snell state that the method can be used for the range 0·2 to 10·0 μg cm^{-3} P.

Kurtz and Arnold (1946) used 7·4 M hydrochloric acid as the acid medium for preparing the ammonium molybdate reagent, and Laverty (1963) modified this to 9·2 M, claiming a more stable blue colour after reduction. Laverty prepared a stock reducing agent powder which was stable for 1 year by grinding together the 1,2,4-amino-naphtholsulphonic acid, sodium pyrosulphite and sodium thiosulphate. For use, a dilute aqueous solution was prepared which was stable for about 3 weeks.

VI / WITH ASCORBIC ACID

Of the many alternative reducing agents ascorbic acid has become favoured owing to the fact that the blue colour of the reduced complex is stable for a long time, enabling reduced solutions to be kept overnight if necessary before measurement.

When first described (Ammon and Hinsberg, 1936), the method suffered from the disadvantage of requiring heat (7 minutes at 37°C) to develop the colour. The procedure, however, was adopted by several workers, Duval (1962) for instance, and Frei *et al.* (1964), who found that a 0·01 M solution of ascorbic acid gave the best results providing that interference from nitrogen compounds was eliminated by adding sulphamic acid.

Sakadi and co-workers (1965) described a method for using ascorbic acid without the necessity of heating and which was designed especially for estimating phosphorus in lactate extracts of soil. In this method the reducing solution is actually a mixture of ascorbic acid and tin(II) chloride.

A disadvantage of the ascorbic acid method of reduction is that during the long period necessary for colour development, some organic phosphorus may be hydrolyzed. To overcome this, Murphy and Riley (1962) proposed the use of a single reagent for reduction containing sulphuric acid, ammonium molybdate, ascorbic acid and antimony ammonium tartrate. This reagent gives a stable blue colour (24 hours) within 10 minutes and is not affected by organic phosphorus. As originally developed, the single reagent was for measuring phosphorus in sodium hydrogen carbonate extracts of soil, but Watanabe and Olsen (1965) showed that the procedure is suitable for determining phosphorus in aqueous extracts of soil.

VII / WITH VANADOMOLYBDATE

If an excess of molybdate ions are added to an acid solution of vanadate and orthophosphate, a yellow-coloured complex of uncertain composition is formed. The yellow colour has been attributed (Jackson, 1958) to a substitution of oxyvanadium and oxymolybdenum radicals for the oxygen of phosphate.

The colour is more stable than molybdenum blue and the method is less subject to interference from iron(III) and silicon than the molybdenum blue method; it also covers a wider range of phosphorus concentration, 1 to 20 μg cm^{-3} P. Best results are obtained if a constant temperature is maintained. In some published procedures the reagent is made up in nitric acid but perchloric acid is preferable. The final acid concentration of the test solution plus reagents should be around 0·5 M and the extinction is measured at 470 nm.

The increasing use of auto-analysis has permitted rapid determination of phosphorus in solution by colorimetric procedures (e.g. Lacy, 1965; Colwell, 1965).

12:3 RECOMMENDED METHODS

12:3:1 Phosphorus in solution

Colorimetric methods

I / WITH CHLOROSTANNOUS ACID IN SULPHURIC ACID (PONS AND GUTHRIE)

Reagents

Standard phosphate solution: Dissolve 0·2195 g of dried (40°C) potassium dihydrogen phosphate in about 500 cm³ of water and add 25 cm³ of 3·5 M sulphuric acid. Dilute to 1 dm³ when the solution contains 50 μg cm⁻³ P.

Ammonium molybdate solution: Dissolve 50 g ammonium molybdate in 400 cm³ of 5 M sulphuric acid and 500 cm³ of water; dilute to 1 dm³.

Sulphuric acid, 0·5 M: 28 cm³ dm⁻³ conc. H_2SO_4.

Tin(II) chloride solution: Dissolve 10 g $SnCl_2$ in 25 cm³ conc. HCl and dilute 1 cm³ to 200 cm³ for use.

Butan-2-ol.

Procedure

Pipette an aliquot of test solution containing from 0·005 mg to 0·045 mg P (i.e. up to 1 μg cm⁻³ when diluted to 50 cm³) into a 125-cm³ separating funnel marked at 20 cm³. Add 5 cm³ of ammonium molybdate solution and dilute to 20 cm³. Add 10 cm³ of butan-2-ol, shake for 2 minutes, discard the aqueous layer and wash the organic layer once with 10 cm³ of 0·5 M sulphuric acid. Add 15 cm³ of diluted tin(II) chloride solution and shake for 1 minute. Discard the aqueous layer and transfer the organic phase to a 50-cm³ volumetric flask, the funnel being rinsed into the flask with ethanol.

Dilute to volume with ethanol and read the transmission at 730 nm from 40 minutes to 18 hours later. A standard curve should be prepared to cover the range 0–1 μg cm⁻³ P.

II / WITH CHLOROSTANNOUS ACID IN HYDROCHLORIC ACID

Reagents

Ammonium molybdate solution: Dissolve 48·0 g ammonium molybdate in 500 cm³ water at 60°C. Cool and mix with 1120 cm³ 10 M HCl. Dilute to 2 dm³ and store in a brown glass bottle containing 100 g boric acid (Laverty).

Tin(II) chloride solution: Weigh 0·25 g metallic tin into a 50-cm³ conical
flask. Add 5 cm³ conc. HCl and 2 drops of 4% $CuSO_4$ solution. Place
a small funnel in the neck of the flask and gently heat the mixture on a
steam-bath until all the tin has dissolved. Dilute the solution, cool and
filter into a 25-cm³ volumetric flask and dilute to volume. The reductant
should be prepared daily.

2,4 Dinitrophenol solution: 0·5% w/v aqueous.

Standard phosphate solution: As in section 12:3:1I.

Procedure

Pipette an aliquot of test solution containing up to 0·050 mg P (1 μg cm⁻³
when diluted to 50 cm³) into a 50-cm³ volumetric flask and dilute to about
20 cm³. Add 1 drop of dinitrophenol solution and 1:1 ammonium hydroxide until a slight permanent yellow colour is produced; discharge the
colour by adding 1 drop of 2 M hydrochloric acid. Add 5 cm³ of ammonium
molybdate solution and dilute to volume. Add 5 drops of tin(II) chloride
solution from a burette exactly 15 minutes before reading the transmission
at 655 nm. The calibration curve should cover the range 0–1 μg cm⁻³ P.

III / WITH VANADOMOLYBDATE

Reagents

Ammonium vanadate solution: Dissolve 2·345 g anhydrous ammonium
  vanadate in 400 cm³ of hot water. Add 17 cm³ of 60% perchloric acid
  and dilute to 1 dm³.

Ammonium molybdate solution: Dissolve 25 g ammonium molybdate in
  400 cm³ of water at 50°C, cool, filter if necessary and dilute to 500 cm³.
  Store in a brown glass bottle.

Standard phosphate solution: As in section 12:3:1I.

Procedure

Pipette an aliquot of test solution containing 0·15–1·0 mg P (or up to
20 μg cm⁻³ when diluted to 50 cm³) into a 50-cm³ volumetric flask, add
5 cm³ of 60% perchloric acid and dilute to about 30 cm³. The acidity
should be equivalent to 8·5–9·0 cm³ of 60% perchloric acid (7–8 cm³ of
72%) and the addition of acid should be adjusted accordingly. Add 5 cm³
of ammonium vanadate solution and if a precipitate of silica forms, filter it
off. Add 5 cm³ of ammonium molybdate solution, dilute to volume and
read the transmission at 470 nm after 30 minutes. The standard curve
should cover the range 0–20 μg cm⁻³ P.

12:3:2 Total phosphorus

Digest 1 g of soil with perchloric acid as described in Chapter 16, section
16:2:3II. During the evaporation of the digest there is no need to drive

11+T.S.C.A.

off completely the perchloric acid; take up the filtered digest in dilute hydrochloric acid and dilute to 100 cm³.

Analyze aliquots of digest for phosphorus either by method 12:3:1ɪɪ using chlorostannous acid, or by method 12:3:1ɪɪɪ using the vanado-molybdophosphoric reaction.

12:3:3 Total inorganic phosphorus

Reagents
As in section 12:3:1ɪ; and concentrated hydrochloric acid.

Procedure

Weigh 1 g of <0·5-mm air-dry soil into a centrifuge tube and mix with 10 cm³ of hydrochloric acid. Heat on a steam-bath for 10 minutes and then add a further 10 cm³ of acid. Allow to stand at room temperature for 1 hour, add 50 cm³ of water and centrifuge. Decant the solution into a 250-cm³ volumetric flask and make to volume.

Determine the phosphorus in solution immediately by the method in section 12:3:1ɪ, in order to prevent errors arising from hydrolysis.

12:3:4 Total organic phosphorus

Weigh 1 g of <0·5-mm air-dry soil into a silica crucible and heat at 240°C for 1 hour in an electric muffle furnace. When cool, transfer the ignited soil to a 100-cm³ centrifuge tube. A second 1-g sample of soil is weighed directly into another centrifuge tube.

The procedure is then exactly as in section 12:3:3 for total inorganic phosphorus and the difference between the phosphorus found in ignited soil and non-ignited soil is a measure of total organic phosphorus.

12:3:5 Fractionation of inorganic phosphorus (Chang and Jackson; Petersen and Corey)

Reagents
Ammonium chloride solution, 1·0 M: 53·5 g dm⁻³ NH_4Cl.
Ammonium fluoride solution, 0·5 M: 18·5 g dm⁻³ NH_4F adjusted to pH 8·2 with 4 M NH_4OH solution. Store in a polythene bottle.
Sodium chloride solution, saturated.
Sodium hydroxide solution, 0·1 M: 4 g dm⁻³ $NaOH$.
Sulphuric acid, 0·25 M: 15 cm³ dm⁻³ conc. H_2SO_4.
Sodium citrate solution, 0·3 M: 75 g dm⁻³ tribasic salt.
Sodium dithionite, solid.
Potassium permanganate solution, 0·25 M: 39·5 g dm⁻³ $KMnO_4$.
Reagents for phosphorus in solution: As for section 12:3:1.

Procedure

The method of fractionating soil inorganic phosphorus is presented in Table 12:2.

Table 12:2

Fractionation of soil inorganic phosphorus by the modified procedure of Chang and Jackson and after Petersen and Corey

Soil: Place 1 g of 0·15-mm soil in a 100-cm³ polypropylene centrifuge tube, add 50 cm³ of 1 M NH₄Cl solution, shake for 30 minutes and centrifuge.

Solution: Determine Saloid-bound-P by method 12:3:1ɪɪ.	*Soil:* Add 50 cm³ of 0·5 M NH₄F solution made to pH 8·2 with NH₄OH, shake for 1 hour and centrifuge.						
	Solution: Filter through activated carbon if necessary and determine Aluminium-bound-P by method 12:3:1ɪɪ.	*Soil:* Wash twice with 25-cm³ portions of saturated NaCl solution, centrifuging each time to recover soil. Discard washings. Add 50 cm³ of 0·1 M NaOH solution, shake for 17 hours and centrifuge.					
		Solution: Add 5 drops of conc. H₂SO₄ and centrifuge.	*Soil:* Wash twice with saturated NaCl solution and discard washings. Suspend soil in 25 cm³ of 0·3 M Na citrate solution, add 1 g of Na dithionite and shake for 15 minutes. Heat to 80°C, dilute to 50 cm³, shake for 5 minutes and centrifuge.				
		Solution: Filter through 0·5 g of activated carbon and determine Iron-bound-P by method 12:3:1ɪ or 12:3:1ɪɪ.	Residue: Discard				
			Combined solution and washings: To 3 cm³ add 1·5 cm³ of 0·25 M KMnO₄ to oxidize excess dithionite and citrate. Allow to stand 2 minutes and determine Reductant-soluble-P in an aliquot by method 12:3:1ɪ.	*Soil:* Wash twice with saturated NaCl solution and add washings to previous solution. To soil add 50 cm³ of 0·1 M NaOH solution, shake for 1 hour and centrifuge.			
				Solution: Determine Occluded-P by method 12:3:1ɪɪ.	*Soil:* Wash twice with NaCl solution. Add 50 cm³ of 0·25 M H₂SO₄, shake for 1 hour and centrifuge; discard residue of soil and determine Calcium-bound-P by method 12:3:1ɪ.		

The sum of the various inorganic fractions of phosphorus should be compared with the total inorganic phosphorus as measured in section 12:3:3.

12:3:6 Phospholipids (Hance and Anderson)

Reagents

Hydrochloric acid, 2·5%: 1:14 v/v conc. HCl:H₂O.
Hydrofluoric acid, 2·5%: 1:24 v/v 60% HClO₄:H₂O.
Acetone.
Petroleum ether, b.p. 40–60°C.

Chloroform.
Ethanol–benzene mixture, 1:4 v/v.
Methanol–chloroform mixture, 1:1 v/v.
Ether–pet. ether (40–60°C) mixture, 1:1 v/v.
Magnesium nitrate solution, 12% w/v aqueous.

Procedure

Shake 1 g of 0·15-mm soil overnight with 10 cm^3 of a mixture of equal volumes of the hydrochloric and hydrofluoric acids in a polythene centrifuge tube. Centrifuge and discard the liquid. Repeat the acid extraction once more. Wash the soil residue free from acid with 10-cm^3 portions of water and add 50 cm^3 of acetone. Cover the top of the tube with a piece of polythene kept in place by a rubber band and allow to stand for 4 hours, mixing the contents every half hour. Centrifuge and transfer the liquid to a 500-cm^3 glass-stoppered bottle. Extract the soil in exactly the same way with 50 cm^3 of pet. ether and add the extract to the acetone extract. Repeat the process with firstly 50 cm^3 of ethanol–benzene mixture and finally with 50 cm^3 of methanol–chloroform mixture; all extracts are combined in the bottle.

Evaporate off the solvents under reduced pressure and at low temperature and re-extract the residue with 20 cm^3 of a cold mixture of ether and pet. ether for 2 minutes. Centrifuge and extract the residue successively with 20 cm^3 of cold chloroform for 2 minutes, 20 cm^3 of ether–pet. ether added cold and brought to boiling on a water-bath and decanted, and 20 cm^3 of chloroform added cold and brought to boiling and decanted.

These extracts are bulked and evaporated to dryness on a water-bath in a silica dish. Moisten the residue with 1 cm^3 of ethanol and 1 cm^3 of magnesium nitrate solution and again evaporate. Ignite the residue at 600°C for 12 minutes, dissolve in 15 cm^3 of 1 M hydrochloric acid and again evaporate to dryness. Add 1 cm^3 of 1 M hydrochloric acid and transfer the solution with hot water to a 50-cm^3 volumetric flask. Dilute to volume and measure the phosphorus in solution by the method in section 12:3:1II, and express the result as mg lipid phosphorus/100 g soil.

12:3:7 Inositol hexaphosphate (Cosgrove)

Reagents

Hydrochloric acid, 0·2 M: 18 cm^3 dm^{-3} conc. HCl.
Hydrochloric acid, 0·1 M: 9 cm^3 dm^{-3} conc. HCl.
Hydrochloric acid, 11·0 M: 990 cm^3 dm^{-3} conc. HCl.
Sodium hydroxide solution, 1·0 M: 40 g dm^{-3} NaOH.
Sodium hydroxide solution, 2·0 M: 80 g dm^{-3} NaOH.
Bromine.
Ether.

Iron(III) chloride solution, 60% w/v aqueous.
Activated charcoal.
Dowex AG 50 W (H) resin.

Procedure

Shake 200 g air-dry soil for 1 hour with 1 dm³ of 0·2 M hydrochloric acid, filter and wash with ethanol. When dry, shake the soil with 1 dm³ of 1 M sodium hydroxide solution at 60°C for 4 hours, cool and centrifuge. Acidify the extract to pH 1·0 with hydrochloric acid and centrifuge. Add solid sodium hydroxide to the soluble, fulvic acid fraction until alkaline and add 40 g in excess. Cool the solution to 0–5°C and treat with 20 cm³ of bromine added 2 cm³ at a time; maintain below 5°C for 18 hours.

Heat the solution at 60°C for 1 hour, cool and acidify with hydrochloric acid. Remove excess bromine by ether extraction and adjust the pH to 2·0. Add 20 cm³ of iron(III) chloride solution and coagulate the precipitate on a steam-bath. Cool, centrifuge and wash the precipitate with 0·1 M hydrochloric acid and water; finally suspend the precipitate in water. Add a slight excess of 2 M sodium hydroxide solution and filter off the precipitated iron(III) hydroxide. Concentrate the alkaline solution of phytin to 50 cm³ and add an equal volume of 11 M hydrochloric acid. Heat the mixture on a steam-bath for 1 hour, cool, dilute and decolorize with charcoal. Re-precipitate the phytin with iron(III) chloride by raising the pH to 2·0 with sodium hydroxide and adding iron(III) chloride solution until the supernatant liquid turns deep yellow. Wash the iron(III) salt and again convert to the sodium salt. Adjust the pH of the solution to 7·0 by adding Dowex resin and filter off the resin again.

Total inositol hexaphosphate can now be estimated by determining the phosphorus content of the solution or the solution can be fractionated by chromatography or paper electrophoresis.

12:3:8 Available phosphorus

As discussed in section 12:2:5, no one method, or even several methods, of determining available phosphorus can be recommended. Choice of extractant must depend upon results of field and greenhouse trials using specific crops under specific conditions. Some of the most commonly used extractants have been discussed and full reference given for experimental details. The basic procedure is the same for all extractants and consists in shaking a soil sample with an extracting solution and determining the phosphorus content of the extract.

For some soils the measurement of Schofield's Phosphate Potential or an equilibrium potential (section 12:2:5) may prove to be the best measure of available phosphorus, while for the best approach to an unknown soil, the procedure of Friend and Birch (section 12:2:5) should be considered.

12:3:9 Phosphorus fixation capacity

Reagents

Standard phosphorus solution: $0\cdot0814$ g dm^{-3} Ca(H$_2$PO$_4$)$_2$.H$_2$O. This solution contains 20 μg cm^{-3} P.

Reagents for determining soluble phosphorus as in section 12:3:1ɪ.

Procedure

Shake 1 g of air-dry, 2-mm soil in a centrifuge tube with 20 cm^3 of standard phosphorus solution for 2 hours. Centrifuge, filter if necessary and determine the phosphorus in solution by the method in section 12:3:1ɪ. From the change in phosphorus concentration of the standard solution, calculate the fixation capacity and express as mg P/g soil.

13

Sulphur

13:1 INTRODUCTION

Although sulphur is a major plant nutrient its determination in soils was neglected for many years, partly because its importance was not sufficiently recognized and partly because of analytical difficulties.

Early attempts at analyzing soils for sulphur content were confined to the determination of the sulphate ion; later, total sulphur analyses were made and organic sulphur was reported as the difference between total and sulphate sulphur. This is more or less the position today except that the analytical methods have been improved.

In humid climates soil sulphur is mostly in organic combination and various workers have reported carbon:nitrogen:non-sulphate-sulphur relationships. As yet little is known of the organic sulphur compounds in soils although sulphur is a necessary constituent of all proteins. Cystine, cysteine, methionine, glutathione, trithiobenzaldehyde and sulphate esters are among the compounds that have been found in soils or their hydrolysates but much research on this subject remains to be done.

The nature of the inorganic compounds of sulphur in soils has received more attention. In normal, aerobic soils the inorganic sulphur occurs almost exclusively as sulphate. Any other forms such as thiosulphate, elemental, tetrathionate and so on, will be transitory and in due course converted to sulphate. If a soil is waterlogged, anaerobic conditions may lead to the presence of monosulphides, polysulphides and elemental sulphur. Polysulphides are found also in well-drained soils which have developed from recent marine deposits and elemental sulphur is found in volcanic regions.

In humid regions the inorganic sulphur fraction is found usually in the subsoil, the topsoil being relatively rich in organic sulphur, but in arid regions sulphate, as gypsum, occurs in the topsoil and often forms a surface coating.

Relatively little of the world's sulphur is available to soils and many countries have reported sulphur deficiencies. Fertilizers such as sulphate of ammonia and superphosphate provide a source of soil sulphur, particularly in agricultural land. Other sources of sulphur are rainfall and inundating waters such as on coastal or riverain flood-plains or on irrigated land. Otherwise the soil is reliant upon the organic–inorganic cycle for its sulphate supply. Vegetable or animal material containing organic sulphur compounds decomposes to give sulphate-sulphur which is then taken up by plants and reconverted to organic forms. This is why humid regions have very little sulphate-sulphur in the topsoil; organically bound sulphur in the upper layers is gradually oxidized to sulphate which is then rapidly leached to the subsoil or taken up by plants. Where the vegetation is relatively permanent, as in a forest for example, in addition to the sulphur cycle dependent upon microbiological oxidation of organic sulphur, there is another cycle whereby sulphate is circulated as such between the vegetation and the soil (Hesse, 1957a). In the case of annual crops which are removed from the soil, the land must be fertilized with sulphur to prevent its eventual deficiency.

Available sulphur

In spite of the fact that plants absorb sulphur almost exclusively as sulphate, determination of the sulphate content of a soil is of little use as a measure of sulphate availability. Moreover, owing to the mobility of the sulphate ion, such determinations can give vastly different results according to the time of sampling. Total sulphur measurements similarly are limited in their usefulness for investigating sulphur availability.

Attempts have been made to determine the rate at which sulphate is formed from organic matter in the same way as nitrification rates are measured. Some workers have reported significant results with certain kinds of soil, but in general, little success has been achieved owing to the very slow rate of breakdown. In a forest soil from East Africa, for example, only 5 μg g^{-1} sulphate-sulphur were produced on incubating the soil under ideal conditions for 6 months although it contained 400 μg g^{-1} organic sulphur (Hesse, 1957a).

The mineralization of organic sulphur in a soil depends upon the sulphur content or, more probably, upon the nitrogen–sulphur ratio, and it should be remembered that any sulphate formed may be fixed against extraction, particularly if much iron or barium is present or if the soil is very acid.

The biological breakdown of organic sulphur compounds is complex and can involve many different organisms. Certain bacteria, particularly those belonging to the genus *Proteus*, break down proteins to give, amongst other products, hydrogen sulphide and mercaptans. Such compounds are then converted to sulphate bacterially and chemically, or if anaerobic conditions prevail, the protein disintegration products will form sulphides and elemental sulphur.

The sulphur cycle in nature is summarized in Figure 13:1, adapted from that of Butlin (1953). In addition there exist organisms (e.g. *Aspergillus niger*) capable of converting sulphur amino-acids directly to sulphate.

As incubation techniques have not been able to provide information regarding sulphur availability, several arbitrary forms of sulphur have been determined for the purpose. Williams and Steinbergs (1959) measured a so-called 'heat-soluble' sulphur and obtained good correlation of their results with plant uptake of sulphur. Kilmer and Nearpass (1960) extracted soil with 0·5 M sodium hydrogen carbonate solution at pH 8·5 and

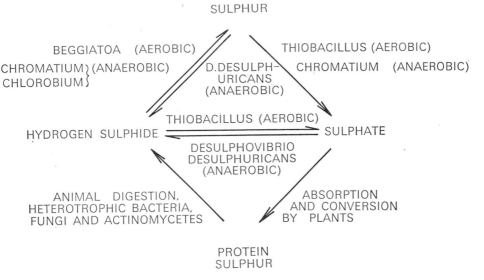

Figure 13:1 The sulphur cycle in nature (from Butlin)

measured the sulphur extracted. The results of Kilmer and Nearpass correlated well with sulphur 'A-values', an r-value of 0·89 being obtained for thirty soils.

'Easily-oxidizable' and 'active-oxidizable' forms of sulphur have been measured by Yomeda (1961). The easily-oxidizable sulphur is that measured as the increase in water-soluble sulphate after oxidizing the soil with hydrogen peroxide, and the active-oxidizable sulphur is calculated from the sulphuric acid equivalent of the titrateable acidity after oxidation of the soil. Barrow (1967) selected a 0·01 M solution of calcium monophosphate as extractant for available sulphur. For most soils the phosphate solution displaced maximum quantities of adsorbed sulphate at a soil–solution ratio of 1:5. Barrow found that the phosphate solution gave clear, easily filterable extracts and did not extract non-available forms of sulphur.

A more direct method of determining available sulphur is that dependent upon the growth rate of mycelia of *Aspergillus niger* (Grigg, 1953).

The main function of sulphur in plant growth is in the synthesis of protein, possibly via amino-acids. For example, Thomas and co-workers (1950) recovered radioactive cystine-sulphur from tomato leaves 10 minutes after supplying $^{35}SO_4$ to the soil. Excess sulphate is stored as such within a plant except that, according to Gilbert (1951), the tea plant is unable to store sulphate and thus is very susceptible to lack of available sulphur in a soil.

Sulphur also aids in the formation of chlorophyll and can stimulate legume growth by affecting the nodular nitrogen fixation. Anderson and Spencer (1950) showed that sulphur was necessary for nitrogen fixation

11*

and that it influenced the conversion of nitrate into protein. Thus if nitrate is applied to a sulphur-deficient soil, it will accumulate as such in the plant; subsequent treatment of the soil with sulphate results in the nitrate being converted to protein-nitrogen.

Effects of sulphur and its compounds upon soil properties

Apart from providing plant-available sulphur, the element or its compounds can be applied to soils for amelioration purposes. The use of gypsum and elemental sulphur for reclaiming alkali soils has been discussed in Chapter 5 (section 5:3). A subsidiary effect of reducing the pH of soils by sulphur treatment is to increase the availability of certain elements, particularly the minor elements, and a sulphur-deficient soil should always be suspected as being trace-element-deficient also.

Quastel and co-workers (1948) compared the effects of sulphur with those of thiosulphate as a means of remedying manganese deficiency. They found that both substances increased manganese uptake by plants from the soil but that the thiosulphate acted more quickly than sulphur and did not change the soil pH so much. They concluded that the action of elemental sulphur was due partly to the formation of thiosulphate.

Another important effect of adding elemental sulphur to a soil is that exerted upon nitrification processes. To a certain extent elemental sulphur will increase the nitrogen uptake of a crop but heavy dressings will depress nitrogen availability. Deleterious effects of elemental sulphur are most marked with base-deficient soils as the sulphuric acid formed is not fixed and the resulting acidity inhibits the nitrifying organisms. The ammonifying organisms are not affected to the same extent and thus an accumulation of ammonium, as the sulphate, occurs. If, on the other hand, the soil is alkaline, elemental sulphur increases nitrification as too high a pH also inhibits the nitrifyers.

Organic sulphur compounds can inhibit nitrification in a soil but this is due to toxic decomposition products. Thus Quastel and Schofield (1949) showed that methionine inhibited nitrification during its breakdown in a soil and only when the methionine itself had completely decomposed did nitrification recommence. Jensen and Sorensen (1952) subsequently suggested that the methionine itself was not toxic but rather that its intermediate disintegration products, such as methyl mercaptan, are responsible. Thiourea was also found to be toxic and in general the toxicity of sulphur compounds appears to be related to the presence of free S—H groups.

Acid sulphate soils

In many parts of the world an excess of inorganic sulphur compounds has proved toxic to plant growth or has induced toxicity. Certain physiological diseases of rice, for example, have been traced to the presence of hydrogen sulphide in reduced layers of periodically submerged soils. Waterlogged, sulphur-rich soils become extremely acid if allowed to dry owing to oxidation of sulphur. Acid sulphate soils, sometimes called 'cat-clays', are commonly found on marine flood-plains and if dried, and thus oxidized, characteristic yellow streaks of basic iron(III) sulphate appear. The pH value of such soils has been known to drop from 6·7 to below 2·0 during oxidation (Tomlinson, 1957) and apart from toxicity arising from acidity *per se*, toxic amounts of aluminium and iron can be liberated.

A considerable amount of investigation has been made into the properties of acid sulphate soils, notably in the Netherlands (Harmsen *et al.*, 1954), West Africa (Tomlinson, 1957; Hart, 1959, 1962, 1963; Hesse, 1961a) and the Far East (Moorman, 1961). The immediate cause of acidity is bacterial oxidation of elemental sulphur and is limited by the rate of formation of elemental sulphur from the soil polysulphides. The polysulphides are sometimes deposited from marine clay suspensions and sometimes partly formed from iron(II) sulphide. Apart from any polysulphides deposited, the soils accumulate sulphur as monosulphide by bacterial reduction of the sulphate in sea water in the presence of actively decomposing organic matter. The characteristic basic iron(III) sulphate found in cat-clays is formed by the hydrolysis of iron(III) sulphate; it varies in composition from $3Fe_2O_3 . 4SO_3$ to $4Fe_2O_3 . 5SO_3$.

It is also possible to have a sulphur deficiency in submerged soils, even though they may be rich in sulphur compounds. Lake muds, for example, are often rich in organic sulphur compounds and polysulphides, yet the lake water can be severely deficient in sulphur. Thus in East Africa the water of Lake Victoria exhibited sulphur deficiency despite a sulphur content of nearly 1% in the bottom mud (Hesse, 1957b); a similar occurrence was found in Japan (Koyama and Sugarawa, 1953).

13:2 DETERMINATION: BACKGROUND AND THEORY

13:2:1 Total sulphur

The determination of the total sulphur in a soil can be made by one of two methods. In one method the sulphur is oxidized to sulphate and in the other it is reduced to sulphide. Sometimes the sulphate is determined as such and sometimes it is reduced to sulphide before determination.

Oxidation to sulphate has been achieved by various reagents and techniques and with varying degrees of efficiency. Fusion of the soil in a platinum or nickel crucible with oxidizing agents is a commonly used procedure. Among the earlier methods was fusion of the soil with sodium carbonate (e.g. Evans and Rost, 1945) but this simple fusion has not generally been found satisfactory although Bardsley and Lancaster (1960), who used this method as well as fusion with sodium hydrogen carbonate, reported that organic sulphur was oxidized and retained by both reagents. Robinson (1945) used a fusion mixture of sodium carbonate and sodium nitrate, Steinbergs *et al.* (1962) fused with sodium hydrogen carbonate mixed with silver oxide and Heathcote (private communication) used sodium hydrogen carbonate mixed with sodium peroxide. Mann (1955) and Marwick (Metson, 1961) fused soil with sodium peroxide alone and Lowe and De Long (1961) obtained satisfactory results by fusing with sodium peroxide in a Parr bomb (1934). About 1 g of 0·25-mm soil mixed with benzoic acid was ignited in the bomb with sodium peroxide under 1520 kN m^{-2} (15 atmospheres) of oxygen.

Butters and Chenery (1959) investigated the efficiency of sodium peroxide and of magnesium nitrate (recommended by Swanson and Latlaw, 1922) for oxidation of sulphur and found peroxide useless and nitrate little better. They showed, however, that the failure of the magnesium nitrate fusion was due to the lack of purity of the commercially available salt, and by preparing their own reagent from 'Specpure' magnesium they obtained good and reproducible results. The average coefficient of variation was 3·1%. Chaudhry and Cornfield (1966) found that heating at 550°C for 3 hours with a potassium nitrate–nitric acid mixture gave satisfactory recovery of total sulphur.

Wet digestion methods of recovering total sulphur from soil include oxidation of the soil with hydrogen peroxide, but although this method has been used by many workers it is not recommended. The results are greatly affected by the quality of reagent which often contains appreciable amounts of sulphate and reproducibility is poor. Similarly a simple acid digestion is useless as volatile sulphur compounds are lost. More accurate results have been obtained (Hesse, unpublished) by adapting the nitric acid digestion method of Revol and Ferrand (1935) to soil analysis. In this method the sample is refluxed with nitric acid and the fumes conducted through a second condenser into bromine water. Later hydrogen peroxide is added and finally the distillate is returned to the digestion flask and evaporated to dryness; sulphate is determined in the residue.

Perchloric acid digestion has been used with some success (Hesse, 1957b) but it is considered that the method as it stands is not sufficiently reliable except as a rapid, routine procedure. Chapman and Pratt (1961) recommended digestion of soil with a 1:1 mixture of perchloric and nitric acids but apart from the proportions of acids, this is essentially the same method as that already mentioned.

Bloomfield (1962) adapted the method of Larsen *et al.* (1959) and oxidized soil sulphur with vanadium peroxide in an atmosphere of nitrogen. The sulphur trioxide produced was passed over reduced copper and hot copper oxide and the sulphur dioxide absorbed in a solution of sodium tetrachlormercurate. An aliquot of the resulting solution was then treated with *p*-rosaniline hydrochloride and dilute formaldehyde solutions. The extinction of the coloured solution obtained was read in a spectrophotometer at 550 nm. The method was applicable to the determination of from 2 to 25 μg of sulphur in 50 cm^3 of final solution; the apparatus (Figure 13:2) is rather complex and each determination takes more time than is desirable if many samples are to be analyzed.

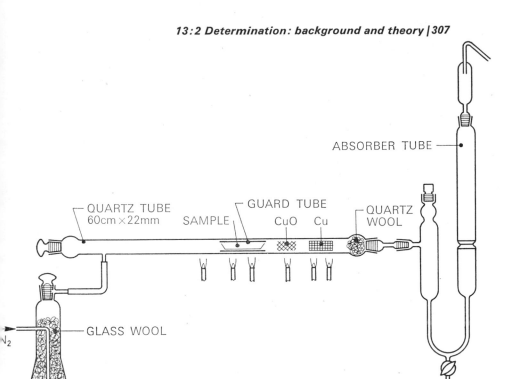

Figure 13:2 Apparatus for determining the total sulphur content of soil (Bloomfield, 1962; *Analyst*)

Methods of total sulphur analysis involving reduction of the sulphur to sulphide before measurement include that of Steinbergs *et al.* (1962) already described in part in section 13:1. After fusion of the soil, the moistened melt is reduced with a mixture of hydriodic, formic and hypo-phosphorous acids in an atmosphere of nitrogen; the sulphide produced is measured as methylene blue (section 13:2:3). Smittenburg *et al.* (1951) heated soil with reduced iron in a current of nitrogen and iodimetrically determined the sulphide produced. Little (1957) developed a method of electrically heating the soil with reduced iron; phosphorus did not inter-fere up to 1% and the method was specific for sulphur.

The choice of method from those later described in detail will depend upon the apparatus available and whether or not the determination is to be rarely or routinely made.

13:2:2 Sulphate-sulphur

Once in solution, soil sulphate can be determined gravimetrically, volu-metrically, colorimetrically, conductometrically, turbidimetrically or nephelometrically. Alternatively the sulphate can be reduced to sulphide and the reduced form determined.

Extraction and determination

Soil sulphate has been determined as water-soluble sulphate and Saalbach *et al.* (1962) obtained a good correlation of sulphate soluble in boiling water with plant uptake of sulphur. As a general procedure, however, measurement of water-soluble sulphate is not satisfactory as not all the inorganic sulphate in soils is soluble in water and as all soils retain sulphate to some extent. Colloidal oxides of aluminium and iron are active in adsorbing sulphate in acid soils, and Aderikhi (1960) found that sulphate adsorption was affected by the concentrations of Ba > Fe > H > Ca > Na in that order. Whilst increasing acidity increases sulphate adsorption, increasing temperature decreases the sorption. Williams and Steinbergs (1962) found that sulphate adsorption is influenced by the type of clay mineral and is a maximum at pH values of 2–4. The same workers examined naturally occurring calcium carbonate and calcareous soils and found that an insoluble sulphate was associated with the carbonate and may comprise an important fraction of the soil sulphate. One soil, for example, had 93% of its total sulphur in this form. The insoluble sulphate was not adsorbed and, in fact, adsorption was negligible at pH greater than 6·5. Thus an acid extraction of sulphate can lead to low results and Williams and Steinbergs recommended that after an acid extraction of soil a second extraction should be made following the addition of a little calcium carbonate.

Studies of sulphate adsorption have been made using isotopic techniques; thus Chao *et al.* (1962) with ^{35}S-gypsum found that soils rich in free iron and aluminium oxides and with high extractable aluminium can fix sulphate more strongly than can other soils. Using ^{35}S-potassium sulphate and acid red soils Liu and Thomas (1961) found that sulphate adsorption increased with time of equilibrium and that the adsorbed sulphate exchanges with some of the hydroxyl ions on the hydroxy-iron (or aluminium) polymers. Freney *et al.* (1962) reported that the sulphate adsorption capacity of clay minerals is related to total surface area. Appreciable amounts of barium in a soil can render sulphate extremely insoluble and calcium has been shown to be involved in sulphate adsorption (Hesse, 1958). In oxidized layers of mud, sulphate can be held against water extraction by adsorption on iron(III) complexes and is released when the mud is reduced (Mortimer, 1941). In the mud of Lake Victoria in East Africa, sulphate was present to about 70 μg g^{-1} yet the water filtered from the mud contained less than 1 μg cm^{-3} sulphate (Hesse, 1957b).

Thus water is limited in use as an extractant for soil sulphate and several alternative extractants have been suggested. Bardsley and Lancaster (1960) recommended the use of ammonium acetate–acetic acid solution, Williams and Steinbergs (1962) used potassium hydrogen phosphate solution and Chesnin and Yien (1950) used a buffered solution of sodium acetate–acetic acid at pH 4·5. Ensminger (1954) investigated the efficiency of several extractants on different soils and obtained similar results from neutral sodium acetate solution, sodium acetate solution at pH 4·8 and potassium hydrogen phosphate solution. Dilute (0·1 M) hydrochloric acid was found to be most unsatisfactory as an extractant of sulphate. Little (1958) measured what he termed 'readily soluble sulphate' by extraction with 0·001 M hydrochloric acid and gave correlations of this with pC measurements (see section 6:2:1). Freney (1958) measured water-soluble sulphate after removing the adsorbing colloids with a Seitz bacterial filter; this determination was specifically of water-soluble and not total sulphate. Barrow (1967) extracted adsorbed plus soluble sulphate with dilute calcium monophosphate solution.

Some soils may contain organic sulphates which are easily converted to inorganic sulphate by heating, grinding or hydrolysis. This was shown by Freney (1961) who compared the results of sulphate and reducible-sulphur analyses of soil extracts made with boiling 0·1 M sodium hydroxide, with 0·1 M hydrochloric acid followed by 0·1 M sodium oxalate solution and with 0·1 M hydrochloric acid followed by 0·1 M sodium pyrophosphate solution. Infra-red spectrophotometric analysis of an acid methanol extract of soil showed the presence of covalent sulphate groups.

Gravimetric method

The gravimetric determination of sulphate involves the precipitation under controlled conditions of barium sulphate which is subsequently weighed. The method, however, is of use only when appreciable quantities of sulphate are present in the soil as otherwise experimental errors become too large. In cases of low sulphate content one can employ the technique of adding a known amount of sulphate to reduce the experimental error, but in general it is preferable to use an alternative method of analysis. Provided that sufficient sulphate is present the gravimetric procedure can be made fairly rapid by using small quantities of soil in centrifuge tubes.

The precipitation of sulphate as the barium salt is subject to several errors and conditions must be rigorously standardized. Although the solubility of barium sulphate increases with acidity, the precipitation must be made under acid conditions to prevent the co-precipitation of barium carbonate or phosphate. The solubility of the sulphate can be reduced by having an excess of barium ions present and it has been found that a pH value of between 1 and 2 of the solution is optimum although values up to 4 give satisfactory results. Barium sulphate also tends to co-precipitate with other ions, giving erroneous results. For example, if barium nitrate is co-precipitated, subsequent ignition before weighing will convert it to the oxide and this will be additive to the weight of sulphate. Co-precipitation of certain metals such as calcium, iron and aluminium will lead to low results (Vogel, 1962). Interfering metals must therefore be removed if possible before precipitation and the effect of the remaining ions is minimized by digesting the precipitate in the diluted solution. Errors due to chlorides are reduced by carrying out the precipitation in hot solution.

Volumetric methods

Weakly acid barium solutions give a red complex with sodium rhodizonate solution and compared with barium sulphate this complex is relatively soluble. Thus if sulphate is added to a solution containing barium ions and sodium rhodizonate, complete precipitation will be indicated by disappearance of the red colour. The converse, that is adding barium ions to sulphate in solution, cannot be done owing to the instability of the indicator. Consequently, to determine sulphate, a known amount of barium is added and the excess back-titrated with standard sulphate solution. The method gives most reliable results with concentrated sulphate solutions, about 0·25 M, which limits its usefulness in soil analysis.

The principle has been applied to analysis of soil extracts by Little (1953) who claimed an accuracy of $\pm 0\cdot 5$ mg $SO_4/100$ g soil if the soil contained about 10 mg $SO_4/100$ g. For higher concentrations of sulphate the errors were less than 5%. Little improved the end-point of the titration by screening with bromcresol purple. The rhodizonate method has received much criticism and is not recommended for soil analysis unless the sulphate content exceeds 100 μg g^{-1}.

Another volumetric method utilizes ethylenediamine tetra-acetic acid, EDTA. Barium sulphate is precipitated and dissolved in a known excess of EDTA solution containing ammonium hydroxide; excess EDTA is then titrated with standard magnesium solution. Munger *et al.* (1950) modified the basic procedure so as to avoid separating the barium sulphate from solution and their method can be used to determine from 2 μg cm^{-3} to 75 μg cm^{-3} sulphate-sulphur. In the procedure described by Bond (1955) an aliquot of solution containing not more than 0·15 m.e. of calcium and magnesium is acidified with hydrochloric acid, mixed with 0·005 M barium chloride solution and boiled. After cooling ammonium chloride, ammonium hydroxide and Eriochrome Black T are added together with enough magnesium chloride to bring the concentration of magnesium to at least 0·02 milliequivalents; the solution is then titrated with EDTA solution. Calcium and magnesium are separately determined with EDTA and the sulphate is found by deducting the amount of EDTA used from the amount equivalent to the calcium, magnesium and barium present.

Sulphate can be titrated with barium solutions using the sodium salt of tetrahydroxyquinone as indicator, but reproducible results are obtained only with high sulphate concentrations. Cantino (1946) used a method whereby an acid solution of barium chromate is added to sulphate solution in excess and the barium not precipitated as sulphate is precipitated with alkali as chromate. The barium chromate is removed and titrated with thiosulphate. Saalbach *et al.* (1962) used a modification of this method.

Sulphate can be estimated by high frequency titration with barium chloride and potentiometrically or amperometrically with lead nitrate. Accurate results can be obtained only for solutions containing down to about 30 μg cm^{-3} $SO_4{}^{2-}$-S (Vogel, 1962).

Fiske (1921) developed a method of titration of sulphate utilizing its reaction with benzidine. In this method a slightly acid solution of benzidine $(C_{12}H_8(NH_2)_2)$ is added to the sulphate solution followed by acetone. After standing the benzidine sulphate is filtered off, heated in water and titrated with sodium hydroxide. Although still described in several books the method is not suited to routine analysis of soils.

Colorimetric method

Sulphate in solution at concentrations from 2 μg cm^{-3} to 400 μg cm^{-3} (that is, from <1 to about 150 μg cm^{-3} S) can be determined with the barium salt of chloranilic acid $(BaC_6Cl_2O_4)$. Sulphate reacts with this reagent to give barium sulphate and the acid chloranilate ion, $C_6Cl_2O_2(OH)_2$, which is purple in colour and at pH 4 has a peak of optical density at 530 nm. If the ultra-violet range is used at 332 nm, a more intense absorption occurs and sulphate-sulphur can be detected to 0·02 μg cm^{-3}. Interfering cations are removed by passing the sulphate solution through an acidic ion exchange resin (e.g. Amberlite IR-120). The effluent is adjusted to pH 4 with dilute hydrochloric acid or ammonium hydroxide and diluted to standard volume. Aliquots are mixed with a buffer solution at pH 4 (0·05 M potassium hydrogen phthalate) and ethanol. Barium chloranilate is added and the excess centrifuged off. The optical density of the supernatant liquid is then measured.

After separation of sulphate with benzidine as described for the volumetric methods, a variety of colorimetric methods exist for completing the determination (Snell and Snell, 1949). For example, the benzidine sulphate can be diazotized and coupled with phenol to give a yellow solution (Kahn and Lieboff, 1928); or the precipitate can be treated with iodine, ammonium hydroxide and potassium iodide, giving a measurable brown colour (Fiske, 1921), or the benzidine absorption can be read directly at 250 nm in the ultra-violet (Andersen, 1953).

By precipitating sulphate with barium chromate the soluble chromate remaining can be determined colorimetrically with diphenylcarbazide and the sulphate calculated (Snell and Snell, 1949).

Conductometric method

Bower and Huss (1948) developed a method whereby sulphate is precipitated as the calcium salt in presence of acetone. After centrifuging, the precipitate is dissolved in water and its electrical conductivity measured. Sulphate concentration is then obtained by reference to a standard curve prepared by measurement of known calcium sulphate solutions. The method has been discussed in Chapter 6, section 6:2:3, as being suitable for determining gypsum in soils.

Turbidimetric method

Sulphate may be determined by the turbidity of suspended barium sulphate. The principle has been adapted to soil analysis by Sheen *et al.* (1935), improved by Chesnin and Yien (1950) and modified by Hesse (1957c).

The procedure by which barium sulphate is precipitated must be carefully controlled as the properties of the suspension are influenced by the velocity of reaction. The barium chloride should be added to the sulphate solution in the solid state as crystals of definite size, and not as solution. The size of the crystals will determine their rate of solution which, in turn, determines the rate of reaction with sulphate. Thus, with a standard shaking procedure, it is possible to obtain reproducible results more easily than by adding barium chloride solution. The optimum pH should be maintained to reduce the effect of other ions and a stabilizer is added to keep the precipitate in suspension. Various stabilizers have been suggested such as glycerol and certain gums, but that proposed by Chesnin and Yien, gum acacia, gives satisfactory results. Jouis and Lecacheux (1962) used a solution of the commercial Tween-20.

In the method of Chesnin and Yien, sulphate is extracted from the soil with a solution of sodium acetate and acetic acid buffered to pH 4·5. After shaking with excess barium chloride, added as graded crystals, gum acacia solution is added in amount according to the approximate sulphate content. The turbidity is then measured with a colorimeter using a blue filter.

Although the turbidimetric method is rapid and convenient it is liable, as are all turbidimetric determinations, to give erroneous results if the soil contains much organic matter (Hesse, 1957c). The extracting solution also extracts colloidal organic matter which can affect the results in two ways. When low sulphate concentrations (0–10 μg cm^{-3} S) are involved the organic matter acts as a protective colloid and causes low results, whereas if the sulphate content is high, organic matter is co-precipitated with barium sulphate, making the precipitate more bulky and causing the results to be high. The latter error increases with sulphate concentration and thus will be of particular importance when analyzing organic soils with high sulphur content.

By adding known amounts of sulphate to a sulphate-free soil and analyzing by the method of Chesnin and Yien, the results shown in Figure 13:3 were obtained. It was subsequently shown that the effect was due entirely to colloidal organic matter, the high results not having been caused by colour of extract, excess calcium or co-precipitation of phosphates, bases or nitrate. Very small quantities of organic matter can result in errors of this kind and co-precipitation was observed in the presence of even 0·01 μg cm^{-3} organic nitrogen.

It follows that colloidal organic matter must be removed before precipitation of barium sulphate; it cannot of course be destroyed *in situ* as oxidizable sulphur may be involved. Investigation of the problem resulted in a modification of the Chesnin and Yien procedure whereby colloidal organic matter is co-precipitated with iron(III) hydroxide and removed by filtration or centrifuging prior to addition of barium chloride. The modification results in the removal of not only organic matter but of iron(III) and soil colour as well, both of which interfere with the determina-

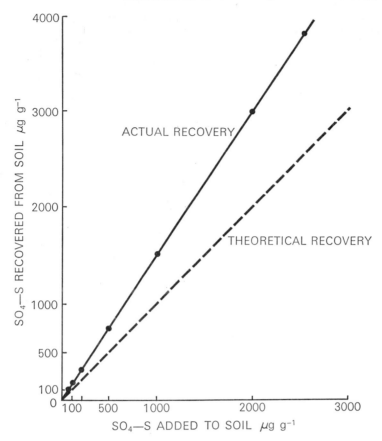

Figure 13:3 Effect of colloidal organic matter upon turbidimetric sulphate determination with increasing sulphate content

tion. Experiment showed that loss of sulphate by adsorption on the hydroxide was negligible. The turbidity was finally measured with a spectrophotometer at 460 nm. Freney (1958) removed colloids with a bacterial filter and also suggested the use of sodium chloride followed by centrifuging or of copper nitrate and calcium oxide followed by centrifuging.

Recent publications show that some investigators using the turbidimetric method of determining sulphate have added the stabilizing gum solution before adding the barium chloride (Bardsley and Lancaster, 1960; Massoumi and Cornfield, 1963). The above findings show that this practice is most undesirable as considerable amounts of colloidal organic matter are thus introduced. If the same procedure is followed when preparing the standard curve, errors may be less appreciable, but the procedure would not take into account the variable colloid content of soil extracts. It must be admitted that Massoumi and Cornfield recovered as little as 2 μg sulphate-sulphur from soils but, as a routine, it is far safer to add the gum solution immediately after precipitation. In the method given by Bardsley and Lancaster (1965), stabilizing gum solutions are omitted altogether when the quantities of sulphate are very small. Bardsley and Lancaster recommend shaking an ammonium acetate–acetic acid extract of soil with activated carbon and filtering before adding the barium chloride, in order to remove colour and colloidal material. This has been found unsatisfactory for some soils as not all colloidal material is necessarily removed and in some cases sulphate is adsorbed.

Steinbergs (1953) found that more reproducible results could be obtained by using slightly impure barium chloride owing to lead sulphate crystals acting as nuclei for barium sulphate precipitation. Butters and Chenery (1959) found that violent shaking of the solutions caused errors and that mixing by slow inversion of the tubes is preferable.

The only difference between turbidimetric and nephelometric measurement of sulphate is in the measurement of the final turbidity. Whereas turbidimetric analysis measures the intensity of transmitted light, nephelometric analysis measures the intensity of light scattered at right angles to the incidence beam.

Reduction method

In several methods of determining sulphate it is first reduced to sulphide, and some methods of achieving this have already been discussed in section

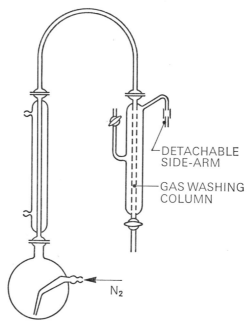

Figure 13:4 Apparatus of Johnson and Nishita for determining sulphate by reduction to sulphide (Reprinted from *Analytical Chemistry*, 1952, Vol. 24, p. 736. © 1952, The American Chemical Society. Reprinted by permission of the copyright owner)

13:2:2. A widely used method is that of Johnson and Nishita (1952) in which sulphate in solution is treated with hydrogen iodide, red phosphorus and formic acid in an atmosphere of nitrogen. The apparatus used by Johnson and Nishita is shown in Figure 13:4. The method was slightly modified by Kilmer and Nearpass (1960) and, in order to be specific for water-soluble sulphate, by Freney (1958). Dean (1966) found the methylene blue finish to the method of Johnson and Nishita tedious and, in general, too sensitive. Dean recommended absorbing the liberated hydrogen sulphide in sodium hydroxide solution, reacting this with a bismuth solution and measuring the extinction of the colloidal bismuth sulphide at 400 nm.

13:2:3 Sulphide-sulphur

Sulphides are normally measured either colorimetrically or volumetrically. The usual colorimetric method is to liberate hydrogen sulphide from the sulphide by acid treatment, absorb this in zinc acetate solution and treat the mixture with *p*-aminodimethylaniline in the presence of iron(III) chloride to give methylene blue,

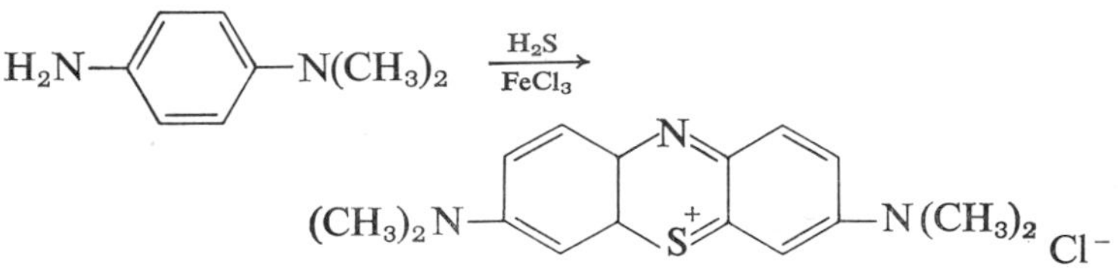

Other colorimetric methods for determining sulphide include conversion to bismuth sulphide with subsequent measurement of light transmittance (Field and Oldach, 1946); with Bindschedler's Green (Snell and Snell, 1949); and as thiocyanate (Allport, 1933).

Volumetric methods of determining sulphide depend upon iodimetry utilizing the reaction,

$$I_2 + H_2S \rightleftharpoons S + 2I^- + 2H^+ \qquad [13:1$$

and in which small amounts of sulphide are added to excess iodine in acid solution. Unused iodine is then determined with thiosulphate.

The various iodimetric methods differ mainly in the manner of absorbing the hydrogen sulphide. Monosulphides in soil are treated with dilute hydrochloric acid in an oxygen-free atmosphere to liberate hydrogen sulphide; this can then be absorbed in a mixture of cadmium and zinc acetates (e.g. Smittenburg *et al.*, 1951) or in potassium hypochlorite (Kitchener *et al.*, 1952) or directly in iodine solution. The precipitated sulphides are then treated with acidified potassium iodide solution. As an alternative to thiosulphate titration of the liberated iodine using starch as indicator, Monand (1953) used the 'Dead-stop' end-point procedure of Foulk and Bawden (1926). This method depends upon the depolarization of electrodes when excess iodine is present.

Liberated hydrogen sulphide can be reacted with arsenic(III) oxide, the excess of which is then determined by titration with iodine (Vogel, 1962; Buriel and Jimenez, 1955).

Monosulphides occur almost exclusively in reduced soils such as muds which need special sampling and handling techniques; this aspect of the analyses is discussed in Chapter 18. Consequently monosulphides are measured usually in wet soils unless precautions are taken to prevent oxidation during the drying process. Hart (1961), for example, dried river muds in vacuo over phosphorus pentoxide in the presence of alkaline pyrogallol.

Polysulphides such as pyrites and marcasite are more difficult to analyze than monosulphides. Interference from monosulphide is easily removed by boiling with hydrochloric acid and then polysulphides can be converted to hydrogen sulphide by heating in concentrated hydrochloric acid with metallic tin (Smittenburg *et al.*, 1951). However, this treatment will also convert elemental and organic sulphur to hydrogen sulphide and in the kinds of soil analyzed for reduced forms of sulphur, elemental and organic forms of sulphur are usually present in appreciable amounts. The elemental sulphur could be removed by solvent extraction but no satisfactory separation of polysulphides from organic sulphur compounds has been advanced. For this reason, Smittenburg called the fraction of soil sulphur measured by reduction with tin and hydrochloric acid 'total oxidizable sulphur'. Sugarawa and co-workers (1954), having removed monosulphides from mud samples, eliminated silica by heating with hydrofluoric and phosphoric acids and hydrolyzed organic matter by boiling with hydrochloric acid. On centrifuging they obtained a black residue of organic matter and pyrites; the polysulphide was determined by X-ray analysis.

The best procedure would seem to be determination of monosulphides and elemental sulphur separately and to record the rest as polysulphide plus organic sulphur. Possibly a closer approximation could be obtained by following the flow sheet in Figure 13:10.

13:2:4 Elemental sulphur

Elemental sulphur can be extracted from soils with acetone, ether, pyridine and certain other solvents. Shorey (1913) isolated free sulphur from soil with steam and extracted the distillate with chloroform. Vallentyne (1952) extracted pond muds with ether and isolated elemental sulphur as crystals.

Romm (1944), having extracted elemental sulphur from soil with acetone, reacted it with potassium cyanide to give thiocyanate which was then titrated with silver nitrate in the presence of formaldehyde to bind residual cyanide ions. Sugarawa *et al.* (1953) used a modification of Morris's (1948) method to determine elemental sulphur that is based upon the reaction of sulphur with sodium sulphite to give thiosulphate.

Elemental sulphur reacts with Schoenberg's reagent, N-(4,4'-dimethoxybenzohydrilidene)-benzylamine, to give a blue thioketone which can be measured colorimetrically at 590 nm and this method has been used by Ory *et al.* (1957). However, as pointed out by Hart (1961), the method would be tedious if adapted to soil analysis as a routine. Hart developed a rapid analytical procedure suitable for routine determination of elemental sulphur present in soils to an extent greater than 0·1%. Hart's method is to extract the soil with acetone, precipitate the sulphur by exchange of solvent by running the acetone solution into water and finally to measure the optical density of the colloidal solution obtained. Hart's procedure is over twenty times as sensitive as the thioketone method and, in the complete absence of interfering organic matter, would be suitable for determining microgramme quantities of sulphur. When applied to the analysis of soil extracts the method is subject to interference from acetone-soluble organic matter and the procedure should not be used uncritically for highly organic soils.

13:2:5 Other inorganic sulphur

Gleen and Quastel (1953) measured polythionates and thiosulphate in soil percolates by methods described by Starkey (1935). This involved the conversion of polythionates to sulphite and thiosulphate which could then be determined by iodine titration.

13:2:6 Organic sulphur

Organic sulphur in soils has been determined almost exclusively as the difference between total and inorganic sulphur. Thus sulphate is measured before and after treatment of the soils with various oxidizing agents, the increase being assumed to be due to oxidation of organic sulphur. For certain soils care must be taken to remove other oxidizable forms of sulphur such as sulphides.

Evans and Rost (1945) determined total organic sulphur by leaching soil with water, followed by dilute hydrochloric acid, and then again with water to wash out chloride. The soil was then oxidized with hydrogen peroxide and sulphate-sulphur determined.

The sulphur content of various organic fractions of a soil (separated as described in Chapter 11) can be determined, but this would be for some specific purpose as the results would be too arbitrary to be considered as total organic sulphur. Thus Evans and Rost measured what they called 'humus sulphur' by shaking the soil leached as described above with ammonium hydroxide and evaporating the filtrate; the residue was then fused with magnesium nitrate and the sulphur measured as sulphate. Evans and Rost found that the ratio of nitrogen to humus sulphur was approximately constant for any particular soil type.

De Long and Lowe (1962) suggested that carbon-bonded sulphur might be determined by a method dependent upon its reduction with Raney nickel. Later De Long and Lowe (1963) detailed such a method, in which hydrogen sulphide liberated from the nickel sulphide produced is estimated as methylene blue. The procedure involves desulphurization in the digestion apparatus of Johnson and Nishita (Figure 13:4) and measures all forms of organic sulphur except alkyl sulphones and ester sulphates. The reagents are the same as those used in the Johnson and Nishita process except that the reducing solution is not required; the usual flask of the apparatus is replaced with one of 150-cm³ capacity. The Raney nickel catalyst is prepared from a 50% nickel–aluminium alloy. From 0·1 g to 0·5 g of soil is weighed into the flask and 0·1 g of catalyst added followed by 5 cm³ of a 5% solution of sodium hydroxide and 25 cm³ of water. The soil is then digested for 30 minutes in a stream of nitrogen, the digest cooled and an excess of 1:1 hydrochloric acid added to liberate hydrogen sulphide. The hydrogen sulphide is then determined as in the routine Johnson and Nishita procedure, being distilled for 30 minutes. As elemental sulphur and mineral sulphides are recovered also, their presence must be taken into account.

Analytical procedures have been developed for the determination of several specific organic sulphur compounds in soil. Shorey (1913) separated an organic fraction of soil which contained trithiobenzaldehyde; the soil sample was repeatedly extracted with boiling glacial acetic acid, the extract evaporated and the residue extracted with ether. Several methods exist for determining the amino-acids cystine, cysteine and methionine, the determinations usually being made upon soil hydrolysates. Once in solution methionine can be determined by reaction with hydrogen iodide to give volatile iodides (Baernstein, 1932) or by demethylation and oxidation of the resulting homocysteine with sodium tetrathionate (Baernstein, 1936). The procedure found most suitable for determining methionine in soil hydrolysates is that given by Williams (1955) and which is based on the McCarthy–Sullivan (1941) colorimetric method in which methionine is reacted with nitroprusside to give a coloured compound. Ion exchange resins are used to purify the hydrolysates and the extinction of the coloured solution is measured with a spectrophotometer.

Cysteine, and cystine via cysteine, can be determined by measuring the sulphur content of precipitated cysteine copper(I) mercaptide (Vickery and White, 1933). A method found suitable for the analysis of soil hydrolysates is that given by Block and Bolling (1951) and is based on Vassel's (1941) use of the Fleming reaction (1930). Fleming discovered that cysteine, when warmed with dimethyl-*p*-phenylendiamine in the presence of iron(III) chloride, gives a blue coloured compound which resembles methylene blue (formed from the amine and hydrogen sulphide). Vassel thought that the end-product is a substituted benzothiazine instead of a phenothiazine as in methylene blue.

Sulphur-containing amino-acids can be estimated *inter al.* by paper chromatography (Dent, 1947; Bremner, 1950). Sowden (1955) determined the amino-acids in soil hydrolysates using column chromatography and concluded that soil materials contain cystine and methionine in very small amounts not detectable by paper chromatography. There is some uncertainty as to whether methionine, and the sulphoxide derived from it, occur naturally in soil or are formed during hydrolysis and subsequent treatment during chromatographic analysis. Stevenson (1954) used ion exchange chromatography to determine the amino-acids in soil hydrolysates.

13:3 RECOMMENDED METHODS

13:3:1 Total sulphur

I / BY OXIDATION WITH MAGNESIUM NITRATE (AFTER BUTTERS AND CHENERY)

Reagents

Magnesium nitrate solution: Dissolve 25 g of Specpure magnesium metal in 400 cm³ of conc. HNO_3 in a 1500-cm³ flask. If the metal does not dissolve completely add a further 50 cm³ of acid. Dilute with 100 cm³ of water to dissolve any crystals of magnesium nitrate, cool and dilute to 500 cm³.

Nitric acid, 25% v/v.

Acetic acid, 50% v/v.

o-Phosphoric acid, concentrated.

Procedure

Mix 1 g of 0·5-mm soil with 2 cm³ of magnesium nitrate solution and evaporate to dryness at 70°C on an electric hotplate. Heat the residue overnight in a stainless steel oven at 300°C. Cool, add 5 cm³ of 25% nitric acid, cover the beaker and digest on a water-bath for $2\frac{1}{2}$ hours; avoid loss of acid.

After cooling, dilute the mixture with water and filter into a 50-cm³ graduated flask. Make to volume and determine sulphate in aliquots of the solution. Place the aliquot in a 50-cm³ volumetric flask, add 5 cm³ of acetic acid, 1 cm³ of phosphoric acid and barium chloride crystals. The phosphoric acid serves to decolorize any iron(III) and the determination is completed as described in section 13:3:2II.

The factor for conversion of sulphate to sulphur is 0·334.

II / BY REDUCTION TO HYDROGEN SULPHIDE (LITTLE)

Apparatus

A heating block is designed to heat the top of the sample first, and to maintain it at red heat whilst the lower part is heating. This is to prevent volatile sulphur compounds escaping before they have reacted with reduced iron. The heating block is depicted in Figure 13:5 and is made from two coils of nichrome wire (0·25 mm diam.) embedded in heat-resisting cement. Lengths of wire 1·5 m and 4·5 m are used to prepare the coils, both of which

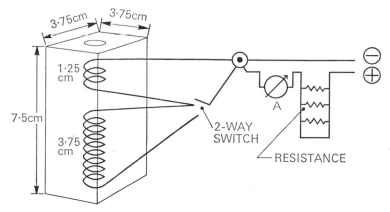

Figure 13:5 Electric circuit and heating blocks for total sulphur analyses (Little, 1957; *J. Sci. Fd. Agric.*)

are 1 cm in diameter; the top coil is 1·3 cm long and the bottom coil is 3·8 cm long (they are embedded in line). The wires protruding from the block are protected from overheating by wrapping with a low-resistance wire. The top coil should give a temperature of 700°C ± 50°C within 10 minutes of switching on the current, and with both coils on for a further 10 minutes, they should both give a temperature of 800–900°C. Little

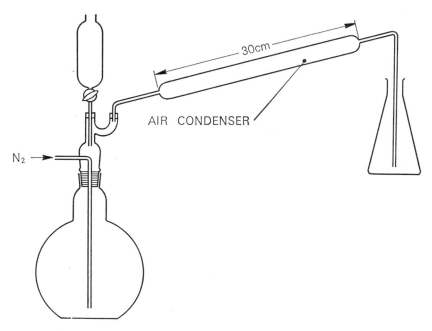

Figure 13:6 Apparatus for liberating and absorbing hydrogen sulphide during the analysis of soil for total sulphur (Little, 1957; *J. Sci. Fd. Agric.*)

found that for these results the top coil needed 1·3 amps at 240 volts and both together needed 1·2 amps. The heating unit, or a series of them, is connected into the circuit shown which includes a variable resistance. Little used electric light bulbs as a resistance but this is a matter of choice.

The apparatus for liberating and absorbing hydrogen sulphide is shown in Figure 13:6. Combustion tubes are made of fused silica, are transparent and are 5 cm long by 1 cm bore; they are closed and rounded at one end.

Reagents

Reduced iron powder: This can be prepared by reducing iron oxide with hydrogen at 700–800°C or by heating iron powder to a dull red heat in a current of hydrogen.

Reagents for determination of sulphide: section 13:3:4.

Procedure

Grind 1 g of 0·15-mm soil in an agate mortar with 5 g of reduced iron powder. Weigh approximately half the mixture into a combustion tube and place about 2 g more of reduced iron on top of the mixture as a guard layer. Place the tube in the heating block so that the boundary of the mixture and guard layer is just below the bottom of the top coil. The tube is best suspended in a stiff wire support so that its position can be varied.

Switch on the top heating coil and 10 minutes later switch on the second coil. Heat for a further 10 minutes, switch off the coils and allow the apparatus to cool.

Place the combustion tube horizontally in the reaction flask, add 50 cm³ of dilute hydrochloric acid and complete the digestion as described for monosulphides (section 13:3:4). Carry out a blank determination for each new batch of reagents and do a standard determination using potassium sulphate as source of sulphur.

13:3:2 Sulphate-sulphur

Extraction

For most purposes the extractant of Chesnin and Yien is recommended, namely a solution of sodium acetate and acetic acid buffered to pH 4·5. Prepare this by dissolving 100 g crystalline sodium acetate in water, adding 30 cm³ of glacial acetic acid and diluting to 1 dm³.

For some specific purposes it may be desired to measure water-soluble sulphate and in such cases, filtration of the extract through a Seitz bacterial filter is advisable. This was recommended by Freney in order to remove traces of sulphate adsorbed on colloidal material.

The extraction is made by shaking 1 part of soil with 5 parts of extractant for 30 minutes and then centrifuging or filtering.

I / GRAVIMETRIC DETERMINATION (USDA HANDBOOK NO. 60)

This method is of particular use when large numbers of samples containing appreciable amounts of sulphate are to be analyzed.

Reagents

Barium chloride solution, 122 g dm^{-3} BaCl$_2$.
Hydrochloric acid, 1·0 M, 90 cm^3 dm^{-3} conc. HCl.
Methyl orange indicator solution.
Ethanol.

Procedure

Pipette a suitable aliquot of soil extract into a 15-cm^3 conical centrifuge tube of known weight and dilute or evaporate, as the case may be, to about 5 cm^3. Add 2 drops of methyl orange solution and then hydrochloric acid until pink; add 1 cm^3 of acid in excess. Heat the tube in a water-bath and add 1 cm^3 of barium chloride solution, drop by drop with shaking.

Keep the tube in the boiling water for 30 minutes and then allow to cool. Centrifuge at 200 rev/s for 5 minutes, decant and discard the liquid and drain the tube by inversion on a pad of filter paper. Wipe the rim of the tube with filter paper and rinse down its sides with 5 cm^3 of ethanol from a pipette. Mix the precipitate with the ethanol, centrifuge and discard the liquid; do not drain the tube this time. Repeat the ethanol washing once more, dry the outside of the tube and leave overnight at 105°C. Cool the tube in a desiccator, weigh and calculate the quantity of barium sulphate.

Calculation

$$SO_4{}^{2-}\text{-S m.e. dm}^{-3} \text{ of extract} = \frac{\text{mg BaSO}_4 \times 2\cdot87}{\text{cm}^3 \text{ in aliquot}} \qquad [13:2$$

II / TURBIDIMETRIC DETERMINATION

This method may be used as a routine for unknown soils and those containing small quantities of sulphate.

Reagents

Barium chloride crystals: BaCl$_2$ crystals ground in a mortar to pass a 0·5-mm sieve and be retained on a 0·25-mm sieve.
Gum acacia solution: 0·25% w/v, aqueous.
Iron(III) chloride solution: 1% w/v aqueous.
Sodium hydroxide: 40% w/v aqueous.
Acetic acid, concentrated.
Standard sulphate solution: 0·5370 g dm^{-3} CaSO$_4$.2H$_2$O. This solution contains 100 μg cm^{-3} S.

Procedure

Add 1 cm³ of iron(III) chloride solution to a suitable aliquot of soil extract, followed by 1 cm³ of sodium hydroxide solution added dropwise with shaking. Filter the solution through a small Whatman No. 41 paper into a volumetric flask. The original flask, precipitate and paper are washed with water, and to the combined filtrate and washings, add 1 cm³ of glacial acetic acid; dilute to volume. As an alternative to filtration, centrifuging may be found more convenient.

Pipette a suitable aliquot of the clear extract (to contain about 5 μg cm⁻³ sulphur when diluted to 25 cm³) into a 25-cm³ graduated flask and dilute if necessary to about 15 cm³. Add approximately 1 g of the graded barium chloride crystals from a standard cup and gently mix the reagent by inverting the flask several times. Add 2 cm³ of gum acacia solution and dilute to 25 cm³. Again invert the flask several times and measure the turbidity with a spectrophotometer at 490 nm or with a blue filter. The calibration curve should cover the range 0–10 μg cm⁻³ S.

III / REDUCTION METHOD (JOHNSON AND NISHITA)

Apparatus

The apparatus is shown in Figure 13:4 which is self explanatory.

Reagents

Reducing agent: Mix 300 cm³ HI, 75 cm³ 50% hypophosphorous acid and 150 cm³ formic acid. Bubble nitrogen through the solution while it is boiled at 115°C for 10 minutes. The solution is stable for 2 weeks.

Nitrogen wash fluid: Add an excess (about 10%) mercury(II) chloride to a 2% solution of potassium permanganate.

Wash solution: Dissolve 10 g sodium dihydrogen phosphate and 10 g pyrogallol in 100 cm³ of copper- and sulphur-free water while nitrogen gas bubbles through. Prepare freshly each day.

Absorbing solution: Dissolve 50 g zinc acetate and 12·5 g crystalline sodium acetate in water, filter and dilute to 1 dm³.

p-Aminodimethylaniline: Dissolve 2 g of reagent in 1500 cm³ of water and add 400 cm³ of conc. H_2SO_4. Dilute the cold solution to 2 dm³.

Ammonium iron(III) sulphate solution: Dissolve 25 g of the salt in 200 cm³ of water containing 5 cm³ of conc. H_2SO_4.

Sulphur-free lubricant: Boil 5 g of silicone grease with 10 cm³ of HI and hypophosphorous acid for 45 minutes under reflux. Discard the liquid and wash the grease with water.

Standard sulphate solution: 5·434 g dm⁻³ K_2SO_4. This contains 1 000 μg cm⁻³ S.

Procedure

Assemble the apparatus and lubricate all joints with sulphur-free grease. Place 10 cm³ of wash solution in the gas washing column. Dilute 10 cm³ of absorbing solution with 70 cm³ of purified water in a 100-cm³ volumetric flask and connect the flask to the side tube of the gas washing column.

Pipette 2 cm³ of soil extract into the boiling flask, add 4 cm³ of reducing solution, quickly attach the flask to the condenser and start the nitrogen flow. Boil the mixture gently for 1 hour using a micro-burner under an asbestos pad, or a small electric hotplate. Remove the volumetric flask together with the detachable side arm, add 10 cm³ of p-aminodimethyl-aniline solution, stopper the flask and shake. Add 2 cm³ of ammonium iron(III) sulphate solution and again shake. Stand for 10 minutes, remove the side tube and rinse it into the flask and dilute to 100 cm³.

Mix and measure the optical density at 670 nm.

13:3:3 Available sulphur

As a general extractant of available sulphur that of Kilmer and Nearpass, namely a 0·5 M solution of sodium hydrogen carbonate at pH 8·5, is recommended. 10 g of soil are extracted with 40 cm³ of solution for 1 hour and sulphate determined in the extract. Results must be correlated with plant response in the usual manner.

An alternative method giving satisfactory results is the biological method of Grigg (1953).

13:3:4 Sulphide-sulphur

Apparatus

A suitable assembly is shown in Figure 13:7 which is self explanatory and easily put together from common laboratory glassware.

Reagents

Hydrochloric acid: 1:1 v/v with water.

Gas washing liquid: As described in section 13:3:2III for the Johnson and Nishita method.

Nitrogen gas.

Potassium hypochlorite solution:
> *Stock solution.* Add 25 cm³ conc. HCl dropwise on to 3·2 g $KMnO_4$ and pass the evolved chlorine into 1 dm³ of 0·5 M KOH solution (28 g dm⁻³ KOH).
> *0·02 M solution.* Add 200 cm³ of stock solution to 500 cm³ water, add 18 g solid KOH and dilute to 1 dm³.

Sulphuric acid: 50 cm³ conc. H_2SO_4/490 cm³ water.

Sodium thiosulphate solution, 0·02 M: Prepare an approximately 0·02 M solution by dissolving 5 g of reagent in water, adding 0·1 g Na_2CO_3 and diluting to 1 dm³. Standardize with potassium iodate.

Potassium iodide solution: 10% w/v, aqueous.

Starch solution: Make 0·25 g soluble starch into a paste with a little cold water and pour it into 50 cm³ boiling water. Boil the solution for 1 minute, cool and mix in 0·25 g KI. Prepare daily.

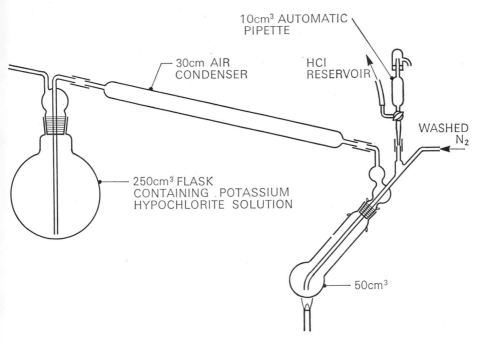

10cm³ AUTOMATIC
PIPETTE

30cm AIR
CONDENSER

HCI
RESERVOIR

WASHED
N₂

250cm³ FLASK
CONTAINING .POTASSIUM
HYPOCHLORITE SOLUTION

50cm³

Figure 13:7 Apparatus for determining monosulphides in soil

Procedure

i. Standardization of sodium thiosulphate solution (Vogel). Dry analytical grade potassium iodate at 120°C and after cooling in a desiccator, accurately weigh 0·14 g to 0·15 g into a 100-cm³ conical flask. Add 25 cm³ of cold, boiled out distilled water to dissolve the iodate and add 2 g of pure, iodate-free potassium iodide followed by 5 cm³ of 1 M sulphuric acid. Titrate the liberated iodine with the thiosulphate solution with constant swirling. When the colour becomes pale yellow add 2 cm³ of starch solution and continue the titration until the blue colour disappears.

The concentration of the thiosulphate solution is calculated from,

$$1 \text{ cm}^3 \text{ } 1\cdot0 \text{ M Na}_2\text{S}_2\text{O}_3 = 0\cdot03567 \text{ g KIO}_3 \qquad [13:3$$

ii. Determination of sulphide-sulphur. Weigh approximately 1 g of soil into the reaction flask which is then attached with small springs to the rest of the apparatus. Pass in nitrogen gas immediately. Place 25 cm³ of 0·02 M hypochlorite solution and 25 cm³ of water in the absorbing flask.

Add 10 cm³ of 1:1 hydrochloric acid to the reaction flask by means of the automatic pipette, and when visible reaction has ceased, heat the mixture for about 45 minutes passing nitrogen all the time. Finally pass nitrogen more rapidly for a few minutes and remove the absorption flask.

Add 5 cm³ of potassium iodide solution and 5 cm³ of sulphuric acid solution to the liquid in the flask and titrate the liberated iodine with thiosulphate using starch as indicator. A blank determination should be made.

Calculation

$$\text{mg S absorbed} = \frac{16\cdot03}{4} \times \text{concentration of } Na_2S_2O_3$$

$$\times \text{ (blank titre } - \text{ actual titre)} \quad [13:4$$

13:3:5 Elemental sulphur (Hart)

Reagents

Standard sulphur solutions: The standard solutions should cover the range 0–10 μg cm^{-3} S and are prepared from recrystallized sulphur and pure, redistilled acetone. Store the solutions in glass-stoppered volumetric flasks and make up solvent evaporation loss as necessary.

Alkaline pyrogallol: Mix 40% aqueous KOH solution with 25% aqueous pyrogallol solution in the ratio of 4:1 as required.

Phosphorus pentoxide.

Redistilled acetone.

Procedure

Dry the soil sample in vacuo over phosphorus pentoxide and in the presence of alkaline pyrogallol; this prevents oxidation of the sulphur before determination.

Grind the dry sample in an agate mortar and accurately weigh a suitable sub-sample into a 100-cm^3 centrifuge tube. Add 25 cm^3 of acetone, stopper the tube and shake for 5 minutes. Centrifuge at 400 rev/s for 15 minutes.

Take a suitable aliquot of the clear supernatant liquid, containing from 100 μg to 700 μg sulphur, and run it into about 80 cm^3 of water in a 100-cm^3 volumetric flask; gently agitate the flask during the process. Colloidal sulphur begins to form immediately after the solvent has been exchanged. Dilute the liquid in the flask with water to 100 cm^3, mix and allow to stand for 3 hours before measuring the optical density at 420 nm.

Calculate the sulphur content of the original sample by reference to a standard curve. If the concentration of sulphur in the extract exceeds 200 μg cm^{-3} the extraction should be repeated using a lower ratio of soil to acetone.

13:3:6 Organic sulphur

Total organic sulphur

As a routine the recommended method is to determine sulphate before and after oxidation of the soil. As a first step, however, any elemental sulphur and/or sulphide-sulphur should be estimated separately. It is likely that the result will be influenced by polysulphides, particularly if monosulphides or elemental sulphur are present and this should be taken into account when interpreting the results. The method of oxidation may be any of those described for total sulphur determination; the use of hydrogen peroxide is not recommended.

Methionine in soil extracts or hydrolysates (Williams)

Reagents

Hydrochloric acid, 6 M: 540 cm³ dm⁻³ conc. HCl.
Amberlite IR-4-B resin.
Amberlite IR-C-50 resin.
Sodium nitroprusside solution: 0·6% w/v, aqueous; prepare freshly.
Standard methionine solution: 1 g dm⁻³ *d,l*-methionine, aqueous. This
   solution contains 1 mg cm⁻³. The standard curve should cover the
   range 100–1 000 μg cm⁻³ methionine.
Acetate buffer solution: 53 g Na₂CO₃ (anhydrous) and 115 cm³ glacial
   acetic acid per dm³.

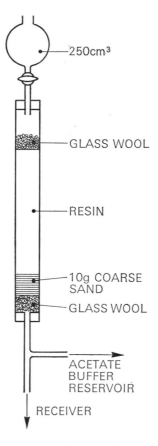

Figure 13:8 Apparatus for use of Amberlite resin during determination of methionine in soil hydrolysates

Procedure

i. Preparation of Amberlite IR-4-B *resin.* Place 70 cm³ of resin into a 400-cm³ beaker and wash three times by decantation with 75-cm³ portions of 4% sodium carbonate solution. Then wash three times with water, three times with 4% hydrochloric acid, three times with water, and again, three times with sodium carbonate solution. Allow the last portion of carbonate solution to remain with the resin for at least 1 hour. Finally wash the resin three times with water and store under water; before use, filter off excess water by suction.

ii. Preparation of Amberlite IR-C-50 *resin.* Set up a 60-cm length of glass tubing 2·5 cm diameter as shown in Figure 13:8 and fill with resin. Allow 250 cm³ of 4% hydrochloric acid to percolate slowly through the resin, and follow with 250 cm³ of water. Next percolate 250 cm³ of 4% sodium hydroxide solution and again wash with water. After draining completely, allow acetate buffer solution to flow in from the bottom of the tube until the resin is submerged. Drain off the buffer solution and repeat the process twice more with fresh buffer solution, allowing the final portion to remain in the tube for 30 minutes. Finally wash the resin with 250 cm³ of water and store under water, in the column, until required.

iii. Preparation of soil hydrolysate. Reflux 1 g of soil for 24 hours with 100 cm³ of 6 M hydrochloric acid. Filter into a 500-cm³ Kjeldahl flask and evaporate to near dryness under reduced pressure; a suitable set-up for this

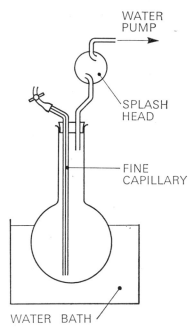

Figure 13:9 Apparatus for concentrating soil hydrolysates

operation is shown in Figure 13:9. The syrup remaining in the flask is dissolved in water and filtered into a beaker. Add about 10 g of Amberlite IR-4-B resin to the solution and stir; sufficient resin should be added to raise the pH to 5. Filter by suction, wash the resin with water and return it to stock for regeneration. The final filtrate should be about 250 cm³. Adjust the pH to 4·7 and percolate the solution through the c-50 resin at 10 cm³ per minute. Elute the column with three 80-cm³ portions of water and concentrate the leachate to about 20 cm³ on a water bath. Filter and make the solution to 25 cm³ in a volumetric flask.

iv. Measurement of methionine. To a series of test-tubes add 1-cm³ aliquots of standard solutions or soil hydrolysates and dilute with 4 cm³ of water. Add 2 cm³ of 5 M sodium hydroxide solution and 1 cm³ of freshly pre-pared sodium nitroprusside solution. Allow the mixtures to stand for 1 hour, add 2 cm³ of *o*-phosphoric acid, shake, allow to stand 5 minutes and read the optical density at 420 nm or with a blue filter. Measure a reagent blank also and for soil hydrolysates a colour blank is measured substituting 1 cm³ of water for nitroprusside solution.

Cystine and cysteine in soil hydrolysates (Block and Bolling)

Reagents

Hydrochloric acid, 5 M: 445 cm³ dm^{-3} conc. HCl.

Iron(III) alum solution: 20 g ammonium iron(III) sulphate/100 cm³ 0·5 M
  H_2SO_4.

Zinc dust.

Dimethyl-*p*-phenylenediamine hydrochloride solution (dye solution): 0·035 g reagent dissolved in 100 cm³ 3 M H_2SO_4 (163 cm³ dm^{-3} conc. H_2SO_4). Store in a refrigerator and prepare every 2 weeks.

Procedure

i. Determination of cysteine. Mix 0·165 g of zinc dust, 3 cm³ of dye solu-tion and 2 cm³ of iron(III) alum solution in a test-tube and allow to stand for 10 minutes. Place the tube in boiling water for 15–30 minutes until all the zinc has dissolved. Cool and add 1 cm³ of soil hydrolysate (made to standard volume with 5 M HCl) and 3 cm³ more of iron(III) alum solution. Mix the solutions and heat in a boiling water-bath for 45 minutes, cool and dilute to 25 cm³. Read the optical density at 580 nm.

ii. Determination of cysteine plus cystine. Pipette 1 cm³ of soil hydrolysate into a test-tube and add 3 cm³ of dye solution and 0·165 g of zinc dust. Stand for 3 minutes, add 2 cm³ of iron(III) alum solution and stand for 45 minutes. Add 3 cm³ more of iron(III) alum solution and heat the stoppered tube in boiling water for 45 minutes or until all the zinc has dissolved. Cool, dilute to 25 cm³ and read the optical density at 580 nm.

Reagent blanks must be read for all determinations.

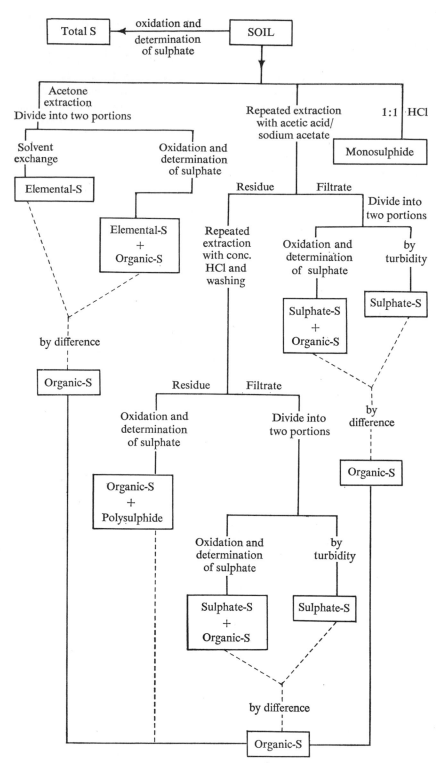

Figure 13:10 Flow sheet for fractionation of soil sulphur

Calculations

The content of cysteine can be estimated directly from a standard curve prepared by measurement of known solutions. Cystine is then calculated as the difference between cysteine plus cystine and cysteine.

A more generally applicable method of estimating the sulphur amino-acids is by paper or column chromatography.

14

Iron, aluminium and manganese

14:1 IRON

14:1:1 Introduction

Occurrence of iron

Iron present in primary minerals is usually associated with igneous rocks as iron ores, biotitic micas and other ferromagnesian silicates. The primary minerals are normally concentrated in the larger sized fractions of soil. In the clay fraction, iron, when it occurs, is usually in the form of oxides, organic complexes and insoluble salts, although vermiculite and biotite can also occur in the clay fraction. Iron oxides are common in sedimentary rocks and iron(II) carbonate (siderite) is often associated with calcite, particularly in a reducing environment.

'Free' or 'active' iron is predominantly in the form of oxides except in some podsolic soils where organic complexes account for most of it. Iron salts usually do not occur to any great extent except in poorly-drained, acid soils where basic iron(III) sulphate is characteristic, generally as a yellow mottling but sometimes as a surface crust.

Compounds of iron in soil are indicators of drainage and weathering conditions and, apart from affecting structure, are one of the principal causes of soil colour. Thus poorly-drained (gley) soils, where reducing conditions prevail, are characteristically coloured greenish-grey to grey by iron(II) compounds. Often the colour does not rapidly change to brown on exposure and the precise nature of the iron(II) compounds is not known. In calcareous soils iron(II) carbonate could be present which is greenish-grey and turns yellow on exposure by oxidation to limonite. Hem (1960) suggested the following equilibrium to be envisaged for iron(II) carbonate,

$$FeCO_3 + H^+ \rightleftharpoons HCO_3^- + Fe^{2+} \qquad [14:1$$

and in a well-drained soil where carbon dioxide is in excess the equilibrium shifts to the left, resulting in a high concentration of hydrogen ions; whereas, if carbon dioxide is restricted and sufficient iron is present, the shift will be to the right. When the soil is subject to alternate oxidizing and reducing conditions iron tends to accumulate in a layer and eventually forms what is known as an 'iron-pan'.

Iron can contribute towards the acidity of a soil by hydrolysis. Hydrolysis of iron(III) in montmorillonitic systems was reported by Bower and Truog (1940) who found that iron was sorbed to an extent of about twice the cation exchange capacity. As they extracted the iron with an acid (pH 3) solution, however, some non-exchangeable iron was probably involved. Thomas and Coleman (1964) used neutral, normal chlorides to investigate hydrolysis of iron(III) in soils. Potentiometric titration of replaced acidity showed that some of that acidity was due to iron. These authors showed that exchangeable iron(III) is unstable, tending to hydro-lyze and the hydrogen ions produced dissolve the lattice cations. Normally the hydrolysis does not proceed to the stage of iron(III) hydroxide forma-tion.

Free iron oxides (Oades, 1963)

The oxide goethite ($HFeO_2$) occurs mostly in temperate regions and is probably formed by the slow precipitation of iron(III) ions at low pH. This is the characteristic mineral of bog iron ore, usually together with limonite ($FeO(OH)_n . H_2O$). In drier regions hematite (α-Fe_2O_3) predominates and this oxide is seldom found in humid regions. In waterlogged soils oxidation of iron(II) hydroxide in the absence of carbon dioxide results in the forma-tion of lepidocrocite (γ-$FeO(OH)$) and this appears in the form of mottling. Two other oxides are distinguished, magnetite (Fe_3O_4) which contains iron(II) and maghemite (γ-Fe_2O_3), which may form as an alteration product of magnetite by weathering and, to a lesser extent, from other oxides by high temperatures in association with organic matter.

The free iron oxides occur in soils as concretions, lateritic crusts and in a state of fine division. If concretions are absent the free iron oxides are concentrated in the clay fraction. Clay pans in poorly drained soils hold up descending iron oxides, giving a concentration of iron in the pan. The presence of iron oxides in soil is generally considered to exert important effects upon their physical properties, but Deshpande and co-workers (1964) found that for certain physical properties this is not necessarily the case. It has been reported (Lutz, 1937; McIntyre, 1956) that a positive cor-relation exists between free iron oxides and the degree of aggregation of clay and silt particles, but Deshpande *et al.* found that the stability of macro-aggregates was not dependent upon the presence of iron oxides. From the results of permeability, wet-sieving, mechanical analyses and swelling determinations, the authors concluded that iron oxides in the soils they examined were present as discrete crystals and did not cement the soil particles. They consider that the widespread belief that iron oxides are important in soil aggregation is due to the methods used for removal of iron oxides not being specific, as they also remove organic matter and aluminium oxides.

Concretionary iron oxide is formed by precipitation under conditions of alternate oxidation and reduction and also by oxidation of plant roots, when the oxides first appear as tubes along the root channels. Free iron oxides are characteristically present in all forms of laterite and are respon-sible for the hardening of that material on exposure. Laterite (Alexander and Cady, 1962) is a strongly weathered material with a high content of iron, aluminium or both. Bases are almost completely absent and primary silicates are usually low, although some laterite may contain large quantities of quartz and kaolinite.

The formation of laterite requires enrichment of soil material with iron and this may arise from removal of other constituents as happens in volcanic and basic rocks, or alternatively, iron may be brought in from other sources by water movement. If subjected to drying, laterite hardens, the process being complex but most probably due to the growth of continuous crystals of iron oxide through the soil (Sivarajasingham *et al.*, 1962). The hardest forms of laterite contain crystalline goethite and hematite and the proportion of hematite increases with age and hardness.

Movement of iron

The solubilities of iron oxides are very low in the oxidized state and movement of iron is more common in the reduced, iron(II) state. The necessary reducing agents come from organic decomposition products, but a more important factor in the mobilization of iron is the formation of organic complexes or chelates. Leachates of leaves and forest litter are very active in the mobilization of iron. Coulson *et al.* (1960) found that polyphenols were specifically involved in complexing iron and Bloomfield (1956) found that simple aliphatic acids and reducing substances are active in this respect. Bremner and co-workers (1946) considered that leaf extracts may be chelating agents and percolation of soil columns with EDTA and other chelating agents produced profiles resembling podsols (Atkinson and Wright, 1957). Duchafour (1963) measured complexed iron in ten different kinds of humus and found that its content was relatively low in forest mulls and in organic matter of rendzinas, but high in other forms of humus. Duchafour found that the iron complexed by fulvic acids was mobile and accumulated in the B-horizons of podsols, whereas iron complexed by humic acids is insoluble. Movement of iron can be expected to be least in chernozemic mull and rendzinas, medium in acid forest mulls and greatest in podsols.

Soluble iron leached through a soil is precipitated at lower depths by adsorption on clays and unchanged oxides, or possibly by an increase in pH which reduces the positive charge acquired by the colloidal particles (Russell, 1961). Hem and Cropper (1959) and Hem (1960) investigated the forms of iron in natural groundwaters and found them to depend upon prevailing pH and redox conditions. Forms of iron found in solution were Fe^{3+}, $FeOH^{2+}$, $FeOH^{+}$, $Fe(OH)_2^{+}$ and Fe^{2+}. In alkaline water $Fe(OH)_3$ may be present. In the pH range of 5 to 8 oxidized iron cannot remain in solution but only in suspension or as complexes. The rate of oxygen diffusion governs the oxidation and precipitation of iron(III) hydroxide, although in acid solution (pH 1·6) no iron(II) was oxidized during 3 months. Hem and Cropper also found that in strongly acid solution the complexing of iron(III) by chloride and fluoride was important.

The availability of dissolved oxygen controls the occurrence of dissolved iron, and at depths where reducing environments occur, a decrease in available oxygen increases the amount of iron(II) in solution. Iron(II) is less strongly complexed that iron(III) and occurs mostly as the simple Fe^{2+} ion.

The precipitation of iron(III) from iron(II) solutions is by hydrolysis, that is,

$$Fe^{2+} + 3H_2O = Fe(OH)_3 + 3H^+ + e^- \qquad [14:2$$

and thus oxidation of iron(II) can result in a drop in pH.

Iron is also brought into solution in the reduced form by bacteria although Oades (1963) considers that this is probably due to reaction with organic compounds produced during bacterial metabolism.

Available iron

Iron is one of the nutrient trace elements essential for plant growth, although needed by most plants in small amount. Iron deficiency shows as chlorosis in the plant and this disease is most common on calcareous soils in which iron may be present but unavailable. The deficiency can be induced by excess phosphorus and high amounts of zinc or copper but a lime-induced deficiency is more usual.

Plant-available iron exists as the iron(II) ion and as the complex iron(III) hydroxide ion, $[Fe(OH)_n]^{(3-n)+}$. As the element occurs in more than one valence state the availability of iron is affected by the state of oxidation of a soil. For this reason iron is normally readily available to swamp rice, for example, and instances have been reported of iron toxicity arising when soils are flooded. Iron deficiency can be controlled by spraying the plants with iron chelates in solution and this is commonly done with vines. The effect is to increase the chlorophyll content and to decrease acidity. In trees, iron is often supplied by injecting iron(II) sulphate pills directly into the trunk.

14:1:2 Determination: background and theory

Analytically we can distinguish total iron, free iron oxides, iron(II) and (III) and plant-available iron.

Total iron

Total iron in a soil is brought into solution by acid digestion or by fusion with sodium carbonate as described in Chapter 16. There is little to choose between the two methods regarding efficiency but it should be borne in mind that excess perchloric acid can interfere with certain colorimetric methods of measuring the dissolved iron, and that hydrofluoric acid may be a preferable reagent. Having decided upon the method for determining the iron in solution, the analyst should read section 16:2 before bringing the total iron into solution.

12*

Total iron(II)

The determination of total iron(II) in soil is not often necessary but it can be brought into solution with a mixture of sulphuric and hydrochloric acids (Kolthoff and Sandell, 1945). The soil is heated in a closed crucible with the acids and evolved water vapour protects iron(II) from oxidation. Mitsuchi and Oyama (1963) modified the procedure by lowering the temperature of decomposition to 80°C which prevents reduction of iron(III) by organic matter. The increased possibility of oxidation of iron(II) was counteracted by introducing carbon dioxide into the system. Walker and Sherman (1962) boiled soil with sulphuric acid and then added hydrofluoric acid. The suspension was poured into a solution of boric and sulphuric acids and filtered. Aliquots of solution were diluted to standard volume with dilute sulphuric acid and analyzed for iron with bathophenanthroline.

Free oxides

Measurement of the free or active iron in soils usually involves reduction as a first step. Tamm (1922) developed an ammonium oxalate extraction of soil, and Mattson (1931) used hot aluminium chloride solution. Drosdoff and Truog (1935) saturated soil suspensions with hydrogen sulphide under alkaline conditions, when the oxides were converted to sulphides which were then extracted with dilute acid. The hydrogen sulphide method was found to be non-applicable to oxides in biotite, basalt and granite. Later, Truog *et al.* (1936) modified the sulphide method by adding oxalic acid and sodium sulphide to the soil suspensions; this released nascent hydrogen sulphide and the resulting iron sulphide was dissolved in the excess oxalic acid.

Several workers have used nascent hydrogen to reduce iron oxides; thus Jefferies (1941) immersed an aluminium cylinder in a boiling suspension of soil and oxalic acid. Later, Jefferies (1945) improved the technique by using magnesium instead of aluminium in the presence of excess oxalate to prevent precipitation of insoluble oxalates. Dion (1944) boiled the soil suspension in 10% ammonium tartrate solution in an aluminium vessel containing an aluminium ribbon spiral; ammonia which volatilized on boiling, kept the pH at about 6·4. Haldane (1956) found Jefferies's magnesium method unsatisfactory and obtained better results by using zinc. The best chemical method, however, appears to be the use of sodium dithionite $Na_2S_2O_4$, and which was first proposed by Galabutskaya and Govorova (1934). The technique was improved by Mackenzie (1954) who recommended heating the soil suspension in boiling water before adding the reagent in order to prevent precipitation of black, insoluble iron compounds. Kilmer (1960) shook the soil suspension with dithionite for 16 hours at room temperature and recommended determining the iron with dichromate, instead of colorimetrically, as the range is unlimited and dilution of solutions is unnecessary. Coffin (1963) simplified the dithionite procedure considerably and claimed that the results were as good, if not better than those of earlier methods. In Coffin's procedure the soil is mixed with a citrate buffer at pH 4·75 which permits the most rapid extraction of iron without precipitation of sulphur. The dithionite reagent is added and the mixture shaken for 30 minutes at 50°C. The filtrate is then digested with perchloric acid and analyzed for iron and aluminium.

Allison and Scarseth (1942) used a biological method of reduction by anaerobically decomposing sugar with the soil. Bromfield and Williams (1963) investigated this biological method and compared it with three chemical methods. They found that by applying the biological method 3 weeks after hydrogen peroxide treatment of the soil it could be made five times more specific for iron than the dithionite method, eight times more specific than the nascent hydrogen sulphide method and fourteen times more specific than the acid ammonium oxalate method. The main conclusion of Bromfield and Williams, however, was that the long incubation and fermentation control necessary gives the method no advantage over more rapid chemical methods.

To remove free iron oxides for the purpose of mineralogical analysis, or for separation of the soil into fractions for elemental analysis, the choice is limited by the need to avoid damage to the silicate minerals. Furthermore, reagents should not contain large amounts of alkali salts (Le Riche and Weir, 1963). Thus hydrochloric acid and sodium dithionite are unsuitable reagents and Le Riche and Weir found that the most suitable method was de Endredy's (1962) modification of Tamm's ammonium oxalate extraction. In this procedure the soil is treated with ammonium oxalate solution at pH 3·3 in ultra-violet light and the extract ignited. Le Riche and Weir found that a single extraction removed most of the free iron oxides without damaging the silicate minerals. They further found that in air-dry soils the solubility of the oxides was decreased, and to obtain correct results, the soils had to be soaked in oxalate solution for at least 5 days before ultra-violet irradiation.

The various free iron oxides cannot be distinguished chemically one from another although the colour is a rough guide (Waegeman and Henry, 1954). The most reliable methods of such analysis are X-ray diffraction (Rooksby, 1961), differential thermal analysis (Mackenzie, 1957) and magnetic susceptibility (Oades and Townsend, 1963).

Available iron

As discussed in section 14:1:1, the plant-available iron in soils is the exchangeable iron(II) and (III) although water-soluble forms should be included. Extraction of the iron(III) form is only complete if the pH is lowered to 3, but under such acid conditions, some non-exchangeable iron is also extracted. Jackson (1958) gives a procedure for extracting soil for what may be termed 'total available iron' which includes the doubtful dilute-acid-soluble forms. In this method the soil is extracted with 1 M ammonium acetate solution buffered to pH 3 with hydrochloric acid.

Usually, however, exchangeable iron(II) together with water-soluble iron is extracted as available iron. Freshly sampled soil is extracted as quickly as possible to avoid oxidation of iron; once in solution, oxidation is of no importance and the iron can be determined by any of the methods described. Jeffery (1961), working with paddy soils, immersed the soil under water in an atmosphere of nitrogen. Aliquots of solution were then pipetted directly into a solution of α,α'-dipyridyl, centrifuged and the colour measured to determine water-soluble iron(II). Ignatieff (1941) found that aluminium chloride solution extracted more iron(II) from soils than does water or potassium chloride solution. The aluminium solution did not reduce iron(III) present during the time of extraction. Ignatieff recommended extracting 20 g of fresh soil with 50 cm³ of 3% aluminium chloride solution immediately after sampling. The mixture is then centrifuged, using a hand-operated centrifuge, and aliquots of extract are pipetted into a buffer solution at pH 4·6 containing α,α'-dipyridyl. Iri (1957) sampled soil directly into a flask containing dilute aluminium chloride solution and transported the flask to the laboratory whilst protecting from oxidation; thereafter the iron was measured with α,α'-dipyridyl. Walker and Sherman (1962) found that organic matter interfered with the determination of iron(II) due to its oxidation. They developed a procedure using bathophenanthroline which gave satisfactory results in the presence of organic matter.

Paddick (1948) extracted available iron from alkaline soils with calcium thioglycollate solution with subsequent treatment of the filtrate with thioglycollic acid and ammonia to produce a measurable purple colour. Paddick regarded available iron in calcareous soils as that occurring mostly as hydroxide and organic compounds. The test that he developed is unaffected by the presence of calcium carbonate.

Bao and co-workers (1964) used a solution containing 0·5% α,α'-dipyridyl and 1 M barium chloride buffered at pH 7 with 3% barium acetate to extract water-soluble, exchangeable and chelated iron(II) from soils. Two extractions for 30 minutes at a ratio of 1:20 were recommended; the solution does not extract iron(III).

Iron in solution

Once in solution, iron can be determined in several ways. Gravimetric analysis is not recommended as the methods do not readily eliminate the effects of aluminium and titanium.

Volumetrically, iron can be determined with potassium permanganate or ceric sulphate after reduction; with titanium(III) or hypovanadous salts after complexing with thiocyanate; with mercury(I) nitrate; iodimetrically and with potassium dichromate. The dichromate titration is considered by Robichet (1957) to be one of the best methods of determining iron.

The traditional colorimetric reaction for determination of iron is with thiocyanate when a red-coloured complex is formed, and the sensitivity of the method can be increased by solvent extraction of the complex. Similarly, solvent extraction of iron as the 8-hydroxyquinolate can be used.

Snell and Snell (1949) describe the use of over thirty reagents for the colorimetric determination of iron and among those commonly used in soil analysis are thiocyanate, α,α'-dipyridyl, ortho-nitroso-R-salt, Tiron, sodium salicylate and orthophenanthroline. After investigation, Robichet (1957) recommended *o*-phenanthroline as being the most suitable reagent, although, as pointed out by Jackson (1958), this reagent is of particular use for very small quantities of iron, and for the relatively large amounts extracted from soils the Tiron reagent is more suitable. Bathophenanthroline, 4,7-diphenyl-1,10-phenanthroline, is about twice as sensitive as *o*-phenanthroline and was used to determine total iron(II) in soils by Walker and Sherman (1962). Jefferies and Johnson (1961) measured the optical density of iron(II) oxalate solutions in a potassium oxalate–oxalic acid buffer and found this to be proportional to the iron content; they developed their technique to measure iron oxides in soil.

For reaction with *o*-phenanthroline the iron must first be reduced and this is accomplished by use of hydroxylamine hydrochloride. The reaction must take place at a buffered pH of 2·8 as reduction is slow at higher pH values and below pH 2 the complex is not formed. There is no interference from other ions likely to be present except from very large amounts of phosphate. If much silica is present the buffer should not be ammonium acetate as the gel of silicic acid which forms may interfere with the colour measurement (Snell and Snell, 1949). Jackson adjusted the pH with hydrochloric acid or ammonium hydroxide using separate aliquots for measurement with a pH meter. Nickel and zinc, if present in large amounts, may interfere but this is overcome by adding excess reagent and reading the transmission within 10 minutes, or by solvent extraction of the complex as perchlorate with nitrobenzene (Snell and Snell, 1949). Orthophenanthroline is a convenient reagent if both iron(II) and (III) are to be measured in the same aliquot, as the reddish iron(II) complex absorbs at 512 nm and both complexes have an identical absorption at 396 nm, the effect being additive. The iron(III) complex is yellow in colour.

Tiron, disodium 1,2-dihydroxybenzene 3,5-disulphonate, gives a purple colour with iron but the colour varies with pH. A buffered pH of 4·7 gives the most satisfactory results (Yoe and Armstrong, 1947). The reagent also reacts with titanium giving a yellow colour which can be measured by destroying the purple colour after reading with dithionite, and thus permitting the estimation of titanium and iron in the same sample. There is hardly any interference from other ions, except perchlorate and fluoride which may influence the choice of method for bringing the iron into solution.

Nitroso-R-salt gives a stable green colour with iron(II) and is more sensitive than either α,α'-dipyridyl or *o*-phenanthroline. Many other elements interfere with the reaction but Dean and Lady (1953) eliminated interference by using hydrazine as reducing agent followed by heating at 80°C at pH 5–7 when the colours of other complexes are destroyed.

With salicylic acid, iron(III) gives an amethyst colour which is suitable for measurement and which follows Beer's Law over a wide range of concentration. As the colour is sensitive to changes in pH, the solution to be measured must be buffered. The method is suitable for routine work if an accuracy of about 3% is acceptable.

Gramms and Kagelmann (1962) determined iron and calcium in the same sample by complexometric titration with Chelaplex III. Polarographic and spectrographic methods can also be used.

14:2 ALUMINIUM

14:2:1 Introduction

Occurrence of aluminium

Aluminium is one of the most commonly occurring elements, being next to oxygen and silicon in abundance. It occurs in many silicate rocks as micas and feldspar and in clays. Aluminium oxide, Al_2O_3, occurs as corundum and emery, and the hydroxide $Al(OH)_3$ occurs as gibbsite. The ore bauxite contains a mixture of aluminium oxides and is formed by prolonged weathering of aluminium-bearing basic rocks in tropical and sub-tropical regions. Other sources of soil aluminium are diaspore, $HAlO_2$ and cryolite, Na_3AlF_6.

Aluminium occurs in laterites and free oxides in association with iron. Norrish and Taylor (1961) extracted almost pure goethites from kaolinitic clays and found them to contain from 15 to 30 mol per cent of aluminium. This was considered to be the explanation for the aluminium brought into solution by dithionite during estimation of free iron oxides. The ease with which goethite is dissolved by dithionite was found to be inversely related to the aluminium content. X-ray measurements revealed that the cell dimensions of soil–goethite spacings are lower than in pure goethite and that aluminium commonly replaces iron in these compounds. Recent work by Deshpande and others (1964) has suggested that the cementing action of free iron oxides upon soil particles is likely to be due to their content of aluminium oxides (section 14:1:1).

In podsols aluminium is deposited together with iron as hydrated oxides or hydroxides. It has been postulated, but not proved (Russell, 1961), that the aluminium passing into the B-horizon reacts with soluble silica, giving rise to the formation of kaolinitic or illitic clays. The mobilization of aluminium may be brought about by organic acids and its re-precipitation by a rise in pH but, as in the case of iron, the movement of aluminium in soils is still not clearly understood. Reducing conditions do not make aluminium soluble in the same manner as iron and manganese and thus soluble aluminium is not necessarily found in poorly drained soils and, indeed, aluminium more characteristically moves in well drained soils.

Raupach (1963) investigated the solubility of aluminium compounds in soil over a pH range of 3·5 to 10 and found that they dissolved to give $AlOH^{2+}$, $Al(OH)_2^+$ and $Al(OH)_4^-$. Values for the aluminium content of soil extracts and displaced soil solutions at different pH levels were compatible with this finding. Thomas (1960) treated aluminium-saturated cation exchangers with calcium hydroxide and obtained $Al(OH)_3$, $Al(OH)_2{}^+$ and $AlOH^{2+}$ according to experimental conditions and type of exchanger.

Humus can retain trivalent ions which are not leached out by neutral salt solutions. Aluminium held in such a manner can be extracted with acid oxalate or pyrophosphate solutions as complexes or chelates (Williams *et al.*, 1958; Martin and Reeve, 1958). Over 10% of the total aluminium in the soil of a West African mangrove swamp was found to be complexed with organic matter (Hesse, 1963).

Exchangeable aluminium

The possible function of aluminium as an exchangeable ion has been extensively studied. Paver and Marshall (1934) and Mukhurjee *et al.* (1947) studied the role of aluminium ions in acid clays and concluded that aluminium does not participate in direct exchange reactions. When montmorillonitic clays were subjected to successive salt and acid treatments, the salt solutions contained increasing quantities of aluminium and this was considered to be the result of breakdown of the clay lattice with simultaneous reduction in cation exchange capacity.

The extraction of aluminium layers increases the cation exchange capacity of clays and conversely, the fixation of aluminium decreases the exchange capacity. Shen and Rich (1962) considered that such decrease in exchange capacity is due to positively charged Al—OH groups occupying exchange sites, thus blocking the pathways to the sites.

The effect of so-called exchangeable aluminium upon soil acidity has been discussed in Chapter 4 and there is little doubt that extractable aluminium influences the pH of a soil. In acid soils any toxic effects on plant growth are very often due to aluminium rather than to acidity *per se*. Vlamis (1953) worked with solutions displaced from limed and unlimed soils and concluded that aluminium toxicity is the main factor limiting plant growth on acid soils, and Chamura (1962) estimated crop tolerance to highly acid soils by measuring the absorption of aluminium by the plant roots.

Aluminium and plant growth

Aluminium appears to affect root growth in particular, and excess aluminium accumulating in roots reduces their capacity for translocating phosphorus. Thus even in soils containing ample phosphorus, excess aluminium will cause a phosphorus deficiency. Amelioration of such soils involves suppression of aluminium activity, for example by liming, and not addition of phosphorus.

The toxic amount of aluminium in a soil will depend upon other soil properties such as pH and phosphorus content and upon the plant to be grown. In culture solutions very small quantities of aluminium can be toxic. Ligon and Pierre (1932) found that as little as 1 μg cm^{-3} aluminium was detrimental to the growth of barley roots. On the other hand, Hackett (1962) found that from 2 μg cm^{-3} to 25 μg cm^{-3} aluminium in solution stimulated root growth from seeds. The growth of subsequent shoots was unaffected, except when aluminium was added to water cultures of seedlings when the shoots and not the roots were stimulated. Later Randall and Vose (1963) using radioactive phosphorus as a tracer found that low levels (1.85×10^{-4} M) of aluminium increased the phosphorus content of roots and shoots of rye grass, whereas high levels (18.5×10^{-4} M) decreased the total phosphorus content. These workers suggested that the observed increase in phosphorus uptake caused by traces of aluminium is a metabolic process due to stimulation by aluminium of cytochrome reduction. In spite of increased uptake of phosphorus the plants may still show phosphorus deficiency symptoms due to immobilization of the phosphorus by aluminium within the plant.

Rice grown in acid soils takes up large quantities of aluminium resulting in restricted root development, and Hesse (1963) found that if rice plants growing in such soils were fertilized with phosphorus, the uptake of phosphorus was increased but the nutrient was precipitated with aluminium in the plant roots.

That aluminium toxicity is largely bound up with phosphorus fixation has been repeatedly demonstrated. Pratt (1961) treated soils with large amounts of calcium phosphate and found that on subsequent acidification to pH 4.7 no aluminium toxicity occurred until all the available phosphorus had been converted to aluminium phosphate and, to a lesser extent, iron phosphate. Williams and co-workers (1958) investigated phosphorus sorption in relation to other soil properties and they found that aluminium extracted by acid oxalate solution gave highly significant correlations. Except for granitic soils where loss on ignition is best, this oxalate-extractable aluminium was considered the best single criterion of phosphorus sorption.

The beneficial effect of applying phosphorus to acid soils can be due largely to fixation of aluminium (Pierre and Stuart, 1933), although the effect may be short-lived owing to the tendency for phosphorus to transfer from aluminium to iron in the course of time. This phenomenon has been discussed in Chapter 12 (section 12:2:6) and appears to be due to the lower solubility product of iron phosphate compared to that of aluminium phosphate. More recently Hsu and Rennie (1962) found that the initial fixation of phosphorus by X-ray amorphous aluminium hydroxide was by absorption, and this was followed by a slow decomposition process during which aluminium ions were again released into solution. The response of rice growing in an aluminium-rich soil to phosphorus fertilization was considered (Hesse, 1963) to be due to fixation of aluminium during the period of early growth.

A great many plants are resistant to aluminium toxicity, some to a surprising degree. Chenery (1949) published a list of what he called 'aluminium accumulators' and his criterion for a plant to be included was that the oven-dry leaves should contain a minimum of 0.2% Al_2O_3 (or 100 µg g^{-1} Al). Hesse (1963) added the mangrove *Rhizophora harrisonii* to the list of aluminium accumulators and it is possible that the large proportion of organically complexed aluminium in the soil of the mangrove swamp previously referred to was due to the decay of mangrove roots containing 0.2% aluminium.

14:2:2 Determination: Background and theory

Total aluminium

The aluminium in soils can be brought into solution by almost any of the methods described in Chapter 16, although fusion with sodium carbonate with subsequent digestion with perchloric acid is most often used. As iron, manganese, titanium and silica are often measured with aluminium, a digestion or fusion procedure should be chosen convenient for all the required determinations.

Available aluminium

Available aluminium can be measured as water-soluble aluminium, although the usual procedure is to take exchangeable or extractable aluminium as being the plant-available form. As found by Plucknett and Sherman (1963), however, the availability of extractable aluminium is dependent upon the method of extraction. The most suitable procedure is extraction with neutral potassium chloride solution as described in the next section.

Exchangeable aluminium

So-called exchangeable aluminium has been extracted from soils with neutral normal ammonium acetate solution in the same manner as for other exchangeable ions. McLean and co-workers (1959) used ammonium acetate solution buffered to pH 4·8 and other workers have attempted to extract the aluminium at the prevailing soil pH by buffering the ammonium acetate solution accordingly. Coleman *et al.* (1959) used normal, unbuffered potassium chloride solution and Yuan (1958) used normal, unbuffered barium chloride solution. Pratt and Bair (1961) investigated these various extractants on acid soils and found that, except for soils containing very little aluminium, barium and potassium chlorides extracted about the same amount of aluminium from any one soil. If the pH of the soil exceeded 4·8, then ammonium acetate buffered to that pH extracted more aluminium than unbuffered solutions, whereas if the pH was less than 4·8, the unbuffered solutions extracted most aluminium. Indeed, the extraction is a continuous function of pH and with buffered solutions the pH of extraction is standardized, whereas with unbuffered, neutral salt solutions the pH of the soil controls the extraction. It is considered that the use of an unbuffered solution is preferable as being less arbitrary, particularly when the extracted aluminium is to be referred to as exchangeable, although Kruptskii *et al.* (1961) insist that standardization by using buffered ammonium acetate should be achieved.

As long as a soil is about neutral in reaction the aluminium extracted by neutral salt solutions may be regarded as exchangeable, but if a soil or the extractant is acidic then other forms of aluminium may be dissolved. For this reason it is preferable to refer to the aluminium extracted from soils as 'extractable' rather than exchangeable; it may still be regarded as available. This can be particularly important when a soil changes radically in reaction during its preparation for analysis. For example, the mangrove swamp soil previously referred to became excessively acid on being air-dried, its pH changing from about 7 to below 2, and thus, although the 0·02% aluminium removed from freshly collected, wet soil might be considered as exchangeable, the 0·2% aluminium found in the dried soil (both quantities were calculated on an oven-dry basis) was largely brought into solution by acid. In both cases, however, the extracted aluminium could be considered as mostly available.

Pratt and Bair also investigated the effect of time of extraction with neutral potassium chloride solution upon extractable aluminium. They found that as time of extraction increased the amount of aluminium extracted increased, and the titrateable acidity first decreased and then increased again (Table 14:1). The difference between the titrateable acidity

Table 14:1

Extractable aluminium, titrateable acidity and displacement of hydrogen ions as affected by time of extraction of a soil with neutral, 1 M potassium chloride solution (Pratt and Bair)

Time of extraction Hours	Al extracted m.e./100 g soil	Titrateable acidity m.e./100 g soil	H⁺ displaced m.e./100 g soil
0·00	7·75	9·80	2·05
0·25	8·10	9·65	1·55
0·50	8·30	9·73	1·43
1·00	8·55	10·09	1·54
2·00	8·80	10·42	1·62
4·00	8·90	10·50	1·60

and the aluminium was due to hydrogen ions and hence some aluminium hydroxide, which is a non-exchangeable form of aluminium, must have dissolved. Thus, to obtain the best measure of exchangeable aluminium, it is best to extract for the minimum time. Pratt and Bair recommended extracting for zero time, this being defined as the time necessary to mix 50 cm³ of solution with 10 g of soil and to filter the suspension.

Aluminium in solution

The classic and most commonly used procedure for measuring aluminium in solution is the colorimetric method employing 'Aluminon'.

Aluminon, aurin tricarboxylic acid, gives a red lake with aluminium in the presence of an acetic acid–acetate buffer (Hammett and Sottery, 1925). Chromium forms a similar lake but this is destroyed if the solution is made alkaline with ammonium hydroxide whereas the aluminium lake is not. Ammonium hydroxide, however, does reduce the intensity of the colour of the aluminium lake and hence it is preferable to remove the interfering chromium, although for most soils such interference is negligible. Iron(III) also gives a similar lake and in this case the interference is more important and must be counteracted. This can be done by extraction into isopropyl ether (Craft and Makepeace, 1945), as the sulphide, as the cupferron complex (Strafford and Wyatt, 1947) and by solvent extraction as thiocyanate. The simplest procedure, however, is to suppress the iron with thioglycollic acid (Mayr and Gebauer, 1938), a procedure adapted for the use of aluminon by Chenery (1948). Chenery by this method measured up to 10 µg aluminium per 5 cm³ in the presence of 200 µg iron, 200 µg phosphate and more than 2 000 µg of magnesium. Thioglycollic acid reduces iron(III) ions to iron(II) which are then strongly complexed; copper is complexed also by the reagent.

Page and Bingham (1962) removed interfering metals by passing the solution through an exchange resin. They used this method for measuring the aluminium extracted by ammonium acetate solution and emphasized that all acetate must be destroyed (by evaporation and digestion with peroxide) before the resin is used, as otherwise, iron will pass through as an acetate complex.

Phosphate and silicate interfere with the method leading to low results and calcium, magnesium, nickel, manganese and cobalt cause high results. Jackson (1958) separates calcium and magnesium by use of ammonium hydroxide and measures the calcium and magnesium separated. A subsequent sodium carbonate treatment removes practically all other interfering ions. Frink and Peech (1962) examined the Aluminon method and found it satisfactory, provided that calcium was removed. One millimol of calcium per 50 cm³ depresses the colour of the lake at low aluminium levels and was found to intensify the colour at high aluminium levels. Frink and Peech also found that fluoride must be absent and recommend its destruction by repeated evaporation with nitric acid.

Apart from specifically interfering ions the Aluminon method is affected by pH during and after colour development, the volume of solution during colour development, the concentration of aluminium, temperature and the presence of colloids. Hsu (1963) made a study of the precision and accuracy of the method as affected by pH, phosphate and silicate. These effects are interrelated and are due to complexes formed with aluminium under certain pH conditions. Hsu found that if the test solution has a high initial pH, results will be poor even though the final pH is adjusted, and this is particularly so in the presence of phosphate or silicate. The colour intensity of solutions as affected by pH of development was plotted (Figure 14:1) and it is apparent that the optimum pH is from 3·7 to 4·0.

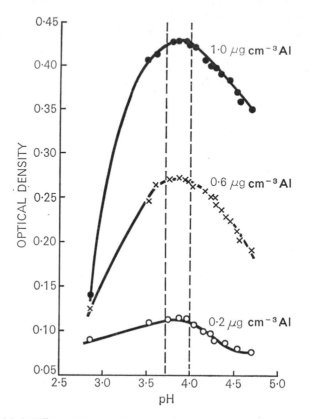

Figure 14:1 Effect of final pH upon colour intensity at 530 nm of aluminium–Aluminon lake suspensions (from Hsu, *Soil Science* © 1963)

The variability of colour appeared to be due to changes in effective concentration of aluminium in solution and in the optimum pH range the colour formation is essentially an ionic reaction, rendering addition of protective colloids unnecessary.

Hsu emphasized the need for controlling the pH of the test solution before adding the buffer solution. This is because hydroxy-aluminium polymers are formed when sodium hydroxide is added to aluminium solutions, and as they are but slowly decomposed at room temperature with dilute acid, they are not destroyed by merely adjusting the final pH.

As a result of his experiments, Hsu developed a modification of the Aluminon method in which the test solution is given a preliminary acid and heat treatment which converts all the aluminium to the ionic state. The colour development with Aluminon is then carried out at optimum pH and at room temperature; subsequent heating of the complex is unnecessary. Heating the test solution after addition of the reagent admittedly hastens colour development but complicates the reaction.

As a result of their investigation Frink and Peech (1962) concluded that a better method of determining aluminium is by solvent extraction of aluminium hydroxyquinolate in chloroform. Iron interference is eliminated by complexing with orthophenanthroline, excess calcium and magnesium are removed and fluoride is eliminated by evaporation with nitric acid. The method can determine from 2 μg to 20 μg of aluminium. A modification of this method in which iron is not complexed with *o*-phenanthroline permits the simultaneous determination of iron and aluminium (Goto *et al.*, 1958). In this method the extinction of hydroxyquinolate extract is read at two wavelengths, 395 nm and 580 nm; iron oxinate absorbs almost to the same extent at both wavelengths whereas aluminium oxinate absorbs negligibly at 580 nm. Thus the iron content can be found from the extinction at 580 nm and a corresponding correction applied to the extinction at 395 nm to obtain the aluminium concentration.

Other colorimetric methods of determining aluminium include the reaction with Eriochrome cyanin-R in a solution buffered at pH 6 to give a violet coloured complex and with Ferron, 8-hydroxy-7-iodo-5-quinoline sulphonic acid, in an acetate buffer at pH 5. The Ferron reaction also permits the simultaneous determination of iron which gives a complex with two absorbancies at 370 nm and 600 nm. Aluminium gives a complex with maximum absorbance at 370 nm.

Jones and Thurman (1957) adapted a method described by Hill (1956) for measuring aluminium with Eriochrome cyanin-R.A. The method was reported as being twice as sensitive as the Aluminon method and is unaffected by the presence of titanium; the accuracy was of the order ± 0.5 μg Al. Iron(III) interferes up to a maximum of 100 μg, after which no increased effect is obtained, and thus a standard addition of more than 100 μg Fe^{3+} is made to test solutions and standards to compensate for interference. No heat is required but the pH of the solution is critical and must be between 5·9 and 6·1. This is because the reagent itself has a high absorbance if the pH is less than 6, but at higher pH values the colour is formed very slowly. The colour is stable for 30 minutes and it is essential to use very pure Eriochrome cyanin.

Adsorption of coloured dyes on colloidal aluminium hydroxide, that is, formation of 'lakes', is an unsatisfactory method of determining aluminium. Far superior results are obtained by using xylenol orange which forms a soluble coloured complex with aluminium (Korbl and Přibil, 1956). Xylenol orange is not specific for aluminium but the only serious interference normally encountered in soil extracts is from iron. Pritchard (1967) developed the xylenol orange method for determining aluminium in soil extracts and used EDTA to eliminate iron interference. In the absence of EDTA the colour produced by the aluminium and iron complexes is stable for several days. The presence of EDTA causes the colour to fade and the method depends on the difference in rates of fading of the two complexes. Pritchard found that the colour produced by 1000 μg of iron fades completely within 1 hour, whereas that produced by aluminium fades slowly enough to allow reproducible readings to be made. The solution is buffered to pH 3·8 and a rectilinear calibration is obtained over the range 0 to 60 μg of aluminium.

14:3 MANGANESE

14:3:1 Introduction

Occurrence of manganese

Manganese occurs in primary, and particularly in ferromagnesian, rocks. By solution from the rocks and subsequent redeposition manganese appears in several minerals, the most abundant of which is pyrolusite, MnO_2. Another common mineral, manganite, $MnO(OH)$, is found frequently in association with granitic, igneous rocks and can change with time into pyrolusite.

In soils we can distinguish water-soluble, exchangeable, reducible and active manganese. Active manganese consists mainly of easily-reducible manganese, unless the pH is less than 6 when exchangeable manganese can be considered as active (Johansson, 1962).

Bacterial action upon hydrated manganese dioxide can form manganese(II) ions (Sherman and Harmer, 1943) which are considered as being plant available. Thus, as shown by Leeper (1947), alternate waterlogging and draining of a soil will increase the divalent available manganese supply. The reverse action of manganese(II) ions converting to insoluble oxides can also be brought about bacterially and this seems to occur in the immediate neighbourhood of plant roots, thus rendering the manganese unavailable, although, as will be discussed later, some doubt exists as to the tenability of this theory.

As oxides manganese occurs commonly together with iron in nodules, iron-pans and laterite, and Drosdoff and Nikiforoff (1940) showed that whereas in small concretions iron predominates, as the nodules become larger the proportion of manganese increases. In poorly drained soils, however, it has been noticed that the smaller concretions contain the most manganese. Concretions form under conditions of alternate wetting and drying of soil as described for iron and the darker their colour the higher their manganese content. The manganese oxide sometimes occurs as a black coating around an iron-rich nodule.

The nature and distribution of manganese in soils is largely dependent upon pH and redox potential. An increasing pH increases oxidation of manganese, whether the process is chemical or biological, and thus at high pH, manganese is relatively unavailable to plants. The deficiency of manganese often encountered in acid organic soils is due to leaching of the mobile manganese(II) ion.

Manganese and plant growth

It is most probable that plants can utilize only divalent manganese although the non-use of the trivalent ion has not yet been conclusively proven. In plants, manganese is a constituent of enzymes, is involved in protein synthesis and, although the evidence is not conclusive, it may also be involved in nitrogen metabolism. Manganese has been shown to increase the translocation of sugars through the stems and leaves of the cotton plant (Abutalybov and Aliev, 1961), to improve the fibre strength of cotton and to favour the accumulation of ascorbic acid in grape leaves and of sugar in grapes.

Excess manganese can cause chlorosis and necrosis of leaves and the element can accumulate in plant tissues. Thus Chamberlain (1956) found approximately 0·9% manganese in sweet potato plants that were severely mottled by chlorosis. Manganese toxicity can usually be remedied by liming the soil, although Johnson (1916) cured chlorosis of pineapple plants by adding iron(II) sulphate and explained this as having the effect of raising the ratio of iron to available manganese. Rubin and co-workers (1962) found that excess manganese in nutrient solutions depressed the carotenoids and chlorophyll in plants and increased the uptake of iron. The harmful effects were removed by treatment with EDTA.

Page *et al.* (1962) obtained a highly significant correlation between water-soluble manganese (x) in soils and manganese (y) in plants grown in those soils. The correlation followed the equation $y = a + bx$ where the constant a was proportional to the total manganese in the soil and where b was a coefficient varying with the place of experiment. Page found that an increase in pH decreased water-soluble manganese within 2 hours and he plotted graphs of pH against pMn (negative logarithm of water-soluble manganese). From the rapidity of decrease in water-soluble manganese and from the curves obtained, he concluded that biological oxidation of manganese is not the cause of reduced manganese availability, as formation of oxides implies a linear relationship between pH and pMn. Page suggested that manganese becomes unavailable with increasing pH because of the pH-controlled formation of organic complexes.

Passioura and Leeper (1963a, 1963b) investigated the manganese deficiencies characteristically occurring in soils of pH 6·5 to 8·0 and sometimes occurring in peaty and sandy soils. Two conflicting explanations previously suggested for these deficiencies were discounted by these workers. One explanation was that the manganese, whilst remaining divalent, becomes covalently bound to organic colloids (e.g. Bremner *et al.*, 1946). Passioura and Leeper disproved this theory by preparing synthetic systems in which manganese and calcium were competitively adsorbed on organic colloids and found little difference between the two metals in this respect. It was concluded that organically bound manganese is no more likely to be found in a non-exchangeable form than is calcium. The second explanation, that the manganese becomes biologically oxidized to higher insoluble oxides, was claimed to be untenable as manganese deficiencies have been remedied by applying manganese oxides, including those oxides formed by biological oxidation (Jones and Leeper, 1951; Bromfield, 1958). Passioura and Leeper suggested that the availability of manganese may be related to the degree of compaction of a soil and, using two different soils, both of which were showing severe manganese deficiency to oats, they obtained up to fifty-fold increases in yield merely by compacting the soils; this was done with laboratory experiments and in the field. Furthermore, a soil which had been treated with manganese dioxide 27 years previously, and in which the beneficial effect of the manganese had disappeared 10 years previously, showed a residual manganese effect when compacted. Passioura and Leeper also discounted the conventional explanation for this compaction effect, that is, carbon dioxide accumulation and reduced aeration causing a drop in pH and a reduced microbial activity. Measurements made on compacted and loose soils indicated equal quantities of divalent manganese, and the drop in pH found was insufficient to explain the increased manganese availability. Moreover, if carbon dioxide was circulated through the loose soil no effect was obtained and oxygen diffusion measurements in the soils showed no differences. Passioura and Leeper's explanation of the compaction effect is that the reserve of available manganese in neutral soils is reactive manganese dioxide, insoluble in water but soluble in the mild reducing agents of plant roots, and that compressing the soil increases the contact area with the plant roots.

Lamm and Maury (1962), using isotopically labelled manganese sulphate, found that manganese available to rye plants was increased by nitrogen application, and this effect was also found by Parker (1962), who further noted that mulching seedlings depressed manganese uptake.

In general, the subject of manganese availability is somewhat neglected; the information published is scrappy and often contradictory.

14:3:2 Determination: Background and theory

Total manganese

The total manganese in a soil is best brought into solution by acid digestion using a mixture of nitric, sulphuric and perchloric acids. This is discussed in Chapter 16, section 16:2.

Available and exchangeable manganese

Available manganese can be estimated biologically by measuring the increase in mycelium growth of *Aspergillus niger* and as the growth of spore-forming bacteria in soil is stimulated by manganese, Casida (1961) proposed that microbiological assay of available manganese be made by agar plug procedures.

Chemically, available manganese has been determined as water-soluble manganese (Page *et al.*, 1962; Cornfield and Pollard, 1950), as that extracted by pyrophosphate (Weir and Miller, 1962) and as that extracted by acetic acid (McCool, 1934). Hoff and Mederski (1958) found that out of eight solutions investigated, ammonium hydrogen phosphate was the most satisfactory extractant for available manganese. Exchangeable manganese is also considered as available and is normally extracted with neutral, 1 M ammonium acetate solution. Johansson (1962) found that exchangeable manganese extracted with 0·5 M magnesium sulphate solution and, more especially, that extracted with 0·05 M cobalt nitrate solution, constituted a high percentage of the total acid-soluble manganese in soils, and that the yield increase obtained from foliar application of manganese sulphate was highly correlated with active manganese and manganese extracted with cobalt nitrate. Weir and Miller (1962) using ^{54}Mn found that the manganese extracted by zinc sulphate solution was in equilibrium with the soluble manganese added to the soil.

Exchangeable manganese increases in amount when a soil is air-dried and thus, available manganese should really be determined in freshly collected samples (Sherman and Harmer, 1943). The increase in exchangeable manganese on drying a soil has been explained as being due to oxidation of iron(II) with simultaneous reduction of higher oxides of manganese and Fujimoto and Sherman (1946) found that drying releases manganese(II) ions from MnO_{1-2}. Timonin (1947) showed that the bacterial oxidation of manganese to the hydrated oxide could be decreased by partially sterilizing a soil, and as Fujimoto and Sherman (1948) found that steam sterilization of soils increases the exchangeable manganese content, it would appear that the drying effect is predominantly biological, as in the case of nitrogen mineralization (Chapter 10, section 10:2:5).

Active or free manganese

Free manganese can be regarded as water-soluble plus exchangeable plus reducible manganese. It can be extracted from soils with sodium dithionite as is free iron (Daniels *et al.*, 1962) and easily reducible manganese has been extracted with a solution of ammonium acetate and hydroquinone, although Jones and Leeper (1950) found that water and hydroquinone was a more satisfactory extractant.

Manganese in solution

The usual and most satisfactory method of determining manganese in solution is to oxidize it with periodate and to measure colorimetrically the permanganate formed,

$$2Mn^{2+} + 5IO_4^- + 3H_2O = 2MnO_4^- + 5IO_3^- + 6H^+ \quad [14:3$$

Oxidizing agents other than periodate have been used; for example, Alten and Weiland (1933) used persulphate in the presence of silver chloride, the silver being necessary to prevent precipitation of manganese dioxide, but the procedure is tedious. Sodium bismuthate is often recommended and gives accurate results for small quantities of manganese, provided that excess reagent is removed before measuring the optical density. Iyer and

Rajagopalan (1936) used sodium bismuthate as an oxidizing agent after reducing the manganese with sodium sulphite and eliminating the effect of chlorine with mercury(II) oxide.

The periodate oxidation of manganese has been investigated a great deal in attempts to improve its accuracy and convenience. In particular the acid medium has caused some controversy. As first introduced by Willard and Greathouse (1917), oxidation was made in the presence of phosphoric acid. With large quantities of phosphoric acid, manganese iodate and periodate are not precipitated and the effect of iron is negligible; also, calcium is not precipitated as when sulphuric acid is used. On the other hand, titanium phosphate may be precipitated and, for this reason, Groves (1951) recommended the use of sulphuric acid after standardizing the iron effect by adding a known excess of iron. Nitric acid may be present in small amount, but with too much nitric acid, formation of colour is retarded and iron(III) periodate may precipitate.

Reducing substances must be absent during the periodate oxidation of manganese and these are destroyed with nitric acid, or nitric acid and hydrogen peroxide, before development of permanganate. Chloride can be tolerated to some extent (Peech *et al.*, 1947), particularly if phosphoric acid is used; if perchloric acid has been used to evaporate the solution there will be little interference from chloride. If sulphuric acid is used, excess calcium must first be removed (Richards, 1930). The degree of acidity is not too critical provided that it lies within certain broad limits (Mehlig, 1939).

Jackson (1958) considers that sodium paraperiodate, $Na_3H_2IO_6$, is a superior reagent to potassium metaperiodate, KIO_4, due to its requirement of acid in an acid-rich system and its non-requirement of water in a water-poor system.

Standard manganese solutions can be prepared by dilution of standardized potassium permanganate solution. The permanganate colour can then be read directly or, as is more usual, the solutions can be reduced with sulphite and treated with periodate in the same way as the test solutions. Alternatively, a standard solution of manganese(II) sulphate can be prepared from a known weight of the salt. In cases where great accuracy is demanded, the blank solution is prepared by destroying the permanganate colour of each test solution with dilute hydrogen peroxide and measuring the background colour.

Other colorimetric methods of determining manganese include the reaction with formaldoxime in alkaline solution, giving a brown-coloured complex (Sideris, 1937). Iron, which gives a similar reaction, is first removed by precipitation as the hydroxide and, if necessary, interfering heavy metals can be removed by solvent extraction. Sideris (1940) modified the method for solutions rich in phosphate by precipitating the phosphate with lead in acetate solution.

Purdy and Hume (1955) oxidized manganese with bromate to a trivalent sulphate complex; the complex is stable for several days and its formation is relatively free from interference by other ions except those forming insoluble sulphates.

Cornfield and Pollard (1950) used the methane base method for determining manganese in solution in the range 0·1 μg to 1·0 μg Mn. The method depends upon the reaction· of manganese with tetramethyl-diamino-diphenylmethane in the presence of acetic acid and periodate to give a deep blue complex (Harry, 1931). Cornfield and Pollard found that the initial blue colour was too fugitive for accuracy, but noted that a stable green colour formed after about 5 minutes and that its intensity was proportional to the amount of manganese present. The highest sensitivity of measurement was at 440 nm but Beer's Law was best followed at 640 nm. The method is of particular use for water-soluble manganese, but ammonium acetate extracts of soil can be evaporated with aqua regia to destroy organic matter, and a water extract made of the residue. Acetate is necessary to prevent precipitation of the methane base but as the extinction value decreases with increasing acidity, the test solution should be kept neutral. Phosphate is added to remove interference from iron; in the absence of phosphate as little as 2 μg iron(III) interferes with the reaction. Interference from other common ions is dealt with by the procedure and duplicate analyses were found reproducible to within 10%.

Manganese can be determined polarographically, spectrographically and by flame photometry. The potentiometric titration of manganese as the manganese(II) ion with potassium permanganate is remarkably free from interference (Vogel, 1962), although volumetric and gravimetric procedures are not commonly used for soil manganese determination. Recently, however, Totev and Kulev (1965) determined exchangeable manganese by extracting soil with potassium chloride solution, removing organic matter with activated charcoal or nitric acid and titrating the solution potentiometrically with permanganate in saturated pyrophosphate solution.

14:4 RECOMMENDED METHODS

14:4:1 Total iron, aluminium and manganese in the same sample

Either of two methods may be used to bring the elements into solution; fusion with sodium carbonate as described in section 16:2:4I, or digestion with acids as described in sections 16:2:4II and 16:2:4III. If acid digestion is used the excess of acid must be driven off before measuring iron, and for manganese determination chloride must be driven off by evaporation with perchloric acid.

14:4:2 Total iron

1 g of 0·15-mm soil is fused with anhydrous sodium carbonate and brought into hydrochloric acid solution after separation of silica as detailed in section 16:2:4I. The iron in solution is determined as in section 14:4:7.

14:4:3 Total iron(II) (Kolthoff and Sandell; Mitsuchi and Oyama)

Reagents

Sulphuric acid, 9 M: 500 cm³ dm⁻³ conc. H_2SO_4.
Hydrofluoric acid, 48%.
Boric acid solution, saturated.
o-Phosphoric acid, concentrated.
Potassium dichromate solution, M/60: As in section 14:4:7I.
Barium diphenylamine sulphonate solution: As in section 14:4:7I.

Procedure

Accurately weigh about 0·5 g of 0·15-mm soil into a 50-cm³ polythene bottle and moisten it with a few drops of water. Add 10 cm³ of 9 M sulphuric acid and displace the air from the bottle with a stream of carbon dioxide. Close the bottle and heat on a boiling water bath for 5 minutes. Remove the lid and add 5 cm³ of hydrofluoric acid; replace the lid and heat in the water-bath at 80°C for 10 minutes. Transfer the contents of the bottle to a 400-cm³ beaker containing 10 cm³ of saturated boric acid solution and 10 cm³ of 9 M sulphuric acid in 200 cm³ of water. The boric acid is to complex the fluoride and prevent the formation of an iron(III) fluoride complex. Stir the mixture using a magnetic stirrer, add phosphoric acid and barium diphenylamine sulphonate and titrate the iron with standard dichromate as described in section 14:4:7I.

14:4:4 Available iron(II) (after Sherman, in Jackson)

Reagents

Ammonium acetate solution, 1·0 M: See section 7:3:1.
Aqua regia: $HCl:HNO_3$, 4:1 v/v.
Hydrochloric acid, 1·0 M: 89 cm³ dm⁻³ conc. HCl.
Standard iron solution: 0·7022 g dm⁻³ $Fe(NH_4)_2(SO_4)_2.6H_2O$ in water containing 5 cm³ conc. H_2SO_4. This solution has 100 µg cm⁻³ Fe.

Procedure

Freshly sample the soil from the field and immediately shake for 30 seconds with ammonium acetate solution using a ratio of ten parts solution to one part soil. Filter at once, using a portable suction filter as described in section 6:2:1 (Plate 4, Figure 6:2) and leach the soil in the funnel with three 50-cm³ portions of ammonium acetate solution. The whole procedure should be completed within a few minutes of sampling the soil to minimize oxidation of iron(II). A second sample of soil should be taken at the same time for moisture determination.

Evaporate the filtrate on a water-bath in a glass basin; treat the residue with 10 cm³ of aqua regia and again evaporate to remove traces of organic matter. Take up the residue in 1 cm³ of 1 M hydrochloric acid and dilute to 100 cm³ in a volumetric flask. Determine the iron content of the solution by removing aliquots for analysis by one of the methods given in section 14:4:7.

14:4:5 Free iron oxides (Coffin)

Reagents

Citrate buffer solution: 10·5 g dm⁻³ citric acid monohydrate and 147 g sodium citrate, tribasic.
Sodium dithionite, powder.
Perchloric acid.

Procedure

Weigh 0·5 g or less (depending upon iron content) of 0·15-mm soil into a 15-cm³ centrifuge tube and add 10 cm³ of citrate buffer solution. Add 0·5 g sodium dithionite, stopper the tube and shake in a water-bath at 50°C for 30 minutes. Centrifuge and digest an aliquot of supernatant liquid in a tall-form beaker with 2 cm³ of perchloric acid until nearly dry. Take up the residue in dilute hydrochloric acid, dilute to standard volume and analyze for iron as in section 14:4:7.

If it is desired to express the free iron in terms of oxides, the factor $Fe_2O_3 = Fe \times 1\cdot14$ is used.

14:4:6 Removal of free iron oxides for soil fractionation and elemental analysis (Le Riche and Weir after de Endredy)

Reagents

Ammonium oxalate solution, pH 3·3, 0·275 M in oxalate: This solution must bc free from trace elements. Distil the required amount of ammonia into a solution of recrystallized oxalic acid.
Potassium chloride solution, 0·1 M: 7·5 g dm⁻³ KCl.

Procedure

Mix 3 g of 2-mm soil with 150 cm³ of ammonium oxalate solution and allow the mixture to stand for 5 days. After standing, continuously stir the suspension for 90 minutes under a Hanovia u.v. lamp (Model 16). Centrifuge and wash the residue twice with 50 cm³ of ammonium oxalate solution and twice with 0·1 M potassium chloride solution. Dry the soil at 60°C.

14:4:7 Iron in solution

I / WITH POTASSIUM DICHROMATE

Reagents

Potassium dichromate solution, M/60: Dry pure potassium dichromate at 120°C and cool in a desiccator. Dissolve 4·904 g in water and dilute to 1 dm³.

Barium diphenylamine sulphonate solution: 0·16% w/v, aqueous.
o-Phosphoric acid, concentrated.
Tin(II) chloride solution: 12 g pure tin dissolved in 100 cm³ conc. HCl and
   diluted to 200 cm³.
Mercury(II) chloride solution, saturated: about 5%.

Procedure

The iron must be in the divalent state and if necessary must be reduced.
A Jones reductor can be used or tin(II) chloride solution as described
below.

Take a 25-cm³ aliquot of test solution containing about 0·1 g iron,
make approximately 5 M with respect to hydrochloric acid and heat to
70–90°C. Add tin(II) chloride solution from a burette with stirring until
the yellow, iron(III) colour has almost, but not quite, disappeared. Thereafter add a diluted tin(II) chloride solution (1 volume solution to 5 volumes
5 M HCl) dropwise with shaking until the solution becomes faintly green.
Protect from the atmosphere and cool rapidly to about 20°C; add 10 cm³
of mercury(II) chloride solution when a slight white precipitate should
form. If the precipitate is heavy or dark in colour, too much tin(II) chloride
has been used and the experiment must be repeated.

Allow the reduced solution to stand for 5 minutes, dilute with 200 cm³
of water containing 5 cm³ sulphuric acid, add 5 cm³ of phosphoric acid
and 0·5 cm³ of indicator solution. Titrate with standard dichromate solution to a permanent violet-blue colour.

$$1 \text{ cm}^3 \text{ M}/60 \text{ K}_2\text{Cr}_2\text{O}_7 \equiv 0\cdot005585 \text{ g Fe} \qquad\qquad [14:4$$

II / WITH TIRON (YOE AND ARMSTRONG)

Reagents

Hydrogen peroxide, 3%.
Buffer solution: Equal volumes of 1 M acetic acid (60 cm³ dm⁻³) and 1 M
   sodium acetate (82 g dm⁻³ anhyd. salt.) adjusted to pH 4·7 with acetic
   acid or sodium hydroxide.
Tiron reagent: 4% w/v, aqueous. Prepare freshly.
Standard iron solution: As in section 14:4:4.

Procedure

Treat the iron in very dilute hydrochloric acid solution with 1–2 drops of
hydrogen peroxide to oxidize to the iron(III) state. Titanium can be
measured in the same aliquot (section 15:5:3). Place 10 cm³ of buffer
solution in a 50-cm³ volumetric flask and dilute to about 30 cm³. Add
5 cm³ of Tiron reagent and an aliquot of test solution containing about
0·05 mg to 0·25 mg of iron. Dilute to 50 cm³ and read the optical density
at 565 nm after 5 minutes.

The standard curve should cover the range 0–10 µg cm⁻³ Fe.

III / WITH *o*-PHENANTHROLINE

Reagents

o-Phenanthroline solution: 1·5% w/v, in ethanol.
Hydroxylamine hydrochloride: 10% w/v, aqueous.
Standard iron solution: As in section 14:4:4.

Procedure

Take an aliquot of test solution containing up to 0·05 mg Fe (or up to
2 µg cm^{-3} when diluted to 25 cm^3) in very dilute hydrochloric acid, dilute
to about 5 cm^3 and adjust to pH 1·5–2·7 with dilute hydrochloric acid
or ammonium hydroxide. The pH can be monitored by using a pH meter
with a separate sample or with 2,4-dinitrophenol as an external indicator
on a spot-plate.

Add 2 cm^3 of hydroxylamine hydrochloride solution and 1 cm^3 of
o-phenanthroline solution. Dilute to 25 cm^3 and read the optical density
at 508 nm. The standard curve should cover the range 0–2·0 µg cm^{-3} Fe.

14:4:8 Qualitative field test for iron(III) and (II)

Reagents

Hydrochloric acid, 1·0 M: 98 cm^3 dm^{-3} conc. HCl.
Potassium thiocyanate solution: 10% w/v, aqueous.
Potassium hexacyanoferrate(III) solution: 0·5% w/v, aqueous.

Procedure

Place two small portions of freshly sampled soil in cavities of a white tile
and moisten with dilute hydrochloric acid. To one sample add a drop of
thiocyanate and to the other a drop of hexacyanoferrate(III). Iron(III) is
indicated by a red coloration with thiocyanate and iron(II) by a blue
coloration with hexacyanoferrate(III).

14:4:9 Total aluminium

Fuse the soil sample with sodium hydroxide and digest the melt with
perchloric acid as described in section 16:2 (16:2:1 and 16:2:4I). Analyze
aliquots of the digest for aluminium as in section 14:4:11.

14:4:10 Available, extractable and exchangeable aluminium (Pratt and Bair)

Reagent

Potassium chloride solution, 1·0 M: 74·6 g dm^{-3} KCl.

Procedure

Shake 10 g of 2-mm soil for a few seconds with 50 cm^3 of neutral, 1 M
potassium chloride solution and filter. Leach the soil on the filter paper

with five 10-cm³ portions of potassium chloride solution and dilute the extract to 100 cm³ in a volumetric flask. Analyze aliquots of extract for aluminium as in section 14:4:11.

14:4:11 Aluminium in solution (Pritchard)

Reagents

Standard aluminium solution: 1·757 g dm⁻³ K₂SO₄.Al₂(SO₄)₃.24H₂O. Dilute 10 cm³ to 1 dm³ giving a solution containing 1 μg cm⁻³ aluminium.
Buffer solution, pH 3·8: Dissolve 136 g of hydrated sodium acetate in water, adjust to pH 3·8 with HCl and dilute to 1 dm³.
Xylenol orange solution: 0·15% w/v, aqueous.
EDTA solution, 0·05 M: 18·6 g dm⁻³ EDTA disodium salt.

Procedure

Transfer an aliquot of test solution containing up to 30 μg of aluminium to a 50-cm³ graduated flask. Add 10 cm³ of buffer solution and 5 cm³ of xylenol orange solution. Place the flask in a water-bath at 40°C for 1½ hours. Cool the solution, add 2 cm³ of EDTA solution, dilute to 50 cm³ and allow it to stand at room temperature for 1 hour. Measure the optical density at 550 nm. The calibration graph should cover the range 0 to 1 μg cm⁻³.

Organic ions such as citrate, oxalate and acetate often present in soil extractants should be destroyed by heating with 40% hydrogen peroxide and evaporating to fuming with 4·5 M sulphuric acid (10 cm³ per 25 cm³ of soil extract). Hydrofluoric acid should be removed before developing the colour and pyrophosphate extracts should be hydrolyzed to orthophosphate.

14:4:12 Total manganese

Reagents

Nitric acid, concentrated.
Sulphuric acid, concentrated.
Perchloric acid, 60%.
o-Phosphoric acid.
Sodium *p*- (or *m*-) periodate.
Standard manganese solution: Standard solutions can be made either from potassium permanganate or from manganese(II) sulphate.

Preparation of standard manganese solution from potassium permanganate
Add approximately 3 g of potassium permanganate to 1 dm³ of distilled water and gently boil for 20 minutes in a covered beaker. Cool and filter through a sintered glass funnel into a brown glass bottle.

Standardize the solution with arsenic(III) oxide (Vogel, 1962) and dilute a suitable aliquot to give a solution containing 100 µg cm^{-3} Mn. Dilute portions of this solution to cover the range 0·5 µg cm^{-3} to 10·0 µg cm^{-3} Mn.

If it is desired to use the standard in the reduced form and submit it to the same treatment as the test solution, acidify with 1:1 sulphuric acid and add sufficient dilute oxalic acid solution to discharge the colour before making to volume.

Preparation of standard manganese solution from manganese(II) sulphate

Dry manganese(II) sulphate at 110°C in an electric oven and cool in a desiccator. Dissolve 0·3076 g of the dry salt in water and dilute to 1 dm³, giving a solution containing 100 µg cm^{-3} Mn.

Procedure

Digest 1 g of 0·15-mm soil in a 250-cm³ beaker or 100-cm³ Kjeldahl flask with 20 cm³ of nitric acid, 10 cm³ sulphuric acid and 2 cm³ perchloric acid. When dense white fumes of acid occur, avoid further loss by covering the beaker and digest for another 30 minutes. Cool the digest, dilute with water and filter into a volumetric flask. Wash the residue with water and dilute the filtrate to volume. Transfer suitable aliquots of solution to small beakers and add 2 cm³ of *o*-phosphoric acid and 0·5 g of sodium periodate; heat the mixture on a hotplate until just below boiling point. After the appearance of the permanganate colour, continue heating for a further 30 minutes adding more periodate if the colour fades. Cool the solution and transfer to a 100-cm³ graduated flask and make to volume. Read the transmittance at 545 nm. The calibration curve should cover the range 0–10 µg cm^{-3} Mn.

14:4:13 Exchangeable manganese

As a routine, exchangeable manganese is best determined in the ammonium acetate extract of soils as this will have been prepared for the determination of other exchangeable cations. For specific purposes other extractants may be used (section 14:3:2) but the subsequent treatment of the extract remains substantially the same. For precise measurement of exchangeable manganese, freshly sampled soil must be extracted (section 14:3:2) and in such cases interpretation of results should take into account other existing field conditions such as pH, redox potential and so on.

The soil leachate must be wet-ashed with nitric and hydrochloric acids as described in section 7:3:3. Take a suitable aliquot of the treated soil extract so as to give a solution containing up to 10 µg cm^{-3} Mn when diluted to 100 cm³. Evaporate to dryness in a 150-cm³ beaker on a water-bath. To the residue add 10 cm³ of concentrated nitric acid and 2 cm³ of perchloric acid. Cover the beaker with a watch glass of the Speedy-vap type and evaporate on a hotplate until dense white fumes appear. Cool and take up in water containing 2 cm³ of phosphoric acid. Add 0·5 g of

13+T.S.C.A.

sodium periodate and develop the permanganate colour as described in section 14:4:12. Dilute the cool solution to 100 cm³ and read the transmittance at 545 nm.

If water-soluble manganese is determined also, its amount should be deducted from the measured exchangeable manganese.

14:4:14 Water-soluble manganese (Cornfield and Pollard)

Reagents

Sodium acetate buffer solution: Dissolve 15·0 g of NaOH in 180 cm³ of water, add 58 cm³ glacial acetic acid and dilute the cooled solution to 250 cm³. This solution is 1·5 M with respect to sodium acetate and 2·7 M with respect to acetic acid.

Potassium phosphate solution: 0·48 g potassium dihydrogen phosphate in 250 cm³ water.

Potassium periodate solution: 0·2 g KIO_4/100 cm³ water. Prepare freshly every 48 hours.

Methane base reagent: 0·5 g 4,4'-tetramethyl-diaminodiphenylmethane dissolved in 6 cm³ of 2 M HCl and the solution diluted to 100 cm³. Prepare every 2 days.

Standard manganese solution: Dissolve 0·575 g $KMnO_4$ in 50 cm³ water. Add 2 cm³ of 1 M H_2SO_4 and bubble SO_2 through the solution to decolorize it. Boil out excess SO_2 and dilute to 1 dm³. The solution contains 20 µg cm⁻³ Mn and aliquots are diluted as necessary.

Procedure

Shake 25 g of 2-mm soil with 125 cm³ of water for 1 hour and filter through a Whatman No. 42 paper. The manganese in the clear filtrate can then be measured with periodate as described for exchangeable manganese (section 14:4:13), but normally it will be necessary to use larger aliquots of extract than for exchangeable manganese. To avoid evaporation of large volumes of liquid the more sensitive methane base method is recommended.

Pipette a suitable aliquot of test solution (containing up to 1 µg Mn) into a 15 cm × 2 cm test-tube marked at 21 cm³. Add 2 cm³ of acetate buffer, 1 cm³ of potassium phosphate solution and dilute to 21 cm³. Shake the tube and warm in a water-bath at 20°C ± 0·5°C for 15 minutes. Add 3 cm³ of periodate solution, shake well and add 1 cm³ of methane base reagent. After 10 minutes ±1 minute, read the transmission at 640 nm.

The calibration curve should cover the range 0–1·0 µg cm⁻³ Mn.

14:4:15 Easily-reducible manganese (Jones and Leeper)

Shake 25 g of soil for 1 hour with 250 cm³ of 0·05% w/v aqueous hydroquinone solution. Filter the suspension through a Buchner funnel and evaporate the filtrate to dryness with 10 cm³ concentrated nitric acid and

2 cm³ of perchloric acid. Treat the residue as described in section 14:4:13 for measurement of exchangeable manganese.

14:4:16 Qualitative field test for manganese

Manganese is detected in the field by the presence of small, hard, characteristically dark-coloured nodules. To test these nodules for manganese add a few drops of hydrogen peroxide solution when a vigorous effervescence indicates its presence.

15

Silicon, titanium and sesquioxides

15:1 INTRODUCTION

Silicon, titanium and sesquioxides have been included together as a logical sequel to the discussion of iron, aluminium and manganese as for many purposes these quantities are considered in relation to each other.

Silica–alumina and silica–sesquioxide ratios are used as a method of expressing the nature of clays; thus a high silica–sesquioxide ratio indicates a siliceous clay and a low ratio indicates a sesquioxidic clay, and many soil properties will depend upon such a distinction. For example, colloidal properties are more developed in siliceous clays, whereas in sesquioxidic clays anion exchange may be a more important factor. In soil survey, the changes down a profile in the silica–alumina and silica–sesquioxide ratios assist in explaining the development of that profile.

The chemical weathering of silicate minerals is predominantly by hydrolysis when silicic acid and its associated bases are liberated. The alkali and alkali-earth bases are leached away as salts, for example hydrogen carbonates, and the sesquioxides remain. In humid regions the silicic acid is also leached away and eventually may be re-precipitated in river sediments. Extreme cases of such a process result in the formation of laterite as described in section 14:1:1. In arid regions the silicic acid is not removed by drainage and can form secondary complexes with the sesquioxides.

Jones and Handreck (1963) found that the amount, kind and crystallinity of free sesquioxides in soils at the same pH, influence the levels of silica in solution. For example, in two sandy-loam soils with the same amount of clay and the same pH, one contained large quantities of sesquioxides and 6 μg g^{-1} silica in solution, whereas the other contained 67 μg g^{-1} silica in solution and hardly any free sesquioxides.

15:2 SILICON

15:2:1 Background and theory

Silicon is the second most abundant element (oxygen being the most abundant) and is present in nearly all minerals. Most of the minerals in igneous rocks are silicates and hence soil is largely composed of silicates. Different classes of silicates—such as Phyllosilicates (e.g. talc) and Tectosilicates (e.g. quartz)—six of which have been proposed by Struntz (1957), depend upon the different arrangements in space of SiO_4^{2-} tetrahedra. This aspect of soil chemistry is discussed in most mineralogy textbooks.

Weathering and vegetation decay can result in amorphous silica and this has been found as a cement binding together particles in some sandstones (Nikiforoff and Alexander, 1942). In Russia, powdered silica was found in the lower horizons of heavy, undrained soils and was probably precipitated during the freezing of silica-saturated soil solutions. A pinkywhite deposit of fine silica is a characteristic of solodized-solonetz profiles and was generally attributed to hydrolysis of sodium-saturated clay with removal of iron and aluminium. Hallsworth and Waring (1964), however, who investigated a solodized-solonetz in New South Wales in which the clay was predominantly kaolinitic, found no accumulation of aluminium oxide as would be expected. As the pH was such as to render aluminium almost insoluble, it appeared that either strong chelating activity was responsible for removal of the liberated aluminium, or else the deposit of silica came from a source other than the clay. Hallsworth and Waring suggested that at periods when the concentration of calcium and sodium ions was high in the moisture film in the clay at the top of the B-horizon, soluble silica is at its maximum concentration and is absorbed by plants and subsequently deposited on the soil.

A knowledge of the movement of silica in soils assists in the interpretation of chemical reactions relating to the genesis of soils. Miller (1963) found that the solubility of silica in soils is rapid and proportional to the water content. Most common soil salts had no effect upon the solubility of silica but calcium carbonate decreased the solubility of simple silica forms and acidification of calcareous soils increased the proportion of soluble silica.

In East Africa, soil fertility as shown by yield, was found to be significantly related *inter alia* to the water-soluble silica content of the soil (Birch, 1955). Water-soluble silica was found to be also related to available phosphate, one of the main factors governing fertility in East Africa. Birch further found that a function of the ratio of water-soluble silica to clay was very significantly and directly related to the silica–sesquioxide ratio, and very significantly and inversely related to the kaolin content of the clay. Significant relationships were also found with the total exchangeable bases and the cation exchange capacity of the soils. It appeared that the amount of water-soluble silica depended mainly upon the kind and amount of clay minerals in a soil, the highest quantities of silica being associated with montmorillonitic soils and the lowest with kaolinitic. Birch thus suggested that determination of water-soluble silica would be very valuable in soil fertility investigations and for soil classification.

All plants contain silicon to some extent, but it is found in greatest quantities in grasses. At one time it was supposed that the silicon in grass improved the strength of the straw but this was disproved at Rothamsted Experimental Station (Russell, 1961). Silicon-deficient plants become very susceptible to disease and take up excessive amounts of iron and manganese. The main effect of silicon in plants appears to be to increase the availability of phosphorus (Fisher, 1929). Soluble silicates added to soils increase plant uptake of phosphorus and this is particularly noticeable with soils having high phosphorus fixation power.

The rice plant is sensitive to a deficiency of silicon and Uchiyama and Onikura (1955) found that when a paddy soil is allowed to dry the soluble silica gels, and is rapidly lost on subsequent leaching. Furthermore, the excess of aluminium left behind is detrimental to the next rice crop.

15:2:2 Determination of silicon

Total silicon

The total silicon content of a soil is almost invariably measured gravi-metrically as silica, after fusion or digestion of the soil; methods of fusion and digestion are discussed in Chapter 16. If an acid digestion is made, once the other elements have been brought into solution, the silica is dehydrated by evaporation with perchloric acid, filtered off and washed. Perchloric acid is a powerful dehydrating agent due to the formation of the azeo-tropic mixture at 203°C. Crude silica can then be weighed directly after ignition to constant weight. For more accurate results the residue is heated with a mixture of sulphuric and hydrofluoric acids to volatilize the silica and the ignited residue is again weighed: the loss in weight represents pure silica.

For the fusion procedure the soil is best fused with sodium hydroxide, the melt taken up in dilute acid and aliquots of solution analyzed for silicon.

Water-soluble silicon

The extraction of water-soluble silicon from soils was investigated by McKeague (1962) who found that the concentration of silica in the extracts of shaken samples was larger than if the soil and solution had attained equilibrium by prolonged standing. McKeague explained this as being due to abrasion caused by shaking the soils. McKeague found also that the amount of silica extracted from soils decreased with increasing pH and de-creasing temperature. Successive extractions showed evidence of a de-sorption process.

Further investigation by McKeague showed that monosilicic acid is effectively removed from solution by precipitated hydroxides of poly-valent metals and fairly effectively by bauxite and ferruginous soils. Ab-sorption is increased as the pH rises from 4 to 9 but decreases again at higher pH levels and depends on the surface area of the added material.

Silicon in solution

Silicon reacts with molybdate in acid solution in a manner similar to phosphorus, giving a yellow coloured complex (Dienert and Wandenbulke, 1923). Provided that the acid concentration is sufficiently low the reaction is more sensitive with silicon than with phosphorus. Iron interferes but this can be remedied by adding an excess of phosphoric acid after reaction with molybdate, when the colour due to iron is destroyed.

Interference from phosphate can be eliminated by correct adjustment of pH (Kahler, 1941), or the phosphomolybdate can be destroyed with citric acid (Schwartz, 1934) or with certain other organic acids. If both phosphate and iron are interfering, sodium fluoride can be added to destroy the unwanted colours (Weihrich and Schwartz, 1941).

Silicomolybdate can be reduced in the same way as phosphomolybdate to give molybdenum blue and thus provides a more sensitive method of measurement. If the silicon in solution is less than $1–2$ μg cm^{-3}, the molybdenum blue method is preferable, but at higher concentrations it is better to measure the yellow colour and avoid the use of reducing agents.

Several reducing agents have been recommended for the preparation of molybdenum blue from silicomolybdate. 1,2,3,4-aminonaphtholsulphonic acid (Bunting, 1944; Straub and Grabowski, 1944) has perhaps received most attention and has been recommended for soil analysis by Metson (1961). Sodium sulphite is a more convenient reductant and permits the measurement of silicon in concentrations much higher than when the sulphonic acid is used. However, the blue colour produced with sodium sulphite conforms to Beer's Law only after several hours (Woods and Mellon, 1941) and for reproducible results, readings must be taken after a definite period of time from adding the reagent.

15:3 TITANIUM

15:3:1 Background and theory

Titanium occurs as ilmenite, $FeTiO_3$, in igneous rocks and as rutile, TiO_2, in granites, gneiss and metamorphic limestone. Rutile is an important, although minor, constituent of some shales in which it is probably formed by decomposition of biotite which sets free titanic acid. Titanium can replace aluminium in six co-ordination and hence sometimes it appears in biotite and also in hornblende. In soils, titanium is typically associated with ferruginous latosols and is thus most commonly found in quantity in the tropics.

15:3:2 Determination

Gravimetrically titanium can be determined as TiO_2 or as the 5,7-dibromoxinate, although the latter procedure is not easily applicable to soil analysis owing to interference from iron. The applicability of the titanium(IV) oxide method depends upon the reagent used. Precipitation with tannin and antipyrine permits the separation of titanium from iron, aluminium, manganese and certain other ions and can be made in the presence of phosphates and silicates. Thus the method is well suited for the determination of titanium in soil digests and quantities of titanium down to 2 mg can be so determined (Vogel, 1962).

Normally a colorimetric procedure is employed. The standard colorimetric method depends upon the formation of a yellow colour when titanium sulphate is oxidized with hydrogen peroxide (Weller, 1882). In soils the important interfering elements are iron and fluorine, iron because it gives a coloured solution and fluorine because even in small amount it bleaches the yellow pertitanic acid colour. Alkali sulphates and large amounts of phosphate also interfere and the colour intensity increases with temperature.

Iron interference can be overcome by adding excess phosphoric acid provided that a like amount is added to the standards; or it can be separated by a preliminary treatment with cupferron (Snell and Snell, 1949), or electrolytically using a mercury cathode (Silverman, 1948). Jackson (1958) reduced the iron in a Jones reductor before measuring titanium. Fluoride must be removed by repeated evaporation with sulphuric acid and the bleaching effect of alkali sulphates is overcome by developing the colour in 1 M sulphuric acid. The presence of perchloric acid does not seriously interfere with the method (Weissler, 1945).

Titanium gives a stable yellow-coloured complex with excess ascorbic acid (Ettori, 1936; Hines and Boltz, 1952) and this is free from the interference of most common ions including iron(III). Fluoride must be removed but phosphate and silicate can be tolerated in concentrations of up to $100 \ \mu g \ cm^{-3}$.

The use of Tiron, disodium 1,2-dihydroxybenzene-3,4-disulphonate, for the simultaneous determination of titanium and iron has been described in section 14:1:2. In soil analysis it is exceptional to determine titanium without also determining iron and so the Tiron method is particularly suitable. Perchloric acid and fluoride interfere and must be removed by evaporation.

15:4 SESQUIOXIDES

15:4:1 Background and theory

Sesquioxides are determined to assist in the classification of soils and clays and are invariably measured gravimetrically after fusion or digestion of the soil. After removal of silica the filtrate is treated with ammonium hydroxide solution in the presence of ammonium chloride until the pH is about 7·5 as indicated by methyl red. If an excess of ammonium ions is present whilst precipitating and during washing of the precipitate, adsorption of other ions is kept at a minimum; to purify the sesquioxide precipitate it is redissolved in dilute acid and then re-precipitated. Finally the precipitate is washed, by decantation as far as possible, with ammonium nitrate solution until all the chloride is removed, and is then ignited to constant weight in a muffle furnace. Chloride must be absent as iron(III) chloride may be volatilized during ignition.

Once the total sesquioxide content of the soil has been determined, the individual oxides of iron, aluminium and titanium can be determined if required. For this the residue is taken up in hydrochloric acid and the individual cations determined. It should be noted that the method includes all the iron, aluminium and titanium in a soil as sesquioxidic; for more accurate results other forms of these elements must be removed before digesting the soil.

Calculations

Silica–alumina and silica–sesquioxide ratios are most conveniently expressed on a molecular basis. The percentages of oxides must be divided by their molecular weights which are 60 for SiO_2, 160 for Fe_2O_3, 102 for Al_2O_3 and 80 for TiO_2, before calculating the ratio value. To obtain a molecular silica–sesquioxide ratio the individual percentages of each oxide must be known. For example the percentage of silica as mol in an ignited clay is given by,

$$\frac{\text{wt. } SiO_2 \times 100}{\text{wt. clay} \times 60}$$

and then SiO_2/Al_2O_3, Fe_2O_3/Al_2O_3 and $SiO_2/(Fe_2O_3 + Al_2O_3)$ ratios are calculated.

15:5 RECOMMENDED METHODS

15:5:1 Total silicon

The soil is digested as described in section 16:2:4II. Heating for 15 minutes with fuming perchloric acid is the best method of dehydrating the silica but the residue should not be allowed to become dry as this leads to incomplete separation.

Filter the silica from the diluted digest with an ashless filter paper of coarse porosity (e.g. Whatman No. 41) and wash with warm, dilute (1:20) hydrochloric acid; wash the residue free from chloride with water.

For more accurate results, evaporate the combined filtrate and washings to dryness and heat the residue at 110°C for 1 hour. This is to recover the small quantity of silicic acid still in solution; extract the residue with dilute hydrochloric acid, filter and wash as before.

Cautiously ignite the combined residues and filter papers to constant weight in a platinum crucible using a muffle furnace and record the result as crude silica.

For many purposes this result is sufficient, but if greater precision is required, the small quantities of co-precipitated iron, aluminium and titanium oxides must be separated. Moisten the residue in the platinum crucible with water, add 3 drops of concentrated sulphuric acid and 5 cm³ of pure hydrofluoric acid. Cautiously evaporate the mixture in a fume cupboard until all the hydrofluoric acid has been driven off. Continue heating to drive off all the sulphuric acid and ignite the residue to constant

13*

weight in a furnace. The loss in weight of the original crude silica is a measure of the pure silica.

SiO_2 figures can be converted to Si by the factor 0·4672.

15:5:2 Water-soluble silicon

I / SILICOMOLYBDATE METHOD (KNUDSON ET AL.)

Reagents

Standard silicon solution: Analyze a commercial solution of sodium silicate gravimetrically by evaporating with hydrochloric acid, filtering and weighing the ignited residue as described in section 15:5:1. Then dilute a suitable aliquot to give a solution containing 100 μg cm^{-3} Si.

Ammonium molybdate solution: 10% w/v, aqueous.

Sulphuric acid, 2 M: 110 cm^3 conc. H_2SO_4 added to about 800 cm^3 of water containing 8 g sodium tetraborate; dilute to 1 dm^3.

Citric acid: 10% w/v, aqueous.

Procedure

Weigh 1 g of 2-mm soil into a 150-cm^3 polythene bottle and add 50 cm^3 of distilled water. Shake for a few minutes to thoroughly mix the contents and leave the bottle standing overnight.

Filter the suspension through a Whatman No. 42 paper and pipette 25 cm^3 of clear filtrate into a 100-cm^3 volumetric flask. Add 1 cm^3 of ammonium molybdate solution and then immediately 1 cm^3 of the sulphuric acid solution, or sufficient to give a pH of 1·5 to 2·0. Allow to stand for 10 minutes and dilute to about 90 cm^3. Add 2 cm^3 of citric acid solution and dilute to 100 cm^3. Immediately after mixing, read the transmittance at 410 nm.

II / MOLYBDENUM BLUE METHOD

Reagents

Standard silicon solution: As in section 15:5:2I.

Ammonium molybdate solution: 10% w/v, aqueous.

Hydrochloric acid, 0·25 M: 22 cm^3 dm^{-3} conc. HCl.

Sodium sulphite solution: 17 g dm^{-3} anhyd. salt.

Procedure

Extract the soil with water as in section 15:5:2I and filter. Pipette 25 cm^3 of clear filtrate into a 150-cm^3 conical flask and add 15 cm^3 of dilute hydrochloric acid. Add 15 cm^3 of ammonium molybdate solution and, after 6 minutes, 2 cm^3 of citric acid and 25 cm^3 of sodium sulphite solutions. Mix and read the optical density at 650 nm exactly 1 minute after adding the reductant. The standard curve should cover the range 0–5 μg cm^{-3} Si.

15:5:3 Titanium and iron (Yoe and Armstrong)

Reagents

All reagents as for section 14:4:7ɪɪ.

Sodium dithionite, powder.

Standard titanium solution: 0·1668 g of ignited TiO_2 is fused with $K_2S_2O_7$ and dissolved in 10 cm³ of 6 M HCl. Add 50 cm³ of 6 M HCl and dilute to 1 dm³. Dilute 10 cm³ of this to 100 cm³ with 0·4 M HCl to give a solution containing 10 μg cm⁻³ Ti.

Procedure

Follow the procedure given in section 14:4:7ɪɪ up to the measurement of the iron in solution. To about 20 cm³ of the same solution add 5 mg sodium dithionite, stopper the tube and mix the contents. After 10 minutes read the optical density at 400 nm. The standard curve should cover the range 0–2·0 μg cm⁻³ Ti.

15:5:4 Silica–sesquioxide ratios

Digest 1 g of soil or clay in a tall beaker with 20 cm³ of concentrated nitric acid to destroy excess organic matter. Cool and add 30 cm³ of 60% perchloric acid; digest until dense white fumes of acid are formed. Cover the beaker and continue the digestion for 15 minutes to dehydrate the silica; do not allow the mixture to become dry. Cool, dilute with 25 cm³ of warm water, filter through an ashless filter paper and wash the residue well with 0·5 M hydrochloric acid. Determine the silica as described in section 15:5:1.

Dilute the filtrate in a 400-cm³ beaker to about 150 cm³, add 5 g of ammonium chloride and heat to near boiling. Add a few drops of methyl red indicator solution (0·2% in ethanol) and then 1:1 ammonium hydroxide solution dropwise until the colour remains yellow. Boil for 2 minutes and filter while still hot through a Whatman No. 41 paper; wash the residue three times with hot 2% ammonium nitrate solution.

Place the filter paper and precipitate in the original beaker and macerate with 10 cm³ of hot, 1:1 hydrochloric acid. Dilute to about 150 cm³ and re-precipitate with ammonium hydroxide as before; filter through a new Whatman No. 41 paper leaving the bulk of precipitate in the beaker. Thoroughly wash the precipitate by decantation with hot 2% ammonium nitrate solution until free from chloride. Finally transfer the whole of the precipitate to the paper and place in a Main-Smith silica crucible. After drying in an electric oven, cautiously ignite the paper and precipitate in a muffle furnace to constant weight. Transfer the residue quantitatively to a beaker and dissolve in 1:1 hydrochloric acid. Alternatively, a known weight of the weighed residue can be taken. Filter the solution into a volumetric flask and dilute to volume. Use aliquots of this solution to measure iron, aluminium and titanium by one of the recommended methods.

Convert the percentages of each element to the oxide and express on a molecular basis; the required ratios can then be calculated (see section 15:4:1). If the total sesquioxide content is to be recorded as R_2O_3, phosphate must be determined in the ignited residue and subtracted from the weight.

16

Total (elemental) analysis and some trace elements

16:1 INTRODUCTION

The relative proportions of the variety of elements composing a soil will differ according to the nature of the parent rock and the degree of weathering. Sesquioxides, for example, are more resistant to weathering than, say, calcium or magnesium which are more easily transformed. As the various mechanical fractions of a soil differ in mineralogical composition they differ also in chemical composition (Failyer *et al.*, 1908).

Complete, total analysis of all elements present was originally used in geological work on rocks and minerals. As now applied to soil investigation, the principal function of total analysis remains connected with mineralogy and assists in determining the soil genesis and classification. Wells and Taylor (1960) arranged soils from different parent rocks in sequences according to degree of weathering and found their related differences by elemental analysis. High contents of iron, magnesium, calcium, titanium, phosphorus, chromium, nickel and cobalt were associated with weakly weathered soils from basalt, whereas these elements were in small quantities in soils from rhylotite. Strongly weathered soils from acidic rocks contained residual silicon, titanium and zirconium and from basic rocks contained small amounts of sodium, potassium, magnesium, calcium, strontium, boron and much less titanium.

Trace elements, using the term in its wider sense to include all elements occurring in small amount, occur in soils in water-soluble forms, as exchangeable ions, in organic complexes, in association with oxides of iron, aluminium and manganese and as constituents of other minerals. Le Riche and Weir (1963) investigated the distribution of several elements in each of the above forms in two soils. They found that the extracted oxides (by ammonium oxalate at pH 3·3 under ultra-violet light) were richest in trace elements. The sand fractions contained relatively small amounts of trace elements except for strontium, tin and zirconium.

Many workers have observed an increase of most trace elements, except strontium, tin and zirconium which decreased, concomitant with an increase in clay content of soil. Chamberlain (1956) studied the distribution of elements down two profiles of the same soil (Kikuyu red loam) which occurred on sites with markedly different drainage. In the well drained soil beryllium, chromium, lead, nickel, tin, vanadium and zinc decreased rapidly with depth for about 1 m and then decreased less rapidly for 6 m Cobalt, on the other hand, increased with depth and titanium after decreasing for 2 metres, increased again. In the poorly drained soil there was hardly any change with depth to 6 metres of any element; moreover, the poorly drained soil contained noticeably smaller quantities of certain elements than did the well drained soil.

Total analysis is of little value for soil fertility investigations as the availability of elements in a soil to plants is poorly correlated with the total amounts present. The only use of total analysis in this respect is to study long term changes in a soil, although sometimes determination of the total amount of an element will reveal its inadequacy. However, total analysis does give a better idea of trace element availability than of macro-element availability due to the greater range of concentration in different soils of trace elements.

Practically every known element has been determined in soil for one reason or another and Swaine (1962) gives a useful listing of the so-called 'minor elements'. Those elements known to be essential for plant growth include boron, calcium, carbon, chlorine, copper, iron, magnesium, manganese, molybdenum, nitrogen, oxygen, phosphorus, potassium, silicon, sodium, sulphur and zinc. There are, in addition, certain elements non-essential to plant growth but essential to animals living upon plants; cobalt is an example of this, its deficiency in grass causing disease in cattle. Iodine deficiency in soils is associated with the incidence of goitre. Some elements such as barium, lithium and vanadium may or may not be essential in agricultural soils and some such as arsenic and chromium are toxic to plants.

With the growing interest in radioactive contamination of soils, elements such as uranium, caesium and yttrium are determined with particular reference to their rate of disappearance, their fixation and their uptake by plants.

A certain amount of confusion can arise from the terms 'trace element' and 'minor element'. Some workers take the logical view and include as trace elements all those that are present in trace amounts, but often both terms are taken as referring to those elements essential for plant growth but in very small concentrations and elements such as lead and chromium are not included. Thus, although total analysis involves the determination of trace elements, a 'trace element' analysis, as sometimes interpreted, does not necessarily involve the total analysis of a soil. As the essential trace elements are concerned with fertility, it is usually their 'available' form that is required rather than total amount. To avoid ambiguity the term 'trace nutrient element' will be used when this is meant.

A shortage of one or more of the essential trace nutrient elements nearly always affects the appearance of plants and many such symptoms have been described (e.g. Wallace, 1961). Often, however, the symptoms are not sufficiently developed or characteristic and only by chemical or biological analysis can a deficiency be detected. To remedy a trace nutrient element deficiency it is usual to apply the element in solution directly to the plant as a spray or injection and not by adding it to the soil where, more likely than not, it would become fixed in an unavailable form. If a soil is definitely known to be deficient in a certain element, the deficiency can sometimes be overcome by soaking plant seeds in a solution of that element before sowing. In general, however, trace nutrient elements are only deficient in their available form and can be rendered available by some treatment of the soil such as lowering the pH.

Different plants take up and utilize trace elements in different ways and this is further discussed in the sections dealing with each element. Chamberlain (1956) found that the trace elements in corresponding parts of different plants grown in the same soil differed markedly; copper, for example, varied from 2 μg g^{-1} to 50 μg g^{-1}, vanadium from 0·02 μg g^{-1} to 1·00 μg g^{-1}, molybdenum from 10 μg g^{-1} to 100 μg g^{-1} and manganese from 100 μg g^{-1} to 20 000 μg g^{-1}.

16:1:1 Total analysis

For the total analysis of a soil by chemical procedures, all the elements have to be brought into solution. Two main methods of achieving this are fusion of the soil with subsequent digestion of the melt and digestion of the soil with strong acids. It is not possible to recommend one method as being better than the other as the suitability of either depends upon the elements to be determined. For the determination of every element it would be necessary to prepare portions of a soil sample in more than one manner.

For fusion of soils, sodium carbonate is most commonly used but potassium pyrosulphate, sodium hydroxide and sodium peroxide are used in certain cases. For example, zirconium is not brought into complete solution by sodium carbonate fusion and potassium pyrosulphate must be used. The acids used for wet digestion of soils include perchloric, nitric, hydrochloric, hydrofluoric, phosphoric and sulphuric and the suitability of an acid or an acid mixture depends upon the elements to be determined.

Probably the most widely used method for the total analysis of soils is emission spectrophotometry using a spectrograph. With an arc or spark spectrograph, theoretically all the elements can be measured; in practice, certain non-metallic ions are difficult to determine by this means. Most spectrographic techniques for the total analysis of soil do not require the prior solution of the sample and the soil can be used directly, in a carbon electrode for example. A detailed discussion of spectrographic analysis is beyond the scope of this book and the reader is referred to standard works on the subject and particularly to Mitchell (1964). One of the attractive features of the method is the automatic, permanent record of analysis obtained in the form of a photographic plate. Thus the estimation of a particular, but previously ignored, element in a sample can often be made at any future date without further analysis other than reading the appropriate portion of the plate.

Particular care and special precautions must be taken when preparing soil samples for total analysis or for the determination of trace elements. Metal trays must not be used for drying the soil; if a suitable plastic tray is not available then sheets of paper should be placed on the trays. Sieves should be of polythene or aluminium and an agate mortar must be used for grinding. Spatulas should be of platinum or non-metallic material. Subsequent treatment of the sieved, dry soil depends somewhat upon the elements to be determined but as a general rule, rubber and metal apparatus should be avoided and water (for reagents and washing) should be demineralized and not merely glass-distilled. All reagents should be tested and purified if necessary; Mitchell (1964) describes the purification of the more common reagents.

16:1:2 Trace nutrient elements

For available trace nutrient element determinations arbitrary methods of extraction are employed or else a biological method is followed. A considerable number or reagents have been investigated in this connection and many have proved satisfactory in specific instances. The most generally useful extractant is accepted as being 2·5% acetic acid, although the diammonium salt of EDTA at pH 5·2 is gaining in popularity. For a few elements, notably molybdenum, boron, manganese and selenium, dilute acetic acid is not an effective extractant and as far as molybdenum is concerned, neutral ammonium acetate solution is preferable. For those elements which occur as anions the water-soluble form is probably the best measure of availability.

Whatever extractant is used the results must be considered in relation to other factors such as climate, crop (past, present and future), relief, parent material, pH, land management and so on. Laboratory investigations into trace nutrient element availability should always be augmented by pot-tests and field experiments.

16:1:3 Elements in solution

Once in solution, the elements can be determined by several different means. For practically every element a colorimetric procedure has been developed and these are fairly rapid and, providing that interference can be overcome, are accurate to micro or semi-micro levels. Certain elements such as iron, silicon and barium can be determined gravimetrically and others like calcium, magnesium and manganese, volumetrically. Polarographic analysis is convenient and accurate for metals such as zinc, copper and manganese, whereas for calcium, potassium and sodium, flame photometry is convenient. With a flame spectrograph routine determinations of calcium, sodium, potassium, manganese, strontium, iron and some other elements can easily and rapidly be made provided that they are in solution.

16:1:4 Calculations

Various methods of reporting the results of a total analysis are followed. The most correct way of calculating is upon the oven-dry weight of soil, the water and —OH groups being separately determined. As the combined water and organic matter are often not considered in total analysis, calculation upon the ignited weight of the soil is often done, but it should be remembered that certain elements such as sulphur and fluorine may be partially lost during ignition. In order to obtain a tidy figure of near 100% for a total analysis, elements can be calculated as oxides on an ignited weight basis, although, apart from serving as a check upon analytical results, there seems to be little point in such calculations.

16:2 FUSION AND DIGESTION OF SOILS FOR TOTAL ANALYSIS

16:2:1 Fusion

Main sources of error in sodium carbonate fusion

The errors peculiar to sodium carbonate fusion of soils in platinum crucibles are largely due to the nature of the particular soil being analyzed. A soil containing much manganese oxide will produce manganates on fusion and subsequent treatment of the melt with hydrochloric acid liberates chlorine which attacks the platinum. Errors of this kind are largely avoided by removing the bulk of the melt from the crucible before acidifying and can be eliminated altogether by treating the melt remaining in the crucible with a little ethanol in 5 M hydrochloric acid to reduce the manganese before digesting in acid.

If a soil contains much iron oxide the iron may combine with the platinum during fusion. This can be minimized by digesting the soil in a beaker with aqua regia, diluting the digest, filtering, washing and drying the residue before it is fused (Jackson, 1958); the final solutions are combined.

Certain other elements also attack platinum, for example bromides and iodides will liberate free halogen if fused in the presence of nitrates and are then acidified. Easily reducible elements and their compounds (e.g. silver, lead, antimony, etc.) may, if reduced, alloy with the platinum, but this type of error is more likely to occur when the crucible is heated by gas and is unlikely to be caused by electric heating.

Fusion with other materials

Potassium hydrogen sulphate or its anhydrous form, potassium pyrosulphate, is employed rather than sodium carbonate for fusion of a soil when certain elements are to be determined. These elements include antimony, cobalt, zinc and zirconium. The fusion is best carried out in a platinum crucible in a manner similar to that for sodium carbonate fusion and the melt is extracted with hydrochloric and perchloric acids. In the case of zirconium to be measured gravimetrically as the silicate, fusion is made in a quartz crucible.

Sodium peroxide is sometimes used for fusion of soils and as this reagent violently attacks platinum, nickel crucibles are used. Similarly nickel crucibles must be used for fusions with sodium or potassium hydroxide as is done in the estimation of arsenic and barium.

16:2:2 Acid digestion

Provided that hydrofluoric acid is not used, the wet oxidation of soils is best carried out in tall-form beakers heated on an electric sand-tray or in electrically heated 100-cm³ Kjeldahl flasks.

Perchloric–nitric acids

The soil sample is first digested with nitric acid alone and then with a mixture of nitric and perchloric acids. Pre-treatment with nitric acid is essential in order to reduce the danger of explosion which may occur with hot perchloric acid in contact with organic matter. If the soil is suspected to contain much antimony it would be advisable to use some other method of decomposition as antimony compounds are particularly explosive if heated with perchloric acid. It is presumed that the analyst is familiar with the dangers of perchloric acid and the precautions to be observed (see Vogel, 1962).

Perchloric–hydrofluoric acids (Jackson)

This digestion is carried out in platinum crucibles and for soils rich in organic matter the sample should be ignited before digestion, remembering, however, that certain elements may be lost during ignition; in some cases it may be best to digest with nitric acid rather than to ignite the soil.

16:2:3 Procedures in general

Special precautions to be observed or modifications to be introduced in the fusion or digestion procedures for determination of certain specific elements are indicated in the sections dealing with those particular elements. Jackson (1958) gives two flow sheets for the determination of some of the more common elements in soil at semi-micro level. In one flow sheet a 0·100-g sample is digested with perchloric–hydrofluoric acids and used to measure iron, titanium, potassium, sodium, calcium and magnesium, with the optional determination of aluminium and manganese. In the other, a similar sample is fused with sodium carbonate and used for the measurement of silicon, aluminium and manganese, with the optional determination of iron, calcium, mangesium and titanium. A flow sheet for the determination of silica, aluminium, iron, titanium, calcium, magnesium, manganese, potassium, sodium, phosphorus and sulphur using a 1·00-g sample of soil is also given by Jackson.

Of the total elements determined in soils, calcium and magnesium have been discussed in Chapter 8, potassium and sodium in Chapter 9, nitrogen in Chapter 10, carbon in Chapter 11, phosphorus in Chapter 12, sulphur in Chapter 13, iron, aluminium and manganese in Chapter 14 and silicon and titanium in Chapter 15. All other elements considered are in this chapter and for convenience of reference are discussed in alphabetical sequence.

Recommended analytical procedures are not given in detail for every element but adequate references are given in each case to enable the required determination to be made.

16:2:4 Recommended methods

I / FUSION WITH SODIUM CARBONATE

Accurately weigh about 1·0 g of 0·15-mm oven-dry soil into a platinum crucible and mix with about six times the weight of anhydrous sodium carbonate. Add a little more carbonate as a layer on top of the mixture. Partly cover the crucible with its lid and heat in an electric muffle furnace to 1200°C. When the melt is liquid, swirl the crucible, using platinum-tipped tongs, to spread the melt in a thin layer round the sides and continue heating for a further 15 minutes. After cooling somewhat, but while the melt is still liquid, repeat the swirling process so that the melt finally solidifies in a thin layer.

When cold, place the crucible inside a beaker and cover it with water. Warm the beaker on a steam-bath until the contents of the crucible have disintegrated. Separation of the melt from the crucible can sometimes be helped by means of a flattened glass spatula but manual deformation of the crucible must be avoided. Remove the crucible and lid and rinse them into the beaker; place the crucible and lid in a second beaker and digest with 5 M hydrochloric acid containing a little ethanol.

The disintegrated material in the first beaker is cautiously acidified by adding about 10 cm³ of concentrated hydrochloric acid and 10 cm³ of 60% perchloric acid. When the violent reaction has subsided, add the solution resulting from digestion of the crucible, cover the beaker with a Speedy-vap (see Vogel, 1962) cover glass and evaporate to fuming on an electric sand-tray. Continue heating for a further 15 minutes to dehydrate the silica, cool the solution and dilute to about 25 cm³ with warm water.

Filter through a Whatman No. 41 paper into a volumetric flask and wash the residue of silica with 0·5 M hydrochloric acid; dilute to volume. Determine the silica as described in section 15:5:1 and use the filtrate for determining the other elements.

II / DIGESTION WITH PERCHLORIC–NITRIC ACIDS

Accurately weigh about 1·0 g of oven-dry, 0·15-mm soil into a 250-cm³ tall-form beaker and add 20 cm³ of concentrated nitric acid. Cover the beaker and cautiously heat to oxidize organic matter. Add 10 cm³ of 60% perchloric acid and digest the mixture (using a Speedy-vap cover) until dense white fumes of acid appear. Use a little extra perchloric acid to wash down the sides of the beaker as necessary. Continue the procedure as described in section 16:2:4I for the dehydration and separation of silica.

Heating should not be such that the residue becomes dry as this will lead to incomplete separation of silica.

III / DIGESTION WITH PERCHLORIC–HYDROFLUORIC ACIDS

Accurately weigh about 0·1 g of oven-dry, 0·15-mm soil into a platinum crucible and moisten with a drop or two of water. Add 5 cm³ of hydrofluoric acid and 0·5 cm³ of perchloric acid, nearly cover the crucible with its lid and heat on a sand-tray at 200°C until the liquid has evaporated. Any unoxidized organic matter deposited on the upper part of the crucible or its lid can be oxidized with the flame of a Meker burner (Jackson, 1958).

After cooling, add 5 cm³ of 6 M hydrochloric acid and half fill the crucible with water. Place the crucible in an electrothermal bunsen and gently boil the contents for 5 minutes. If this treatment does not completely dissolve the solids, evaporate to dryness and repeat the acid digestion.

When the residue has been quite dissolved, transfer to a 100-cm³ volumetric flask and dilute to volume.

16:3 ANTIMONY

16:3:1 Background and theory

Although widely distributed in minerals, for example stibnite, Sb_2S_3, antimony is hardly mentioned in the literature of soil science but has been reported as being less toxic than arsenic to plants.

Once in solution, antimony can be determined colorimetrically with Rhodamine-B (Ward and Lakin, 1954), Brilliant Green (Stanton and MacDonald, 1962a), potassium iodide and with pyridine (Snell and Snell 1949). It is perhaps best determined spectrographically.

16:3:2 Recommended method (Ward and Lakin)

Reagents

Potassium hydrogen sulphate.
Hydrochloric acid, 1:1 v/v.
Sodium sulphite solution, 1% w/v, aqueous.
Cerium(IV) sulphate solution, 3·3% w/v of anhydrous cerium(IV) sulphate in dilute (1:35 v/v) sulphuric acid.
Hydroxylamine hydrochloride solution, 1% w/v, aqueous.
Isopropyl ether.
Rhodamine-B solution, 0·02% w/v in 1:10 hydrochloric acid.
Standard antimony solution: Dissolve 0·100 g Sb metal in 25 cm³ of hot conc. H_2SO_4. Dilute to 100 cm³ with water and then to 1 dm³ with 1:3 H_2SO_4. The final solution contains 100 μg cm⁻³ Sb.

Procedure

Fuse up to 0·5 g of 0·15-mm soil in a tube with 1·5 g of potassium hydrogen sulphate until all organic matter has decomposed and the tube is filled with white fumes. Cool the tube and rotate it to form a thin layer of melt round the wall. Add 6 cm³ of 1 : 1 hydrochloric acid and heat until the salts are in solution; do not boil. Add 1 cm³ of sodium sulphite solution to reduce the antimony to trivalent ions and 3 cm³ of 1 : 1 hydrochloric acid. Shake, filter and wash the precipitate twice with 3-cm³ portions of hot 1 : 1 hydrochloric acid and once with 2 cm³ of hot water. Cool to below 25°C, add 3 cm³ of cerium(IV) sulphate solution to oxidize the antimony again and 10 drops of hydroxylamine hydrochloride solution to destroy excess cerium(IV) sulphate. Add 45 cm³ of water and again cool.

Add 5 cm³ of isopropyl ether and shake for 30 seconds. After standing for 5 minutes, drain off the aqueous layer as completely as possible and add 2 cm³ of a 1 % solution of hydroxylamine hydrochloride in 1 : 10 hydrochloric acid to ensure the elimination of iron(III). After shaking, discard the aqueous layer and repeat the hydroxylamine treatment.

Add 2 cm³ of Rhodamine-B solution and shake for 10 seconds; separate the ether layer through a plug of cotton wool and read its transmission at 550 nm.

If much arsenic or tin is present it must be removed and a method for this is given by Snell and Snell (1949). The calibration curve prepared from dilutions of the standard should cover the range 0–30 µg cm⁻³ Sb.

16:4 ARSENIC

16:4:1 Background and theory

A widespread arsenic mineral is arsenopyrite, AsFeS, and arsenic is present in small amounts in nearly all soils and is widely distributed in plants, certain of which can accumulate the element in their roots. Arsenic is sometimes added to soils in the form of arsenical sprays used to combat fungi and other plant diseases. If the soil is sandy the applied arsenic can be very toxic to plants but toxicity is less likely with clay soils, particularly if they contain much iron.

Everett (1962) showed that increasing the pH value of grassland soils decreased the plant uptake of arsenic, although fertilizing with phosphorus had the opposite effect. Everett further found that a high phosphorus content in the grass itself eliminated injury due to arsenate and decreased injury due to arsenite. The best method of countering arsenic toxicity in soil is to add lime or hydrated iron(III) oxide. It has been demonstrated that arsenic can stimulate the activity of some of the nitrifying organisms in soil.

Determination of arsenic

Small and McCants (1961) developed a colorimetric method for determining arsenic in soils depending upon its reduction and reaction with ammonium molybdate. The standard method, and that adopted by the Association of Official Agricultural Chemists, is the Gutzeit method in which the arsenic is reduced to arsine which is then reacted with mercury(II) bromide sorbed on paper. The dark stain produced is compared with those obtained from known quantities of arsenic. Almond (1953) suggested fusing the soil in a nickel crucible with potassium hydroxide and taking up the melt in hydrochloric acid before applying the Gutzeit method.

Arsenic can be measured spectrographically although the method is not sufficiently sensitive for very small quantities of the element.

16:4:2 Recommended method (Small and McCants)

Apparatus

A digestion apparatus is set up as shown in Figure 16:1. The funnel trap (Corning, MW 90210) contains cotton wool saturated with lead acetate solution.

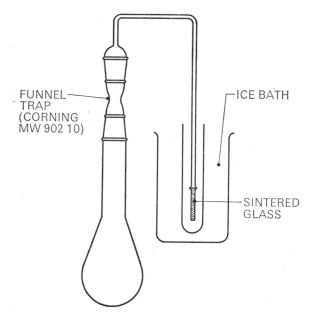

Figure 16:1 Apparatus for the determination of arsenic in soil (Small and McCants, 1961; Soil Science Society of America, *Proceedings*)

Reagents

Sulphuric acid, concentrated.
Perchloric acid, 72%.
Hydrochloric acid, concentrated.
Potassium iodide solution, 15% w/v, aqueous.
Tin(II) chloride solution, 40% w/v, aqueous.
Zinc metal, mossy.

Iodine solution, 0·001 M: Dissolve 2·54 g iodine and 8 g KI in 25 cm^3 of water, dilute to 1 dm^3 and store in the dark. This solution is 0·02 M and the 0·001 M solution is prepared from it by dilution immediately before use.

Ammonium molybdate solution: Dissolve 5 g ammonium molybdate in 300 cm^3 of water and add 70 cm^3 of conc. H_2SO_4; dilute to 500 cm^3.

Hydrazine sulphate solution, 0·15%, w/v.

Standard arsenic solution: Place 1·320 g As_2O_3 in 500 cm^3 of water and add 10 cm^3 of 35% w/v NaOH solution. When all the oxide has dissolved, neutralize the solution with 70% sulphuric acid and dilute to 1 dm^3. This solution contains 1000 µg cm^{-3} As.

Procedure

Heat 1 g of 0·15-mm soil in the Kjeldahl flask of the apparatus until fuming with 2 cm^3 (or more if necessary) of concentrated sulphuric acid. When the organic matter has been destroyed and the soil assumes a grey colour, cool the flask and add 3 cm^3 of perchloric acid and a few boiling chips. Boil the mixture for 2 hours, the acid condensing on the neck of the flask. Cool, add 20 cm^3 of water, 10 cm^3 of concentrated hydrochloric acid, 2 cm^3 of potassium iodide solution and 0·5 cm^3 of tin(II) chloride solution. Allow the flask to stand for 15 minutes, add 5 g of mossy zinc and join the flask on to the rest of the apparatus. The fritted end of the delivery tube dips into 30 cm^3 of the iodine solution. Distil the mixture for 1 hour.

Add 2 cm^3 of ammonium molybdate solution to the distillate and then 0·8 cm^3 of hydrazine sulphate solution. Place the tube in a boiling water-bath for 10 minutes, cool and read the transmission at 840 nm. The colour is stable for 24 hours.

Prepare a calibration curve to cover the range 0–1 µg cm^{-3} As. The graph is linear if lg% transmission is plotted as ordinate and quantity of arsenic as abscissa.

16:5 BARIUM

16:5:1 Background and theory

Barite, $BaSO_4$, is a common mineral and occurs together with calcite in limestone and is associated with ores of copper, silver, lead, cobalt and manganese. As the carbonate, barium occurs in witherite. The element is widely distributed in soils and plants and can replace bases on the soil colloids.

Barium, even in small concentration, can be toxic to plants although the toxicity is reduced if excess calcium is present. Barium can reduce plant uptake of sulphur by precipitating the highly insoluble barium sulphate.

Determination of barium

Mitsui and co-workers (1960) determined barium in soils by neutron radioactivation but the standard methods are by emission spectrophotometry. Barium can be determined turbidimetrically as sulphate and by flame photometry. Barium forms few coloured compounds and its colorimetric determination is confined to the method depending upon formation of the chromate (e.g. Middleton and Westgarth, 1964).

16:5:2 Recommended method (Middleton and Westgarth)

Reagents

Nitric acid, 5 M: 315 cm^3 dm^{-3} conc. HNO$_3$.
Ammonium hydroxide solution, 0·5 M: 36 cm^3 dm^{-3} conc. NH$_4$OH solution.
Ammonium nitrate solution, 10% w/v, aqueous.
Potassium chromate solution, 0·73% w/v, aqueous.
o-Phosphoric acid, concentrated.
Standard barium solution: 1·7787 g dm^{-3} BaCl$_2$.2H$_2$O. This contains 1000 μg cm^{-3} Ba.

Procedure

Pipette a suitable aliquot of soil digest or extract into a tall-form 125-cm^3 beaker. Add 1 drop of 5 M nitric acid and neutralize with dilute ammonium hydroxide solution using phenolphthalein as indicator. Add 5 cm^3 of ammonium nitrate solution and heat to 90°C on a water-bath. While hot, add 5 cm^3 of potassium chromate solution slowly with swirling. Leave the beaker at room temperature for 3 hours and filter through a Whatman No. 1 paper into a 50-cm^3 volumetric flask containing 2·5 cm^3 of phosphoric acid. Wash the precipitate twice with water by decantation and twice whilst on the paper. Dilute to volume and read the transmission at 490 nm or with a blue filter.

Measure a reagent blank and prepare a calibration curve to cover the range 0–1000 μg cm^{-3} Ba.

16:6 BERYLLIUM

16:6:1 Background and theory

Beryllium is found particularly in those soils derived from beryl, Be$_3$Al$_2$(Si$_6$O$_{18}$), which is fairly well distributed and is found commonly in granitic rocks.

Determination of beryllium

Beryllium is brought into solution by alkaline fusion or digestion with strong acids and once in solution, can be determined colorimetrically with *p*-nitrobenzene-azo-orcinol with which it forms a lake. Aluminium and zinc do not interfere with this method but copper must be removed. Sandell (1940) recommends the use of morin, a tetrahydroxyflavanol, which gives a fluorescence measurable under ultra-violet light and which can be made specific for beryllium. Among other reagents used are quinalizarin which gives a blue lake (Fischer, 1922) and curcurmin which gives a red colour (Kolthoff, 1928). With the curcurmin method potassium, sodium, lithium, calcium and barium do not interfere but iron and aluminium must be absent. Sunderasan and Sankar Das (1955) precipitated beryllium as the ammonium phosphate and determined the associated phosphorus; interfering elements were eliminated by complexing with EDTA.

16:6:2 Recommended method (Sunderasan and Sankar Das)

Reagents

Di-ammonium hydrogen phosphate solution, 10% w/v, aqueous.
Bromcresol green solution, 0·05% w/v in ethanol.
EDTA solution, 10% w/v of sodium salt, aqueous.
Hydrogen peroxide.
Hydrochloric acid, 1:1 v/v.
Ammonium acetate solution, 2 M: 154 g dm^{-3} of the salt.
Ammonium nitrate solution, 2% w/v, aqueous.
Nitric acid, 1:4 v/v.
Sulphuric acid, 1 M: 28 cm^3 conc. H_2SO_4/500 cm^3.
Sodium molybdate solution, 10% w/v, aqueous.
Standard beryllium solution: Standardize a stock solution of beryllium chloride by the oxide method (Vogel, 1962) and dilute to give a solution containing 10 mg Be/100 cm^3.

Procedure

Pipette an aliquot of soil digest or extract to contain up to 1 mg Be into a 40-cm^3 centrifuge tube. Dilute to 30 cm^3 and add 1 cm^3 of ammonium hydrogen phosphate solution and 1 cm^3 of EDTA solution. Add a few drops of hydrogen peroxide and 3 drops of bromcresol green solution. Acidify with 1:1 hydrochloric acid and place the tube in a boiling water-bath; add ammonium acetate solution dropwise until the colour is blue. Leave the tube in the water-bath for 30 minutes, cool, centrifuge and filter. Wash the precipitate three times with ammonium nitrate solution made to pH 3 with ammonium acetate and then dissolve it in hot, 1:4 nitric acid. Evaporate to dryness and re-dissolve in 5 cm^3 of 1 M sulphuric acid; add 5 cm^3 of sodium molybdate solution. Dilute to 50 cm^3 in a volumetric flask and read the transmission of the solution at 420 nm.

Prepare a calibration curve to cover the range 0–1 mg Be by taking aliquots of standard solution up to 10 cm^3.

16:7 BORON

16:7:1 Background and theory

Boron occurs as the borates: borax; boracite, $Mg_3B_7O_{13}Cl$; kernite, $Na_2B_4O_3$; ulexite, $NaCaB_5O_9 \cdot 8H_2O$; and colemantite, $Ca_2B_6O_{11} \cdot 5H_2O$. It is most abundant in arid regions and its deposition is associated with the evaporation of water.

Boron is remarkable in soils by its very narrow range between deficiency (for plant growth) and toxicity. Less than 1 $\mu g\ g^{-1}$ of boron may mean a deficiency, yet 3 $\mu g\ g^{-1}$ may be toxic. The element affects the actively dividing plant tissues and its deficiency is recognized by rotting of the softer parts. It is associated with calcium uptake by plants and in fertility studies it is often useful to measure boron–calcium ratios.

It has been shown that a boron deficiency decreases the permeability of the plasma membrane and boron-deficient plants may accumulate nitrogen as nitrate. The toxicity of excess boron to plants is connected with transpiration; thus Oertl and Kohl (1961) found necrotic spots on plants to have a high concentration of boron. Oertl and Kohl attributed the narrow range of boron-deficient–toxic levels to the localization of boron in plants. Oertl (1962) later showed that boron in leaves is mobile, moves with the transpiration stream and can largely be removed by guttation. One cause of boron toxicity in soil is irrigation with water containing appreciable amounts of the element (Eaton, 1935). This is a common occurrence in arid regions and, as pointed out by Singh and Kanwar (1963), the problem is intensified in saline alkali soils where sodium is the predominant cation and insufficient calcium is present.

To remedy boron deficiency in soils without causing toxicity has presented a problem for many years. Adding the nutrient as magnesium borate has been recommended as large amounts of this compound appear to have no toxic effects and the boron is not lost by leaching.

Very often a total analysis will reveal an adequate amount of boron in a soil and yet it is in a form unavailable to plants. Olson and Berger (1947) maintained that the most useful measure of available boron is the water-soluble form. The content of water-soluble boron in soils is influenced by pH, organic matter and amount of colloids. Thus the amount of boron soluble in water increases with increasing clay, silt and organic matter unless the pH simultaneously increases. In saline alkali soils an increase in pH increases the water-soluble boron and in such soils the element may become toxic.

Determination of boron

The total boron in a soil can be brought into solution by fusion with sodium carbonate or with pyrophosphate. Acid digestion does not recover all compounds of boron. Scharrer and Gottschall (1935) fused soil with potassium hydroxide; Whetstone (1942) recommended sodium dihydrogen phosphate.

Once in solution boron can be determined titrimetrically, colorimetrically and spectrographically. A biological assay method of determining available boron has been developed by Colwell (1946).

The titrimetric method of measuring boron (Bobko and Matveeva, 1936; Whetstone *et al.*, 1942) involves conversion of boron to methyl borate by distilling the soil extract with methanol and sulphuric acid. The distillate is re-distilled in the presence of alkali and the ignited residue taken up in dilute acid. After expulsion of carbon dioxide the solution is titrated with either sodium or barium hydroxide to neutrality. Addition of mannitol to the neutral solution liberates acid and the titration of this is a measure of the boron present; the titration can be direct or electrometric.

Numerous reagents have been used for the colorimetric determination of boron, quinalizarin and curcurmin being the most popular. Quinalizarin reacts with boric acid forming a chelate ring and gives a blue-coloured complex. The reaction takes place in strong sulphuric acid and the precise strength of the acid influences the intensity of colour. Temperature must also be controlled and oxidizing agents must be absent. Strmeski (1959) used a solution of quinalizarin in a 1 : 1 acetic–sulphuric acid mixture to determine micro quantities of boron in soils. Ellis *et al.* (1949) pointed out that there is considerable overlapping of absorption bands of reacted and free quinalizarin, thus restricting the range; the use of 1,1'-dianthrimide in concentrated sulphuric acid was recommended as a preferable reagent.

The curcurmin reagent, 1,7-bis(4-hydroxy-3-methoxyphenyl)-1,6-heptadiene 3,5-dione, is more convenient than quinalizarin as concentrated sulphuric acid is not involved. The reaction takes place in ethanol solution and small temperature changes do not affect the colour. Nitrates and excessive amounts of iron and molybdenum must be removed; a generally applicable method of separating boron from interfering elements is to distil as methyl borate.

Carmine also forms a strongly coloured complex with boron in a manner similar to that of quinalizarin, the reaction taking place in concentrated sulphuric acid (Hatcher and Wilcox, 1950). No interference is experienced from silica, fluorine, ammonium, calcium, molybdenum, magnesium, sodium or potassium and there is no temperature effect between 20°C and 35°C. Hatcher and Wilcox found the standard error of the mean difference to be ± 0.1.

Ráb (1962) described the photometric measurement of boron with Algol blue-3R which gives reliable results for minute quantities of boron but here also, concentrated sulphuric acid is involved. Dean and Thompson (1955) used a flame photometer for measuring boron as a solution of methyl borate. Stanton and McDonald (1966) proposed a colorimetric procedure using methylene blue. The sample is treated with a mixture of sulphuric and hydrofluoric acids and an aliquot of digest mixed with methylene blue solution. The coloured fluoborate compound is extracted into 1,2-dichloroethane and its extinction measured at 640 nm.

In all methods of boron analysis, boron-free glassware must be used;

not, for example, Pyrex which is a borosilicate glass. It is advisable to predetermine the reactivity of the glassware used; this can be done by a 16-hour agitation with ammonium acetate solution at pH 4 (Holden and Bontoyan, 1962).

16:7:2 Recommended methods

Total boron (after Wear, 1965; Naftel, 1939)

Reagents

Sodium carbonate, anhydrous (or sodium dihydrogen pyrophosphate).
Sulphuric acid, 2 M: 102 cm^3 dm^{-3} conc. H$_2$SO$_4$.
Bromthymol blue indicator solution.
Phenolphthalein indicator solution.
Ethanol.
Hydrochloric acid, 0·1 M: 9 cm^3 dm^{-3} conc. HCl.
Calcium hydroxide suspension, 0·4 g Ca(OH)$_2$/100 cm^3 water.
Curcurmin reagent: Dissolve 0·4 g curcurmin and 5·0 g oxalic acid in 100 cm^3 ethanol. Store in a brown bottle in a refrigerator.
Standard boron solution: 0·5716 g dm^{-3} boric acid. For use dilute 10 cm^3 to 1 dm^3 giving a solution containing 1 μg cm^{-3} B.

Procedure

Fuse 1 g of 0·15-mm soil with sodium carbonate (or with the pyrophosphate) as described in section 16:2:4ɪ. Take the melt up in water and add 2 M sulphuric acid to neutralize the solution to bromthymol blue, used as an external indicator. Transfer the solution to a 500-cm^3 volumetric flask and dilute to about 400 cm^3 with ethanol. Add sodium carbonate solution until the solution is alkaline and dilute to 500 cm^3 with ethanol. Mix and filter the suspension and pipette 100 cm^3 of solution into a 250-cm^3 beaker; add 50 cm^3 of water and concentrate to about 15 cm^3 on a water-bath. Transfer the concentrated solution to a platinum dish and evaporate to dryness. Heat the dish to destroy organic matter and take up the residue in 5 cm^3 of 0·1 M hydrochloric acid.

Add 4 cm^3 of curcurmin reagent, heat the dish in a water-bath at 55°C ±3°C and leave in the water-bath for 15 minutes after the liquid has evaporated. Moisten the residue with a little ethanol and transfer with more ethanol to a 15-cm^3 centrifuge tube. After centrifuging, decant the liquid into a 25-cm^3 volumetric flask and make to volume with ethanol. Alternatively, transfer the whole contents of the dish to the flask with ethanol and filter after diluting to volume.

Measure the optical density of the solution at 540 nm; the colour is stable for 2 hours.

To prepare the calibration curve, pipette aliquots (0–5 cm^3) of the dilute standard boron solution into evaporating dishes and evaporate to dryness with 5 cm^3 of calcium hydroxide suspension. After cooling, add 1 drop of phenolphthalein solution and then 2 M hydrochloric acid until the pink colour is discharged; add $0 \cdot 5$ cm³ of acid in excess. Add 4 cm³ of curcurmin solution and follow the procedure described for the test solution.

Provided that the boron in the aliquot analyzed is between $0 \cdot 2$ μg and $0 \cdot 8$ μg, Wear reports that the coefficient of variation is less than 3%.

Water-soluble boron

Reagents

Barium chloride solution, 10% w/v, aqueous.
Hydrochloric acid, 2 M: 178 cm³ dm⁻³ conc. HCl.
Calcium hydroxide suspension.⎫
Curcurmin solution. ⎬As for total boron.
Standard boron solution. ⎭

Procedure

Gently reflux 20 g of air-dry, 2-mm soil for 5 minutes with 40 cm³ of water containing $0 \cdot 5$ cm³ of barium chloride solution (Berger and Truog, 1939). After cooling, decant the suspension into a centrifuge tube and centrifuge. Pipette a suitable aliquot, containing up to 5 μg boron, into an evaporating dish, add 5 cm³ of calcium hydroxide suspension and evaporate to dryness. After cooling, add 1 drop of phenolphthalein solution and enough 2 M hydrochloric acid to discharge the colour; add $0 \cdot 5$ cm³ of acid in excess. Add 4 cm³ of curcurmin reagent and proceed as described for total boron.

16:8 BROMINE

16:8:1 Background and theory

Bromine occurs in certain silver minerals such as bromyrite, AgBr, and embolite, Ag(Br.Cl), and can be deposited from volcanic gases. Bromine is normally present in minute amounts in soils but is not retained against leaching and so is easily lost; it tends to accumulate in organic soils. Bromine has been found toxic to plant growth but only during late stages of development.

Determination of bromine

If chlorides are absent, bromides can be estimated nephelometrically as silver bromide, but general methods of analysis involve displacement of bromine by chlorine. The iodine content of the sample must be low as compared with the bromine content, less than 30%. The red colour of the liberated bromine can be measured directly or, in order to increase the sensitivity, the bromine can be extracted into carbon tetrachloride before estimation.

Fuchsin reacts with bromine to give a violet coloured compound that can be extracted into iso-amyl alcohol (Indovina, 1935) and the reaction of bromine with fluorescein to give eosin is free from chlorine interference.

Bromides can be estimated in the presence of chlorides by reaction with potassium persulphate and strychnine (Brinkley, 1948).

Atamanenko *et al.* (1962) determined bromine and iodine polarographically. The soil is heated at 600°C to remove organic matter and then extracted with normal hydrochloric acid; a stationary platinum electrode was used in the determination.

Conway (1950) describes micro-diffusion methods for determining bromine, based on oxidation to the elementary form with subsequent absorption in potassium iodide solution and the determination of the iodine liberated. The method provides for an accurate distinction between bromine and chlorine.

16:8:2 Recommended method (Conway; Stout and Johnson)

Reagents

Starch–iodide solution: Dissolve 3 g of KI in 100 cm^3 of starch solution.

Starch solution: Grind 4 g of soluble starch into a smooth paste with water and then grind in 0·01 g of mercury(II) iodide. Pour the paste into 1 dm^3 of boiling water with stirring.

Potassium chromate solution, 40% w/v, aqueous.

Sulphuric acid solution: 800 cm^3 dm^{-3} conc. H_2SO_4.

Standard bromide solution: 1·4891 g dm^{-3} KBr. This contains 1000 μg cm^{-3} Br.

Procedure

The Conway units shown in Figure 10:14 can be used, or one of their modifications. Starch–iodide solution is placed in the centre well, and in the outer chamber, place 2 cm^3 of test solution which is usually a water extract of soil. Add 1 cm^3 of potassium chromate solution to the test solution and then 1 cm^3 of sulphuric acid. Seal the unit and gently agitate for 1½ hours; draw off the contents of the centre well and wash the well several times with water, adding the washings to the main solution. Make the iodine solution to standard volume (5 cm^3) and read its transmission at 540 nm.

Chlorine does not interfere up to 2% although iodine must be absent. Stout and Johnson (1965) emphasize that the concentration of the sulphuric acid is critical.

A special tray designed to shake Conway units, and having an aspiration device for transferring solutions from the centre well to volumetric flasks, is described by Johnson *et al.* (1958).

16:9 CADMIUM

Cadmium is widely distributed in small amounts in soils and occurs as the mineral greenockite, CdS, usually in association with zinc ores.

Determination of cadmium

Apart from spectrographic determination which is not particularly sensitive for cadmium, the only described method for soil analysis is that of Popea and Gutman (1961) using dithizone (diphenylthiocarbazone), which is adapted from Sandell's (1939) procedure for mineral analysis. The method is subject to interference from other ions and Popea and Gutman proposed the extraction of cadmium dithizonate into carbon tetrachloride after elimination of iron and manganese. The complex is not very stable in carbon tetrachloride, however, and Snell and Snell (1949) advocate chloroform as solvent. The procedure measures up to about 1 $\mu g \, cm^{-3}$ cadmium in solution.

16:10 CAESIUM

Caesium occurs in the rare mineral pollucite, $Cs_4Al_4Si_9O_{26}.H_2O$, in pegmatites. The chief interest in the determination of caesium in soils lies in the fact that ^{137}Cs is a fission product liable to contaminate agricultural land.

Experiments in Russia using ^{137}Cs showed that the element is strongly adsorbed by soils. The experiments showed that if the caesium content was less than 10^{-7} mol/g in the soil, it was specifically sorbed; if the concentration was between 10^{-7} and 10^{-5} mol/g the caesium was in combined forms and if the concentration was 10^{-4} mol/g then exchangeable caesium was present.

The general behaviour of caesium is like that of potassium. Coleman and co-workers (1963) examined the exchange displacement behaviour of vermiculite and of heated potassium montmorillonite. Their findings suggested that the sorption of caesium in interlayer spaces leads to interplanar distances admitting potassium or ammonium but restricting the entry of calcium.

Residual ^{137}Cs in soils can seriously affect plant growth, particularly in sandy soils; on the other hand sandy soils permit the leaching of radioactivity. ^{137}Cs appears to accumulate in the vegetative parts of plants. In Japan it was found that rice was taking up excessive amounts of caesium from lowland soils, particularly when ammonium or rubidium was present and that potassium depressed the uptake of caesium. Investigations using ^{137}Cs showed that addition of rubidium or ammonium decreased the amount of carrier-free ^{137}Cs absorbed by the soil and increased its uptake by plants. Similarly uptake of rubidium was increased by adding caesium. The high uptake of caesium by rice was explained as being due to the available nitrogen being in the form of ammonium ions.

Haghiri (1962) demonstrated that the uptake of ^{137}Cs by maize was affected by temperature, being greatest at 21°C and least at 10°C. Chelating agents which normally increase plant uptake of minor elements have no effect upon uptake of caesium (Essington *et al.*, 1962).

Analysis of soil for ^{137}Cs from fifteen sites in Great Britain, made by the UK Atomic Energy Authority, showed that the isotope was present mostly in the surface mat of decomposed vegetation and that its concentration in soil below a depth of 3 inches was negligible. The amounts of ^{137}Cs found were related directly with rainfall.

Determination of caesium

The usual method of determining ^{137}Cs in soil is by γ-spectrometry and it can be extracted from soil with 6 M nitric acid at 100°C.

Provided that phosphorus is removed and potassium known, caesium can be determined in solution using a flame photometer at a wavelength of 852 nm (8521 Å). Caesium can also be determined in the same way as potassium by forming the cobaltinitrite complex which is then colorimetrically measured with α-naphthylamine and sulphanilic acid (Snell and Snell, 1949).

16:11 CHLORINE

Chlorine occurs in silicate rocks but most soils receive their chlorine content from rain. Rothamsted in England, for example, receives an average annual deposition of 17·6 kg chlorine per hectare and the Hebrides in Scotland receive as much as 6328 kg chlorine per hectare every year.

Chlorine is essential for photosynthesis in green plants (Warburg, 1949, for instance); it is involved in non-cyclic photophosphorylation and in the riboflavine phosphate pathway of cyclic photophosphorylation (Arnon, 1959). Warburg further found that chlorine is essential for oxygen evolution by isolated chloroplasts and could be replaced for this function only by bromine. Arnon found that chlorine was not, however, essential for the vitamin-K pathway of cyclic photophosphorylation which is the only other kind of photophosphorylation shared by chloroplasts and bacterial particles. Thus, in the absence of chlorine, chloroplasts act like bacterial chromatophores and can carry out the vitamin-K pathway only. One of the main functions of chlorine, and in which it is not specific, is to regulate osmotic pressure and cation balance in plant cells.

The problem of excessive amounts of chlorine in soils has been discussed in Chapter 6. Chlorine is easily leached from soils to the lower layers but is drawn to the surface again during dry periods and the maximum accumulation of chlorides is found usually in the top 5 cm of a soil.

Determination of chlorine

The determination of water-soluble chlorides has been discussed and described in Chapter 6.

For the determination of total chlorine in soil the sample can be fused with very pure sodium carbonate, but however pure the reagents, a blank determination is essential.

Small amounts of chlorine are best determined by micro-diffusion methods (Conway, 1950; Stout and Johnson, 1965) in which the combined form is oxidized to elemental chlorine with potassium permanganate and absorbed in potassium iodide solution; the liberated iodine is then determined. Bromine is simultaneously liberated and absorbed and must be determined separately and subtracted. The bromine is determined by micro-diffusion as described in section 16:8:2 and chlorine does not interfere. Iodine is also determined by this procedure and must be allowed for if present in large amount; normally the small quantities of iodine in soil can be ignored.

16:12 CHROMIUM

Chromium as chromite, $FeCr_2O_4$, and magnesiochromite, $MgCr_2O_4$, is a constituent of periodite rocks and their derived serpentines. In soils except for those formed from serpentines or other ferromagnesian rocks, chromium occurs in extremely minute quantities. The element has been found to occur in clay colloids isomorphous with the oxides of aluminium and iron.

The mineral chromite is quite inert, unless the soil is acid in reaction when chromium, particularly as chromate, is toxic to plant growth. Such toxicity can be remedied by liming the soil. Ng and Bloomfield (1962), during their experiments upon mobilization of minor elements in soil by anaerobic conditions, found that chromium was unreactive in this respect.

Determination of chromium

Chromium can be determined gravimetrically as chromium(VI) oxide, as barium chromate or as lead chromate. Volumetrically it can be determined by oxidizing to dichromate, adding a known excess of iron(II) and back-titrating with potassium dichromate solution. Alternatively, after oxidation to dichromate the solution can be potentiometrically titrated with ammonium iron(II) sulphate solution.

In alkaline solution as chromate or dichromate, chromium can be determined colorimetrically, the transmittance being read at 365–370 nm. With S-diphenylcarbazide chromates in acid solution give a violet colour almost specific for chromium (Urone and Anders, 1950), although a few elements such as iron and vanadium interfere and must be removed. Blundy (1958) oxidized chromium to chromate with ammonium hexanitratocerate in hot acid solution and extracted with isobutyl methyl ketone at low temperature prior to reacting with S-diphenylcarbazide. By this method Blundy estimated from 10 μg to 100 μg chromium in the presence of 100 μg nickel and iron. Blundy obtained a mean recovery of 100·3% with a standard deviation of ±0·8%. For the spectrometric measurements Blundy used a violet filter of wavelength 366 nm (3660 Å).

14 + T.S.C.A.

16:13 COBALT

16:13:1 Background and theory

Cobalt occurs in the minerals cobaltite (Co, Fe)AsS; skutterudite, (Co, Ni, Fe)As$_3$; linnaeite, Co$_3$S$_4$; and erythite, Co$_3$(AsO$_4$)$_2$.8H$_2$O. Kidson (1938) obtained evidence that the cobalt content of many soils is related to the magnesium content of the parent rocks. Thus serpentine gives soils with high cobalt content whereas soils derived from granite are low in cobalt. There is a correlation with some soils between cobalt and total iron and clay content. Peaty soils are usually very low in cobalt content.

Excess cobalt in soil can interfere with the uptake of iron by plants and causes the chlorophyll content to decrease. On the other hand, excess cobalt can increase the protein content of plants. The chief importance of cobalt in agricultural soils, however, is connected with the fact that ruminants have an essential need for the element; cobalt deficiency in pasture soils has led to incidence of disease in sheep and cattle.

Determination of cobalt

The total cobalt in a soil can be brought into solution by fusion with potassium hydrogen sulphate, with sodium carbonate or by digestion with strong acids. Plant-available cobalt is best extracted with dilute, 2·5%, acetic acid at pH 2·5.

Once in solution cobalt can be determined colorimetrically by several procedures. Stanton and McDonald (1962b) developed a method that depends upon the blue colour formed with cobalt and tributylamine and by which from 1 μg cm^{-3} to 250 μg cm^{-3} cobalt can be measured in the presence of 20% iron.

Many nitroso reagents have been used to determine cobalt, for example nitroso-R-salt (Kidson, 1938), orthonitrosophenol (Cronheim, 1942), orthonitrosocresol (Ellis and Thompson, 1945) and α-nitroso-β-naphthol (Nicholl, 1953). Of the nitrosocresol complexes, only those of cobalt, iron(III) and palladium are soluble in ligroin (petroleum ether, b.p. 90–100°C) and this forms the basis of isolation of cobalt. Palladium can be ignored in most soils and iron(III) is separated by a dithizone extraction, any traces remaining being reduced with hydroxylamine hydrochloride.

A classic colorimetric method is to complex cobalt with thiocyanate, when a blue colour is formed. Copper and iron give similar reactions but can be suppressed by reduction; the transmission is read at 625 nm or at 312 nm in the ultra-violet range. If desired, the method can be made more sensitive by converting the complex to tetraphenylarsonium-cobaltothiocyanate and extracting into chloroform (Potratz and Rosen, 1949).

Apart from spectrographic measurement in which a method involving platinum electrodes is recommended (Puig *et al.*, 1960), cobalt can conveniently be determined polarographically.

16:13:2 Recommended methods

I / TOTAL COBALT (HOLMES; ELLIS AND THOMPSON)

Reagents

Standard cobalt solution: Heat cobalt sulphate at 250°C for 8 hours and cool in a desiccator. Dissolve 0·2630 g $CoSO_4$ in 50 cm^3 of water containing 1 cm^3 conc. H_2SO_4 and dilute to 1 dm^3. This solution contains 100 μg cm^{-3} Co. For use, dilute 5 cm^3 to 1 dm^3, giving a solution with 0·5 μg cm^{-3} Co.

Ammonium citrate buffer solution, 25%: Dissolve 225 g ammonium citrate in water and dilute to 1 dm^3. Place the solution in a large separating funnel and add 40 cm^3 to 45 cm^3 of strong ammonium hydroxide solution to give a pH of 8·5. Extract the solution with 20-cm^3 portions of dithizone in carbon tetrachloride until the organic phase remains green and the aqueous phase orange. Then extract the orange colour with pure CCl_4.

Bromthymol blue indicator solution, 0·04% w/v, aqueous.

Dithizone solution: Dissolve 0·25 g of diphenylthiocarbazone in 1 dm^3 of pure CCl_4. Add 2 dm^3 of 0·02 M ammonium hydroxide solution and shake. Discard the CCl_4 phase and extract the solution with 50-cm^3 portions of pure CCl_4 until the extract is clear green. Add 500 cm^3 of CCl_4 and 50 cm^3 of 1·0 M HCl; shake and discard the aqueous layer. Dilute the CCl_4 with more CCl_4 to 2 dm^3 and store in a brown glass bottle. Add 25 cm^3 of water partly saturated with sulphur dioxide.

Hydrochloric acid, 0·02 M: 2 cm^3 dm^{-3} conc. HCl.

Nitric acid, concentrated.

Perchloric acid, 60%.

Sodium borate buffer solution: Dissolve 20 g boric acid in 1 dm^3 of water containing 22 cm^3 of 1·0 M NaOH solution. The pH should be 7·5.

Sodium nitrosocresol reagent: Dissolve 6 g hydroxylamine hydrochloride and 15 g $CuCl_2$ in 900 cm^3 of water in a large separating funnel. Add 5 cm^3 of *m*-cresol and 15 cm^3 of 30% H_2O_2 and shake the mixture. After standing for 2 hours add 25 cm^3 of conc. HCl and shake; add 100 cm^3 of pet. ether, b.p. 40–60°C, and shake for 1 minute. Place the aqueous phase in a clean funnel and extract with pet. ether until all the reagent has been removed. Wash the petroleum extract three times with water and shake with four 100-cm^3 portions of 1% copper acetate solution. Filter off the solid and store the solution in a refrigerator.

The sodium reagent is prepared from the copper compound as required. Place 75 cm^3 of the above reagent in a separating funnel and shake with 10 cm^3 of conc. HCl and 500 cm^3 of pet. ether. Collect the organic phase and wash twice with 100 cm^3 of 0·01 M HCl and twice with water. Add 50 cm^3 of sodium borate buffer solution and 2 cm^3 of 1·0 M NaOH solution. After shaking, separate and retain the aqueous phase.

Ligroin (petroleum ether b.p. 90–100°C).

Copper acetate solution, 4% w/v, aqueous (or 4% $CuCl_2$ solution).

Hydroxylamine hydrochloride solution: Dissolve 10 g hydroxylamine hydrochloride and 9·5 g sodium acetate in 500 cm^3 of water. The pH should be 5–5·2.

Pure carbon tetrachloride: Dry the solvent by shaking with anhydrous calcium chloride and filter into a Pyrex still. Distil the liquid and collect the fraction boiling at 76·7°C.

Procedure

Fuse 1 g of soil with sodium carbonate as described in section 16:2:4I or digest with perchloric acid as in section 16:2:4II. Take the residue up in dilute hydrochloric acid and dilute to 100 cm^3.

Pipette 50 cm^3 of digest into a 125-cm^3 separating funnel, add 5 cm^3 of ammonium citrate buffer and adjust the pH to 2·5 with ammonium hydroxide using 3 drops of the acid form of bromthymol blue as indicator. When the colour turns yellow, add 5 cm^3 of dithizone solution and shake for 5 minutes. Boawn (1962) has described a suitable set-up for mechanically shaking a number of funnels. The carbon tetrachloride phase is discarded and the extraction repeated with 5 cm^3 more of dithizone solution. Cobalt, lead and zinc now remain in the aqueous phase and copper is in the discarded organic phase (section 16:14:2).

Add 5 cm^3 of citrate buffer to the cobalt solution and adjust the pH with ammonium hydroxide, using phenolphthalein, to 8·5. Add 10 cm^3 of dithizone solution and shake for 2 minutes. Run the dithizone phase into a clean funnel and re-extract the aqueous phase with more dithizone until the extract is green. Finally, extract the aqueous phase with 10 cm^3 of pure carbon tetrachloride; completeness of extraction is indicated by the carbon tetrachloride being clear green. Keep the aqueous phase in the funnel.

Add 25 cm^3 of 0·02 M hydrochloric acid to the combined dithizone extracts and shake for 2 minutes; run the dithizone phase into a 100-cm^3 tall-form beaker.

Add the aqueous, acid phase to the first aqueous extract kept in the first funnel and use the solution (if so desired) to determine lead (16:19:2).

Add 0·5 cm^3 concentrated nitric acid to the dithizone extract in the beaker and evaporate until just dry. Add 5 cm^3 perchloric acid and again evaporate; do not continue heating after the residue is dry as this causes loss of cobalt. After cooling, add 10 cm^3 of 0·02 M hydrochloric acid, warm to aid solution and transfer the contents of the beaker into a clean separating funnel using a little water as necessary. Add 10 cm^3 of sodium borate buffer solution to bring the pH to 7·5 and add 0·5 cm^3 of sodium nitrosocresol reagent. If, after shaking, an orange colour does not dominate over the violet colour, add more reagent. Add 10 cm^3 of ligroin and shake for 5 minutes. Discard the aqueous phase. Add 10 cm^3 of 4% copper acetate (or chloride) solution and shake for 1 minute; discard the aqueous phase. Add 25 cm^3 of water, shake and discard the aqueous phase. Add

5 cm³ of hydroxylamine hydrochloride–sodium acetate reagent, shake for
1 minute and discard the aqueous phase.

Dry the stem of the funnel with filter paper, run the ligroin phase
through a plug of cotton wool into a spectrophotometer cell and read its
transmission at 360 nm.

A blank determination must be carried through all stages of the determination. The calibration curve should cover the range 0 to 1·5 μg cm⁻³
Co; the accuracy of the method is to within 0·01 μg cm⁻³ Co.

II / AVAILABLE AND EXTRACTABLE FORMS OF COBALT

Follow the procedure of section 16:13:2ɪ for determining cobalt in any
chosen extractant, unless dithizone has been used as part of the extracting
procedure when this step can be omitted.

16:14 COPPER

16:14:1 Background and theory

Occurrence

Copper is widely distributed in soils and minerals; one of the more important copper minerals is chalcocite, Cu_2S, and another is chalcopyrite,
$CuFeS_2$.

As a plant nutrient, copper occurs in the enzyme polyphenol oxidase
which is involved in more than one function, depending upon the plant. In
the potato, for instance, the enzyme is involved in respiration whereas in
spinach it contributes to photosynthesis. Copper seems to be involved in
the synthesis of cytochrome oxidase and possibly increases chlorophyll and
carotenoids in plants. Yates and Hallsworth (1963) investigated the role of
copper in the metabolism of nodulated clover and found its effect to vary
with the supply of combined nitrogen. When copper was deficient, plants
receiving nitrate-nitrogen accumulated amino-acids, whereas plants relying
upon symbiotic fixation of nitrogen showed a continuous increase in soluble
amino-acids correlated with the level of copper. Yates and Hallsworth
obtained evidence suggesting that copper is directly concerned in the formation of γ-amino-n-butyric acid in the plant nodules and as this acid is a particular constituent of nodule protein the effect of copper appears to be a
unique requirement.

Uptake of copper by plants is hindered by aluminium and Hiatt and
co-workers (1963) showed that as little as 0·1 μg g⁻¹ Al seriously reduced
copper uptake by wheat roots. This effect was overcome by increasing the
available copper and thus it appears that copper and aluminium were com-
peting for common binding sites at the root surfaces.

Excess copper can be toxic to plants and this is liable to occur where
copper compounds are applied as fungicides, particularly on acid soils.
Drouineau and Mazoyer (1962) found copper toxicity in vineyards a few
years after spraying with copper sulphate; at pH values less than 6 in the

soil, 100 μg g^{-1} exchangeable copper was toxic and 50 μg g^{-1} were toxic if the pH dropped below 5. In the copper-affected plants, sodium, calcium and magnesium were reduced in amount and potassium was increased. Copper also prevents uptake and translocation of iron, thus inducing chlorosis. Coffee plants found to contain high amounts (300 μg g^{-1}) of copper in their leaves were exhibiting magnesium-deficiency symptoms in spite of adequate magnesium in the soil (Chamberlain, 1956).

Generally copper remains isotopically exchangeable in acid soils and is fixed in complex form in alkaline soils. Copper added to soil is easily sorbed by clay acids. Henkens (1961) found that copper penetration in a clay soil was about 2·5 cm during the year following application, and had reached 5 cm several years later. Menzel and Jackson (1950) showed that kaolinite is particularly active in sorbing copper as the $(CuOH)^+$ ion; at least 50% of sorbed copper was found in this form.

Copper is rapidly and strongly fixed by organic matter and copper deficiency is commonly found in highly organic soils. Such a deficiency is almost a feature of newly prepared peat soils. In Florida, USA, as much as 113 000 kg per hectare (per 10^4 m²) of copper sulphate has been added to peat soils with no toxic effects and in Carolina peaty soils receive 226 kg/hectare copper sulphate when first prepared and 57 kg/hectare annually (Green, 1952).

Sorteberg (1962) found that cultivated peat soil induced copper deficiency when mixed with uncultivated peat; this effect was nullified by drying the peat at 105°C for 4 hours.

Copper sulphate added at 45 kg/hectare to a highly organic soil under mangrove vegetation had beneficial effects upon the early stages of growth of rice and resulted in a 40% increase in yield of paddy over that from untreated soil (Hesse, 1962). An additional point of interest is that large quantities of copper are fixed by organic soils in a non-toxic form. In many such soils, particularly if used for rice cultivation, copper sulphate is used to combat fungal diseases such as piricularia, and effectively large doses can be given without fear of causing copper toxicity.

In some peat soils addition of copper can cause an iron chlorosis unless iron is simultaneously applied, and in the same way manganese and boron uptake may be hindered. The C/N balance is important when copper is deficient, and nitrogen fertilizers can aggravate a copper deficiency. In soils other than peats, increasing acidity usually increases the availability of copper but in peat soils the contrary has been found.

The nature of the organic–copper complexes formed in peat soils has been investigated by Ennis (1962) who prepared complexes by saturating the soils with copper sulphate solution followed by washing with dilute hydrochloric acid and water. Ennis found that 60% of the copper exchange capacity was accounted for by phenolic hydroxyl groups and 30% by carboxylic groups. Thiol groups appeared to form particularly stable complexes with copper and the number of stable complexes increased with the sulphur content of the soil.

Tobia and Hanna (1961) displaced soil solutions with ethanol and separated organic copper complexes from free copper by means of exchange resins. They found that the complexes were proportional in amount to the amounts of organic matter present, but obtained no relationship with total carbon contents.

Total copper

Total copper in soil is best brought into solution by decomposition with hydrogen fluoride as in some soils all the copper is not released by perchloric acid digestion (Jackson, 1958).

Available copper

Many reagents have been proposed for the extraction of plant-available copper. Steenbjerg and Boken (1948) used dilute (0·1 M) hydrochloric acid, a reagent which extracts organically bound copper and basic copper compounds. Antipov-Karataev (1947) used 0·5 M nitric acid and for calcareous soils, Morgan's buffered acetic acid solution has been recommended. Viron (1955a; 1955b) extracted soil with 0·05 M EDTA solution, the extract being subsequently ignited and the residue evaporated with hydrofluoric and hydrochloric acids. Henkens (1961) compared EDTA and dilute nitric acid extractions with a method of biological assay and concluded that while all three methods were equally satisfactory, the nitric acid extraction was quickest.

Fiskell and Westgate (1955) recommended extraction of available copper with neutral ammonium nitrate solution and ammonium acetate-extracted copper has been reported as giving the best correlation with plant uptake. Ammonium acetate at pH 4·8 has been used for extraction of copper at fixed pH.

What was termed 'potentially available' copper in soil was extracted by Lundblad *et al.* (1949) with perchloric acid after the soil had been treated with nitric acid. The technique was modified by Kittrick (Jackson, 1958) who digested the soil with perchloric and sulphuric acids, and by Henricksen and Jensen (1958) who digested the soil with perchloric acid after igniting at 500°C for 4 hours. These determinations are, in effect, of total copper.

Copper in solution

Copper in solution can be determined gravimetrically as copper(I) thiocyanate, copper benzoinoximate, copper salicylaldoxime and with ethylenediamine (Vogel, 1962). Volumetric methods using cerium(IV) sulphate, sodium thiosulphate or potassium iodate can be used and a high frequency titration using EDTA has been developed (Vogel, 1962).

Apart from spectrographic and polarographic measurements, the standard procedure for determining copper in soil extracts is a colorimetric method using sodium diethyl-dithiocarbamate and Sandell (1950) has developed a procedure permitting the estimation of copper and zinc in the same extract. The reaction has a wide tolerance to pH conditions and interfering compounds are removed by extraction into dithizone. The golden-coloured complex is usually extracted into carbon tetrachloride for spectrophotometric measurement and Beer's Law is complied with up to 2·5 µg cm^{-3} Cu. Cheng and Bray (1953) used EDTA to remove all interfering metals with the exception of bismuth. Andrus (1955) used zinc dibenzyl-dithiocarbamate in carbon tetrachloride solution for determining copper in plant digests, and the method was found applicable to soil extracts by Friend (1958) who reported that the results agreed well with those of spectrographic analysis. The procedure, having obtained the copper in solution, was merely to shake a diluted aliquot with the reagent, filter and measure the extinction at 440 nm. Andrus found no interference from other elements up to 1000 µg iron, 100 µg cobalt, 1000 µg manganese and 500 µg of molybdenum.

Tobia and Hanna (1961) removed interfering organic matter from soil extracts by passing the extract through an exchange resin which retained free copper and allowed organically bound copper to pass.

Breckenridge and co-workers (1939) introduced 2,2'-biquinoline as a specific reagent for copper. The reagent reacts with copper(I) ions to give a purple complex soluble in ethanol. The procedure was adapted for soil analysis by Cheng and Bray (1953). The copper is reduced with hydroxylamine hydrochloride which also reduces interfering iron and manganese. Interference with the method comes not so much from metallic ions but from various anions such as oxalate and silicate and from organic matter. The main interfering cation is ammonium.

16:14:2 Recommended methods

Total copper

Digest 1 g of 0·15-mm soil in a platinum crucible with a mixture of perchloric and hydrofluoric acids as described in section 16:2:4(III). The final residue is taken up in dilute hydrochloric acid and made to 50 cm^3. Determine copper in aliquots of the solution as described in section 16:14:2I.

Available copper

i. Immediately available copper. For most purposes the immediately available copper in soil is best extracted with neutral, normal ammonium acetate solution together with other exchangeable ions as described in Chapter 7. The extract is evaporated with nitric and hydrochloric acids in the usual manner (section 7:3:3) and then again evaporated with perchloric acid. Finally, take up the residue in dilute hydrochloric acid and determine the copper as in section 16:14:2I.

ii. Potentially available copper. For many purposes the determination of total copper serves as an estimation of potentially available copper. For a less drastic digestion of the soil Kittrick's modification of Lundblad's method can be used.

Digest from 2 g to 5 g of soil with 10 cm³ of a mixture of 100 cm³ of 60% perchloric acid and 10 cm³ of sulphuric acid. The digestion is conveniently done in Pyrex tubes, several of which can simultaneously be heated in an oil-bath (Kittrick recommended a phosphoric acid bath). Continue heating until the residue is white and all the perchloric acid has been evolved; dilute with water, filter and make to standard volume. Analyze aliquots of the solution for copper as in section 16:14:2ɪ.

Copper in solution

ɪ / SODIUM DIETHYL-DITHIOCARBAMATE METHOD (HOLMES AFTER SANDELL)

Reagents

Standard copper solution: 0·3928 g $CuSO_4.5H_2O$ dissolved in water containing 2 cm³ conc. H_2SO_4 and diluted to 500 cm³. The solution contains 200 μg cm⁻³ Cu.

Ammonium citrate buffer solution.
Dithizone solution.
Pure carbon tetrachloride.
Bromthymol blue indicator solution.
⎫ As in section 16:13:2ɪ.

Sodium diethyl-dithiocarbamate solution: Dissolve 0·2 g of reagent in 100 cm³ of water and filter into a separating funnel. Shake with 5 cm³ portions of pure CCl_4 to remove any copper. Store in the dark.

Ammonium hydroxide solution, 1·0 M: Distil strong ammonia solution from Pyrex and collect in a cold polythene bottle. Determine the concentration and dilute aliquots to give solutions 1·0 M.

Ammonium hydroxide solution, 0·01 M: Prepare from the 1·0 M solution.

Potassium iodide solution: Dissolve 10 g KI in 490 cm³ of water and add 5 cm³ conc. HCl and 5 cm³ 0·05 M sodium sulphite solution. Shake with 10-cm³ portions of dithizone solution until no discoloration of reagent occurs. Remove excess dithizone by shaking with pure CCl_4.

Procedure

Pipette an aliquot of test solution containing up to 2 μg cm⁻³ Cu into a 125-cm³ separating funnel. Add 5 cm³ of ammonium citrate buffer and bring to pH 2·5 with ammonium hydroxide using 3 drops of the acid form of bromthymol blue as indicator. Add 5 cm³ of dithizone solution and shake the mixture for 5 minutes. Run the carbon tetrachloride phase into a clean funnel and repeat the extraction of the sample. Combine the two extracts. The carbon tetrachloride phase will contain the copper and the aqueous phase contains any zinc which may be present (see section 16:34:2).

Add 10 cm³ of potassium iodide solution to the copper solution and shake for 2 minutes to remove interfering ions such as bismuth. Again run the organic phase into a clean funnel and remove excess dithizone by shaking with 50 cm³ of 0·01 M ammonium hydroxide solution for 1 minute.

14*

Transfer the copper solution to a 100-cm³ tall-form beaker and evaporate with 0·5 cm³ of concentrated nitric acid. Add 0·5 cm³ of 60% perchloric acid and re-evaporate until nearly, but not quite, dry. Cool, add 10 cm³ of water and boil off any traces of nitric acid. Dissolve the residue in a little hot, 1 M hydrochloric acid and place in a clean separating funnel. Using phenolphthalein as indicator, titrate the solution with ammonium hydroxide solution to pH 8·5. Add 5 cm³ of carbamate solution and 10 cm³ of carbon tetrachloride. Shake for 3 minutes, discard the aqueous phase, dry the stem of the funnel with filter paper and run the carbon tetrachloride phase through a plug of cotton wool into a spectrophotometer cell. Read the transmission at 400 nm or with violet filters.

Prepare a calibration curve from aliquots of diluted standard copper solution to cover the range 0–2 μg cm^{-3} Cu.

II / 2,2′-BIQUINOLINE METHOD (CHENG AND BRAY)

Reagents

Hydroxylamine hydrochloride.
Sodium acetate solution, 1·0 M: 136 g dm^{-3} cryst. salt.
Biquinoline reagent: Dissolve 0·2 g of 2,2′-biquinoline in 900 cm³ of iso-amyl alcohol with gentle warming. When cool, dilute to 1 dm³ with more of the alcohol.
Standard copper solution: As for carbamate method.

Procedure

Pipette an aliquot of test solution containing up to 60 μg Cu into a separating funnel; add a few milligrammes of hydroxylamine hydrochloride to reduce the copper. Adjust the pH to 4–5 with sodium acetate solution and add 10 cm³ of biquinoline solution. Shake for 1 minute, run the alcohol phase through a plug of cotton wool into a dry spectrophotometer cell and measure the transmission at 540 nm. If the solution is a little cloudy it can be cleared by adding a few drops of methanol. The calibration curve should cover the range 0–60 μg cm^{-3} Cu.

Separation of extracted copper from organically complexed copper (Tobia and Hanna)

Prepare a column 10 cm by 0·3 cm² of cross-linked Dowex 50 resin in the sodium form and of 0·15 mm to 0·07 mm (100–200 mesh). Percolate the soil extract slowly through the resin which retains uncomplexed copper ions. When about 25 cm³ of extract has passed, wash the resin with two 5-cm³ portions of water and add the washings to the leachate.

If desired, the organically bound copper can be determined in the leachate after digestion with nitric and perchloric acids. The resin is eluted with 50 cm³ of 1 M sodium chloride solution and copper determined in the leachate.

16:15 FLUORINE

16:15:1 Background and theory

Occurrence

Fluorine as the mineral fluorite, CaF_2, is widely distributed, usually in association with other minerals such as calcite, dolomite, gypsum and barite. Fluorine commonly finds its way into soils from phosphate rocks and in volcanic regions is deposited from the atmosphere via rain. In soils fluorine is found from minute trace amounts up to about 0·1 %. Fertilizers such as superphosphate and limestones can inadvertently increase the fluorine content of soils, probably in the form of fluorapatite, $Ca_5F(PO_4)_3$.

Deterioration of teeth is related to fluorine deficiency in soils, but the element does not appear to be essential in any way for plant growth. According to Specht and MacIntire (1961), an excess of fluorine in soils of humid regions has no ill effects upon plants or their consumers.

Total fluorine

The decomposition of fluorine compounds in soil is best done by treatment with hot (165°C) sulphuric acid and steam distillation as hydrofluosilicic acid (Willard and Winter, 1933). A suitable apparatus is shown in Figure 16:2. Sodium hydroxide is added to the distillate which is then evaporated to small volume. Sulphate and phosphate which interfere with the subsequent determination of fluorine are removed by steam distilling from perchloric acid at 135°C. Silver perchlorate solution is added to prevent volatilization of hydrogen chloride and the distillate is collected in a flask containing 1 drop each of 50% sodium hydroxide solution and *p*-nitrophenol indicator solution.

Jefferies (1951) slightly modified the above procedure by adding potassium permanganate before the first distillation to oxidize sulphides and avoid hydrogen sulphide in the distillate. The fluorine in the distillate is then determined.

Available fluorine

For the estimation of plant available fluorine in soils, Brewer (1965) recommends a water extraction at a ratio of 1:1 or 1:2 w/v. Wider ratios of soil to solution should be avoided so that the relatively insoluble calcium fluoride is not dissolved.

For routine analyses, fluorine in the water extract can be measured colorimetrically without further treatment, but for more accurate results the fluorine must be distilled out with perchloric acid and the distillate titrated with thorium nitrate.

Fluorine in solution

Fluorine in solution can be titrated with thorium nitrate with which it forms thorium tetrafluoride. If the titration is done in the presence of a suitable organic indicator dyestuff, at the end-point excess thorium ions produce a coloured lake (Milton, 1949). Willard and Horton (1950) considered that the two most satisfactory dyestuffs were Chrome Azural-S and Alizarin red-S. Of these two indicators Brewer (1965) prefers the former.

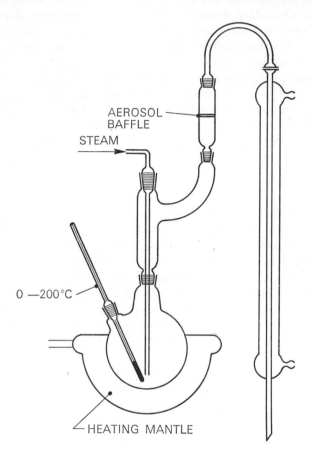

AEROSOL BAFFLE

STEAM

0 —200°C

HEATING MANTLE

Figure 16:2 Steam distillation apparatus for the determination of fluorine in soil

Several colorimetric procedures for estimating fluorine have been proposed. Megregian (1954) developed a method which depends upon the decolorizing effect of fluorine upon zirconium Eriochrome cyanine-R lake and Shaw (1954) exploited the same decolorizing effect upon zirconium alizarin lake.

The coloured lake produced by fluorine with thorium nitrate can be measured spectrophotometrically (e.g. Dean *et al.*, 1957).

A polarographic method of determining fluorine has been described by Shoemaker (1955) and a turbidimetric method was introduced by Brandt and Duswalt (1958).

16:15:2 Recommended methods

Total fluorine (Willard and Winter; Brewer)

Apparatus

The distillation apparatus used is shown in Figure 16:2.

Procedure

Weigh 1 g of oven-dry, 0·15-mm soil into the distillation flask and add 50 cm³ of sulphuric acid diluted with 25 cm³ of water; a 500-cm³ conical flask acts as receiver. Heat the mixture until the temperature reaches 150°C and then admit steam; continue heating to 165°C and adjust the rate of distillation to about 6–8 cm³ per minute. Stop the distillation when nearly 500 cm³ of distillate have been collected; add 10 cm³ of 10% sodium hydroxide solution to the distillate and evaporate to near dryness. Add sufficient water to the residue to bring the volume to about 30 cm³.

Transfer the solution to a clean distilling flask with a mixture of 50 cm³ perchloric acid and 25 cm³ of water. Add a 15% w/v solution of silver perchlorate to a slight excess after precipitation of any chloride present. Add glass beads to prevent bumping, connect the flask to the apparatus and place a 500-cm³ conical flask marked at 400 cm³ as receiver. Place in the receiver 1 drop of 50% sodium hydroxide solution and 1 drop of 0·5% *p*-nitrophenol solution. Heat the flask to 128°C, admit steam and adjust to 135°C. Distil at about 4 cm³ per minute, maintaining the volume of liquid in the distilling flask. When 400 cm³ of distillate have been collected, change the receiver and collect 100 cm³ more of distillate. Both distillates are analyzed for fluorine, the second portion acting as a check on complete recovery. Make the distillates to standard volume and analyze aliquots as in the following section.

Fluorine in solution (Milton; Willard and Horton; Brewer)

Reagents

Standard fluorine solution: 2·211 g dm⁻³ NaF (oven-dried at 100°C). Store in a polythene bottle. The solution contains 1000 μg cm⁻³ F.

Sodium hydroxide solution, 0·1 M: 4 g dm⁻³ NaOH.

Perchloric acid, 1·0 M: 8·5 cm³ 70% $HClO_4$/100 cm³.

Thorium nitrate solution, 0·025 M: 13·8 g dm⁻³ $Th(NO_3)_4.4H_2O$. 0·0025 M: By dilution of 0·025 M solution. 0·0005 M: By dilution of 0·025 M solution.

Chrome Azural-S indicator: 0·016 g reagent/100 cm³ of redistilled water.

Procedure

Add 1·0 M perchloric acid dropwise to a suitable aliquot of test solution until the yellow colour of the *p*-nitrophenol solution present is discharged.

Adjust the pH of solution to 3·3 using 0·1 M perchloric acid or 0·1 M sodium hydroxide solution and a pH meter.

Prepare a colour blank by adding 1 cm³ of indicator solution and 0·1 cm³ of 0·0005 M thorium nitrate solution to 50 cm³ of water at pH 3·3 (adjusted with 1·0 M perchloric acid and a pH meter) and contained in a Nessler comparison tube.

Titrate 50-cm³ aliquots of test solution with 0·0005 M thorium nitrate solution using 1 cm³ of indicator solution until the colour matches the colour blank. If more than 2 cm³ of nitrate solution are required, repeat the titration using the stronger, 0·0025 M solution and multiply the titre by five. A fresh colour blank must be made every 15 minutes.

Pipette duplicate 2-cm³ portions of standard fluorine solution, diluted to contain 10 μg cm⁻³ F, into Nessler tubes and dilute to 50 cm³ (each tube will thus contain 20 μg F). Add 1 cm³ of indicator solution and titrate with 0·0005 M thorium nitrate solution to match the colour blank. Prepare and titrate a reagent blank in the same manner.

Calculation

Deduct the reagent blank from the test solution titre giving titre A.

Subtract 0·1 cm³ from the titre of standard fluorine solution and multiply the result by 20 giving titre B. Then,

$$A \times B = \mu g \text{ F in the 50 cm}^3 \text{ of solution titrated.} \qquad [16:1$$

16:16 GERMANIUM

Germanium has been found in trace amounts in soils and is associated with the sulphide minerals of silver, lead, tin and zinc.

Germanium in soil can be brought into solution by fusion with alkali or by digestion with acids. Separation from other elements can be effected by distillation as tetrachloride in a stream of carbon dioxide or more simply by extraction into carbon tetrachloride from an acid (10 M HCl) solution.

Once the germanium has been separated from other elements it can be determined colorimetrically by the molybdenum blue method in a manner similar to that used for phosphorus (Hybbinette and Sandell, 1942). A more rapid method of analysis is by reaction with phenylfluorone and measurement of the transmission of the resulting solution at 510 nm (Schneider and Sandell, 1954). After digestion of soil and separation of silica the residue is taken up in 10 M hydrochloric acid. The germanium is extracted into carbon tetrachloride and the organic layer extracted with water. Aliquots of the aqueous layer are then reacted with a methanol solution of phenylfluorone, the pH being buffered to 5 with sodium acetate–acetic acid solution.

16:17 GOLD

16:17:1 Background and theory

Gold occurs in several minerals, such as calverite, $AuTe_2$; petzite, Au_2Te; nagyatite, a sulphotelluride of gold; and sylvanite, $AuAgTe_2$. It is widely distributed in soils in trace amounts and can be accumulated by plants.

Determination of gold

Gold is brought into solution by digesting the soil with aqua regia. The digest is then evaporated to dryness and the residue taken up in dilute hydrochloric acid.

Most colorimetric methods of determining gold depend upon reduction of the metal to the colloidal state, although as little as $0 \cdot 1$ μg cm^{-3} can be determined by reaction with *p*-diethylamino-benzylidene-rhodanine with which a violet colour is produced (Poluektov, 1943).

Stanton and McDonald (1964) determined gold by its reaction with Brilliant Green. The procedure has many steps of evaporation, solution and filtration and needs careful attention.

16:17:2 Recommended method (Stanton and McDonald)

Reagents

Digestion acid: 400 cm^3 conc. HCl, 100 cm^3 conc. HNO_3 and 500 cm^3 of water.

Hydrochloric acid, 2 M: 178 cm^3 dm^{-3} conc. HCl.

Hydrochloric acid, 0·5 M: By dilution of the 2 M acid.

Copper solution: 4 g $CuSO_4 . 5H_2O$/100 cm^3 of 2 M HCl.

Tellurium chloride solution: 0·21 g $TeCl_4$/100 cm^3 of 2 M HCl.

Tin(II) chloride solution: 20 g $SnCl_2$/100 cm^3 of 2 M HCl.

Brilliant Green solution, concentrated: 1 % w/v in absolute ethanol.

Brilliant Green solution, dilute: Dilute 5 cm^3 of the 1 % solution to 100 cm^3 with toluene just before use.

Standard gold solution: 0·504 g sodium chloroaurate dissolved in 25 cm^3 of 1·0 M HCl and diluted to 50 cm^3. This solution contains 5 mg cm^{-3} and is diluted for use with 0·5 M HCl.

Procedure

Digest 20 g of soil with 100 cm^3 of digestion acid and evaporate to dryness. Add 100 cm^3 of 2 M hydrochloric acid, boil for 15 minutes and filter through a Whatman No. 42 paper. Wash the soil residue with 2 M hydrochloric acid and dilute the cold filtrate to 200 cm^3 with 2 M hydrochloric acid.

Dilute a suitable aliquot of solution to 50 cm^3 with 2 M hydrochloric acid, add 1 cm^3 of copper solution, 2 cm^3 of tellurium solution and 10 cm^3 of tin(II) solution. Boil the solution and maintain nearly boiling for 30 minutes. Add 2 drops of tin(II) chloride solution to ensure complete precipitation. Filter while hot with a filter stick of porosity 3; wash the beaker and precipitate with 2 M hydrochloric acid and dissolve in 1 cm³ of hot aqua regia. Rinse the precipitation beaker with 1 cm³ of hot aqua regia, and pass the solution through the filter stick and wash it with a little water. Evaporate the solution to dryness in a test-tube and dissolve the residue in 2·5 cm³ of 2 M hydrochloric acid with gentle warming.

Transfer the solution to a small separating funnel and dilute to 50 cm³; it should now be 0·05 M in HCl. Add 5 cm³ of dilute Brilliant Green solution and shake for 30 seconds. Discard the aqueous layer, add 50 cm³ of 0·5 M hydrochloric acid and 0·25 cm³ of 1 % Brilliant Green solution and shake for 2 minutes. Discard the aqueous layer and filter the organic phase into a dry spectrophotometer cell. Measure the absorbance at 650 nm.

The calibration curve should cover the range 0–2 μg cm⁻³ Au.

16:18 IODINE

Iodine occurs in a few minerals such as iodyrite, AgI, and iodobromite, Ag(Cl,Br,I), which are related to the silver mineral cerargyrite, AgCl. The element has been found in certain magnesium limestones and dolomites, in coal and in rock-salt but the basic source of iodine in soils is the oceans via the atmosphere. Iodine is more prevalent in organic soils than in other kinds, and bottom deposits of lakes are often rich in iodine due to its accumulation by phytoplankton.

Borst-Pauwels (1961) found that the vegetative growth of plants was favourably influenced by traces of iodine, although some, such as oats and turnip, were adversely affected. On further investigation, Borst-Pauwels found that potassium iodide depressed the growth of oats more than did potassium iodate and that the iodine uptake was greater from iodide than from iodate. A quantitative relationship was established between plant growth and iodine content of roots.

Normally a deficiency of iodine in soils does not greatly affect plant growth but it can affect the animals dependent upon the plants. In Switzerland, parts of America and in Russia, high incidence of goitre is associated with soils of low iodine content. Iodine-deficient soils are commonly acid in reaction and iodine cannot be retained by highly leached clays.

Determination of iodine

McHargue et al. (1932) volatilized iodine from soils by heating in a closed tube in a furnace. The volatile products were absorbed in potassium carbonate solution and after destruction of organic matter, iodine was extracted into ethanol. Godfrey and co-workers (1951) determined iodine polarographically.

Several volumetric and gravimetric techniques are available for determining iodine in solution, but for soil extracts a colorimetric procedure is preferable. Iodine is liberated with an oxidizing agent such as bromine water, extracted into an organic solvent and the optical density of the solution measured. The sensitivity of the method can be increased by oxidizing the iodine to iodate which is then used to liberate iodine from hydriodic acid.

Proskuryakova (1964) utilized the catalytic effects of iodine upon the reaction of arsenic(III) acid and cerium(IV) sulphate and upon the reaction between sodium nitrite and iron rhodanide in the presence of nitric acid to estimate the element.

Iodine can quantitatively be estimated in the presence of chloride and bromide by a micro-diffusion method (Conway, 1950) provided that it is present in sufficient amount.

16:19 LEAD

16:19:1 Background and theory

Lead occurs commonly as the sulphide mineral galena, as the carbonate cerussite, and as the sulphate anglesite. It is nearly always present in soils although in small amount. The solubility of lead in soil increases with acidity and in acid soils it is accumulated by plants: food plants may acquire toxic (to the consumer) amounts of lead. Nowadays there is a danger of land near motorways becoming contaminated with lead from exhaust fumes (Parsons, private communication).

Determination of lead

For the determination of total lead in soil a spectrographic procedure is best. The standard chemical method of determining lead in small quantities is colorimetrically with dithizone. As the reagent gives coloured complexes with many metals, the main problem has been elimination of interference. Sandell (1937) developed a suitable technique for determining lead in rock samples. A carbon tetrachloride extraction of the dithizone complex at low pH (2·5) separates other ions from lead, cobalt and zinc, and subsequent treatment with dithizone under alkaline conditions separates the lead from cobalt and zinc.

16:19:2 Recommended method (Holmes)

Reagents

Standard lead solution: Dissolve 0·1599 g of oven-dried (110°C), recrystallized $Pb(NO_3)_2$ in 1% nitric acid and dilute to 1 dm^3. This solution contains 100 μg cm^{-3} Pb.

Other reagents as for the determination of cobalt, section 16:13:2I.

Procedure

Fuse 1 g of 0·15-mm soil with sodium carbonate as in section 16:2:4I, or digest with acids as in section 16:2:4II; after separating the silica, take up in dilute hydrochloric acid.

Pipette an aliquot of soil digest or extract corresponding to about 0·5 g of soil into a separating funnel. Follow the procedure given for the determination of cobalt (section 16:13:2I) up to the point where the aqueous extracts are combined ready for the determination of lead.

To the aqueous extract add 5 cm³ of ammonium citrate buffer solution and 10 cm³ of 10% potassium cyanide solution. Raise the pH to 9·3 with ammonium hydroxide using thymol blue as indicator. Add 10 cm³ of dithizone solution and shake for 3 minutes. Transfer the carbon tetrachloride phase to a clean funnel, wash it with 50 cm³ of 0·01 M ammonium hydroxide solution and filter through a cotton-wool plug into a spectrophotometer cell. Measure the optical density of the solution at 520 nm.

The calibration curve should cover the range 0–1 µg cm⁻³ Pb and a reagent blank must be measured.

16:20 LITHIUM

16:20:1 Background and theory

Lithium is widely distributed in small amounts in association with mica-type minerals. Unlike potassium, lithium is not adsorbed on soil colloids and, except in very small amount, is toxic to plant growth (Haas, 1929; Aldrich *et al.*, 1951). Bradford (1963) investigated the lithium content of natural waters in California and found high lithium associated with low magnesium and/or high sodium. If the kind of soil favoured lithium accumulation and if the lithium content exceeded 0·5 µg cm⁻³ in water, toxic effects were observed.

Determination of lithium

Total lithium in soil can be brought into solution by alkali fusion or acid digestion as described in section 16:2:4, and available lithium is extracted from soil with ammonium acetate solution.

Lithium is conveniently measured by flame photometry where the slight interference from sodium and potassium can be compensated for by adding these elements to the standard solutions. Bradford and Pratt (1961) purified soil extracts by passing them through a cation exchange resin before determining lithium by flame photometry.

Provided that iron is removed, 1 µg to 10 µg of lithium can be accurately determined in 10 cm³ of solution by reaction with Thoron, *o*-(2-OH-3, 6-disulpho-1-naphthylazo)-benzenearsonic acid. This reagent gives an orange-coloured complex with lithium and up to fifty times the lithium content of sodium may be present (Kivznetsov, 1948).

Patassy (1965) determined lithium in solution with an atomic absorption spectrophotometer.

16:20:2 Recommended method (Bradford and Pratt)

Apparatus

Prepare glass columns from 25-cm^3 burettes of 1 cm diameter. Plug the bottom of the burette with glass wool for 2·5 cm; place 38 cm of exchange resin on top of the glass wool and join the top of the burette with tubing to a separating funnel.

Reagents

Hydrochloric acid, concentrated.

Methanol.

Standard lithium solution: Dry lithium chloride overnight at 110°C and cool in a desiccator. Weigh a small quantity into a beaker by calculating the difference in weight of a weighing bottle before and after removal. 0·6109 g dm^{-3} LiCl contains 100 μg cm^{-3} Li. Dissolve the salt in water and dilute to 1 dm^3.

Resin: Use Dowex-50 w-x8 resin of 0·15–0·07 mm (100–200 mesh). Flush the resin in the column with 100 cm^3 of 3 M HCl and then with 100 cm^3 of water. The same treatment is used to regenerate the resin after use.

Procedure

Total lithium is brought into solution by alkali fusion or acid digestion and taken up in dilute hydrochloric acid. Extractable lithium is removed from soil by extracting 10 g of soil with 50 cm^3 of 1 M ammonium acetate solution for 4 hours; after filtering, leach the soil with 100 cm^3 more of solution. Evaporate the leachate to dryness and ignite at 500°C. Add 10 cm^3 of water and sufficient 3 M hydrochloric acid to neutralize the residue. Dilute, filter and make to 150 cm^3.

Pass the lithium solution through the column of resin, being careful to avoid air bubbles; discard the leachate. Leach the column with 100 cm^3 of 0·2 M hydrochloric acid in 30% v/v methanol, discard the leachate and leach with 200 cm^3 of the acid–methanol solution. Collect the final leachate and evaporate to dryness on a steam-bath. Add 5 cm^3 of the acid–alcohol solution and transfer the solution to a small beaker.

Analyze the solution with a flame photometer using an oxygen–acetylene flame; the wavelength should be 670·8 nm.

Prepare a calibration curve to cover the range 0–5 μg cm^{-3} Li.

16:21 MERCURY

16:21:1 Background and theory

The natural occurrence of mercury in soils is of negligible importance but the element is often added to soil as a fungicide, usually in the form of organic complexes such as mercaptobenzothiazole. The fact that mercury is an accumulative poison and is taken up by plants necessitates its occasional determination in soil.

Determination of mercury

Stock (1938) heated soil at 800°C for several hours and the volatilized mercury was collected and deposited electrolytically upon copper from which it was then distilled. The size of the condensed droplets was measured and the mercury estimated.

Polley and Miller (1955) treated a 1-g sample of soil with 5 cm³ to 15 cm³ (according to organic matter content) of concentrated sulphuric acid and, without heating, added hydrogen peroxide dropwise until reaction ceased. The mixture was then gently heated and dropwise addition of peroxide continued; the mixture was finally boiled, cooled and diluted. Potassium permanganate was added to slight excess and the excess destroyed with hydroxylamine hydrochloride. Mercury in the filtered solution was determined with dithizone.

Popea and Jemaneaunu (1960) digested soil with potassium nitrate and sulphuric acid, eliminated interfering elements with potassium thiocyanate and EDTA and determined mercury with dithizone. Vasilevskaya and Shcherbakov (1963) sublimed mercury from soil in Penfield tubes and dissolved it in nitric acid before determining with dithizone.

Pickard and Martin (1963) investigated the known procedures and found severe interference from copper. Pickard and Martin overcame copper interference by distilling mercury from soil in hydrogen chloride gas after digestion of organic matter. The mercury was then extracted from the distillate with dithizone. Recovery of added mercury to soil, as mercury(II) chloride, was good and no losses occurred at any stage of the experiment.

Mercury(II) ions react with dithizone at low pH to give a yellow-coloured complex which is extractable into carbon tetrachloride.

16:21:2 Recommended method (Pickard and Martin)

Digestion apparatus

A two-necked, 500-cm³ round-bottomed flask is fitted with an air condenser (30 cm × 3·5 cm) and a 50-cm³ cylindrical separating funnel. A double-surface water condenser is fitted on top of the air condenser. The flask is heated with a 250-watt heating mantle.

Distillation apparatus

The distillation apparatus is shown in Figure 16:3. The flask used for digestion is attached to the air condenser.

Reagents

Ammonium hydroxide solution, 5%: Concentrated ammonium hydroxide solution of r.d. 0·880 is diluted with water in the ratio of 1 to 19.

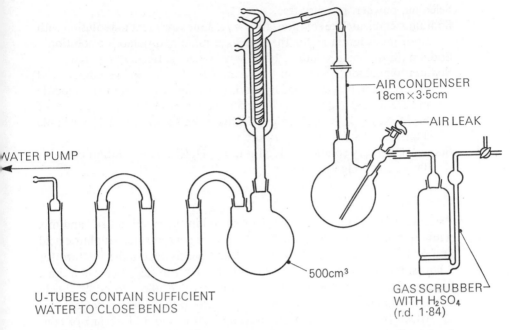

Figure 16:3 Apparatus for the distillation of mercury from soil (Pickard and Martin, 1963; *J. Sci. Fd. Agric.*)

Buffer solution: 1·0 M sodium acetate solution is made by dissolving 136 g of the crystalline salt in water and diluting to 1 dm³. Test the solution with dithizone for freedom from interference and purify if necessary as described for hydroxylamine hydrochloride solution. To 50 cm³ of the solution add 13 cm³ of 1·0 M HCl and dilute to 250 cm³.

Carbon tetrachloride: Redistilled and purified as described for the determination of cobalt (section 16:13:2I).

EDTA solution: 4%, w/v, aqueous solution of disodium salt.

Dithizone stock solution: Dissolve 0·05 g of dithizone in 20 cm³ of carbon tetrachloride and filter. Extract the solution with 100 cm³ of 5% NH₄OH and wash the alkaline layer twice with 5-cm³ portions of carbon tetrachloride. Make the alkaline layer just acid with HCl and add 5 cm³ of hydroxylamine hydrochloride solution. Extract with 100 cm³ of CCl₄ and wash the organic layer twice with 15-cm³ portions of water. Store the solution in a refrigerator.

Dithizone dilute solution: Just before use, dilute the stock solution with ten volumes of CCl₄.

Hydroxylamine hydrochloride solution: Prepare a 20% w/v aqueous solution of the reagent, and extract with several portions of dilute dithizone solution until the extract is colourless after removal of excess dithizone with 5% NH₄OH. Extract any remaining dithizone with CCl₄ until two successive washings are colourless. Filter through a Whatman No. 54 paper to remove traces of CCl₄.

Selenium powder, mercury-free.

Sodium metabisulphite solution, 20% w/v, aqueous: Treat the solution with dithizone solution as for the hydroxylamine hydrochloride solution.

Sodium thiosulphate solution, 1·5% w/v, aqueous, freshly prepared.

Sodium hypochlorite solution: The available chlorine in the commercial 15% solution is determined by mixing an aliquot with potassium iodide and acetic acid and titrating the liberated iodine (Vogel, 1962). 1 cm³ of 0·5 M $Na_2S_2O_3 \equiv 0·03456$ g Cl. Dilute the solution to 5% available chlorine and store in refrigerator.

Standard mercury solution: 0·1354 g dm^{-3} $HgCl_2$. This solution contains 100 μg cm^{-3} Hg.

Procedure

Place a sample of up to 40 g of 0·15-mm soil in the reaction flask and mix with 0·1 g of selenium powder. Add 50 cm³ of concentrated sulphuric acid and 5 cm³ of concentrated nitric acid and gently digest the mixture for 2 hours. If charring occurs, carefully add more nitric acid, 1 cm³ at a time. Simmer the mixture for a further 30 minutes and allow to cool. Slowly wash down the condenser with 40 cm³ of water while swirling the flask.

Attach the flask to the distillation apparatus and add 50 cm³ of concentrated sulphuric acid. When the apparatus is fully assembled and tested for air-tight joints, draw a slow stream of air through by means of a water pump and with the air-leak partly open. Gently heat the flask until about two-thirds of the water present has distilled and then with the air-leak still partly open, pass a slow (about 30 cm³/minute) stream of hydrogen chloride gas through the reaction flask. The air-leak should be adjusted so as to draw a steady stream of gas through the absorption tubes. Distil with hydrogen chloride for 2 hours, allow the apparatus to cool and remove the distillate. Wash the condenser and U-tubes with water, adding the washings to the distillate: the final volume should be about 125 cm³.

Filter the distillate (Whatman No. 541 paper) into a 600-cm³ graduated beaker and neutralize (using litmus paper) with concentrated ammonium hydroxide solution. Add concentrated hydrochloric acid dropwise until just acid and then add 10 cm³ of sodium metabisulphite solution. After mixing, add 25 cm³ of buffer solution and 5 cm³ of EDTA solution; dilute to 250 cm³ and adjust to pH 5 if necessary.

Transfer the solution to a separating funnel and extract with 10 cm³ of dithizone solution. Run the extract into a 100-cm³ separating funnel containing 25 cm³ of 0·1 M hydrochloric acid. Extract twice more with 5-cm³ portions of dithizone, combining all extracts in the 100-cm³ funnel. If the third portion of dithizone changes colour (indicating mercury still present) make another extraction. To the extracts add 5 cm³ of hydroxylamine hydrochloride solution and shake for 1 minute; transfer the organic layer to a second 100-cm³ funnel containing 50 cm³ of 0·1 M hydrochloric acid. Wash the aqueous layer remaining in the first funnel with 5 cm³ of pure carbon tetrachloride and add the washing to the separated organic layer.

Add 2 cm³ of sodium thiosulphate solution to the dithizone extract and shake for 1 minute. Discard the carbon tetrachloride layer and wash the aqueous layer with two 3-cm³ portions of pure carbon tetrachloride. Add 3 cm³ of sodium hypochlorite solution to the aqueous layer, shake for 1 minute, blow out any evolved chlorine and again shake. Extract the solution with two 3-cm³ portions of carbon tetrachloride and discard the extracts.

Add 20 cm³ of buffer solution and 5 cm³ of EDTA solution, mix well and add 5 cm³ of carbon tetrachloride and 1 cm³ of dithizone solution. Shake vigorously and separate the organic layer into a clean 100-cm³ separating funnel. Extract the aqueous layer once more with 5 cm³ of carbon tetrachloride and a few drops of dithizone and add the extract to the first. Remove excess dithizone by shaking twice with 15-cm³ portions of ammonium hydroxide solution.

Dry the carbon tetrachloride extract by running through about 2 g of anhydrous sodium sulphate into a small (25-cm³) volumetric flask of low actinic glass, dilute to volume with pure, dry carbon tetrachloride and measure the transmittance at 490 nm.

Prepare a calibration curve by adding mercury(II) solutions covering the range 0–25 μg Hg to 50-cm³ portions of 0·1 M hydrochloric acid. Add 5 cm³ of hydroxylamine hydrochloride solution, 20 cm³ of buffer solution and 2 cm³ of EDTA solution and extract the mercury with dithizone; continue the analysis as for the test solution.

Instructions for the safe handling of the HCl gas cylinder are obtainable from the suppliers and a three-way stop tap is inserted between the reducing valve and gas scrubber. Pickard and Martin recommend using polyvinyl chloride tubing, not rubber, which, although giving high blanks when new, gives satisfactory service after having been used.

16:22 MOLYBDENUM

16:22:1 Background and theory

Molybdenum is widely distributed as molybdates; it occurs for example, in wulfenite, $PbMoO_4$, and as the sulphide it occurs as molybdenite, MoS_2, an accessory mineral in granites.

Increasing acidity mobilizes molybdenum and Ng and Bloomfield (1962) showed that molybdenum in soils is mobilized anaerobically when the soil is flooded and incubated with organic matter.

Molybdenum is adsorbed by clays as MoO_4, the adsorption being maximum at pH 2, and kaolin ceases to adsorb the element if the pH rises above 8 (Gallego and Jolin, 1960). Adsorption reactions between molybdate and soils can be predicted from the Freundlich equation provided that it is modified to include a pH term (Reisenauer *et al.*, 1962). Reisenauer found the equation applicable over a wide range of soils and pH values covering the physiologically important range of molybdenum content. Analogous patterns of reaction were obtained with hydrous oxides of aluminium, iron and titanium. In the case of absorption of molybdate on $Fe_2O_3 \cdot xH_2O$, it was accompanied by stoichiometric release of one H—OH and two hydroxy groups. Such conditions are satisfied by sodium molybdate reacting with iron(III) hydroxide to yield $Fe(MoO_4)_3$ or by chemisorption of molybdate.

As a trace nutrient element, molybdenum differs from others in that plants can develop satisfactorily in its absence provided that ammonium-nitrogen is present. The molybdenum appears to be essential for the nitrate-reducing enzyme to function and if nitrogen is supplied to a molybdenum-deficient soil exclusively as nitrate, then this accumulates as such in plants. Furthermore, legumes are unable to fix atmospheric nitrogen in the absence of molybdenum.

Karim and Deraz (1961) found that molybdenum increased plant uptake of nitrogen, potassium and calcium in mustard and induced better growth and yield than any other micro-nutrient. Amongst its other beneficial effects in soil, molybdenum has been reported as having promoted germination of maize, counteracting aluminium toxicity and increasing fruit size of citrus. Healy *et al.* (1961) found a significant incidence of dental caries associated with low molybdenum in soil.

The amount of molybdenum present in soils can, however, be critical, an excess being toxic. Wheat, for example, has been known to accumulate molybdenum and be toxic to cattle. Plants growing on cretaceous shales are likely to accumulate molybdenum and give toxic effects similar to those of selenium. In Kenya, coffee plants accumulated molybdenum and in spite of an ample magnesium supply exhibited magnesium deficiency symptoms (Chamberlain, 1956). Some of the coffee plants contained as much as 10 μg g^{-1} Mo which is above the toxic level for cattle. In another investigation Chamberlain found that lupins were accumulating molybdenum to such a degree that the soil molybdenum was liable to be exhausted.

Determination of molybdenum

For determining the total molybdenum content of a soil it is preferable to fuse the sample with sodium carbonate rather than to digest with acids as the acids must be removed prior to measurement of the molybdenum in solution.

Plant-available molybdenum has been extracted with ammonium oxalate solution at pH 3·3 (Grigg, 1953; Gupta and MacKay, 1966) and can also be measured by biological assay (Roschach, 1961), although the water-soluble form is probably the best indication of availability. Lowe and Massey (1965) considered the molybdenum extracted by hot water to be better related to plant uptake than oxalate-soluble molybdenum. Difficulty was found, however, in extracting sufficient molybdenum to give reliable analytical results. Lowe and Massey eventually extracted the soil for 10 hours with water in a soxhlet apparatus.

Spectrographic and polarographic determinations of molybdenum are convenient, a gravimetric method in which molybdenum is weighed as lead molybdate has been used (Stanfield, 1935) and several colorimetric methods exist. The classic method is that dependent upon the thiocyanate complex of molybdenum. The thiocyanate complex when reduced, for example with tin(II) chloride or ascorbic acid, gives a coloured solution, the transmission of which can be measured directly or after extraction into an organic solvent. Rhenium gives the same reaction but this can normally be ignored in soil analysis. Silica must be absent and the colour of the complex fades rapidly. Purvis and Peterson (1956) and Lowe and Massey (1965) used a procedure (Marmoy, 1939) whereby molybdenum is reacted with ammonium thiocyanate in the presence of iron(III). After reduction with tin(II) chloride the orange-coloured complex is extracted into isopropyl ether. Iron must be present for full colour development but the formation of red iron(III) thiocyanate interferes and the iron must therefore be kept in the reduced state when taking the readings. Tin(II) chloride is satisfactory for reduction if the amount of iron is small, but with large amounts of iron present, sodium sulphite must also be added prior to organic extraction of the complex (Hogan and Breen, 1963). Barshad (1949) found that for the analysis of plant material which invariably contained iron, it is more convenient to add a known excess of iron(III) chloride to test solutions and standards alike rather than to reduce the iron. The effect of iron is to intensify the colour of the complex.

Another commonly used reagent for the colorimetric determination of molybdenum is dithiol (4-methyl-1,2-dimercaptobenzene) which gives a green molybdenum complex (Williams, 1955). Bingley (1963) suggested removing interfering copper by means of thiourea; potassium iodide solution is added to the acid test solution and the liberated iodine cleared by adding ascorbic acid. Tartaric acid and thiourea are then added immediately before the dithiol reagent and the colour extracted into iso-amyl acetate. Gupta and MacKay found that the dithiol solution must be prepared immediately before use and they mixed thioglycollic acid with the reagent. A similar reagent, toluene-3,4-dithiol, was proposed by Scholl (1962).

16:22:2 Recommended methods

Total molybdenum

Fuse 1–2 g of 0·15-mm soil in a platinum crucible with sodium carbonate as described in section 16:2:41. Metal grinding machines must not be used to prepare the soil. Separate silica and evaporate the filtrate with hydrochloric acid; take up the residue in dilute hydrochloric acid. Determine the molybdenum in solution by one of the methods given.

Available molybdenum (Lowe and Massey)

Water-soluble molybdenum is recommended as the best measure of plant-available molybdenum in soils.

Extract 25 g of 2-mm soil with 125 cm³ of hot water in a soxhlet apparatus for 10 hours; the water should contain 2 drops of dibutyl citrate. Add 5 cm³ of concentrated nitric acid to the extract and concentrate to about 10 cm³. Transfer the concentrated solution to a beaker and cautiously evaporate to dryness. Rinse the flask with 10 cm³ of nitric acid mixed with 2 cm³ of perchloric acid and add the washings to the beaker. Digest the solution on a steam-bath for 6 hours and then gently evaporate to dryness. Add 2 cm³ of hydrogen peroxide and again evaporate to dryness. Take up the residue in 30 cm³ of water, boil for a few minutes, cool, add 5 cm³ of concentrated hydrochloric acid and filter. Make the filtrate to 50 cm³ and determine the molybdenum in solution.

Note: Do not use rubber stoppers and re-distil all water from borosilicate glass.

Molybdenum in solution

I / THIOCYANATE METHOD (PURVIS AND PETERSON; HOGAN AND BREEN)

Reagents

Redistilled water must be used throughout the analysis.

Ammonium thiocyanate solution, 10% w/v, aqueous.

Iron(III) chloride solution: 49 g dm⁻³ $FeCl_3.6H_2O$.

Sodium nitrate solution: 42·5 g $NaNO_3$/100 cm³.

Tin(II) chloride solution: 20 g $SnCl_2$ dissolved in 20 cm³ conc. HCl with heating, and the solution diluted to 200 cm³. Prepare daily.

Sodium sulphite solution, 20% w/v, aqueous.

Isopropyl ether: To isopropyl ether in a separating funnel add one-tenth of its volume of a mixture of equal parts of tin(II) chloride solution, water and ammonium thiocyanate solution. Shake and discard the aqueous layer. Prepare daily.

Standard molybdenum solution: Dissolve 0·150 g MoO_3 in 10 cm³ of 0·1 M NaOH solution. Make acid with HCl and dilute to 1 dm³. The solution contains 100 μg cm⁻³ Mo.

Procedure

Place 50 cm³ to 100 cm³ of test solution in a 150-cm³ separating funnel. Add 1 cm³ of iron(III) chloride solution, 1 cm³ of sodium nitrate solution, 5 cm³ of ammonium thiocyanate solution, 2 cm³ of tin(II) chloride solution and 1 cm³ of sodium sulphite solution in that order. Shake the solution well after each addition. Add 10 cm³ of isopropyl ether and shake for 1 minute. Dry the organic phase by running it through a cotton-wool plug and read its optical density at 475 nm.

Note: For total molybdenum determinations the addition of iron(III) chloride can usually be omitted as sufficient iron will be present.

A calibration curve is prepared to cover the range 0–1 $\mu g\ cm^{-3}$ Mo.

II / DITHIOL METHOD (WILLIAMS; BINGLEY; GUPTA AND MACKAY)

Reagents

Potassium iodide solution, 50% w/v, aqueous.
Ascorbic acid solution, 5% w/v, aqueous, freshly prepared.
Tartaric acid solution, 10% w/v, aqueous.
Thiourea solution, 10% w/v, aqueous, freshly prepared and filtered.
Dithiol solution: To 0·2 g of melted dithiol (m.p. 51°C) add 100 cm³ of
    1% w/v NaOH solution and keep the mixture at 51°C with stirring for
    15 minutes. Add 1·8 cm³ of thioglycollic acid.
Iso-amyl acetate.

Procedure

Take an aliquot of test solution to contain up to 1 $\mu g\ cm^{-3}$ Mo and treat with an excess of potassium iodide solution. Clear the liberated iodine by adding ascorbic acid solution dropwise. Add 1 cm³ of tartaric acid solution and 2 cm³ of thiourea solution. Add 4 cm³ of dithiol solution and allow the mixture to stand for 30 minutes. Extract the complex into iso-amyl acetate by shaking vigorously for 30 seconds with 5 cm³ of acetate. Dry the organic phase and measure the optical density at 680 nm.

The procedure determines up to 20 μg Mo in solution.

16:23 NICKEL

The principal source of nickel in soils is pentlandite (Fe, Ni)$_9$S$_8$, which occurs in basic rocks. Nickel is present in other minerals, for example garnierite, a nickel magnesium silicate, niccolite and millerite, NiS. The element is present to some extent in all soils, its concentration being highest in clays and mineral soils and lowest in organic soils such as peat. It is more commonly found in the finer fractions of soil.

Except in minute quantities, nickel is toxic to plant growth and is partly responsible for the low fertility of soils derived from ferromagnesian rocks.

Determination of nickel

Spectrographic analysis of nickel in soil is made using platinum electrodes and before its determination the element can be separated from interfering iron by chromatography.

Flame photometric and polarographic methods exist for determining nickel, although the former method is not recommended if calcium is present.

Of the colorimetric methods of determining nickel in solution, the best known is reaction with dimethylglyoxime. More recent reagents include diethyl-dithiocarbamate with which cobalt and copper can be simultaneously determined, and 1,2-cyclohexane-dione-doxime (nioxime). The reaction between nickel and nioxime is similar to that using dimethylglyoxime but the suspension is more stable and little interference is experienced from other ions (Johnson and Simmons, 1946).

Volumetrically nickel can be determined by mixing with an excess of EDTA solution, buffering to pH 9 and back-titrating the excess EDTA with standard magnesium solution (Vogel, 1962) and this method was used for soil analysis by Baker and Burns (1964).

16:24 RUBIDIUM

Rubidium occurs in soils derived from pegmatites and it acts in a soil in a manner similar to that of potassium. It is absorbed by the soil colloids and retained against leaching.

Colorimetrically rubidium can be determined in solution by formation of the silver cobaltinitrite complex. If potassium is determined separately and phosphorus is removed, rubidium can be determined by flame photometry at a wavelength of 795 nm (Whisman and Eccleston, 1955).

Rubidium is encountered in soil science mainly in connection with measurement of self-diffusion coefficients of other ions. Radioactive rubidium, ^{86}Rb, is adsorbed on to a soil which is then placed in contact with a similar soil section containing no tracer. The self-diffusion coefficient is calculated from the amount of radioactive rubidium diffusing in unit time (Schofield and Graham-Bryce, 1960).

16:25 SELENIUM

16:25:1 Background and theory

Selenium occurs as zorgite, crockesite (Cu, Tl, Ag)$_2$Se, in pyrites and as metallic selenides. On weathering, selenides are converted into selenites which are adsorbed on to hydrous iron oxides (Goldschmidt, 1954); selenium can thus be expected to concentrate in clays and iron oxides.

There is mounting evidence that selenium is an essential micro-nutrient for plant growth (Shrift, 1964) but if it is present to excess in soils, and this is usually the case only in soils of cretaceous shales in semi-arid regions, it is toxic either to plants or to animals feeding on the plants. Selenium toxicity is the cause of 'alkali disease' of cattle and it has been found that quantities of selenium as low as 1 μg g^{-1} in the soil, although permitting normal plant growth, result in grain being toxic to cattle (Hardy, 1948). Plants usually show toxicity symptoms if selenium in the soil exceeds 15 μg g^{-1}. Hamilton and Beath (1963) studied the uptake of selenium by many different plants and found that most plants can absorb, metabolize and store selenium in their tissues. All the plants studied could absorb selenium as the selenate or as organic forms and the organic selenium was in equilibrium with the inorganic selenium in plants. Some plants, for example *Neptunia amplexicaulis,* can accumulate selenium and studies using radioactive selenium, ^{75}Se, showed that although the element is converted in most plants to seleno-amino-acids, in *N. amplexicaulis* such conversion does not take place.

Shrift (1964) presented a tentative scheme for a selenium cycle in nature. Selenium accumulator plants can biosynthesize organic selenium compounds, such as selenium methylselenocysteine, and these are released into the soil when the plant decays. The fact that the selenium in soils is predominantly inorganic is considered by Shrift to be due to conversion of organic selenium compounds by micro-organisms.

Sulphate, by acting antagonistically, inhibits selenium toxicity and thus gypsum application can reduce plant uptake of selenium although this has been reported to be not the case for every soil. Iron(III) oxides also reduce plant uptake of selenium by forming insoluble basic iron(III) selenite.

Determination of selenium

There is no simple, reliable spectrographic procedure for determining selenium. Robinson *et al.* (1934) described a method of isolating selenium from soils by distillation as the tribromide. The resulting solution was reduced with hydroxylamine and analyzed colorimetrically. Klein (1941) first treated samples with nitric and sulphuric acids and then converted the selenium present to the tetrabromide with a mixture of hydrogen bromide and bromine. Distillation into hydrobromic acid gave a solution of selenious acid. Elemental selenium was precipitated with sulphur dioxide and hydroxylamine hydrochloride, redissolved and again oxidized with bromine water to selenious acid. The selenious acid was estimated by a thiosulphate–iodine titration using the procedure of Norris and Fay (1896). Klein reported an accuracy of 0·1 mg selenium with selenium contents less than 50 mg and for larger quantities of selenium the accuracy was of the order of 99%.

Plant-available selenium is estimated as water-soluble selenium, the soil sample being refluxed with water for 30 minutes and an aliquot of filtrate being concentrated by evaporation for analysis.

Selenium in solution has been estimated colorimetrically by utilizing the colour produced on reduction of selenium compounds to the elemental form (e.g. Robinson *et al.*, 1934). Lambert and co-workers (1951) converted selenium in solution to selenious acid with bromine and hydrobromic acid and destroyed excess bromine with nitrous acid in an atmosphere of carbon dioxide; excess nitrous acid was in turn destroyed with urea. Interfering iron was oxidized to the trivalent state and then converted to a colourless complex with phosphoric and tartaric acids; formic acid was added to prevent interference from arsenic. The selenious acid was used to liberate iodine from cadmium iodide and the iodine determined. Vogel (1962) describes a method in which selenious acid is reacted with potassium permanganate for its estimation.

A colorimetric method for determining selenium has been described by Hoste and Gillis (1955) using 3,3'-diaminobenzidine with which selenium gives a yellow-coloured diphenylpiazselenol. The method was modified by Cheng (1956) who employed EDTA to mask the effect of polyvalent metals. Strong oxidizing or reducing agents must be absent and excess vanadium interferes. The complex is extracted into toluene at pH 7 and the extinction of the solution measured at 420 nm; Beer's Law is followed over a range of 0.5 μg to 2.6 μg cm^{-3} Se.

16:25:2 Recommended methods

Total selenium (Robinson; Klein; Fine)

Reagents

Nitric–sulphuric acid mixture: Add 1 volume conc. H_2SO_4 to 2 volumes conc. HNO_3.
Mercury(II) oxide solution: 5% w/v in conc. HNO_3.
Bromine–hydrobromic acid (conc.): Mix 15 cm^3 Br$_2$ with 985 cm^3 of constant boiling HBr.
Bromine–hydrobromic acid (dil.) Add 10 cm^3 of saturated bromine water to a mixture of 85 cm^3 water and 5 cm^3 of HBr.
Sulphur dioxide: From siphon.
Hydroxylamine hydrochloride solution: 10% w/v, aqueous.
Phenol solution: 5% w/v, aqueous.

Procedure

Weigh 10 g of 0.5-mm (40-mesh) air-dry soil into a 250-cm^3 round-bottom flask. Add a few boiling chips, 150 cm^3 of the acid mixture and 10 cm^3 of mercury(II) oxide solution. Heat the flask on a steam-bath in a fume chamber for 30 minutes with occasional shaking. If necessary (e.g. for very organic soils) extra nitric acid is added to complete oxidation. Gently heat the flask to expel nitric acid and until dense white fumes of sulphur trioxide appear.

Cool the digest and dilute with 50 cm^3 of water. After again cooling add 25 cm^3 of concentrated bromine–hydrobromic acid mixture and connect the flask to a vertical condenser for distillation. The delivery tube should dip below the surface of 5 cm^3 of the dilute bromine–hydrobromic acid mixture contained in a 125-cm^3 conical flask cooled in ice.

Gently heat the flask until the selenium bromide and most of the hydrobromic acid has distilled over. At the beginning of the distillation free bromine should appear in the receiving flask indicating complete oxidation of organic matter. Quickly cool the digestion flask and add a further 10 cm^3 of concentrated bromine–hydrobromic acid mixture. Con-

tinue the distillation until all the hydrobromic acid has passed over as indicated by bubbling of the sulphuric acid.

Filter the distillate through a fine sintered glass crucible and wash the
crucible with two small portions of water. Saturate the filtrate with sulphur
dioxide and add 2 cm³ of hydroxylamine hydrochloride solution. Cover the
neck of the flask with a small watch glass, or loosely stopper with a stem-
med glass bulb and warm on a steam-bath for 30 minutes. Cool the flask
and leave overnight for the precipitated selenium to settle. Filter off the
selenium through a pad of asbestos in a sintered glass crucible and wash
the precipitate several times with water.

Re-dissolve the selenium by adding 4 cm³ of dilute bromine–hydrobromic acid mixture to the vessel used for precipitation and thence to the
crucible. After 1–2 minutes, slowly suck the liquid through the crucible into a vial. Repeat with a further 2 cm³ of dilute bromine–hydrobromic acid
mixture; if much selenium is present it may be necessary to use additional
amounts of solvent. Wash the asbestos pad with water; the final filtrate
should not exceed 25 cm³.

Add three drops of phenol solution to the filtrate and stir to dispel
excess bromine, then immerse the vial in hot water for 5 minutes to complete the elimination of bromine. The selenium in solution is then measured
by one of the methods given.

Extraction of water-soluble selenium

Boil 100 g of 2-mm, air-dry soil under reflux with 500 cm³ of water for
30 minutes. Filter and concentrate 400 cm³ of filtrate to small volume. The
final solution, or suitable aliquots from it, are analyzed for selenium by the
colorimetric method described in section 16:25:2ɪɪ.

Selenium in solution

I / IODOMETRIC METHOD (NORRIS AND FAY; KLEIN; FINE)

Apparatus

Klein recommended making the titration in a small (25 cm³) Nessler-like
tube having a narrow (1 cm diameter) lower portion of capacity 5 cm³. The
lower part of the tube is painted black and the colour is then viewed down
the length of the tube against a background of unglazed white opal glass.

Reagents

Sodium thiosulphate solution: Dissolve 24·82 g of sodium thiosulphate in
water and dilute to 1 dm³ to give a solution 0·1 M. Store in a brown
glass bottle. Immediately before use dilute an aliquot to give a solution
0·005 M.

Iodine solution: Dissolve 25 g of KI in about 600 cm³ of water and add
6·350 g of iodine. When solution is complete dilute to 1 dm³ giving a

solution 0·05 M. Store in a brown glass bottle and immediately before use dilute an aliquot to give a solution 0·005 M.

Starch indicator solution: 0·5% w/v, aqueous.

Standard selenium solution: Dissolve 2·190 g of pure Na_2SeO_3 (dried at 150°C) in 80 cm^3 of 48% HBr and dilute to 1 dm^3. The solution contains 1000 μg cm^{-3} Se.

Procedure

To duplicate 10-cm^3 portions of water, add 2 cm^3 of iodine solution and 3 drops of starch solution. Titrate with thiosulphate to obtain the thiosulphate equivalent of the iodine solution.

Dilute 10 cm^3 of the standard selenium solution to 1 dm^3 giving a solution containing 10 μg cm^{-3} Se. Add 2 cm^3 of dilute bromine–hydrobromic acid mixture (see 'Total selenium') to 5-cm^3 aliquots of the dilute standard. Add sulphurous acid dropwise until the yellow colour has nearly disappeared and finally decolorize the solution with 3 drops of phenol solution (see 'Total selenium'). Add 5 drops of starch solution and 2 cm^3 of 0·005 M thiosulphate solution. Add 0·005 M iodine solution to slight excess as shown by a blue colour. If less than 1 cm^3 of iodine solution is required, add sufficient thiosulphate so that at least 1 cm^3 of iodine is needed. Titrate with 0·005 M thiosulphate until the blue colour just disappears. The total volume of thiosulphate and the volume of iodine solution added are noted. From the total volume of thiosulphate used, subtract the thiosulphate equivalent of the iodine added to obtain the volume of thiosulphate used in reaction. Then mg of selenium present (i.e. 0·05 mg for a 5-cm^3 aliquot) are divided by the thiosulphate used to give the selenium equivalent in mg cm^{-3} of the thiosulphate.

Suitable aliquots of the test solution are taken and 5 drops of starch solution added. Add an excess of about 50% thiosulphate (0·005 M) estimated from the approximate selenium content and the fact that 1 cm^3 of 0·005 M thiosulphate is equivalent to about 0·1 mg selenium. Then add a slight excess of iodine and titrate with the thiosulphate solution. Subtract the thiosulphate equivalent of the iodine used from the total amount added and multiply by the selenium equivalent to obtain the mg Se in the aliquot of test solution.

II / 3,3'-DIAMINOBENZIDINE METHOD (CHENG)

Reagents

3, 3'-Diaminobenzidine solution: 0·5% w/v of the hydrochloride in water. Store in a refrigerator.

Formic acid, 2·5 M: 140 cm^3 dm^{-3} anhyd. acid.

EDTA solution, 0·1 M: 38 g dm^{-3} (di-Na salt).

Standard selenium solution: As for iodometric method.

Toluene.

Procedure

Pipette an aliquot of test solution, containing up to 50 µg Se and in which strong oxidizing or reducing substances are absent, into a 100-cm³ beaker. Add 2 cm³ of 2·5 M formic acid and dilute to 50 cm³. Adjust the pH to 2–3 if necessary. Add 2 cm³ of the 3,3′-diaminobenzidine reagent and allow to stand for 30 minutes. Add 10 cm³ of EDTA solution to suppress heavy metals and adjust the pH to 6–7 with ammonium hydroxide solution. Transfer the solution to a 125-cm³ separating funnel and add 10 cm³ of toluene. Shake for 30 seconds and measure the extinction of the toluene layer at 420 nm.

Beer's Law is followed over the range 0–2·5 µg cm^{-3} Se in toluene. A reagent blank must be carried out each time.

16:26 SILVER

Argentite, Ag_2S, is the best known silver-containing mineral and is associated with native silver, galena, sphalerite and ores of copper. In soils silver is generally toxic to plant growth but has been known to accumulate in mushrooms.

Determination of silver

Silver is extracted from soil by acid digestion and is classically determined nephelometrically as the chloride. A colorimetric method is to combine the silver with dithizone having separated it from lead, zinc, cobalt and other ions by complexing with EDTA. Any copper present is then separated by decomposing the dithizonate with chloride and re-extracting with dithizone at pH 3. The transmittance of silver dithizonate in carbon tetrachloride is read at 460 nm.

16:27 STRONTIUM

Strontium occurs as the sulphate in celestite and as the carbonate in strontianite. In small quantities it is widely distributed in soils and reacts with soil colloids in a manner similar to that of barium.

The sorption of strontium by soil increases with increasing strontium concentration and increasing pH up to a value of 5; this is due to a decrease in activity of hydrogen and aluminium ions. Above pH 5 strontium absorption by soil is little affected by changes in soil reaction. The absorbed strontium is converted to a non-exchangeable form, the proportion of which increases with organic matter content and cation exchange capacity.

Strontium is absorbed in the upper layers of soil and plant uptake depends upon rooting depth; the element is not essential for plant growth and can be toxic. Sandy soils permit the greatest uptake of strontium which can be decreased by adding calcium to the soil. In calcareous soils strontium

15+T.S.C.A.

exchange reaches a constant equilibrium, and as it is not retained any more than is calcium, information regarding its movement in soil can be obtained by inference by measuring calcium. This procedure has been used to predict long term consequences of accumulation in soils of radioactive strontium, ^{90}Sr. The ^{90}Sr content of oats has been found to be highly correlated with the reciprocal of exchangeable calcium.

The contamination of agricultural land with fission products, which include ^{90}Sr, and the consequent contamination of plants, is a serious possibility and the amelioration of such soils has recently received much attention. Nishita and co-workers (1961) showed that ^{90}Sr is taken up by plants to a much greater extent than are other fission products. Deposition of radioactive strontium was found to be correlated with the rainfall of areas concerned. The isotope can be removed from soil by leaching with EDTA solution although subsequent fertilization with regard to nutrient elements would then be necessary. Deep ploughing and liming of acid soils both reduce strontium uptake (Milbourn, 1960; Romney *et al.*, 1959; Andersen, 1963). Gregers-Hansen (1964) found that heat treatment of soil, either alone or combined with addition of dihydrogen phosphate, reduced strontium extractability from soils.

Hempler (1963) used ^{90}Sr to examine the availability of strontium to plants and he stressed the importance of the discrimination factor:

$$\frac{\text{(Ca in soil solution)} \times \text{(Sr in plant)} \times \text{(Sr in soil solution)}}{\text{(Ca in plant)}}$$

for predicting harmful effects of ^{90}Sr.

The self-diffusion coefficient of strontium in soils was found by Graham-Bryce (1963) to be one-tenth that of the monovalent ions potassium and rubidium.

Determination of strontium

Total strontium in soil has been measured by neutron radioactivation although a spectrographic procedure is generally employed. Available strontium has been extracted from soil with ammonium acetate solution and with ammonium chloride solution and measured with an atomic absorption apparatus at a wavelength of 461 nm (4607 Å) (David, 1962).

Flame photometry can be used to measure strontium in solution (Smales, 1951) but it is best measured at a wavelength of 461 nm, rather than at 675 nm as recommended, as there is less interference from calcium (Whisman and Eccleston, 1955).

16:28 THALLIUM

Thallium occurs principally in the mineral crooksite in which it is associated with selenium, copper and silver and in lorandite, $TlAsS_2$.

Thallium is often added to soils as a constituent of pesticides and is very toxic to plants and animals. The toxic effects are greatest in sandy soils where contents of about 30 μg g^{-1} will completely inhibit plant growth.

Determination of thallium

The classical method of determining thallium is by reduction to the thallium(I) ion which is then employed to liberate iodine from iodides. Iron interferes with the method but can be suppressed (Akeman, 1948).

Colorimetric methods for estimating thallium in solution include reduction of phosphotungstomolybdic acid to molybdenum blue, formation of the brown sulphide and with 8-hydroxquinoline. All these methods need modification to eliminate interference from other ions normally found in soils.

16:29 TIN

Tin as cassiterite, SnO_2, occurs in small amounts in association with granitic rocks.

Tin can be brought into solution from soil by alkali fusion or by acid digestion, and once in solution, can be estimated by several colorimetric methods. With dithiol a red-coloured complex is formed but the reaction is subject to interference from other ions such as copper, iron and cobalt; this necessitates separation of the tin by distillation. McDonald and Stanton (1962) have described a procedure for determining tin in soil in which the element is complexed with pyrogallolphthalein (gallein). In this method the soil is heated with ammonium iodide until sublimation ceases and the residue is leached with dilute hydrochloric acid. The iodine is reduced with hydrazine hydrate and the tin reacted with gallein in the presence of a chloroacetate buffer. The complex is extracted into 1-pentanol and the optical density of the solution measured at 500 nm. The method was used to determine from 0·1 to 2·0 μg cm^{-3} tin in solution.

16:30 TUNGSTEN

16:30:1 Background and theory

Tungsten is found in the minerals wolframite (FeMn)WO_4, and scheelite, $CaWO_4$, both of which are associated with granites.

Tungsten has not been found essential for plant growth but it may partly replace molybdenum if that element is deficient (Bowen and Cawse, 1962) and it has been shown to increase the fixation of atmospheric nitrogen by legumes.

Determination of tungsten

Spectrographic methods are not sufficiently sensitive for the determination of tungsten (Ward, 1951).

The total tungsten in soil is best extracted by alkali fusion. Early methods of determining tungsten in solution depended upon the yellow colour produced when an alkaline solution of tungstate is treated with thiocyanate, acidified and reduced. North (1956) detected down to 4 μg cm^{-3} tungsten with a blue-green dithiol complex extracted into iso-amyl acetate. The dithiol reagent as used by North was found by Bowden (1964) to be unstable and to give an unsatisfactory, high blank. Bowden found that by using a mixture of white spirit and ethanol as solvent, the standards were stable for 48 hours.

The soil sample is fused with a mixture of sodium carbonate, potassium nitrate and sodium chloride and the melt extracted with water. An aliquot treated with tin(II) chloride and the reagent produces a blue-green colour which is extracted into the organic phase and the extinction of the solution is measured at 600 nm. The sensitivity of the method is 2 μg cm^{-3}.

16:30:2 Recommended method (Bowden)

Reagents

Standard tungsten solution: 0·09 g sodium tungstate, $Na_2WO_4.2H_2O$, is dissolved in water and made to 500 cm^3. The solution contains 100 μg cm^{-3} W and is diluted as necessary.

Flux: Grind together 5 parts Na_2CO_3, 1 part KNO_3 and 4 parts NaCl and pass through a 0·18-mm (80-mesh) sieve.

Tin(II) chloride solution: Dissolve 10 g $SnCl_2$ in conc. HCl and dilute with more HCl to 500 cm^3.

Solvent: Mix 50 cm^3 white spirit with 50 cm^3 of absolute ethanol.

Toluene 3,4-dithiol solution: 0·5% w/v. Dissolve 20 g NaOH in 500 cm^3 of water and cool. Break a 5-g vial of reagent under the surface of the alkali and gently warm if necessary to effect solution. Add 10 cm^3 thioglycollic acid and dilute to 1 dm^3. Store in a polythene bottle.

Procedure

Mix 0·25 g of 0·18-mm (80-mesh) soil with 1·25 g of flux in a thick-walled, Pyrex test-tube. Fuse for 5 minutes, cool and add water to 5 cm^3. Digest the melt in a boiling water-bath for 30 minutes, readjust the volume to 5 cm^3, shake the tube and allow to stand for 3 hours. Transfer 0·5 cm^3 of the supernatant liquid to a dry 16-mm × 150-mm test-tube and add 5 cm^3 of tin(II) chloride solution. Heat the tube in a boiling water-bath for 2 minutes and add 1 cm^3 of reagent; heat for a further 10 minutes with occasional shaking. Cool and shake the stoppered tube for 30 seconds with 5 cm^3 of solvent added. Read the optical density of the dried organic phase at 600 nm.

The calibration curve should cover the range 0–1 μg cm^{-3} W.

16:31 URANIUM

The occurrence of uranium in minerals is not common, although even small amounts may be of importance. The principal uranium minerals are uraninite, UO_2, and carnotite, $K_2(UO_2)_2(VO_4)_2 \cdot 3H_2O$.

Paper chromatography is widely used to measure uranium in solution and Thompson and Lakin (1957) have developed this procedure for determining up to 20 μg cm^{-3} uranium. Purushottam (1960) slightly modified the method by using a mixture of citric acid, acetone, acetic acid and nitric acid as solvent. Zagran and Vlasov (1962) used complexes of uranium with sulphate and chloride to separate from 10^{-3} to 10^{-6} g of uranium from soil.

The most practical colorimetric procedure for determining uranium in solution depends upon the absorbance peak of the uranium(IV) ion at 660 nm. Vanadium is conveniently measured in the same solution at 700 nm (Canning and Dixon, 1955). The uranium and vanadium are reduced with iron(II) sulphate in phosphoric acid and the method is subject to little interference from other ions.

16:32 VANADIUM

Occurrence of vanadium in soils

Vanadium occurs with uranium in carnotite and also in vanadinite, $Pb_5Cl(VO_4)_3$, patronite and roscoelite, $K_2V_4Al_2Si_6O_{20}(OH)_4$. It is commonly found in bituminous deposits.

In soils vanadium is known to occur as the vanadate of copper, zinc, lead, iron, manganese, calcium and potassium, and it can replace aluminium in clays. Nakamura and Sherman (1961) found very high vanadium contents in humic ferruginous latosols in Hawaii, and in the surface soils of that island vanadium varied in concentration from 190 to 1520 μg g^{-1}, the average concentration being 450 μg g^{-1}. Nakamura and Sherman considered that vanadium may have been concentrated in the soil by biochemical processes.

Small amounts of vanadium are stimulating to plant growth but if present in amounts greater than about 10 μg g^{-1}, the element is usually toxic. Legumes utilize vanadium in nitrogen fixation and can tolerate higher amounts than can other plants. A few species of plants can accumulate vanadium, particularly if they absorb much calcium as well, when the vanadium is precipitated in the roots (Cannon, 1963).

Determination

Vanadium in solution can be determined colorimetrically together with uranium after reduction with iron(II) sulphate (section 16:31). With hydrogen peroxide, vanadium reacts in acid solution to give peroxyvanadic acid and a reddish-brown colour which can be estimated by its extinction at 460 nm (Wright and Mellon, 1937), although a more sensitive reading can be obtained at 290 nm in the ultra-violet (Telep and Boltz, 1951).

16:33 YTTRIUM

Yttrium is deposited in soils as ^{91}Y, a fission product. Very little investigation has been made into its effect upon soil except that it has been shown that chelates increase its uptake by plants.

Although usually measured spectrographically, yttrium can be determined colorimetrically with hematoxylin (Sarma and Raghava-Rao, 1955) with which it forms a lake. As the lake is stable only at pH 6·0–6·5, the reaction of the test solution must be carefully controlled. Lanthanum forms a similar lake but the elements are differentiated by measuring the optical density at two wavelengths, 600 nm and 650 nm, and applying a simple calculation.

16:34 ZINC

16:34:1 Background and theory

Zinc is widely distributed as the sulphide sphalerite, the carbonate smithsonite and the silicates calamine and hemimorphite, and is associated with pyrites. In Bulgaria a marked accumulation of zinc was found in B-horizons of some podsolized soils.

Zinc is typically deficient in peat soils and twenty of the USA states have been reported as zinc deficient. An excess of copper can induce zinc deficiency and this should be considered when applying copper to peat soils during their reclamation. In Hawaii, Kanehiro (1964) found that zinc-deficient soils were better correlated with acid-extractable zinc contents than with total zinc contents and that such soils were associated with intensely weathered latosols or where subsoils had been exposed. Lucerne is a strong zinc-extracting plant and hence is a good cover crop for zinc-deficient soils.

Shortage of zinc in a soil leads to a decrease in plant protein synthesis and a decrease in uptake of phosphorus and nitrogen; manganese uptake on the other hand increases. Bertrand and Wolf (1961) found that adequate zinc is absolutely necessary for the synthesis of alanine, glycine, proline, threonine, serine, valine, leucine, aspartic acid and glutamic acid, and that it has a specific role in the synthesis of tyrosine, tryptophan and phenylalanine.

It has been shown that zinc increases the fibre strength of cotton and that if cotton seeds are soaked in a zinc solution before sowing, or if zinc is applied to the roots of cotton seedlings, the zinc is translocated to actively growing parts of the plant.

Stewart and Leonard (1963) found that a zinc–EDTA complex was an excellent source of zinc for plants if mixed with a large quantity of soda-ash. Zinc sulphate supplied available zinc provided that it was mixed with calcium chloride or, to a lesser extent, with other salts. Laboratory studies using ^{65}Zn showed that calcium does not replace exchangeable zinc in soils. Studies on zinc availability using corn as an indicator plant showed that the availability may be influenced by soil temperature effects on microbial activity (Bauer and Lindsay, 1963).

An excess of zinc in soil suppresses phosphorus uptake by plants and can cause leaf chlorosis. Brown and co-workers (1962), using sweet corn as an indicator plant in different soils, found a critical level of response to applied zinc sulphate at about 0·5 μg g^{-1} of dithizone-extractable zinc in the soil. About 35% of the zinc remained in the soils after harvesting and showed a residual effect for several years. There appeared to be no downward movement of zinc through the soil.

Bingham *et al.* (1963) investigated the retention of zinc by montmorillonite using the chloride, nitrate, sulphate and acetate salts. Bingham found that when the pH was such that the solubility product of zinc hydroxide was not exceeded, the amounts of zinc retained by the chloride sulphate and nitrate systems were more or less equal and the amount of zinc absorbed was equal to the cation exchange capacity of the montmorillonite as determined with the ammonium ion. In the acetate system, although the pH of the equilibrium solution was below that favouring formation of the hydroxide, zinc was retained in amount in excess of the cation exchange capacity. The retention of zinc acetate was shown to be related to the concentration of undissociated acetic acid rather than to the amount of metal absorbed. Similar results were obtained for copper retention.

Total zinc

Spectrographic determination of zinc in soils is not particularly satisfactory as the element is observed only when present in amounts which would be abnormal in soil. Pometun and Boyarova (1961) distilled zinc from a small electrode crater containing 200 mg of soil mixed with 200 mg of carbon plus 10% cadmium iodide. Using this technique Pometun and Boyarova spectrographically measured zinc in soil from 0·002% to 0·01% with an error less than ±8%.

Total zinc can be brought into solution from soil by fusion with sodium carbonate or potassium pyrosulphate or by digestion with a mixture of perchloric and hydrofluoric acids. In all cases a reagent blank is essential.

Available zinc

Plant-available zinc was extracted by Eve (1955) by shaking 5 g of soil with 50 cm^3 of 0·1 M hydrochloric acid for 45 minutes. The same method was used by Wear and Sommer (1948) who correlated their results with zinc deficiencies. Other extractants used with success, considering specific soils and crops, have been acidified potassium chloride (Hibbard, 1940), acidified ammonium acetate (Lyman and Dean, 1942) and dilute magnesium sulphate solution acidified so as to cause the pH of the extract to correspond to that of the soil (Bergh, 1948). Jackson (1958) recommends extracting the soil directly with dithizone in carbon tetrachloride solution in the presence of ammonium acetate. This serves to remove any zinc dissolved from the soil by the acetate into the organic phase. Nelson *et al.* (1959) used 0·1 M hydrochloric acid and extractions were made with successive portions of acid until the pH of the extract remained unaltered.

Stewart and Berger (1965) considered that as extraction of available zinc is governed by pH a neutral salt solution should be used. Stewart and Berger chose 1 M magnesium chloride solution as the magnesium ion has a similar ionic charge to the zinc ion. 10 g of soil were shaken for 45 minutes with 50 cm^3 of the extractant and then centrifuged. A 25-cm^3 aliquot of the extract was treated in a separating funnel with 10 cm^3 of neutral, 1 M ammonium acetate solution and 25 cm^3 of a 0·01% solution of dithizone in carbon tetrachloride; the organic phase was separated for analysis.

Martens and co-workers (1966) compared 0·1 M hydrochloric acid, dithizone and magnesium sulphate solution as extractants of available zinc with total zinc determination, and a biological method using *Aspergillus niger*. Martens concluded that the biological method and determination of total zinc were of equal importance for estimating available zinc. Some of the zinc extracted by the dilute hydrochloric acid proved to be unavailable to plants.

Zinc in solution

Zinc in solution is conveniently measured polarographically. Iron(III) must be absent and this is managed by a dithizone extraction of the zinc. Cobalt, which is extracted also, does not normally interfere with the analysis, and the copper and nickel extracted can be measured polarographically in the same sample as they have well separated waves.

Zinc has been measured gravimetrically by precipitation from soil extracts with 8-hydroxyquinoline. Various colorimetric procedures have been developed for determining zinc. Bogg and Alben (1936) precipitated zinc from hydrochloric acid solution with hydrogen sulphide, redissolved the precipitate and reacted the zinc with potassium hexacyanoferrate(II). Zinc forms a blue-coloured complex with 'zincon', 2-carboxy-2'-hydroxy-5'-sulphoformazylbenzene (Rush and Yoe, 1954). Interfering elements are removed on exchange resins (e.g. Pratt and Bradford, 1958) and the optical density of the zinc complex solution is read at 600 nm at pH values of 9 and 5·2. The reading at pH 9 includes zinc and copper whereas the reading at pH 5·2 does not include zinc which is then obtained by difference.

The usual procedure for measuring zinc colorimetrically depends upon the reaction of zinc with dithizone (diphenylthiocarbazone) when a red-coloured complex is formed. Interfering ions are removed by pH control and complexing reagents. Verdier *et al.* (1957) separated zinc, lead and copper as dithizonates at pH 9 in a citrate buffer. The lead and zinc were then re-extracted into acid solution.

The first step in the dithizone separation is the extraction of dithizone complexing metals such as zinc, copper, cobalt and nickel from others such as iron and aluminium; this is done in a citrate buffer at high pH. Copper and zinc are then separated under acid conditions and the copper complexed with carbamate before measuring the zinc. Extraction is usually done with mechanical shaking, and Boawn (1962) has described a suitable apparatus for this.

Hubbard and Scott (1943) developed a method of determining zinc using di-β-naphthylthiocarbazone which is more sensitive than dithizone and interfering elements are removed with sodium diethyl-dithiocarbamate. The results obtained by Hubbard and Scott agreed well with those obtained by polarographic analysis.

16:34:2 Recommended method (Holmes; Sandell)

Reagents

Standard zinc solution: Dissolve 0·1 g of pure metallic zinc in 50 cm^3 of water containing 1 cm^3 of conc. H$_2$SO$_4$. Dilute to 1 dm^3, giving a solution containing 100 μg cm^{-3} Zn.

Ammonium citrate buffer solution.
Ammonium hydroxide.
Dithizone solution. }As in section 16:13:2ɪ.
Carbon tetrachloride.
Carbamate solution.

Procedure

Follow the procedure described in section 16:14:2ɪ for the determination of copper up to the separation of the first aqueous phase. Add 5 cm^3 of ammonium citrate buffer solution to the aqueous phase and adjust the pH to 8·3 with ammonium hydroxide solution, using phenolphthalein as indicator. If necessary the precipitation of iron and aluminium can be prevented by adding more buffer solution. Extract the zinc with two 10-cm^3 portions of dithizone solution; the aqueous phase becomes orange in colour when all the zinc has been extracted. Extract the aqueous phase once more, this time with pure carbon tetrachloride.

Discard the aqueous phase and place the combined organic phases into a clean separating funnel. Shake for 2 minutes with 50 cm^3 of 0·02 M hydrochloric acid, discard the organic phase and wash the aqueous layer once with carbon tetrachloride. Add 5 cm^3 of citrate buffer solution and raise the pH to 8·3 with ammonium hydroxide. Add 10 cm^3 of dithizone solution and 10 cm^3 of carbamate solution and shake the mixture for 2 minutes. Transfer the zinc dithizonate phase to a clean funnel and shake with 25 cm^3 of 0·01 M ammonium hydroxide solution.

15*

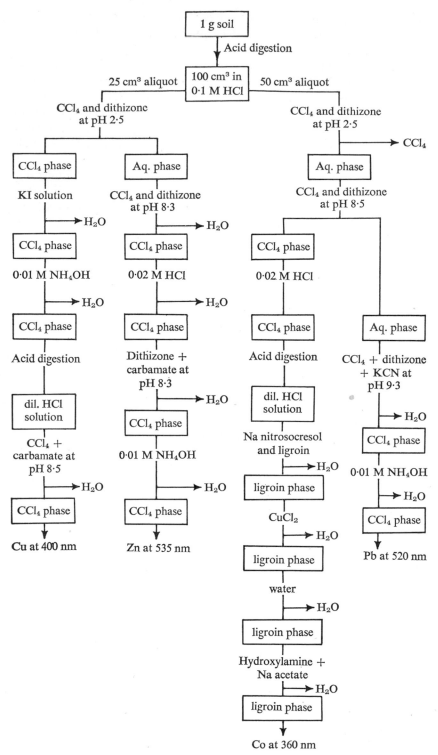

Figure 16:4 Flow-sheet for the determination of cobalt, copper, lead and zinc in soil

Remove 5 cm³ of zinc dithizonate solution by inserting a pipette through the aqueous phase and read the optical density at 535 nm. A blank determination is essential and the calibration curve should cover the range 0–3 μg cm^{-3} Zn. The accuracy of the method is of the order 98–99%.

In Figure 16:4 is shown a flow-sheet summarizing the analysis of a soil sample for cobalt, copper, lead and zinc by the method of Sandell as modified by Holmes.

16:35 ZIRCONIUM

16:35:1 Background and theory

Zirconium occurs naturally as the silicate zircon, $ZrSiO_4$, which is some-times found in alluvial sands and is common in limestone and gneiss. It is also found as the oxide in baddeleyite and as a mixture of the oxide and silicate in zirkite.

The mineral zircon is extremely resistant to weathering and the principal use of zirconium estimations in soil has been as a reference in studies of weathering processes. However, it has been pointed out by Le Riche and Weir (1963) that whereas determination of zircon in sands is a satisfactory measure of weathering, the determination of total zirco-nium in soil cannot be so regarded. This is because a high proportion of the total zirconium in soils has been found in clays and extractable oxides and can thus be translocated.

Alexander and co-workers (1962) measured zirconium by X-ray emission spectrography and found it to be twice as high in concentration in loess as in glacial till. Alexander considered that the analysis provides a means of distinguishing between the two materials.

Determination of zirconium

Willard and Hahn (1949) measured zirconium in soils gravimetrically as the silicate after fusing the soil with pyrosulphate in a quartz crucible, precipitating the zirconium and igniting. Petretic (1951) employed sodium peroxide for the fusion.

Mortimore and Romans (1952) effectively measured zirconium by X-ray fluorescence. The soil was freed from organic matter with hydrogen peroxide, dried and ground; the zirconium was then excited by WX-radiation and its fluorescence spectrum obtained with a rock salt crystal. For quantitative estimation the intensity of ZrK-radiation at 15·9 degrees of 2-theta was used.

Zirconium forms a red lake with alizarin red-S and this reaction has been developed as a colorimetric method for determining zirconium in clay samples by Green (1948). Green found that a simple sodium carbonate fusion of clay did not recover all the zirconium present and he recom-mended a double fusion with sodium hydroxide. The melt of the first fusion was digested with hot water, filtered and the filtrate discarded. The residue was digested with hydrochloric acid, the solution was filtered into a beaker and cautiously neutralized (at room temperature to avoid precipitation of zirconyl hydroxide) with sodium hydroxide solution. After just acidifying with hydrochloric acid, the solution was passed through a silver reductor to eliminate iron interference and was then treated with sodium alizarin sulphonate. The optical density of the solution was measured at 520 nm. Zirconium remaining on the filter paper was recovered by igniting the paper and fusing the residue with sodium hydroxide (the second fusion). Green used the method to determine zirconium up to 0·275 mg with an accuracy of 0·003 mg.

Grimaldi and White (1953) measured zirconium in clay samples by isolating it from interfering metals by precipitation with *p*-dimethylamino-azophenolarsonic acid in 1:1 hydrochloric acid and then developing the yellow-coloured complex of zirconium with quercetin (3,3′,4′,5,7-penta-hydroxyflavone). Iron and tin are the elements which interfere most but fluorine must be absent, and therefore the sample must not be treated with hydrofluoric acid during its preparation.

16:35:2 Recommended method (Grimaldi and White)

Reagents

Standard zirconium solution: Mix 2·17 g $ZrO(NO_3)_2.2H_2O$ with 100 cm³ of 1:1 HCl and gently boil to reduce the volume to 75 cm³. Dilute to 500 cm³ with 1:1 HCl. Standardize by precipitating 20-cm³ aliquots with redistilled ammonium hydroxide solution and igniting the oxide which is weighed.

> Add 5 cm³ of the standardized solution to 395 cm³ of 1:1 HCl and dilute to 1 dm³ with water. 1 cm³ will now contain approximately 10 μg Zr.

Flux: 3 parts K_2CO_3 and 1 part borax.

Potassium hydroxide solution: 100 g KOH/100 cm³ of water.

Iron(III) chloride solution: Dissolve 0·1 g of pure iron wire in hydrochloric acid. Evaporate to dryness and add 1 cm³ of 1:1 HCl and 20 cm³ water to the residue. Digest, cool and dilute to 100 cm³.

p-Dimethylamino-azophenolarsonic acid solution: 0·3% w/v in 1:1 HCl.

Quercetin solution: 0·5 g reagent/300 cm³ ethanol. When dissolved, dilute with more ethanol to 500 cm³.

Procedure

Ignite 0·2 g of 0·15-mm soil in a platinum crucible. Mix 3 g of flux with the residue and fuse for 30 minutes. Leach the melt with 50 cm³ of water, add 10 cm³ of potassium hydroxide solution and then 5 cm³ of iron(III) chloride solution to act as a carrier for the zirconium. Digest the mixture on a steam-bath for 1 hour, filter and wash with 1% potassium hydroxide solution. Dissolve the residue in 10 cm³ of hot 1:1 hydrochloric acid, filter into a beaker and add 5 cm³ of *p*-dimethylamino-azophenolarsonic acid solution. Adjust the volume to 50 cm³ and digest the solution on a steam-bath for 1 hour. Filter whilst still hot, using macerated filter paper accelerators. The temperature must be kept above 70°C or titanium is precipitated with the zirconium. Wash the precipitate well with a hot solution of *p*-dimethylamino-azophenolarsonic acid prepared by mixing 10 cm³ of the reagent with 70 cm³ of concentrated hydrochloric acid and diluting to 500 cm³.

Place the paper and precipitate in a platinum crucible, dry the precipitate and allow to cool. Ignite the cold precipitate at 500°C. Add 10 drops of concentrated sulphuric acid to the residue, cover the crucible and heat for about 30 minutes to dissolve the solids. Evaporate off the acid and remove the last traces by holding the crucible in a bunsen flame at a temperature below 500°C. Add 5 cm³ of 1:4 hydrochloric acid to the residue and warm the crucible to dissolve the zirconium. Transfer the solution to a 25-cm³ volumetric flask, dilute to 15 cm³ with water and cool. Add 5 cm³ of ethanol and 3 cm³ of quercetin solution. Dilute to 25 cm³ and read the optical density at 440 nm.

A blank experiment, including the filter paper digestion, must be made and the calibration curve should cover the range 0–0·5 mg Zr.

17

Oxidation–reduction potentials

17:1 BACKGROUND AND THEORY

17:1:1 Introduction

The process of oxidation is fundamentally that of addition of oxygen but can also be considered as loss of hydrogen. A factor common to both these reactions is loss of electrons, for example:

$$Fe^{2+} \longrightarrow Fe^{3+} + e \qquad [17:1$$

Similarly the process of reduction can be considered as that of acquiring electrons and an oxidation–reduction system is represented as:

$$\text{Oxidized state} + \text{electrons} \rightleftharpoons \text{Reduced state} \qquad [17:2$$

and the number of electrons involved depends upon the particular system. Oxidation of divalent tin for example involves the loss of two electrons,

$$Sn^{2+} \rightleftharpoons Sn^{4+} + 2e \qquad [17:3$$

The importance of the state of reduction or oxidation in a soil was pioneered by Gillespie (1920) who concluded that 'good' soils are more oxidizing than are 'bad' soils. Some of the main inorganic oxidation–reduction systems of importance in soils are:

$$O_2 + 4H^+ + 4e \rightleftharpoons 2H_2O \qquad [17:4$$

$$NO_3^- + 2H^+ + 2e \rightleftharpoons NO_2^- + H_2O \qquad [17:5$$

$$Fe(OH)_3 + 3H^+ + e \rightleftharpoons Fe^{2+} + 3H_2O \qquad [17:6$$

$$Fe_3O_4 + 8H^+ + 2e \rightleftharpoons 3Fe^{2+} + 4H_2O \qquad [17:7$$

The state of oxidation or reduction becomes of prime importance when considering wet soils, the properties of which depend largely upon oxygen availability. Organic matter decomposing in aerobic soils results in synthesis of microbial tissue, disappearance of simple organic compounds and evolution of carbon dioxide. If oxygen is deficient, fatty and certain other organic acids accumulate, microbial synthesis is diminished and respiration decreases. Greenwood (1961) found that the changeover from aerobic to anaerobic metabolism of organic matter takes place at an oxygen concentration of about 3×10^{-6} mol dm^{-3} in widely different kinds of soil.

Fresh organic matter in a soil aids the formation of reducing conditions, and Willis (1936) considered that reducing conditions in waterlogged soils may be due primarily to percolation of water-soluble organic matter from above. Robinson (1930) found high amounts of iron and manganese in waterlogged soil as proto-hydrogen carbonates, formed by carbon dioxide from organic matter. The proto-hydrogen carbonates did not occur in the absence of organic matter. Shapiro (1958) found that flooding acid soils containing organic matter increased the availability of phosphorus; calcareous soils remained unaffected. Bloomfield (1951, 1953) showed that aqueous extracts of leaves were capable of reducing iron(III) in soil. In coastal saline mangrove swamps of West Africa the presence of large quantities of decomposing mangrove roots accelerates reduction of sulphate in the tidal floodwater and plays a considerable part in the formation of sulphur-rich soils (Hart, 1959).

The reduction or oxidation of iron in acid soils is one of the main chemical effects of a changing state of oxidation. Reduced iron is soluble, and for this reason phosphorus availability is increased in reduced soils. Manganese reacts similarly and is more easily reduced than is iron. The soil solution of a non-calcareous, reduced soil has been shown to contain Fe^{2+}, Mn^{2+}, NH_4^+, sulphide, nitrite, hydrogen sulphide, alcohols, aldehydes, ketones, sugars, carboxylic acids and increased amounts of phosphate and exchangeable bases over those found in the oxidized condition of the soil. The process of nitrification will not proceed beyond ammonification in a reducing soil and the ammonium exchanging on the soil complex liberates soluble calcium, magnesium and so on.

Mortimer (1942) demonstrated that the surface layers of lacustrine muds are adsorbing and are capable of locking up added fertilizers. Mortimer suggested that the adsorbing properties of the muds are due to colloidal iron(III) hydroxide. Due to seasonal changes in lakes the colloidal complexes are destroyed by lack of oxygen and on reduction, liberate iron(II) and nutrient bases.

Many organic oxidation–reduction systems play an important part in soils; for example,

$$acetaldehyde \rightleftharpoons acetic\ acid + 2H + 2e \qquad [17{:}8$$

$$succinic\ acid \rightleftharpoons fumaric\ acid + 2H + 2e \qquad [17{:}9$$

$$pyruvate \rightleftharpoons lactate + 2H + 2e \qquad [17{:}10$$

$$cysteine \rightleftharpoons cystine + 2H + 2e \qquad [17{:}11$$

Sometimes soils are encountered in which different horizons are in different states of reduction or oxidation. Soils which have a high water table are normally oxidizing on the surface but reducing in their lower strata, and if much iron is present the differences can be strikingly obvious to the eye. Certain basin soils in East Pakistan, for example, which are annually flooded but exposed during the dry seasons, exhibit strongly coloured profiles. The upper soil is red, lower down the colour is yellow, then green and in the deep subsoil the colour is slaty-blue. McKeague and

Bentley (1960) found oxidation–reduction potentials a better criterion for classifying gley soils than determination of free iron.

Even under a permanent water cover, soils can have an oxidized top layer. Some lake muds, for example, have a thin layer of oxidized material which is an iron(III) humate deposit kept in the oxidized state by oxygen dissolved in the water. In contrast, a newly flooded soil may develop a reducing layer over an oxidized horizon. Formation of reducing conditions in a flooded topsoil will be aided by the organic matter present.

Soils with different oxidizing conditions at different depths must be fertilized with care. The problem arises particularly with lowland rice cultivation. Ammonium-nitrogen, if placed as a surface dressing, will be oxidized to nitrate and if this is leached down into the reducing zone it will be lost by denitrification (see Chapter 10, section 10:2:6). If the fertilizer is placed immediately into the reduced layers of soil the nitrogen will remain available as ammonium. Nitrates as such can be used as a top dressing during the period of rapid growth of the plants when it is taken up more quickly.

17:1:2 Oxidation–reduction potentials

The physical measurement of oxygen in soils is not simple and in practice it is more feasible to measure what is known as the 'oxidation–reduction potential'. Oxidation–reduction potentials, sometimes abbreviated to 'redox potentials', were first measured by Bancroft (1892). Up to a point the potential will depend upon the availability of oxygen, but in more reduced soils other systems are involved.

If an inert, unattackable metal such as platinum or gold is placed in a solution of oxidizing substance it tends to lose electrons to the oxidizing agent and thus acquires a positive charge. We can say that the platinum acquires a positive potential with respect to the solution. Similarly, the metal placed in a reducing solution tends to acquire electrons and has a negative potential with respect to the solution.

The power of a solution to oxidize or reduce must be considered in terms of capacity and intensity. The oxidizing or reducing capacity of a solution depends merely upon the concentration of the oxidizing or reducing agent present. The intensity of oxidation or reduction is more fundamental and is related to the nature of the substances present. The potential assumed by an inert metal placed in a solution is a measure of the oxidizing or reducing intensity.

It follows that for a system to have an oxidation–reduction potential, both oxidized and reduced forms must be present together and the system must be thermodynamically reversible.

Faraday's (1834) second law of electrolysis predicts that if one faraday of electricity passes between two electrodes in an electrolyte, 1 gramme-equivalent of a substance will be liberated at each electrode. Consequently during the reduction of, for example, iron(III), for every gramme-atom of iron reduced 1 faraday of electricity is lost.

For the potential existing between a metal and a solution of its ions, the electrode potential equation derived from the Nernst equation is:

$$E = \frac{RT}{nF} \ln \frac{p}{P}$$
[17:12

where R is the gas constant, T is the absolute temperature, n is the number of electrons involved, F is the faraday (96 500 coulombs), p is the osmotic pressure and P is the electrolytic solution pressure.

Assuming a system containing both reduced and oxidized forms and being thermodynamically reversible, the potential (E) acquired by an inert electrode placed in the system will be given by:

$$E = E_0 + \frac{RT}{nF} \ln \frac{a_{ox.}}{a_{red.}}$$
[17:13

where E_0 is a constant known as the 'standard oxidation–reduction potential' of the system and a refers to activities.

Using known values for R and F and converting the logarithm to base ten, equation 17:13 becomes:

$$E = E_0 + \frac{0 \cdot 0002T}{n} \lg \frac{a_{ox.}}{a_{red.}}$$
[17:14

and as it is usual to use concentrations instead of activities,

$$E = E_0 + \frac{0 \cdot 0002T}{n} \lg \frac{[\text{oxidized state}]}{[\text{reduced state}]}$$
[17:15

The oxidation–reduction potential E thus is dependent upon the ratio of concentrations of oxidized and reduced states. If the concentrations are equal then the logarithm of the ratio becomes zero and E will equal E_0. Thus the standard oxidation–reduction potential of a system is the value of E when the oxidized form is present in the same concentration as the reduced form. For example, the electrode potential of a tin(IV)–tin(II) system is $-0\cdot15$ volts with respect to the hydrogen electrode and that of an iron(III)–iron(II) system is $+0\cdot771$ volts. The higher the value of E_0 the more powerfully the element can oxidize and vice versa.

Standard redox potentials are determined experimentally and can be found tabulated for most redox systems. In mixed systems, for example tin and iron, that with the higher E_0 value will oxidize the other:

$$2Fe^{3+} + Sn^{2+} \longrightarrow 2Fe^{2+} + Sn^{4+}$$
[17:16

This fact is made use of analytically in potentiometric titrations. Addition of an oxidizing agent to a reduced state causes the potential of the latter to rise. The increase in potential occurs quickly at first, then slowly, and again quickly near the end of oxidation. The potential is plotted against per cent oxidized form to give the end-point; such a curve for the iron(III)–iron(II) system is shown in Figure 17:1.

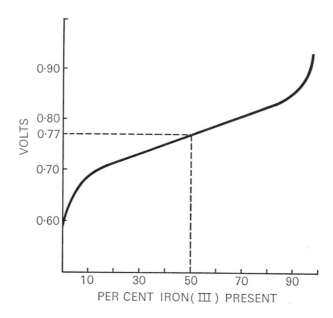

Figure 17:1 Potentiometric curve for the iron(III)–(II) system

Mortimer (1941) gave the following list of oxidation–reduction potentials for important transformations occurring in lake muds at pH 7:

$$NO_3^- \longrightarrow NO_2^- \qquad 450 - 400 \text{ mV}$$
$$NO_2^- \longrightarrow NH_4^+ \qquad 400 - 350 \text{ mV}$$
$$Fe^{3+} \longrightarrow Fe^{2+} \qquad 300 - 200 \text{ mV}$$
$$SO_4^{2-} \longrightarrow S \qquad 100 - 60 \text{ mV}$$

Measurement

Although an inert electrode placed in an oxidation–reduction system assumes a potential, the electrode forms only half a cell and cannot be used by itself to measure the potential. It is necessary to complete the cell with another half-cell of known potential. The standard half-cell is calibrated against the normal hydrogen electrode and all oxidation–reduction potentials measured in this way are referred to the hydrogen electrode and are denoted by the term *Eh*.

The e.m.f. of the cell is measured in volts and is the arithmetical sum or difference of the two electrode potentials depending upon their sign.

17:1:3 *r*H values

If we consider a system in which the reduced form differs from the oxidized form by two atoms of hydrogen, the reversible potential is given by,

$$E = E_0 + \frac{RT}{2F} \ln \frac{[\text{oxidized state}] \, [\text{H}^+]^2}{[\text{reduced state}]} \qquad [17:17$$

$$= E_0 + \frac{RT}{2F} \ln \frac{[\text{ox.}]}{[\text{red.}]} + \frac{RT}{F} \ln [\text{H}^+] \qquad [17:18$$

With an ordinary hydrogen electrode with a gas pressure p, the potential is given by,

$$Eh = \frac{RT}{2F} \ln \frac{I}{p} + \frac{RT}{F} \ln [\text{H}^+] \qquad [17:19$$

and so we can consider an oxidation–reduction electrode of the type involving two atoms of hydrogen as a hydrogen electrode whose gas pressure is determined by the terms E_0 and $\dfrac{[\text{ox.}]}{[\text{red.}]}$.

Clark and Cohen (1923) defined this gas pressure by introducing the term '*r*H' where,

$$rH = -\lg p \qquad [17:20$$

and suggested the term *r*H as a measure of oxidizing intensity.

The general equation 17:18 can be written as,

$$E = \frac{(2 \cdot 302 \, RT)}{2F} \, rH + \frac{RT}{F} \ln [\text{H}^+] \qquad [17:21$$

2·302 being the factor for converting the logarithm to base ten. Then by converting the term $\ln [\text{H}^+]$ to pH,

$$E = \frac{(2 \cdot 302 \, RT)}{2F} \, rH - \frac{(2 \cdot 302 \, RT)}{F} \, pH \qquad [17:22$$

and equation 17:22 can be simplified by putting in known constants and temperature values.

It was later realized (Clark *et al.*, 1928) that if the oxidized or reduced forms of a system are also acidic or basic, the respective dissociation constants must be included in the expression and the system no longer resembles a simple hydrogen electrode. The term *r*H was accordingly discarded by its proposers, although subsequent use of it has been made by other workers.

17:1:4 Relationship between *Eh* and pH

As the *r*H value assumes a relationship between the oxidation–reduction potential and pH, it can be calculated that for each unit drop in pH the potential will increase by a definite amount depending upon temperature. At 20°C, for example, the value is 58 mV.

Even though the rH assumptions are not necessarily valid, correction of potential for pH in this manner is often done as an approximation. Pearsall (1938) calculated the potentials of lake muds to the predominating pH value of 5. As an example of the calculation consider a potential of 180 mV at pH 6. This corrected to pH 5 will be (180 + 58) or 238 mV. Similarly, 250 mV at pH 4 will become 250 − (58 × 3), or 76 mV at pH 7.

Jeffery (1960) added organic matter to an oxidized soil and then imposed anaerobic conditions by flooding in an atmosphere of nitrogen. The flooded soils were incubated for up to 21 days and analyzed for Eh, pH, Fe^{3+} and Fe^{2+}. For the soil alone, that is with no added organic matter, the value of Eh fell gradually to +0·1 volt and ended in an ill-poised* system fluctuating around 0·1 volt (Figure 17:2). For soil with added

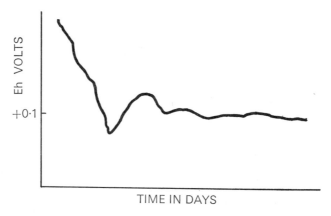

Figure 17:2 Oxidation–reduction potential of soil as affected by flooding without addition of organic matter (Jeffery, 1960; *J. Soil Sci.*)

organic matter an initial period of intense reduction occurred to a value of $Eh = -0·25$ volt, which was followed by the release of systems with higher potentials establishing a well poised system at +0·1 volt (Figure 17:3). During the first stage there was a rapid release of manganese and during the second stage iron was dissolved from oxides and hydroxides. The recovery was due to oxidation of organic matter by the released iron and involved conversion of iron(III) oxide to basic iron(II) hydroxide. The net reaction resulted in a situation where both iron(III) and iron(II) ions were present, which Jeffery considered as a desirable state for the soil to be in.

* The term 'poised' is used for redox systems in the same way as the term 'buffered' is used for pH discussion. A system which has a stabilized redox potential is said to be 'well poised'.

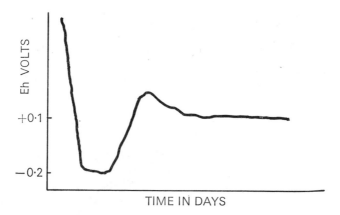

Figure 17:3 Oxidation–reduction potential of soil as affected by flooding after addition of organic matter (Jeffery, 1960, *J. Soil Sci.*)

As the standard potential for the iron(III)–iron(II) system is $+0.771$ volts, the activities of iron(III) and iron(II) during the third stage of slow decline are related to the oxidation–reduction potential by the expression,

$$Eh = 0.77 + 0.058 \lg \frac{a_{Fe^{3+}}}{a_{Fe^{2+}}} \qquad [17:23$$

$$= 0.77 + 0.058 \lg [Fe^{3+}] - 0.058 \lg [Fe^{2+}] \qquad [17:24$$

and Jeffery showed that if one assumes the ionic product of water to be 10^{-14} and the solubility product of iron(III) hydroxide to be 10^{-35} (Evans and Pryor, 1949), the value of $0.058 \lg [Fe^{3+}]$ can be estimated from pH. Thus,

$$\lg [Fe^{3+}] = -3 \, pH + 7 \qquad [17:25$$

and therefore,

$$0.06 \lg [Fe^{3+}] = -0.18 \, pH + 0.42 \qquad [17:26$$

and so equation 17:24 can be written,

$$Eh = 1.19 - 0.18 \, pH - 0.06 \lg [Fe^{2+}] \qquad [17:27$$

or by more precise calculation and at 30°C,

$$Eh = 1.033 - 0.18 \, pH - 0.06 \lg [Fe^{2+}] \qquad [17:28$$

Likely values of the term $0.06 \lg [Fe^{2+}]$ are between $+0.12$ volt and $+0.18$ volt as the concentration of iron(II) is likely to be 10^{-2} or 10^{-3}.

Plots of *Eh* calculated from equation 17:28 and pH values gave linear relationships which agreed with Jeffery's experimental determinations of the third stage of reduction (Figure 17:4).

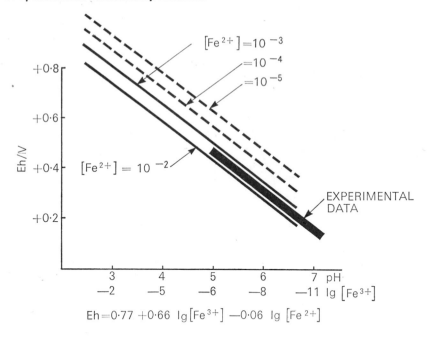

Figure 17:4 Relation between redox potential, pH and concentration of iron(III) and iron(II) ions (Jeffery, 1960; *J. Soil Sci.*)

It was demonstrated by Jeffery that in waterlogged soil the formation of basic iron(II) hydroxide explains the concomitant changes in *Eh* and pH. Subsequently Jeffery (1961) proposed three terms to define the state of reduction of a flooded soil. The terms were: 'oxidizing conditions', those of a normal aerated soil, where rice plants suffer from lack of ammonium-nitrogen and where phosphorus is precipitated by iron(III); 'healthy reducing conditions' where rice plants can flourish, nitrogen and phosphorus are available and both iron(II) and iron(III) are present, and; 'extreme reducing conditions' where the plants cannot flourish as the rhizosphere can no longer protect the roots from, for example, sulphide. Under healthy reducing conditions small oxygen pockets in the rice roots can raise the potential of the soil in the immediate vicinity. From the experimental results of Jeffery, Greene (1963) expressed the states of reduction in the form shown in Figure 17:5.

Further investigation into the relationship between *Eh* and pH was made at the International Rice Research Institute in the Philippines (1963). A large selection of soils differing in properties were flooded. The oxidation–reduction potentials measured immediately after flooding ranged from $+350$ mV to $+620$ mV. These potentials were highly correlated negatively with pH and were expressed by the regression,

$$Eh = 0.851 - 0.056 \text{ pH} \qquad [17:29$$

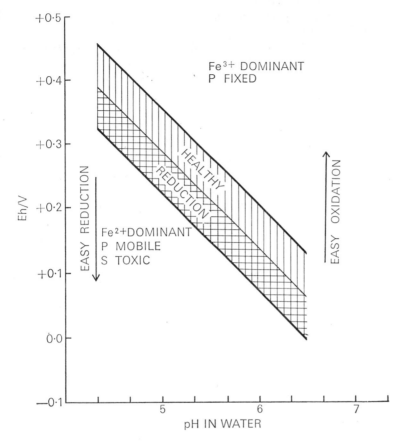

Figure 17:5 Illustrating the three stages of reduction of a paddy soil (Greene, 1963 after Jeffery; *J. Soil Sci.*)

which implies an Eh/pH slope of -0.056 volts per pH unit. This result was compared to the theoretical thermodynamic potential for the oxygen system,

$$O_2 + 4H^+ + 4e \rightleftharpoons 2H_2O \qquad [17:30$$

given by,

$$Eh = 1.25 + 0.015 \lg pO_2 - 0.06 \text{ pH at } 30°C \qquad [17:31$$

and which can be corrected for the irreversibility of the oxygen potential at bright platinum electrodes to,

$$Eh = 0.92 + 0.015 \, pO_2 - 0.06 \text{ pH} \qquad [17:32$$

or, when pO_2 is 20%, to

$$Eh = 0.92 - 0.06 \text{ pH} \qquad [17:33$$

This expression agrees well with the observed values of 0·851 volts and

−0·056 volt and it was concluded that the redox system in aerobic soil is the irreversible oxygen system.

After the soils had remained flooded for 1 day the oxidation–reduction potentials decreased, most strongly in neutral soils and least in acid soils. Further changes varied with the kind of soil and confirmed the role of organic matter in lowering the redox potentials of flooded soils. The results further indicated two poising systems, by manganese in the range $+100$ mV to $−50$ mV and by iron in the range $−50$ to $−200$ mV. The redox potential values, however, could not be correlated with iron or manganese contents. As the amounts of iron or manganese in the soil solution did not govern the redox potential of the soil it was considered whether the redox potential of the soil solution itself was so governed. This hypothesis was examined and it was found that the redox potential values of soil solutions in general were higher than those of the respective soils and tended to increase after 80 days of anaerobic incubation. Air had been rigorously excluded and the pH did not change by more than 0·2 of a unit. It was concluded that redox potentials in soils are, in fact, bacterial potentials.

Bacterial potentials are known to have pH slopes as found in reduced soils and are thought to be caused by systems of bacterial metabolism and activated by enzymes.

The higher redox potentials of the soil solutions were used for correlation with amounts of iron(II) and manganese(II) ions present and the best correlation was that between iron(II) concentration and the potential at pH 7 after 143 days of flooding,

$$E_7 = 0·967 − 0·368 \lg [Fe^{2+}] \qquad [17:34$$

which is different from the theoretical correlation for the iron(III) hydroxide–iron(II) system and so was dismissed as incorrect or fortuitous.

Soils were next anaerobically incubated with and without added organic matter, and pH and Eh values were measured. Also determined were aerobic values of dEh/dpH. The results confirmed the earlier conclusion that aerobic soils have a dominant oxygen system. The aerobic potentials were from 460 mV to 550 mV at pH 7, and Eh/pH gradients of $−50$ mV to $−55$ mV were obtained. These results compared well with theoretical values of 530 mV and $−60$ mV.

On flooding, the presence of organic matter accelerated the drop in oxidation–reduction potentials and it was found that the changes in potential were amenable to mathematical description by the expression,

$$\lg Eh = a + b \lg t + c (\lg t)^2 \qquad [17:35$$

where t is the time of anaerobic incubation.

For a given soil it was found that the value of dEh/dpH varies with time of flooding and is altered at any time by organic matter. The changes in the function bore no relation to the iron(III)–(II) system. For the iron(III)–(II) system to be operating, dEh/dpH should be zero up to a pH value of 3 from the acid side, $−180$ mV between pH 3 to 7 and $−60$ mV from pH 7 upwards. Actual values diverged widely from these figures. For aerobic soils a constant value of -58 mV was obtained over a range of pH 2 to pH 10, and below pH 2 the function tended towards zero, indicating the operation of the iron(III)–(II) system. For soils not under aerobic conditions the function varied widely in the pH range 1 to 6 and between pH 7 to 10 was constant at -110 mV.

The final conclusion from these experiments was that the oxidation–reduction potentials as measured in flooded soils are really bacterial potentials.

In recent years there has been an increasing use of *Eh*–pH diagrams to show and interpret their interrelationship as applicable to other functions. For example Garrels and Christ (1965) show, *inter alia*, how the stability of iron oxides is affected by changes in *Eh* and pH and how carbon dioxide affects iron–water–oxygen relations by means of *Eh*–pH diagrams. These authors give the mechanics of constructing such diagrams in considerable detail.

17:1:5 Reducing power and redox levels

Lamm (1956) considered that it is incorrect to speak of well established soil oxidation–reduction potentials which have been obtained experimentally with a platinum electrode. Lamm suggested an empirical term 'redox levels' giving an expression for the mean reducing or oxidizing activities in dry soil samples. If a soil is treated with a solution containing the reversible redox system, potassium hexacyanoferrate(II)–potassium hexacyanoferrate(III), the e.m.f. of a platinum electrode against a calomel reference electrode will be different from the potential measured in pure solutions. These potential differences, ΔE mV, can be correlated with oxidizing properties of the soil. By treating the same soil with the hexacyanoferrate(II)–(III) system, using different ratios of hexacyanoferrate(II) to (III) and buffering all solutions with ammonium acetate, values of ΔE can be plotted against corresponding redox potentials in pure solutions. An S-shaped curve is obtained which intersects the E axis at a point E'. The point E' represents a solution that does not change its oxidation–reduction potential when in contact with the soil and can be regarded as an expression of the 'redox level' of that soil.

In practice, 10 g of air-dry soil are shaken for 1 hour at room temperature with 50 cm^3 of a series of buffered solutions. The concentrations of potassium hexacyanoferrate(II) and (III) in the solutions are between 10^{-4} and 10^{-2} mol dm^{-3} and E values are all referred to the hydrogen electrode.

Lamm found ΔE values to be reproducible but realized that this could have been caused by absorption or precipitation instead of by a purely redox process. However, when buffered and with acetate ions present at a ratio of 100:1 to hexacyanoferrate(II) ions, such absorption was not likely.

Birch (1949) estimated the total reducing power of a lake mud by titration with acid permanganate, the end-point being determined poten-tiometrically. In soil, however, the degree of total oxidation or reduction is of limited value and Birch wished to find the conditions under which iron would be converted to the trivalent state. Consequently known amounts of iron(III) chloride solution were added to samples of the mud and oxidation–reduction potentials were determined. The presence of iron(III) was tested for after each addition with ammonium thiocyanate solution. Birch did not obtain a positive test for iron(III) until the redox potential reached $+640$ mV, that is, when 98.6% of the added iron(III) had been reduced. This result suggested that the mud could rapidly and quantitatively reduce iron(III) until the redox potential reached 640 mV, after which iron would accumulate in the oxidized state. Birch proposed that the reducing power of muds can be compared by titration to *Eh* 650 mV with 0·1 M iron(III) chloride solution. Curves plotted of *Eh* against added iron changed from the horizontal to the vertical at *Eh* 650 mV and the results were more sharply defined if plotted on a log/log basis.

Scott and Evans (1955) considered that measurement of oxidation–reduction potentials is unreliable as an index of oxygen available to crops in waterlogged soils. Using a special apparatus (Figure 18:3), Scott and Evans measured oxygen present in saturated soils and found that the gas disappeared after 10 hours of flooding. At this stage the soil was reducing, and even if re-oxygenated, the reducing conditions were not easily removed, although redox potentials indicated a rapid rise in the state of oxidation.

Heintze (1934) similarly had found that redox potentials are no criterion of waterlogging or of iron(II) as he obtained oxidizing potentials in gleyed, waterlogged soil containing iron(II). Heintze explained this as being due to alternating conditions of oxidation and reduction in the soil.

17:2 POTENTIOMETRIC MEASUREMENT

17:2:1 Apparatus

The electric circuit necessary to measure oxidation–reduction potentials in soil is the cell,

$$Pt \mid Soil \parallel saturated\ KCl \mid Hg_2Cl_2.Hg$$

liquid junction

connected through a potentiometer. The platinum electrode may be replaced with a gold electrode but this is seldom encountered in soil analysis. Similarly the standard half-cell is invariably a saturated calomel electrode in which a mixture of mercury and mercury(I) chloride are covered with a saturated solution of potassium chloride. The standard potential of the calomel half-cell as determined against a 1 M hydrogen electrode,

$$Hg \mid Hg_2Cl_2.KCl_{sat.} \parallel H^+.H_2 \mid Pt$$

is 0·246 volt at 25°C. In practice, the calomel cell is balanced against a Weston cell of known e.m.f. before use to determine its precise potential.

Various patterns of the calomel cell have been proposed but the most usual is that shown in Figure 17:6 and details of its preparation are given

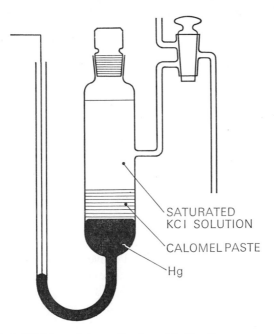

SATURATED
KCl SOLUTION

CALOMEL PASTE

Hg

Figure 17:6 Hildebrand type of calomel half-cell

in section 17:4:2; potassium chloride solution can be renewed at the point of contact by means of the stopcock. A more compact form of the electrode is commercially available for use with pH meters and which incorporates its own salt bridge.

The circuit as set up in practice may be depicted as in Figure 17:7 and the two halves of the cell must be connected by a salt bridge so that the two systems do not become mixed. The electrolyte of the bridge must not react chemically with either system and saturated potassium chloride solution is used when analyzing soils. In the commercial, compact calomel electrode the bridge is formed by saturated potassium chloride solution via a sintered glass disc or a porous fibre plug, and both reference and platinum electrodes may be placed directly in the soil sample. When using a calomel half cell of the general type shown in Figure 17:6, the potassium chloride of the salt bridge is made up in a solution of agar-agar which sets to a jelly in the containing tube. As the exposed surface of the agar–potassium chloride jelly becomes contaminated by soil, a short section of the glass tube is cut off with a file to expose a fresh surface of electrolyte. As a null method is used for measuring the voltage, the length of the salt bridge does not affect the results.

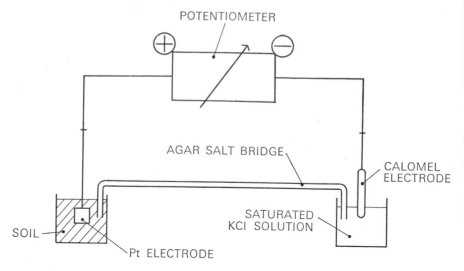

Figure 17:7 Circuit for measuring oxidation–reduction potentials using a salt bridge of the agar jelly type

For fieldwork using a simple potentiometer and calomel cell of the type shown in Figure 17:6, it was found convenient (Hesse, 1950) to make the agar–chloride salt bridge in a flexible length of polythene tubing with short pieces of glass tubing at each end. This gave more manoeuvrability than a rigid glass tube, especially when working in soil pits with the potentiometer at the surface. To expose clean surfaces of electrolyte the small glass tube at the end (needed to enable the bridge to be firmly pushed into the soil) could be pushed further into the plastic tubing and the extruded jelly cut off.

It is usual to use the platinum electrode in a clean, bright condition and the electrode must frequently be cleaned in chromic acid to prevent polarization. When not in use, platinum electrodes are stored under distilled water.

If the system is not well poised or is aerated, a bright platinum electrode may be of no use (Quispel, 1947) as its surface will be coated with oxides. In such cases Quispel advocates a platinum black electrode which catalyzes the reaction velocity of oxidation. Quispel prepared the electrode by placing a platinum wire as cathode in a 3% solution of chloroplatinic acid containing a trace of lead acetate; the anode was another platinum wire. The electrodes were connected to 4 volts for 5 minutes, rinsed, put into a 3% solution of sulphuric acid and connected to 4 volts for 2 minutes. The electrode was kept immersed in water for 24 hours before use. After use the electrode could be re-platinized by placing it as anode in chloroplatinic acid for 5 minutes and then re-plating as before.

Mortimer (1941) built special compound platinum electrodes to permit redox determinations in mud at various depths without disturbing the sample more than necessary. Small strips of platinum foil, 10 mm × 2 mm,

were welded on to separate platinum wires each fused into a piece of glass
tubing (Figure 17:8). Twelve of these electrodes were arranged round a
cork in a spiral to give a gradual increase in depth of penetration. The
platinum foil was welded so that the sharp edge was towards the mud
sample for minimum interference. Mortimer emphasized the necessity for
clearly labelling each electrode. For more simple experiments it is sufficient

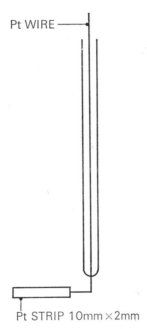

Pt WIRE

Pt STRIP 10mm × 2mm

Figure 17:8 Platinum electrode for measuring redox potentials in lake
muds. Several such electrodes are fixed round a cork in a gradually descending spiral (Mortimer, 1941; *J. Ecol.*)

to prepare two or three separate electrodes fixed together at different
levels. For example, the apparatus shown in Figure 17:9 was found convenient for measuring the potentials of mud samples and their overlying
water; the rubber stopper carrying the platinum electrodes, and either the
salt bridge or compact calomel electrode, could be transferred bodily from
sample to sample contained in similar-sized bottles.

The platinum electrode must be in good contact with the soil. Pearsall
and Mortimer (1939) first cut a slit in the soil with a knife, then inserted
the platinum electrode and gently pressed the soil together. Quispel (1947)
used a platinum wire as electrode and inserted it into the soil via a hole
initially made with a glass tube. The calomel reference electrode or salt
bridge should be placed so as to ensure adequate contact, and in wet
samples, should be placed below the limits of waterlogging.

Once the complete circuit has been made, the potential is read in volts
or millivolts. Normally the platinum electrode is connected to the positive

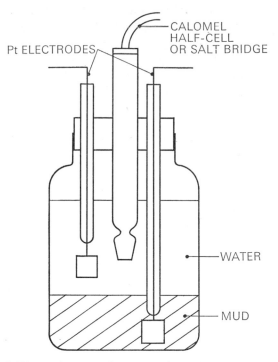

Figure 17:9 Electrode set up for measuring redox potentials in muds and their overlying water without disturbing the samples

pole of the potentiometer, but sometimes when the soil is strongly reducing, it becomes necessary to change over the electrodes in order to obtain a reading.

As the saturated calomel electrode has a potential of 250 mV (at 20°C) this value must be added to, or subtracted from, the observed value according to which terminal of the potentiometer it is connected.

17:2:2 Preparation of soil samples

For determination of oxidation–reduction potentials in the field the electrodes are placed in the soil *in situ* and no preparation is necessary. A certain amount of time must be allowed to elapse before taking the reading in order to allow for the disturbance caused to the equilibrium by inserting the electrodes. The time factor has been found to vary according to soil type and existing conditions. Pearsall (1950) recommended waiting for at least 10 minutes; Pearsall and Mortimer found that 30 minutes were necessary in order to obtain steady readings; Mortimer (1941) and Heintze (1934) waited for 2 hours and Quispel (1947) recommended leaving the electrodes in the soil for 2 days before taking the reading.

Transportation of soil samples to the laboratory for redox potential measurements causes changes due to alteration of aeration, temperature and moisture. Pearsall (1938) tried dropping soil into toluene to keep the potential steady, but with limited success; slightly better results were obtained if the soil was kneaded under toluene to remove water anaerobically.

Birch (1949) found that well poised mud samples, in which the redox potential increased due to sampling and transportation, rapidly reassumed their original state of reduction on standing under water. Even on vigorous aeration of the suspension resulting in an *Eh* value of $+400$ mV, the original low values were soon re-established. This was not so for bleached, iron-poor and ill-poised muds. Mud samples collected from the bed of Lake Victoria were measured immediately after sampling and again in the laboratory several days later; it was found that the original low *Eh* values were rapidly re-established by storing the mud under water in the dark for 24 hours (Hesse, 1957).

For laboratory measurement of oxidation–reduction potentials the soil sample must be mixed with water. Remesow (1930) prepared soil suspensions and allowed them to stand undisturbed for 18 hours. Three platinum electrodes were then placed in the suspension and left overnight before readings were taken. Hertzner (1930) obtained best results by placing dry soil in a Gooch crucible standing in a pan of boiled, distilled water. When the soil was saturated with water a coiled platinum wire electrode was inserted and the potential measured within 1 hour. Heintze (1934) mixed dry soil to a paste with water and Willis (1936) shook soil with water for 3 days in unstoppered bottles at constant temperature. Heintze further investigated the effects of replacing air in the containers with nitrogen but found this to have no effect upon the oxidation–reduction potentials subsequently measured.

Brown (1934), using a soil–water ratio of $1:1$, studied the effect of time of measurement after mixing. Brown found that leaving soil waterlogged for 12 hours or more resulted in a negative shift of potential and Remesow's method, for example, was thus not acceptable. Very short times of standing resulted in unstable potentials. Brown concluded that a waiting time of 2 hours was optimum.

Brown then examined Hertzner's Gooch crucible procedure using five variations of the technique. For example, the electrode was placed in the soil while the latter was dry, after saturating and with different time intervals. All five procedures were discarded. Ten different procedures involving displacement of air with nitrogen were next examined. For example, nitrogen gas was bubbled through the suspension before and during measurement, the pressure was reduced before admitting nitrogen and so on. All procedures using nitrogen were rejected. Six procedures involving heat treatment, centrifuging and reduced pressure were all rejected. Brown eventually tested seventy procedures for the laboratory determination of oxidation–reduction potentials of air-dry soil samples. The final recommended procedure was to first grind the dry soil with a rubber pestle to pass a 1-mm sieve and then to weigh 7 g into a 30-cm³ vial. 4 cm³ of water (a little more for clays) were added and the mixture rubbed with a glass rod. Coiled platinum wire electrodes were cleaned with chromic acid, rinsed and flamed in an ethanol burner before every measurement. The electrode was inserted immediately the soil paste was prepared and was completely submerged but not touching the vial. After standing for 30 minutes the vials were centrifuged for 5 minutes and the potentials immediately measured. Brown claimed that by this procedure a deviation from the mean is no greater than 0·003 volt.

In practice, when oxidation potentials are required for a particular soil or mud, it is far preferable and relatively simple to measure them *in situ* rather than transport a sample to the laboratory. Laboratory measurements should be confined to those forming part of a laboratory experiment where the soil is receiving certain treatments. For example, in the experiments performed by Mortimer (1941) using the apparatus described in Figure 18:2, periodic measurement of redox potentials formed part of the routine. The experiments of Hesse (1957), Jeffery (1961) and those of the International Rice Research Institute were all specific laboratory investigations. McKeague and Bentley (1960) prepared plastic columns (Figure 17:10) to investigate variations of redox potential in soils with varying water tables. The columns were rectangular boxes with

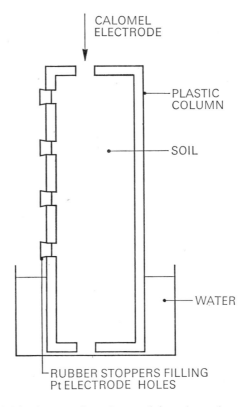

Figure 17:10 Plastic container for studying the redox potential of soil when flooded to different depths (McKeague and Bentley, 1960; *Can. J. Soil Sci.*)

12-mm holes in one side at 100-mm intervals. The column was filled with soil and the holes closed with rubber bungs; the soil was submitted to various conditions of drainage and flooding by placing the column in tanks of water to the desired depth. Platinum electrodes could be inserted through the side holes and the calomel electrode was placed through the hole in the top of the column.

Patrick (1960) devised an apparatus for studying soil properties when the oxidation–reduction potential was kept constant. An automatic titrator was modified to deliver air instead of liquid. The soil was mixed 1:1 with water and 800 g of suspension was kept submerged for several weeks. A sample was then placed in a test-tube which was stoppered with a bung carrying a bright platinum electrode, a salt bridge and an air inlet tube reaching to the bottom of the test-tube. The other end of the salt bridge was connected to a calomel half-cell and the test-tube was wrapped in aluminium foil and kept in a bath at $30°C \pm 0.5°C$.

A selected oxidation–reduction potential was set on the titrator. When the potential fell below this value a solenoid valve opened and air was bubbled through the suspension until the selected potential was again reached, when the valve closed. A recording potentiometer could be used to obtain an *Eh*–time graph. The apparatus worked well over the range -100 mV to $+500$ mV.

As the redox potential of a soil varies with location of electrodes and is affected by such factors as carbon dioxide, Ponnamperuma *et al.* (1967) suggested measuring the more reproducible potential of the soil solution.

17:3 COLORIMETRIC MEASUREMENT

Oxidation–reduction indicators are of two main types, those which are irreversibly oxidized and those which are reversibly oxidized. The latter are of importance for measuring redox potentials.

Certain oxidizing agents undergo colour changes when reduced and so act as their own indicators, for example permanganate and iodine. For general chemical analysis it is only necessary to know the point at which there is a rapid change in potential; the precise value of that potential is immaterial and certain oxidation–reduction indicators are well suited for this purpose. For example, during the estimation of organic carbon in soil (section 11:2:3), diphenylamine, which changes from colourless to violet on oxidation, is used to detect the end-point of the iron(II) titration. Pearsall and Mortimer (1939) used a dilute solution of diphenylamine in sulphuric acid as a general test for oxidizing conditions in muds, a blue colour being obtained at oxidation–reduction potentials of $+350$ mV or above. Another substance which has been used in soil science as a general indicator of oxidizing conditions is ammonium thiocyanate which gives a red colour in the presence of iron(III). Ammonium thiocyanate is not, however, an oxidation–reduction indicator in the true sense, being really

16+T.S.C.A.

an indicator for iron(III). Pearsall and Mortimer shook samples of mud with two volumes of a saturated ethanolic solution of ammonium thiocyanate (Misra, 1938). After standing for 10 minutes the colour was noted and 1 drop of hydrogen peroxide solution added to see if the colour was intensified. Colours were recorded on an arbitrary scale of 0–4; for example, a value of 2/3 denoted a colour of intensity 2 due to iron(III) and a colour of intensity 3 due to oxidation of iron(II), and this gives a rough idea of the state of reduction of the mud. Greenwood and Lees (1960) used the true oxidation–reduction indicator, methylene blue, in alkaline solution to indicate whether or not anaerobic conditions were being maintained during an experiment.

For the determination of actual oxidation–reduction potentials, fully reversible indicators are necessary in which each shade of colour represents a definite potential; for example,

$$Fe^{3+} + \text{reduced indicator} \rightleftharpoons Fe^{2+} + \text{oxidized indicator} \qquad [17\!:\!36$$

The quantity of indicator used must be small so that changes in its oxidation–reduction balance will not affect the main reaction. The term 'potential mediator' refers to a substance which rapidly comes to equilibrium with a redox system, and most oxidation–reduction indicators fall into this classification.

The indicators mostly used for determining oxidation–reduction potentials are those with an equilibrium system like that of quinone–hydroquinone,

$$Q + 2H + 2e \rightleftharpoons 2H_2Q \qquad [17\!:\!37$$

where the potential is given by equation 17:18. Thus the potential of a system containing an indicator is determined not only by the system itself but by the standard potential of the indicator and the hydrogen ion concentration of the medium. Calculation of the potential is therefore complicated and involves dissociation constants. It is thus customary to use the term $E_{0'}$ which includes the terms E_0, $\frac{RT}{F} \lg [H^+]$ and the dissociation constants and refers to a particular pH value. That is,

$$E = E_{0'} + \frac{RT}{2F} \ln \frac{[Q]}{[H_2Q]} \qquad [17\!:\!38$$

Values of $E_{0'}$ for different indicators are obtained by measurement of E using mixtures containing known amounts of oxidized and reduced substances, and then applying equation 17:38. Values of $E_{0'}$ at pH 7 are found tabulated in books of reference and some of particular interest in soil analysis are reproduced in Table 17:1. Each indicator has only a narrow range of usefulness as outside its own range it is either fully oxidized or reduced and will undergo no further colour change.

To measure the redox potential of a solution at known pH a suitable indicator is added and knowing the $E_{0'}$ value at that pH, E is estimated from

Table 17:1

Oxidation-reduction indicators and their E_0, values at pH 7 which are of particular interest in soil science

Indicator	E_0, volts at pH 7
Benzyl viologen	−0·359
Neutral red	−0·325
Rosinduline	−0·281
Diazine green	−0·255
Phenosafranine	−0·252
Potassium indigo disulphonate	−0·125
Potassium indigo trisulphonate	−0·081
Potassium indigo tetrasulphonate	−0·046
Methylene blue	+0·011
Brilliant cresyl blue	+0·047
Thionine	+0·063
Toluylene blue	+0·115
1-Naphthol-2-sodium sulphonate indo-2,6-dibromophenol	+0·119
1-Naphthol-2-sodium sulphonate indo-phenol	+0·123
Carvacrol indophenol	+0·172
Thymol indophenol	+0·174
o-Cresol-indo-2,6-dichlorophenol	+0·181
o-Cresol indophenol	+0·191
m-Cresol indophenol	+0·208
Phenol-indo-2,6-dichlorophenol	+0·217
Phenol blue	+0·224
Phenol indophenol	+0·227
o-Bromophenol indophenol	+0·230
o-Chlorphenol indophenol	+0·233
m-Carboxy-phenol-indo-2,6-dibromophenol	+0·250

the colour produced. As this procedure is not sufficiently accurate, in practice a series of indicators is added to aliquots of solution and their colour changes noted. The required potential is that lying between the values of E_0, of the two borderline indicators. For example, if all indicators having E_0, values above +239 mV are decolourized, whereas those with E_0, values below +235 mV remain unaltered, then the oxidation–reduction potential of the unknown solution is approximately +237 mV.

As many redox indicators are also acid–base indicators the colour changes must not be confused. For instance an indicator which is red when acid will probably become colourless when reduced, whereas for the same indicator in alkaline solution the colour change may be from blue to colourless.

For clear, well poised solutions an approximation of oxidation–reduction potential can be obtained by the above procedure or by comparing the unknown solution plus indicator with a series of dilutions of that indicator, but for soils the matter is not so simple. Indicator solutions added directly to soil are adsorbed or affected by factors not always related to the state of oxidation. Making water extracts of the soil is not a satisfactory procedure and the potential of the extract will show little relationship to the actual redox conditions. An attempt was made to shake mud samples with alcoholic (both ethyl and amyl to reduce adsorption) solutions of indicator and to note the colour of the liquid, but with no success. Soil solutions extracted by conventional means could be measured colorimetrically for redox potential, but no evidence has been traced indicating the validity of such measurements.

On the whole, colorimetric procedures for determining the oxidation–reduction potential of soil are unsatisfactory and are not recommended.

17:4 RECOMMENDED METHODS

17:4:1 Potentiometric: line-operated pH meter method

Preparation of soil sample

The soil sample must be prepared according to circumstances and purpose of analysis and this has been discussed in section 17:2:2. Normally the soil will be saturated or waterlogged and will either form part of a laboratory investigation or will have been suitably transported from the field. To measure the potential of the soil solution this should be displaced into a receiver filled with nitrogen gas.

Apparatus

Most commercial pH meters can be used or adapted to measure millivolts in either the ± 700 mV or the 1400 mV range and the manufacturers' instructions for operating the instrument should be followed. Millivolt measurements are made with the thermocompensator of the instrument disconnected; a manual temperature compensation unit is automatically disconnected when the millivolt scale is being used.

For measurements in the ± 700 mV range the instrument will have a fixed, central zero and the standardizing control will be inoperative. For measurements in the 1400 mV range the zero can be set to any desired point.

Various electrode set-ups have been described in the text but in every case the inert and the reference electrodes are connected to the pH meter in the usual manner, the inert electrode taking the place of the glass electrode used for pH measurement. For the reference half-cell a calomel electrode is used and which incorporates the salt bridge; it is prepared and maintained as for pH measurements. The inert platinum electrodes must be kept clean. Spade-type electrodes can be immersed in chromic acid, rinsed and flamed. Certain commercial platinum electrodes have the metallic surface flush-mounted in glass; these can be cleaned with a mild abrasive or more thoroughly by an electrolytic process.

Procedure

Place the two electrodes in the soil sample so as to ensure good contact and then switch the instrument to the desired mV scale. Allow a definite period of time to elapse before taking the reading; this period must be determined by experiment to obtain reproducible results (see section 17:2:2). The standard potential of the calomel half-cell (+250 mV) is added to the measured millivolt reading to give the redox potential of the sample. At the same time measure and note the pH of the soil; this value must be quoted when reporting the *Eh* value.

17:4:2 Potentiometric: independent salt bridge method

Measurement of redox potentials in soil using a battery powered potentiometer will seldom be made in the laboratory where a pH meter is available; the procedure is of most use in the field.

Preparation of soil sample

Redox potentials will be measured in naturally wet soils in the field and no preparation is necessary.

Apparatus

A commercial galvanometer potentiometer is used with which balance is effected by a main dial and three range multipliers. A standard Weston cell is usually incorporated for standardizing the instrument. The inert platinum electrode is prepared as in section 17:4:1 and in the first place is connected to the positive terminal of the meter. If the compact calomel electrode is not used, the Hildebrand type (Figure 17:6) can be prepared as follows.

The vessel should be thoroughly cleaned and rinsed with distilled water. Place a layer of clean mercury at the bottom of the vessel to make contact with the sealed-in electrode. Prepare calomel paste by grinding together in a glass mortar mercury(I) chloride, mercury and saturated potassium chloride solution. Decant off excess liquid and grind the paste twice more with fresh portions of potassium chloride solution. Place a layer of paste about 1 cm thick on top of the mercury in the vessel and then completely fill the vessel with saturated potassium chloride solution which has been shaken with solid mercury(I) chloride and filtered.

The salt bridge is prepared from an aqueous solution of agar-agar and potassium chloride as follows:

Add 5 g of agar to 100 cm^3 of 3·5 M potassium chloride solution (261 g dm^{-3} KCl) and heat to boiling until the agar has dissolved. Fit a piece of Tenoplast tubing of the required length with short pieces of glass tubing, rounded at the ends. Attach a short length of tubing to one of the glass pieces and fill the tube completely with agar solution by gentle suction, avoiding any air bubbles. Leave the free glass end dipping into the stock solution of agar and close the tubing at the other end with a clamp. When the jelly has set, remove the end tubing, clean the glass ends and cut with a file a small segment off each to leave a clean, smooth surface of jelly exposed. When not in use, leave the tube with its ends dipping into saturated potassium chloride solution.

The fundamental circuit is shown in Figure 17:7.

Procedure

Press the platinum electrode into the soil and insert one end of the agar salt bridge near by, making sure that the soil is sufficiently wet to give good electrical contact. The other end of the bridge tubing should be passed through one hole of a two-holed rubber bung fitted into a small bottle containing saturated potassium chloride solution. Fit the side-arm of the calomel electrode through the second hole of the bung and open the tap on the electrode just sufficiently to ensure electrical contact.

Release the potentiometer needle and adjust to the mechanical zero. Join on the standard (Weston) cell and adjust the dials to give zero deflection at 1018 mV; re-check the zero. Connect the platinum electrode to the positive terminal of the meter and the calomel electrode to the negative terminal. Make contact and turn the dials to give zero deflection; note the voltage. With strongly reducing soils the electrodes may have to be reversed to obtain a reading. Add the potential of the calomel electrode (or subtract if the electrodes have been reversed) to the measured millivolts to give the redox potential of the soil. The standard potential of the calomel electrode is +250 mV.

The accuracy of the instrument can be checked by measuring the potential of a solution containing 0·479 g $FeSO_4$ and 3·694 g of ammonium iron(III) sulphate; the potential of this solution should be +750 mV.

18

Waterlogged soils

18:1 BACKGROUND AND THEORY

18:1:1 Introduction

Theoretically the bottom deposits of oceans, lakes, rivers and ponds are all waterlogged soils and all have occasion for analysis. This chapter, however, is concerned with the analysis of more easily accessible waterlogged soils although lake beds are not ignored entirely.

It is as well to be quite clear as to what is meant by a waterlogged soil. In an unsaturated soil the fine pores may or may not contain water, depending upon the moisture content and the large pore spaces are filled with air. When both fine and large pore spaces are completely filled with water the soil is said to be waterlogged. Even so, there are several different kinds of waterlogged soil to consider:

i. Soils permanently submerged under water of various and varying depths and which include bottom deposits and permanent swamps; the water may be fresh or salt.

ii. Soils periodically waterlogged by submergence; for example, certain coastal regions diurnally flooded with sea water, and riverain floodplains. In East Pakistan, for example, vast tracts of country are flooded annually to considerable depths during the monsoon seasons and yet become dry later in the season. It is not uncommon to have soils dry to a depth of 1–2 metres which will be under 6 metres of water in a few months' time, and vice-versa. In East Africa certain areas alternate between being lakes and dust bowls according to the season.

iii. Soils which perhaps are more truly what is conceived as waterlogged, that is, completely and permanently saturated and with some, but not much surface water such as many swamps and bogs.

iv. Soils with a high and fluctuating water table, which are thus periodically waterlogged but sometimes partly aerobic. Naturally occurring paddy soils fall into this group.

v. Soils periodically and artificially waterlogged by irrigation. This group includes rice-growing soils contained within polders.

There are special cases where waterlogging is associated with other soil features such as salinity or acidity. For example, some years ago the water table of the Indus Plain in West Pakistan was not too high and represented an equilibrium between infiltrating river and rain water from the north and underground streams and evaporation in the south. When the extensive irrigation canals of the country were constructed, excess water seepage destroyed the equilibrium and the water table rose. Now, large areas of what was once agricultural land are waterlogged and non-productive. Where the water table is very close to the surface, capillary rise of salts takes place from the groundwater, and by evaporation at the surface, causes salinity. Sometimes the soil becomes alkali as well which aggravates the problem (see Chapter 6).

The problem of acid–sulphate soils has been discussed in Chapter 13, but to briefly recapitulate: land, usually coastal, which has been subjected to prolonged flooding with sea water accumulates sulphur. The sulphur is present largely as polysulphides and if for any reason the land is allowed to dry, as for example by poldering, strongly acid and toxic conditions are produced which are not necessarily removed by re-waterlogging.

We are not concerned here with the economic problems of waterlogged soils but with how to analyze and investigate them.

It may be that the physical properties of wet soils are of highest importance but the chemical properties which are a natural consequence are also important. The chemical effects resulting from a high or fluctuating water table and of impeded drainage must be considered in relation to the availability of plant nutrient elements and of possible toxic substances. In view of the desperate need to increase the world's available agricultural land it is essential to develop and utilize its wet soils. In West Africa, for example, it is estimated that over 7500×10^6 m^2 (750 000 ha) of fertile, coastal swamp land lie unexploited.

Rice is the principal wet soil crop and nearly all research on waterlogged soils has been directed to the understanding and development of 'paddy' soils. The other important aspect of the subject is usually the concern of hydrologists and biologists and deals with the fertility of lakes and ponds.

18:1:2 Chemistry

The chemical properties of waterlogged soils depend mainly upon the fact that they are oxygen deficient. A deficiency of oxygen leads to the aerobes quickly exhausting the available supply and then disappearing or becoming inactive. Facultive and obligate anaerobes then become dominant in decomposition processes and utilize oxidized substances in the soil, such as nitrates, as hydrogen acceptors. The soil rapidly changes from an oxidizing system to a reducing system. The process is closely interrelated with oxidation–reduction and has been partly discussed in Chapter 17. The study of Chapter 17 is essential for a proper understanding of the problems of analyzing a waterlogged soil.

It is possible for a waterlogged soil to be oxidizing at its immediate surface and this is so if the soil is submerged beneath oxygenated water, as in certain lakes for example. Seasonal changes can deplete the hypolimnion of oxygen resulting in the reduction of the oxidized soil surface with concomitant release of nutrient elements into the water. This is important to plant and animal life in lakes and has been studied by Mortimer (1941) who demonstrated that the nutrients are adsorbed on to iron(III)–humate complexes on the oxidized mud surface; reduction of the iron destroys the complexes.

Sometimes an oxygen-depleted, waterlogged soil contains more oxygen at depth and this is especially so if the soil has been flooded from above. In such event reduced, soluble complexes of iron and manganese are leached downwards and are commonly re-oxidized and precipitated at depth. A fluctuating water table can often be traced by local deposits of iron and manganese.

The most important chemical changes to consider are those concerning reduction of iron, manganese and sulphate, the decomposition of organic matter, nitrogen changes and changes in reaction.

Anaerobic decomposition and nitrogen changes

In the first place, decomposition of organic matter proceeds more slowly than in aerobic soils and the products are different. Plant residues tend to accumulate, as witness peat bogs and fibrous mangrove swamps. Among the decomposition products are ammonia, carboxylic acids, amines, methane, carbon dioxide and reduced sulphur compounds. Ponnamperuma (1955) considered that the objectionable odour of some waterlogged soils is due to one or more of the following organic matter decomposition intermediates: skatole from tryptophan, tetramethylenediamine from arginine, pentamethylenediamine from lysine and mercaptans and hydrogen sulphide from cystine and methionine. While on the subject of 'smelly' muds, it is of interest that the mud of Lake Victoria in East Africa does not have an offensive odour unless it is first boiled (Hesse, 1958); this is presumably a partial sterilization or other effect upon the micro-organisms.

Work carried out at the International Rice Research Station (1963) showed a highly significant correlation between the total organic matter content of a soil and the oxidizable organic material in the waterlogged soil solution, and which competes with plant roots for any available oxygen. The correlation equation obtained was,

$$(ox.) = 0.54 + 2.48(OM) \qquad r = 0.749 \qquad [18:1$$

where $(ox.)$ is the peak oxidizable organic matter in m.e. dm^{-3} of soil solution and (OM) is percentage of organic matter in the soil.

If a waterlogged soil has a low organic matter content the intensity of reduction will be less, and certain chemical transformations may take different paths than when actively decomposing organic matter is present. For example, Subrahmanyan (1927) found that nitrate added to waterlogged soil without organic matter was reduced to ammonium without any reduction in total nitrogen content. If glucose was added also, denitrification occurred and nitrogen was lost. Denitrification takes place more completely if the water table fluctuates and the soil is periodically aerated and this results in high C/N ratios.

16*

The capacity of a soil to produce ammonium during waterlogged incubation is used as an index of plant-available nitrogen (Chapter 10, section 10:2:5). Research at the International Rice Institute led to the kinetics of ammonium formation being described by the equation,

$$\frac{A - y}{A} = e^{-ct} \qquad\qquad [18:2$$

where A is the mean maximum quantity of ammonium-nitrogen in parts per million of oven-dry soil, y is the actual concentration of ammonium-nitrogen t days after flooding and c is a parameter depending upon the kind of soil.

The value A is a characteristic for a given soil at given temperature and was calculated from organic content by the expression,

$$A = 14 \cdot 14 + 38 \cdot 44 \ (\% \ \text{organic matter}) \qquad r = 0 \cdot 810 \qquad [18:3$$

Equally significant was the time required for a soil to liberate half the maximum amount of ammonium-nitrogen.

A secondary effect of ammonia production is the release into solution of exchangeable cations such as calcium, magnesium and potassium by the process of cation exchange. It has been claimed that the often observed rise in pH values of soils newly flooded is due to production of ammonium (Metzger, 1930; Subrahmanyan, 1927), and Subrahmanyan related pH with $\lg [NH_4^+]$.

Iron(III)–iron(II) systems

The theory of changes in the state of oxidation of iron has been discussed in Chapter 17. In waterlogged soils the iron systems nearly always dominate the chemical equilibrium.

Experiments at the International Rice Institute using thirty-one different soils resulted in the derivation of equations enabling prediction to be made of iron(II) concentrations in any given soil at any given time after flooding. For example, for a moderately acid, fine sandy loam with 8% organic matter the equation was,

$$\lg [Fe^{2+}] = 0 \cdot 438 + 2 \cdot 235 \lg t - 0 \cdot 744 \ (\lg t)^2 \qquad [18:4$$

where $\lg [Fe^{2+}]$ is the concentration of iron(II) in $\mu g \ g^{-1}$ soil and t is the time in days of flooding. The equations agreed well with observed values and helped to predict availability or deficiency of iron for rice plants.

It should be noted that not all the iron(II) in a waterlogged soil is plant-available and less than five per cent has been found in ionic form. In calcareous soils which are waterlogged iron(II) often occurs as the carbonate siderite.

Elements other than iron are similarly reduced and made soluble, for example, cobalt, nickel, molybdenum and manganese. The kinetics of manganese transformations are similar to those for iron and its reduction appears to depend mainly upon the concentration of 'active' manganese (section 14:3:1).

Ng and Bloomfield (1962) found that copper was released in water-logged soils but if organic matter was present in quantity the copper(I) ions were complexed out of solution. In general, trace nutrient element deficiencies do not occur in waterlogged soils.

Changes in the iron(III)–iron(II) system lead to secondary chemical effects. One of the most important of these is the release of available phosphorus from iron(III) phosphates in waterlogged, acid soils (Shapiro, 1958). Mortimer (1941) found that reduction of iron(III) hydroxide–silica complexes released water-soluble silica.

Changes in reaction

It is a general observation that acid soils become less acid when water-logged. This has been attributed to release of bases, and the measurement of the amount by which a soil becomes less acid has been used as an index of increased solubility of bases. As already mentioned, some workers have attributed the rise in pH of soils when waterlogged to ammonia formation; others have cited sulphate reduction. Schollenberger (1928) considered that manganese reduction was responsible and Joffe (1935) cited, amongst other things, iron(III) reduction.

Not so commonly reported is an observed decrease in pH values of calcareous and alkali soils when waterlogged.

A study of changes in reaction when acid, calcareous and alkali soils are waterlogged was made by Ponnamperuma *et al.* (1966) and as an overall conclusion, they found that the changes are governed by the partial pressure of carbon dioxide in soil and by the $Fe(OH)_3$–Fe^{2+} system.

In waterlogged soil the diffusion of carbon dioxide is restricted and the gas tends to accumulate. The effect of carbon dioxide upon the reaction of calcareous and alkali soils has been demonstrated by Whitney and Gardner (1943), Yaalon (1957) and Turner (1958) and has been discussed in Chapter 3. Thus in waterlogged soils with high pH, an accumulation of carbon dioxide can be expected to affect the reaction.

Ponnamperuma calculated partial pressures of carbon dioxide from the expression,

$$P_{CO_2} \text{ (atm.)} = \frac{a_{H^+} \times a_{HCO_3^-}}{k K_1} \qquad [18{:}5$$

where k is the solubility coefficient of carbon dioxide in water and K_1 is the first dissociation constant of carbonic acid. The activities were calculated from the Debye–Hückel equation,

$$-\lg f = \frac{A z^2 . \sqrt{\mu}}{(1 + aB \sqrt{\mu})} \qquad [18{:}6$$

It was found that the reduction in pH of calcareous and alkali waterlogged soils could be explained quantitatively by one or more of the systems,

$$
\begin{array}{llll}
Na_2CO_3 & - & H_2O & - & CO_2 \\
CaCO_3 & - & H_2O & - & CO_2 \\
MnCO_3 & - & H_2O & - & CO_2 \\
Fe_3(OH)_8 & - & H_2O & - & CO_2
\end{array}
$$

Equations for each of the systems were calculated to relate the pH changes with P_{CO_2}. For the calcium carbonate system, for example,

$$pH = 6.00 - \tfrac{2}{3} \lg P_{CO_2} + \tfrac{1}{3} \lg f_{HCO_3^-} - \tfrac{1}{3} \lg f_{Ca^{2+}} \qquad [18:7$$

By assuming corrections for ionic strengths (μ) to be small, it was deduced that the pH $-$ lg P_{CO_2} curves should have slopes of -1.0 for alkali soils, $-\tfrac{2}{3}$ for calcareous soils and $-\tfrac{1}{3}$ for iron-rich soils. Practical observations agreed with this deduction.

The experiments of Ponnamperuma showed that nearly all soils except acid sulphate soils, regardless of their initial pH value, assumed neutrality after about 15 weeks of waterlogging. A correlation was obtained between waterlogged pH values and the partial pressures of carbon dioxide and which could be expressed by,

$$pH = 6.45 - 0.36 \lg P_{CO_2} \qquad r = -0.651 \qquad [18:8$$

In general the chemical kinetics of waterlogged soils can be expressed by mathematical functions of the type

$$
\begin{aligned}
y &= a + bx, \\
\lg y &= a + b \,(\lg x) + c \,(\lg x)^2, \\
y &= a + bx + cx^2, \text{ and} \\
\lg (A - y) &= \lg A - cx
\end{aligned}
$$

and the parameters for particular elements can be correlated with the properties of the soils when in the dry condition.

18:1:3 Classification of wet soils

There is a need for a scheme of classification of wet soils to parallel that for dry soils, but to date no completely satisfactory scheme has been devised. Dudal (1957) presented a classification of 'paddy' soils using conventional terms. A system of classification based upon drainage has been proposed but the different conditions of drainage have not been adequately defined. Jeffery (1963) suggested a classification based upon the state of reduction. As an example Jeffery would consider a wet soil as a 'swamp-rice soil' when maintained as an iron(III)–(II) soil where all sulphide has been precipitated and iron(II) ions exist in solution at concentrations greater than 10^{-4} mol dm^{-3}.

18:2 FIELD SAMPLING, STORAGE AND PREPARATION

18:2:1 Introduction

In the investigation of waterlogged soils there is a basic difference between the analysis of naturally occurring waterlogged soil and that of pre-dried, artificially waterlogged soil. For laboratory prepared waterlogged soils no special problems or techniques have to be considered other than those already pertaining to the sampling and preparation of dry soils. The changes in behaviour of a soil caused by drying have been discussed in Chapter 2 and these should be borne in mind when re-wetting dried soils.

18:2:2 Field sampling

It was considered by Beacher (1956) that the time of sampling a waterlogged soil may be more important than the method. This is especially so if the soil in question has a fluctuating water regime but the method of sampling is of importance for certain analytical purposes. The time of sampling will depend largely upon the purpose of analysis and usually samples will be collected at several different times. For soils which have alternating wet and dry conditions, samples should be taken during both periods as this will assist in the interpretation of results.

Beacher also recommended collection of the sample when it is wet but not saturated. Beacher was primarily considering paddy soils; not all waterlogged soils can be collected in this manner and techniques must be evolved and improved for the collection of not only saturated, but submerged soils.

For certain purposes composite sampling may be sufficient or even best, but experience has shown that waterlogged soils can vary considerably over relatively short distances. Thus in most cases analyses of individual samples can be better correlated and interpreted.

If it is at all possible, sampling should be done according to the profile of the soil rather than merely topsoil and subsoil. In waterlogged, and particularly in submerged, soils this is difficult. Preliminary auger borings sometimes help to reveal a profile and in all cases it is a wise precaution to try and sample separately the surface layer of soil as shallowly as possible; oxidized surface layers are usually only 1–2 mm thick.

Wet soils which are to be analyzed while still wet should not be collected in bags—even plastic bags. The only suitable containers are rigid glass or polythene vessels. The container should be filled as completely as possible, leaving no air space at the top; as there is nearly always an excess of water present, the jar or tube should be topped up with water to the very brim. A rubber bung can then be inserted in such a way (by slightly compressing it) as to displace water without admitting air. Screw caps on bottles are not satisfactory except the type which hermetically seal the jar over a glass lid. Chaudhry and Cornfield (1966) suggested using glass containers from which all air had been displaced with carbon dioxide.

This procedure may be a little difficult in practice, but in any case, nitrogen would be better than carbon dioxide for most subsequent analyses.

Collection of samples in this general way will be found fairly satisfactory for a number of analyses; for example, total quantities of elements are not likely to alter and polysulphides, carbon, carbonates, soluble salts and so on will not be appreciably affected. Such analyses as monosulphide will have to be made soon after sampling and, as discussed in Chapter 17, initial changes in reducing conditions caused by sampling are usually overcome by a short time of anaerobic storage. Certain properties, however, such as the immediate nitrate, nitrite or ammonium content, will be irrevocably changed if the soil sample is handled and transported. This difficulty can be partly overcome by on-the-spot treatment; for example, nitrate could be extracted into potassium chloride solution and preserved with toluene, but such analyses are seldom required.

The precise technique of taking the sample will depend upon the size of sample required and upon its accessibility. Upper horizons of soils which are merely saturated or under shallow water can be sampled with a small trowel or spatula. If the soil is sufficiently soft, the receiver itself can be used to scoop out a sample and for smaller samples a cork-borer or glass tube can be used as a small auger.

Deeper samples can be obtained with a post-hole auger and with suitable soils, deep and very deep samples can be obtained with a peat-borer. A peat-borer of the Djos type has been found exceptionally useful for obtaining deep samples of waterlogged soils. The borer is so constructed that the sampler section (which is, or can be, graduated in centimetres) remains closed when the borer is being screwed into the soil. When the sampler has reached the required depth (the extendable shaft is graduated) the handle is turned anti-clockwise. The action opens the sample tube and fills it by the rotary action of a sharp cutting blade. The handle is then turned clockwise again, which shuts the sampler tube and protects the sample from contamination whilst it is brought to the surface. Using a Djos peat borer, undisturbed samples of mud in Lake Victoria were collected to a depth of fifteen metres below the surface of the mud and under nine metres of water (Hesse, 1957). Furthermore, the sample is contained under practically air-tight conditions, and on opening the tube, accurate sub-sampling down the metre-length profile can be done. Measurements such as pH or *Eh* can be made immediately by inserting electrodes at various points without disturbing the sample.

Soft surface horizons of waterlogged soil can be sampled without disturbing the profile by inserting a wide glass tube into the soil and closing its ends *in situ*. The top of the tube should be closed before the bottom to avoid disturbance of the soil in the tube. An automatic sampler of this type (the Jenkin sampler) was devised at the laboratory of the Freshwater Biological Association in England (Plate 11). The apparatus, which is complex and expensive, was designed to sample surface layers of lake deposits and is gently lowered with the glass tube open at both ends to the mud surface. The glass tube penetrates the soft mud, and as soon as the legs of the apparatus take the weight off the lowering cable, a spring is released which very slowly closes the two ends of the tube with covers. Alternatively, a sliding weight can be sent down the cable to release the mechanism at any desired time. The sample enclosed in the tube together with its overlying water is then drawn to the surface.

18:2:3 Storage and preparation

In general, waterlogged soils should not be stored except for a short time. For measurement of oxidation–reduction potentials, storing field-collected samples under water and in the dark for 1–2 days is necessary to re-impose the natural reducing conditions disturbed by sampling and transport. Such storage will also restore the equilibrium of the main oxidation–reduction systems (in well poised soils), permitting analysis for iron(II) for example.

Many investigators have air-dried wet soils and prepared and stored them in the usual manner (Chapter 2). The samples are then analyzed as a dry soil or are re-waterlogged for definite periods of time as necessary. For many purposes drying waterlogged soils is convenient and permissible. For total analyses, carbon analysis, soluble salts and carbonates, little change will be caused by drying and undoubtedly weighing and sub-sampling will be more simple.

For other analyses re-submerging the samples is permissible and the procedure has the advantage that weighing of the sample can be done when it is dry. The method has been extensively used in fertility experiments and correlation of results can be made with field observations. However, air-drying may change the availability of nutrient elements in an inconsistent manner and care must be taken that the relevant irreversible changes do not occur when a waterlogged soil is dried. For example, sulphur-rich wet soils which, when dried, produce acid sulphate soils are not necessarily reduced to the same state on waterlogging again; often large amounts of iron and aluminium are released by the acid conditions. These are the sort of facts that can be determined by simple experiments and in any case it is always of interest to see how a waterlogged soil will be affected by drying. Drying can be done in several ways; normal air-drying, rapid drying with a warm air draught, very slow, controlled air-drying and drying under anaerobic conditions. Fairly rapid drying will probably cause fewer changes than slow drying as certain transformations depending upon micro-organisms will not have time to occur. Small samples can be dried in an oxygen-free atmosphere for specific analyses. Thus Hart (1961) dried river muds over phosphorus pentoxide in a desiccator containing a beaker of alkaline pyrogallol prior to determination of elemental sulphur. This procedure could be made more effective by flushing out the desiccator with nitrogen before evacuation; alternatively, soils could be dried in a current of pure, dry nitrogen.

If a wet soil is dried and then re-waterlogged it should not, when dry, be ground in a metal grinder. Traces of metal are liable to give rise to nascent hydrogen when the soil is again anaerobic, which causes excessive reduction to take place (Moraghan, 1963).

For many analyses, particularly those in connection with fertility studies, it has been shown that, in view of the errors caused by transportation and collection of waterlogged soil, there is no point in subjecting routine samples to submergence before analysis. Basic information of soil factors existing in the natural flooded state are obtained and correlated with the results of dried soil analyses. Thus at the West African Rice Research Station (1961), it was found that for the extraction of available phosphorus from paddy soil it was immaterial whether the fresh wet soil, air-dried soil, finely ground dry soil or dried and re-submerged soil was used.

Robinson (1930) collected waterlogged soil solutions by inserting a filter candle into the soil and connecting to a bottle which could be exhausted with a hand pump. Ponnamperuma *et al.* (1966) mixed air-dry soils with krilium before waterlogging and collected the soil solution by gravity percolation through a sintered glass plug. The solution passed into a flask which contained liquid paraffin and nitrogen.

18:3 LABORATORY SAMPLING OF WET SOILS

Sub-sampling a waterlogged soil for analysis can be extremely difficult and leads to serious errors, particularly as most wet soils contain partially decomposed pieces of organic matter. Sieving is virtually impossible, and in any case undesirable as the sample should be disturbed as little as possible.

As far as can be managed, sub-sampling should be avoided and the complete sample analyzed. For certain experiments this is not too difficult. For example, in nitrogen incubation studies involving determination of total nitrogen (as for denitrification investigation), the incubation should be done in a Kjeldahl flask; the sample can then be digested as a whole without further sampling. Similarly, samples incubated in flasks to study nitrate and ammonium changes or changes in phosphorus solubility should be extracted as a whole in the incubation flasks. When a dry soil, which is re-waterlogged for specific experiments, can be weighed while it is still dry, excellent reproducibility can be obtained by extracting the whole sample.

Brogan and co-workers (1963) made soils to a slurry with water and sampled by volume with a syringe. This procedure was compared with drying, sieving and weighing using thirty-seven soils. Reproducible results for pH determinations were obtained, but not for phosphorus determinations. The inference from this result is that quantitative sampling was not achieved, but the idea of sampling wet soils by volume merits further study.

For samples containing heterogeneous amounts of macro-organic matter which is almost impossible to remove, it may be best to use a high-speed blender to homogenize the sample. Sampling can then be done by weight, using a tared receptacle or by volume. Whichever method is decided upon it is imperative to sample the wet soil for moisture determination at the same time and at every time.

Very wet soils which settle under excess water must be separated from that water and this can be done by decantation or even filtration, although centrifuging is probably best.

The need for replication of experiments and analyses when dealing with waterlogged soils is of even greater importance than with dry soils owing to the difficulty of handling the samples. Experiments should be made to see if dry soil or dried and re-wetted soil can be used instead of the fresh, wet soil.

18:4 SPECIAL ANALYTICAL METHODS

18:4:1 Procedures
Remarkably little research has been made into methodology as applied to wet soils. This may be because, in general, special procedures are not necessary. The Kjeldahl method for total nitrogen is perfectly adequate, although for including certain forms of nitrogen such as nitrite, the Bremner–Shaw modification is to be preferred to the routine salicylic acid method (see section 10:2:1).

Nitrites, nitrates and ammonium should be extracted with neutral potassium chloride solution to avoid loss of nitrite. For the determination of plant-available nitrogen anaerobic incubation of the sample is best (section 10:2:5). Results, however, are dependent upon crumb size and for small scale experiments, on pre-drying of the sample which entails errors due to the drying effect. The effects upon nitrogen (for example) mineralization by pre-drying must therefore be taken into consideration and can be estimated by a separate experiment.

The Walkley–Black procedure is suitable for organic carbon determinations unless the soil contains much organic matter. As subsampling would introduce errors it is best to estimate organic carbon in such soils by the low temperature loss on ignition method (section 11:2:2).

For extraction of exchangeable and easily soluble cations the usual buffered solutions are satisfactory. For example, ammonium acetate extracts can be used for the determination of calcium, magnesium, potassium, sodium, manganese, aluminium and phosphorus. When extracting a wet soil the water already present should be taken into account and should also be analyzed separately. It is often only necessary to add a calculated amount of solid ammonium acetate or potassium chloride to the wet sample instead of a solution. Using wet soils in this way, however, creates problems of obtaining uniform soil–extractant ratios.

Several workers have reported Olsen's hydrogen carbonate method (section 12:2:5) as giving highest correlations between available phosphorus and crop response in waterlogged soils. This is probably a pH effect, and with acid sulphate soils in West Africa, a dilute hydrochloric acid extraction at pH 0·0 was found best (West African Rice Research Station, 1961). Extractants containing acetic acid are not recommended for phosphorus extraction owing to the weak replacing power of the acetate ion.

18:4:2 Calculation and interpretation

The results of an analysis of a waterlogged soil are difficult to interpret. One problem is whether to try and interpret on a wet basis or to convert to an oven-dry basis. It must be remembered that in all cases the moisture content exceeds 100% on a wet basis and results calculated to oven-dry soil may be quite misleading, except for purposes of comparison. Another pertinent factor is the density of the soil. Quite often an organic, waterlogged soil is very bulky for its weight and proposals have been made to express analytical results on a volume basis, the 'm² 15-cm' ('acre six-inch') concept, as being more meaningful.

For specific soils with known characteristics the analysis of the dry soil

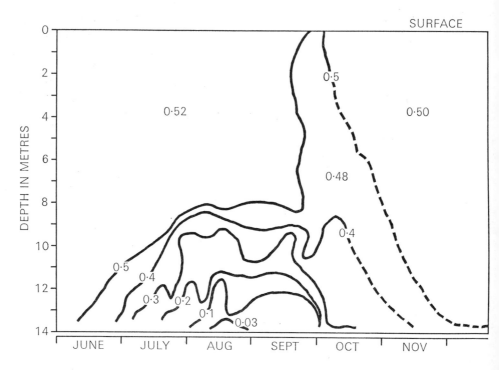

Figure 18:1 An example of a time–depth diagram with plotted *Eh* isopleths

can be correlated with wet soil analyses or with field properties and behaviour. Examples of this are given by the equations derived at the International Rice Research Institute (section 18:1).

Mortimer (1941) used a method of presenting analytical results of submerged soils so as to give time–depth pictures. An example of this is given in Figure 18:1. Mortimer intended this presentation for results of water analysis but there is no reason why it should not be used for wet soil profiles.

18:5 SPECIAL LABORATORY TECHNIQUES FOR STUDY

For a simple investigation of the fertility of waterlogged soils Jeffery (1961) used polythene buckets with no drainage holes. Bulk samples of soil were kept flooded in the buckets for several weeks before experiments commenced. Thereafter water content was kept up by daily weighing and, in some cases, tidal wash was simulated by pouring off the standing water every morning and reflooding in the evening.

Ng and Bloomfield (1962) mixed 2·5 g of air-dry soil with 7·5 cm³ of water in 50-cm³ conical flasks. Treatments were given as required and the flasks flushed out with nitrogen. Before incubation a small vial was placed in the flask containing moist, copper-activated steel wool to remove traces of oxygen; the flasks were then sealed. For many purposes simple incubation under water is sufficient.

For investigating the effect of a fluctuating water table a reservoir (separating funnel) of water can be connected to the bottom of a leaching tube containing the soil sample. The water level can be altered by raising or lowering the reservoir and a refinement is to attach the reservoir to a slowly revolving wheel which can be adjusted to give a regulated flooding and draining of the sample (to imitate diurnal flooding by a river for example).

Mortimer (1941) devised a simple piece of apparatus for studying variations with time of pH, *Eh* and conductivity of lake deposits and their overlying water. Mortimer's apparatus is shown in Figure 18:2 and consists of pairs of tanks, one for aerobic studies and one for anaerobic. The tanks were square battery jars and the rest of the apparatus is self-explanatory. Mud samples collected with a Jenkin sampler were separated into two parts, one being the shallow oxidized surface layer. The lower horizon of mud was poured into the jars to give a layer of 3 cm depth and the jars were filled with tap water. The surface mud was then mixed with water and added to the tanks so that it gently settled back on to the surface of the reduced mud. The water was then siphoned off and the muds left to stand for several days. Lake water was then carefully run in without disturbing the mud and to a depth of 20 cm. The tanks were left for 1 week, shielded from dust but in the light.

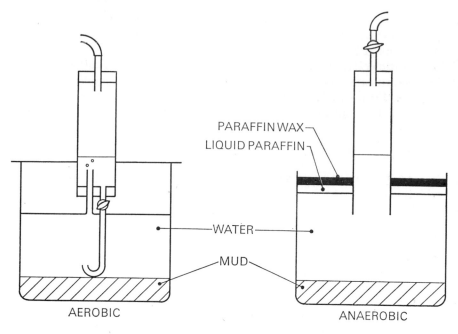

PARAFFIN WAX
LIQUID PARAFFIN

WATER

MUD

AEROBIC

ANAEROBIC

Figure 18:2 Apparatus for studying submerged soils under aerobic and anaerobic conditions (Mortimer, 1941; *J. Ecol.*)

In the aerobic tank the wide glass tube was adjusted so that when suction was applied water travelled up the short (4 mm bore) tube until the level in the main tank dropped below the short tube when air was sucked in. This aerated the water in the wide tube before it flowed back via the J-shaped tube (2 mm bore).

After inserting the wide tube in the anaerobic tank, the water surface was covered with liquid paraffin to a depth of 2 cm. Molten paraffin wax was then poured carefully on top and allowed to harden. Measuring instruments, for example electrodes, could be introduced through the wide tube. After measurements had been made, water was drawn up by suction until no air remained beneath the paraffin and the tap closed. Before the next measurements were taken the tap was opened and the water allowed to drop gently to its own level. Samples of water could be removed for analysis with a small pipette with its tip bent into a J-shape to avoid disturbing the mud surface.

Scott and Evans (1955) designed a special cell for measuring dissolved oxygen in water-saturated soils. The cell which was used in conjunction with a recording polarograph, is shown in Figure 18:3. It is a glass cylinder with five openings along one side and one on the opposite side. A plastic cylindrical screen is fitted between the two openings opposite to each other. The cell is filled with air-dry soil and water of known oxygen content is allowed to flow in from the left hand side; as the soil opposite each opening becomes saturated the openings are stoppered or electrodes are in-

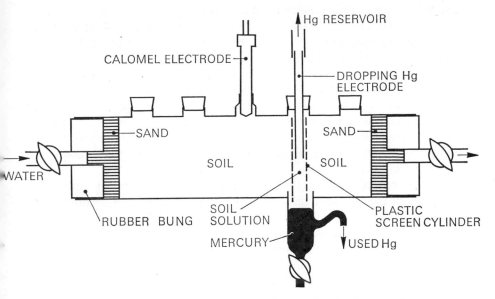

Figure 18:3 Cell used for the determination of dissolved oxygen in water-logged soils (Scott and Evans, 1955; Soil Science Society of America, *Proceedings*)

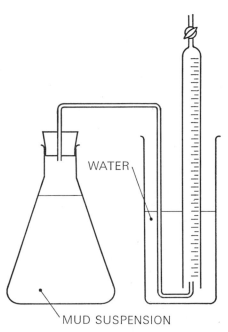

Figure 18:4 Apparatus for measuring and collecting evolved gases during anaerobic decomposition of waterlogged soil (Allen *et al.*, 1953; *J. Hyg.*)

serted. Polarographic measurement is made of dissolved oxygen when the soil is saturated completely, the soil solution acting as base electrolyte. Full experimental details were given by Scott and Evans and can be found also in 'Methods of Soil Analysis' (*Agronomy* 9, 1966). Scott and Evans found that as soon as the soil was completely flooded the oxygen content dropped and completely disappeared within 10 hours. Subsequently introduced oxygenated water was de-oxygenated more quickly than that originally introduced. Armstrong (1967) pointed out that the apparatus of Scott and Evans does not permit measurement of *Eh* and oxygen at the same point. Armstrong thus replaced the dropping mercury electrode with a platinum micro-electrode for this purpose.

Allen and co-workers (1953) used the apparatus shown in Figure 18:4 to measure the volume of gases evolved from an anaerobically decomposing soil. The macro-respirometer of Birch and Friend (section 11:2:7) has been used (Hesse, 1958) to study the course of organic matter decomposition in waterlogged and submerged soils in contact with oxygen.

Appendix I

Element		Atomic weight	Standard solution
Aluminium	(Al)	26·98	1·757 g dm^{-3} Al$_2$(SO$_4$)$_3$ 100 μg cm^{-3} Al
Antimony	(Sb)	121·76	0·100 g Sb in 25 cm^3 hot conc. H$_2$SO$_4$ and dilute to 100 cm^3. Make to 1 dm^3 with 1:3 H$_2$SO$_4$ 100 μg cm^{-3} Sb
Arsenic	(As)	74·91	1·320 g As$_2$O$_3$ in 500 cm^3 water plus 10 cm^3 35% w/v NaOH. Neutralize with H$_2$SO$_4$ and dilute to 1 dm^3 1000 μg cm^{-3} As
Barium	(Ba)	137·36	0·1779 g dm^{-3} BaCl$_2$.2H$_2$O 100 μg cm^{-3} Ba
Beryllium	(Be)	9·01	By dilution of standardized BeCl$_2$ solution.
Bismuth	(Bi)	209·00	0·2321 g Bi(NO$_3$)$_3$.5H$_2$O in 5 cm^3 1:10 HNO$_3$ and dilute to 1 dm^3 100 μg cm^{-3} Bi
Boron	(B)	10·82	0·5716 g dm^{-3} H$_3$BO$_3$ 100 μg cm^{-3} B
Bromine	(Br)	79·92	0·1489 g dm^{-3} KBr 100 μg cm^{-3} Br
Cadmium	(Cd)	112·41	0·1631 g dm^{-3} CdCl$_2$(anhyd.) 100 μg cm^{-3} Cd
Caesium	(Cs)	132·91	0·1267 g dm^{-3} CsCl 100 μg cm^{-3} Cs
Calcium	(Ca)	40·08	0·2497 dry CaCO$_3$ in 5 cm^3 HCl; boil out CO$_2$ and dilute to 1 dm^3 100 μg cm^{-3} Ca
Carbon	(C)	12·01	———
Cerium	(Ce)	140·13	0·1228 g CeO$_2$ in 10 cm^3 1:2 H$_2$SO$_4$ with heat. Dilute to 1 dm^3 100 μg cm^{-3} Ce
Chlorine	(Cl)	35·46	0·1648 g dm^{-3} NaCl 100 μg cm^{-3} Cl
Chromium	(Cr)	52·01	0·283 g dm^{-3} dry K$_2$Cr$_2$O$_7$ 100 μg cm^{-3} Cr
Cobalt	(Co)	58·94	0·2630 g dry CoSO$_4$ in 50 cm^3 water plus 1 cm^3 conc. H$_2$SO$_4$. Dilute to 1 dm^3 100 μg cm^{-3} Co
Copper	(Cu)	63·54	0·1964 g dm^{-3} CuSO$_4$.5H$_2$O 100 μg cm^{-3} Cu
Fluorine	(F)	19·00	0·2211 g dm^{-3} NaF 100 μg cm^{-3} F
Germanium	(Ge)	72·60	0·1441 g GeO$_2$ in water. Add few drops 12% NaOH, dilute, neutralize with 1:1 HCl, add drops HCl in excess and dilute to 100 cm^3 10 μg cm^{-3} Ge
Gold	(Au)	197·00	0·1008 g Na chloroaurate in 25 cm^3 1·0M HCl; dilute to 500 cm^3 100 μg cm^{-3} Au
Hydrogen	(H)	1·01	———
Iodine	(I)	126·91	0·1308 g dm^{-3} KI 100 μg cm^{-3} I

Element		Atomic weight	Standard solution
Iron	(Fe)	55·85	0·7022 g dm^{-3} ammonium iron(III) sulphate plus 5 cm^3 H$_2$SO$_4$ 100 μg cm^{-3} Fe
Lead	(Pb)	207·21	0·1599 g Pb(NO$_3$)$_2$ in 1% HNO$_3$ and dilute to 1 dm^3 100 μg cm^{-3} Pb
Lithium	(Li)	6·94	0·6109 g dm^{-3} LiCl 100 μg cm^{-3} Li
Magnesium	(Mg)	24·32	1·0131 g dm^{-3} MgSO$_4$.7H$_2$O 100 μg cm^{-3} Mg
Manganese	(Mn)	54·94	0·3076 g dm^{-3} dry MnSO$_4$ 100 μg cm^{-3} Mn
Mercury	(Hg)	200·61	0·1354 g dm^{-3} HgCl$_2$ 100 μg cm^{-3} Hg
Molybdenum	(Mo)	95·95	0·150 g MoO$_3$ in 10 cm^3 1·0 M NaOH, acidify with HCl and dilute to 1 dm^3 100 μg cm^{-3} Mo
Nickel	(Ni)	58·71	0·4479 g dm^{-3} NiSO$_4$.6H$_2$O 100 μg cm^{-3} Ni
Nitrogen	(N)	14·01	0·236 g dm^{-3} (NH$_4$)$_2$SO$_4$ 50 μg cm^{-3} NH$_4$-N 0·361 g dm^{-3} KNO$_3$ 50 μg cm^{-3} NO$_3$-N 0·246 g dm^{-3} NaNO$_2$ 50 μg cm^{-3} NO$_2$-N
Oxygen	(O)	16·00	
Phosphorus	(P)	30·98	0·2195 g dry KH$_2$PO$_4$ in 700 cm^3 water plus 25 cm^3 3·5 M H$_2$SO$_4$; dilute to 1 dm^3 50 μg cm^{-3} P
Platinum	(Pt)	195·09	0·1000 g cleaned Pt in 1:3 HNO$_3$-HCl. Evaporate nearly dry, add 10 cm^3 HCl and 0·06 g NaCl. Re-evaporate and dilute to 1 dm^3 100 μg cm^{-3} Pt
Potassium	(K)	39·10	0·1907 g dm^{-3} KCl 100 μg cm^{-3} K
Rubidium	(Rb)	85·48	0·1415 g dm^{-3} RbCl 100 μg cm^{-3} Rb
Selenium	(Se)	78·96	0·1404 g dm^{-3} SeO$_2$ 100 μg cm^{-3} Se
Silicon	(Si)	28·09	By dilution of analyzed sodium silicate solution.
Silver	(Ag)	107·88	0·1575 g dm^{-3} dry AgNO$_3$ 100 μg cm^{-3} Ag
Sodium	(Na)	22·99	0·2541 g dm^{-3} NaCl 1000 μg cm^{-3} Na
Strontium	(Sr)	87·63	0·2415 g dm^{-3} Sr(NO$_3$)$_2$ anhyd. 100 μg cm^{-3} Sr
Sulphur	(S)	32·07	0·5370 g dm^{-3} CaSO$_4$.2H$_2$O 100 μg cm^{-3} S
Tellurium	(Te)	127·61	0·1250 g dm^{-3} TeO$_2$ 100 μg cm^{-3} Te
Thallium	(Tl)	204·39	0·1304 g dm^{-3} TlNO$_3$ 100 μg cm^{-3} Tl
Thorium	(Th)	232·05	0·2379 g dm^{-3} Th(NO$_3$)$_4$.4H$_2$O 100 μg cm^{-3} Th
Tin	(Sn)	118·70	1·000 g Sn in 100 cm^3 1:1 HCl and dilute to 1 dm^3 with 1:1 HCl 1 000 μg cm^{-3} Sn
Titanium	(Ti)	47·90	0·1668 g TiO$_2$ fused with K pyrosulphate. Dissolve in 6 M HCl and dilute to 1 dm^3 100 μg cm^{-3} Ti
Tungsten	(W)	183·86	0·18 g dm^{-3} Na$_2$WO$_4$.2H$_2$O 100 μg cm^{-3} W
Uranium	(U)	238·07	By dilution of analyzed uranyl nitrate solution.

Element		Atomic weight	Standard solution
Vanadium	(V)	50·95	0·2295 g NH_4VO_3 in 100 cm^3 water plus 15 cm^3 1:1 HNO_3. Dilute to 1 dm^3 100 μg cm^{-3} V
Zinc	(Zn)	65·38	0·100 g Zn in 50 cm^3 water plus 1 cm^3 conc. H_2SO_4 and dilute to 1 dm^3 100 μg cm^{-3} Zn
Zirconium	(Zr)	91·22	By dilution of standardized zirconium nitrate solution.

Appendix II

STRENGTH OF COMMON ACIDS AND ALKALIS AND PREPARATION OF STANDARD SOLUTIONS

Reagent	Approximate		
	Relative density	% by weight	Preparation of 1·0 M solution
Acetic acid	1·05	99·5	58 cm^3 dm^{-3}
Ammonia solution	0·90	27	71 cm^3 dm^{-3}
Hydrochloric acid	1·18	36	89 cm^3 dm^{-3}
Hydrofluoric acid	1·15	46	38 cm^3 dm^{-3}
Nitric acid	1·42	70	63 cm^3 dm^{-3}
Perchloric acid	1·66	72	86 cm^3 dm^{-3}
	1·54	60	103 cm^3 dm^{-3}
Potassium hydroxide	—	—	56 g dm^{-3}
Sodium hydroxide	—	—	40 g dm^{-3}
Sulphuric acid	1·84	96	56 cm^3 dm^{-3}

Appendix III

CONVERSION FACTORS, DEFINITIONS AND INTERRELATIONSHIPS

Note: Where applicable, SI units are given first and are followed by non-SI units in parenthesis.

Concentration

1 microgramme $= \mu g = 1 \times 10^{-6}$ g $[= 1 \gamma]$

μg cm^{-3} $[= \gamma$ cm^{-3} = p.p.m. (in solution)]

μg g^{-1} $[=$p.p.m. (mass)]

kg per 10^4 m^2 $[=$kg/hectare $= 1 \cdot 21 \times$ lb/acre (thus lb/acre $= 0 \cdot 11$ kg per 10^3 m^2)]

g dm^{-3} $[=$g/litre]

cm^3 $[=$cc $=$ ml]

dm^3 $[=$litre (l)]

Conductance (*G*) and Conductivity (*σ*)

1 siemens $= 1$ S $= 1/$ohm $[= 1$ mho]

mS cm^{-1} = millisiemens per centimetre $[=$mmhos/cm]

Ion exchange

1 milliequivalent = equivalent weight expressed in milligrammes

Milliequivalents per cent = m.e.% = % $\times$ 1000/equivalent weight

$$\text{Percent base saturation} = \frac{\text{T.E.B.} \times 100}{\text{C.E.C.}} = \frac{S \times 100}{T}$$

where T.E.B. is total exchangeable bases in m.e.%,
C.E.C. is cation exchange capacity in m.e.%,
S-value is T.E.B. in m.e.%,
T-value is C.E.C. in m.e.%.

$T = S - \text{ex.H}^+$

Exchangeable sodium percentage = E.S.P.
$$= \text{ex.Na m.e.} \% \times 100/\text{C.E.C. m.e.} \%.$$

Colorimetry

1 nanometre (nm) $= 10^{-9}$ metre $= 1000$ pm (picometre) $[= 1$ millimicron $(m\mu) = 10$ angstrom $(10$ Å$)]$

Ultra-violet range of spectrum covers 185–400 nm

Visible range of spectrum covers 400–760 nm

Infra-red range of spectrum covers 760–15 000 nm

The Beer–Lambert Law states that $\lg I_0/I_t = \varepsilon c t$

where I_0 is incident light intensity,
  I_t is transmitted light intensity,
  c is concentration in mol dm^{-3},
  t is the thickness of the medium,
  ε is the molecular extinction coefficient.

The term $\lg I_0/I_t$ is known as the Optical Density, the Extinction or the Absorbance of the medium.

Sieve mesh size

Data for sieves showing comparison of US Standard and I.M.M. sieves

US		I.M.M.	
Openings in mm	Mesh No.	Openings in mm	Mesh No.
2·00	10	2·54	5
1·00	18	1·06	12
0·42	40	0·42	30
0·25	60	0·25	50
0·18	80	0·21	60
0·15	100	0·16	80
0·11	140	0·10	120
0·07	200	0·08	150
0·05	300	0·06	200

Notes: 1. The US Standard and the Tyler sieves can be considered as being interchangeable.
2. The openings in mm are calculated as squares.
3. Sometimes sieve sizes are given in microns; for example, a 2000 micron sieve (or 2 mm) and 150 micron (or 0·15 mm), and so on.

Appendix IV

BIBLIOGRAPHICAL NOTE

Books

ASSOCIATION OF OFFICAL AGRICULTURAL CHEMISTS (1955), *Official methods of analysis*, 8th edn. Washington, USA.

BLACK, C. A., Ed. in Chief (1965), *Methods of soil analysis* (*Agronomy* No. 9). Am. Soc. of Agron. Inc., Madison, USA.

CHAPMAN, H. D., and PRATT, P. F. (1961), *Methods of analysis for soils, plants and waters*. Univ. of Calif., Div. of Agric. Sci.

GEDROITS, K. K. (1963), *Chemical analysis of soils*. Israel Prog. for Sci. Transl., Jerusalem.

JACKSON, M.L. (1958), *Soil chemical analysis*. Constable, London.

METSON, A. J. (1961), *Methods of chemical analysis for soil survey samples*. NZ DSIR Soils Bulletin, No. 12.

MITCHELL, R. L. (1964), *The spectrochemical analysis of soils, plants and related materials*. Commonwealth Bureau of Soils Technical Communication No. 44A.

ORGANIZATION FOR EUROPEAN ECONOMIC COOPERATION (1956). *The organization and rationalization of soil analysis*. Paris.

PIPER, C. S. (1942), *Soil and plant analysis*. Univ. of Adelaide.

RICHARDS, L. A., Ed. (1954), *The diagnosis and improvement of saline and alkali soils*. U.S Dep. Agric. Handbook No. 60.

WRIGHT, C. H. (1934), *Soil analysis, a handbook of physical and chemical methods*. Thomas Murby, London.

Journals

There is no journal specifically dealing with soil analysis. The analyst can keep fairly well up to date by reading the abstract journal *Soils and Fertilizers* which has a section devoted to methods of soil analysis. The following is a short list of the most commonly encountered journals in which methodology appears. In the list of References journal titles have been abbreviated according to the *World List of Scientific Periodicals* (Butterworths, London).

Analyst. London.
Analytical Chemistry. Easton, Pa.
Australian Journal of Soil Research. Melbourne.
Canadian Journal of Soil Science. Ottawa.
Empire Journal of Experimental Agriculture. Oxford.
Irish Journal of Agricultural Research. Dublin.
Journal of Agricultural and Food Chemistry. Washington.
Journal of the Association of Official Agricultural Chemists. Washington.
Journal of the Indian Society of Soil Science. New Delhi.
Journal of the Science of Food and Agriculture. London.
Journal of Soil Science. Oxford.
New Zealand Journal of Agricultural Research. Wellington.
Plant and Soil. The Hague.
Proceedings, New Zealand Society of Soil Science. Wellington.
Proceedings, Soil Science Society of America. Madison.
Soil Science. Baltimore.
Soils and Fertilizers. Farnham Royal.

References

Chapter 1

FISHER, R.A. (1958), *Statistical methods for research workers*. Oliver & Boyd, London.
HALL, A. D. (1949), *The soil*. 5th edn., John Murray, London.
HART, M. G. R., and ROBERTS, E. H. (1961), *Nature, Lond.*, 189, 598.
HESSE, P. R. (1968), *Pak. J. Soil Sci.*, 4, 1.
PANTONY, D. A. (1961), *R.I.C. Lect. Ser. No. 2*.
RUSSELL, E. W. (1961), *Soil conditions and plant growth*. 9th edn., Longmans, London.
U.S. Dept. Agric. (1951), Handbook No. 18, *Soil survey manual*.
VERMEULEN, F. H. B. (1953), *Pl. Soil*, 4, 267.
VOGEL, A. I. (1962), *A textbook of quantitative inorganic analysis*. 3rd edn., Longmans, London.

Chapter 2

ATTOE, O. J. (1946), *Proc. Soil Sci. Soc. Am.*, 11, 145.
BARROW, N. J. (1961), *Aust. J. agric. Res.*, 12, 306.
BIRCH, H. F. (1958), *Pl. Soil*, 10, 9.
— (1959), Ibid., 11, 262.
— (1960), Ibid., 12, 81.
FRENEY, J. R. (1958), *Soil Sci.*, 86, 241.
GRIFFITHS, E., and BIRCH, H. F. (1961), *Nature, Lond.*, 189, 424.
NEVO, Z., and HAGIN, J. (1966), *Soil Sci.*, 102, 157.
SCHALSCHA, E. B., *et al.* (1965), *Proc. Soil Sci. Soc. Am.*, 29, 481.
STOTZKY, G., GOOS, R. D., and TIMONIN, M. I. (1962), *Pl. Soil*, 16, 1.
VAN SCHUYLENBORGH, J. (1954), *Neth. J. agric. Sci.*, 2, 50.

Chapter 3

BIILMAN, E. (1924), *J. agric. Sci.*, 14, 232.
BIRCH, H. F. (1953), Ibid., 43, 329.
CLARK, J. S., and HILL, R. G. (1964), *Proc. Soil Sci. Soc. Am.*, 28, 490.
CROWTHER, E. M. (1925), *J. agric. Sci.*, 15, 201.
HUBERTY, M. R., and HAAS, A. R. C. (1940), *Soil Sci.*, 49, 455.
JOHNSTON, J., and WALKER, A. C. (1925), *J. am. Chem. Soc.*, 47, 1807.
KUHN, S. (1930), *Z. PflErnähr. Düng. Bodenk.*, 18A, 309.
MARSHALL, C. E. (1953), *Proc. Soil Sci. Soc. Am.*, 17, 219.
PEECH, M., OLSEN, R. A., and BOLT, G. A. (1953), Ibid., 17, 214.
REED, J. F., and CUMMINGS, R. W. (1945), *Soil Sci.*, 59, 97.
SCHOFIELD, P. K., and TAYLOR, A. W. (1955), *Proc. Soil Sci. Soc. Am.*, 19, 164.
SØRENSON, S. P. L. (1909), *C. r. des trav. du. lab. Carlsberg*, 8, 1.
TURNER, R. C., and CLARK, J. S. (1967), *Trans. int. Soc. Soil Sci. Comm. II and IV* 1966.
— and NICHOL, W. E. (1962), *Soil Sci.*, 94, 58.
ULRICH, B. (1961), *Landw. Forsch.*, 14, 225.
WHITNEY, R. S., and GARDNER, R. (1943), *Soil Sci.*, 55, 127.

Chapter 4

ADAMS, F., and EVANS, C. E. (1962), *Proc. Soil Sci. Soc. Am.*, 26, 355.
ALBRECHT, W. A., and SMITH, G. E. (1952), *Trans. int. Soc. Soil Sci. Comm. II and IV*, 1, 119.

BRADFIELD, R., and ALLISON, W. H. (1933), *Trans. 2nd Comm. int. Soc. Soil Sci.*, A63, 79.

BROWN, I. C. (1943), *Soil Sci.*, 56, 353.

DUNN, L. E. (1943), Ibid., 56, 341.

HARWARD, M.E., and COLEMAN, N. T. (1954), Ibid., 78, 181

JACKSON, M. L. (1958). *Soil Chemical Analysis*. Constable.

KEENEY, D. R., and COVEY, R. B. (1963), *Proc. Soil Sci. Soc. Am.*, 27, 277.

MARKERT, S. (1961), *Albrecht-Thaer-Arch.*, 5, 766.

MCLEAN, E. O., SHOEMAKER, H. E., and HOURIGAN, W. R. (1960), *Trans. 7th int. Congr. Soil Sci.*, 2, 142.

MEHLICH, A. (1938), *Proc. Soil Sci. Soc. Am.*, 3, 162.

— (1948), *Soil Sci.*, 66, 429.

METSON, A. J. (1961), *New Zealand D.S.I.R. Soils Bull. No. 12.*

MIDDLETON, K. R., and WESTGARTH, D. R. (1964), *Soil Sci.*, 97, 221.

PARKER, J. (1929), *J. am. Soc. Agron.*, 21, 1030.

PEECH, M., and BRADFIELD, R. (1948), *Soil Sci.*, 65, 35.

—, COWAN, R. L., and BAKER, J. H. (1962), *Proc. Soil Sci. Soc. Am.*, 26, 37.

PETERS, H., and MARKERT, S. (1961), *Albrecht-Thaer-Arch.*, 5, 655 and 744.

PRATT, P. F., and BAIR, F. L. (1962), *Soil Sci.*, 93, 329.

PURI, A. N., and ANAND, B. (1934), Ibid., 37, 49.

RUSSELL, E. W. (1961), *Soil conditions and plant growth.* 9th edn., Longmans, p. 97.

SCHOLLENBERGER, C. J., and DREIBELIS, F. R. (1930), *Soil Sci.*, 30, 161.

SHOEMAKER, H. E., MCLEAN, E. O., and PRATT, P. F. (1961), *Proc. Soil Sci. Soc. Am.*, 25, 274.

WOODRUFF, C. M. (1948), *Soil Sci.*, 66, 53.

YUAN, T. L. (1959), Ibid., 88, 164.

Chapter 5

BASCOMB, C. L. (1961), *Chem. Ind.*, 45, 1826.

BRUIN, P. (1938), *Versl. Landbouwk. Onderz.*, 44, 693.

COLLINS, S. H. (1906), *J. Soc. Chem. Ind., Lond.*, 25, 518.

DEB, B. C. (1963), FAO report, *High dam soil survey project, Aswan.*

ERIKSON, A. E., and GIESEKING, J. E. (1947), *Soil Sci.*, 63, 451.

HORTON, J. H., and NEWSON, D. W. (1953), *Proc. Soil Sci. Soc. Am.*, 17, 414.

HUTCHINSON, H. B., and MACLENNAN, K. (1941), *J. agric. Sci.*, 6, 323.

LEHR, J. J., and WESEMAEL, J. C. VAN (1961), *Landbouwk. Tijdschr.*, 73, 1156.

PETERSEN, G. W., CHESTERS, G., and LEE, G. B. (1966), *J. Soil Sci.*, 17, 328.

PIPER, C. S. (1942), *Soil and Plant Analysis.* Univ. of Adelaide.

PRATT, P. F., and BAIR, F. L. (1962), *Soil Sci.*, 93, 239.

RUSSELL, E. W. (1961), *Soil conditions and plant growth.* 9th edn., Longmans, p. 108.

SCHOLLENBERGER, C. J. (1930), *Soil Sci.*, 30, 307.

— (1945), Ibid., 59, 57.

SCHOONOVER, W. R. (1952), *Examination of soils for alkali.* Univ. Calif.

SKINNER, S. I. M., HALSTEAD, R. L., and BRYDON, J. E. (1959), *Can. J. Soil Sci.*, 39, 197.

SOKOLOVICH, V. E. (1964), *Khimiya sel. Khoz.*, 9, 20.

TENANT, C. B., and BERGER, R. W. (1957), *Am. Miner.*, 42, 23.

TINSLEY, J., TAYLOR, T. G., and MOORE, J. H. (1951), *Analyst*, 76, 300.

TURNER, R. C. (1960), *Can. J. Soil Sci.*, 40, 219.

— and SKINNER, S. I. M. (1960), *Can. J. Soil Sci.*, 40, 232.

VOGEL, A. I. (1962), *A textbook of quantitative inorganic analysis.* 3rd edn., Longmans.

WEISSMAN, R. C., and DIEHL, H. (1953), *Proc. Iowa Acad. Sci.*, 60, 433.

WILLIAMS, D. E. (1948), *Proc. Soil Sci. Soc. Am.*, 13, 127.

WRIGHT, J. R., HOFFMAN, I., and SCHNITZNER, M. (1960), *J. Sci. Fd. Agric.*, 11, 163.

Chapter 6

BAEYENS, J., and APPELMANS, F. (1947), *Bijz. Nammer Feb.*, p. 222.

BEST, R. J. (1929), *J. agric. Sci.*, 19, 533.

— (1950), *Trans. 4th Int. Congr. Soil Sci.*, 3, 162.

BIRCH, H. F. (1959), *Nature, Lond.*, 183, 1415.

BOWER, C. A., and HUSS, R. B. (1948), *Soil Sci.*, 66, 199.

— and WILCOX, L. V. (1965), *Monog. am. Soc. Agron.*, 9, 933.

BUCHNER, A. (1951), *Z. PflErnähr. Düng.*, 55, 124.

BURD, J. S., and MARTIN, J. C. (1923), *J. agric. Sci.*, 13, 625.

DASHEVSKII, L. I. (1959), *Trudy vsesoyuz. nauch. issled. Inst. sakhar Svekly*, 34, 328.

DAVIS, R. O. E., and BRYAN, H. (1910), *U.S. Dep. Agric. Bur. Soils Bull.*, 61.

DEB, B. C. (1963), FAO report, *High Dam Soil Survey Project, Aswan*.

DELVER, P., and KADRY, L. T. (1960), *Trans. 7th int. Congr. Soil Sci.*, 2, 370.

GRIFFITH, G. AP. (1950), *Nature, Lond.*, 165, 571.

HEIMANN, H. (1958). *4th Congr. Int. Potash Inst.*, p. 173.

JACKSON, M. L. (1958), *Soil Chemical Analysis*. Constable.

LAGERWERFF, J. V., AKIN, G. W., and MOSES, S. W. (1965), *Proc. Soil Sci. Soc. Am.*, 29, 535.

LEO, M. W. M. (1963), *Soil Sci.*, 95, 142.

LONGENECKER, D. E., and LYERLY, P. J. (1964), Ibid., 97, 268.

MAGISTRAD, O. C., REITEMEIER, R. F., and WILCOX, L. V. (1945), Ibid., 59, 70.

MERKLE, F. G., and DUNKLE, E. C. (1944), *J. am. Soc. Agron.*, 36, 10.

METSON, A. J. (1961), *New Zealand C.S.I.R. Soils Bull. No. 12*.

PARKER, F. W. (1921), *Soil Sci.*, 12, 209.

PIPER, C. S. (1942), *Soil and Plant Analysis*. Univ. of Adelaide.

REITEMEIER, R. F. (1946), *Soil Sci.*, 61, 195.

— and RICHARDS, L. A. (1944), Ibid., 57, 119.

— and WILCOX, L. V. (1946), Ibid., 61, 281.

RICHARDS, L. A. (1947), *Agr. Eng.*, 28, 451.

— (1949), *Agron. J.*, 41, 446.

— (1955a), *Soil Sci.*, 80, 55.

— (1955b), Ibid., 79, 423.

RUSSELL, E. W. (1961), *Soil conditions and plant growth*, 9th edn., Longmans, p. 386.

SIMPSON, J. R. (1960), *J. Soil Sci.*, 11, 45.

TABIKH, A. A., and RUSSELL, J. (1961), *Soil Sci.*, 91, 70.

THEMLITZ, R., and ISHAKOHLN, O. (1962), *Z. PflErnähr Düng.*, 96, 126.

TINSLEY, J., TAYLOR, T. G., and MOORE, J. H. (1951), *Analyst*, 76, 300.

USDA (1954), Handbook No. 60, *Diagnosis and improvement of saline and alkali soils*, ed. L. A. Richards.

VOGEL, A. I. (1962), *A textbook of quantitative inorganic analysis*. 3rd edn., Longmans.

WADLEIGH, G. H., GAUCH, H. G., and STRONG, D. G. (1947), *Soil Sci.*, 63, 341.

WANG, C. C. (1953), *Acta pedol. sin.*, 2, 97.

WETSELAAR, B. (1960), *Nature, Lond.*, 186, 572.

WESEMAEL, J. C. VAN, and LEHR, J. J. (1954), *Trans. 5th int. Congr. Soil Sci.*, 2, 273.

WHITNEY, M., and MEANS, T. H. (1897), *U.S. Dep. Agric. Div. Soils Bull.*, 8.

WHITTLES, C. L., and SCHOFIELD-PALMER, E. K. (1951), *J. Soil Sci.*, 2, 243.

WILCOX, L. V. (1951), *Soil Sci.*, 72, 233.

YAALON, D. H. (1963), *Bull. Res. Coun. Israel*, 11G, 105.

YARON, B., and MOKADY, R. (1962), *Pl. Soil.*, 17, 392.

Chapter 7

AMER, F. (1960), *Trans. 7th. int. Congr. Soil Sci.*, 2, 53.

BAKER, A. S. (1962), *Proc. Soil Sci. Soc. Am.*, 26, 213.

BAKER, J. and BURNS, W. T. (1964), *Canad. J. Soil Sci.*, 44, 360.

17 + T.S.C.A.

BASCOMB, C. L. (1964), *J. Sci. Fd. Agric.*, 15, 821.

BERGSETH, H., HAGEBO, F. A., LIEN, H., and STEENBERG, K. (1963), *Soil Sci.*, 95, 97.

BLACK, I. A., and SMITH, E. (1950), *J. Sci. Fd. Agric.*, 7, 201.

BLUME, J. M., and SMITH, D. (1954), *Soil Sci.* 77, 9.

BOLT, G. H., and WARKENTIN, P. P. (1958), *Kolloid-Z.*, *156, Bd. Heft. 1*, 41.

BOWER, C. A. (1960), *Trans. 7th int. Congr. Soil Sci.*, 2.

—, REITEMEIER, R. F., and FIREMAN, M. (1952), *Soil Sci.*, 73, 251.

—, REITEMEIER, R. F., and FIREMAN, M. (1962), Ibid., 93, 343.

BROWN, I. C. (1943), Ibid., 56, 353.

CARLSON, R. M. (1962), *Diss. Abstr.*, 24, 911.

CHAPMAN, H. D. (1965), *Monog. am. Soc. Agron.*, 9, 891.

DE ENDREDY, A. S., and QUAGRAINE, K. A. (1960), *Trans. 7th int. Congr. Soil Sci.*, 2, 132.

DELL'AGNOLA, G., and MAGGIONI, A. (1964), *Agrochimica*, 8, 139.

EDWARDS, A. P. (1967), *Can. J. Soil Sci.*, 47, 140.

ENSMINGER, L. E. (1944), *Soil Sci.*, 58, 425.

GEDROIZ, K. K. (1924), Ibid., 16, 473.

GILL, W. R., and SHERMAN, G. D. (1952), *Pac. Sci.*, 6, 138.

GILLINGHAM, J. T. (1965), *Can. J. Soil Sci.*, 45, 102.

HAAN, F. A. M. DE, and BOLT, G. H. (1963), *Proc. Soil Sci. Soc. Am.*, 27, 636.

HALEVY, E., and TZUR, Y. (1964), *Soil Sci.*, 98, 66.

HELMY, A. K. (1963), Ibid., 95, 204.

HENDRICKS, S. B., and FRY, W. H. (1930), Ibid., 29, 457.

HISSINK, D. J. (1923), Ibid., 15, 269.

KAMPRATH, E. J., and WELCH, C. D. (1962), *Proc. Soil Sci. Soc. Am.*, 26, 263.

KANSHIRO, Y., and SHERMAN, G. D. (1960), *Adv. Agron.*, 5, 219.

KELLEY, W. P. (1948), *Cation exchange in soils*. Reinhold, N.Y.

— (1964), *Soil Sci.*, 97, 80.

— and BROWN, S. M. (1924), *Calif. agric. Exp. Stn. Tech. Pap. 15*.

— , DORE, W. H., and BROWN, S. M. (1931), *Soil Sci.*, 31, 25.

LEGG, W. T. (1963), Ibid., 95, 214.

MARSHALL, C. E. (1949), *Monog. am. Soc. Agron.*, 1, 106.

MAYWALD, H. (1962), *Z. Landw. Vers u UntersWes.*, 8, 521.

MEHLICH, A. (1942), *Soil Sci.*, 53, 1.

— (1953), *J. Ass. Off. agric. Chem.*, 36, 445.

METSON, A. J. (1961), *New Zealand C.S.I.R. Soils Bull. No. 12.*

MIDDLETON, K. R., and WESTGARTH, D. R. (1964), *Soil Sci.*, 97, 221.

MITCHELL, R. L. (1936), *J. Soc. Chem. Ind.*, 55, 267T.

NIJENSOHN, L. (1960), *Trans. 7th int. Congr. Soil Sci.*, 2, 36.

OKAZAKI, R., SMITH, H. W., and MOODIE, C. D. (1962), *Soil Sci.*, 93, 343.

—, —, and — (1963), Ibid., 96, 205.

—, —, and — (1964), Ibid., 97, 202.

OLSON, L. C., and BRAY, R. H. (1938), Ibid., 45, 483.

PEECH, M. (1945), Ibid., 59, 25.

PETER, H. and MARKERT, S. (1960), *Z. Landw. Vers u UntersWes.*, 6, 505.

PIPER, C. S. (1942), *Soil and Plant Analysis*, Univ. of Adelaide.

RAGLAND, J. L. (1962), *Proc. Soil Sci. Soc. Am.*, 26, 133.

RICH, C. I. (1962), *Soil Sci.*, 93, 87.

SCHNITZER, M. (1965), *Nature, Lond.*, 207, 667.

SCHOFIELD, R. K. (1947), Ibid., 160, 408.

SCHOLLENBERGER, C. J. (1948), *Analyt. Chem.*, 20, 1121.

— and SIMON, R. H. (1945), *Soil Sci.*, 59, 13.

SCHÖNBERG, W. (1963), *Albrecht-Thaer-Arch.*, 7, 721.

SHAW, W. M., and MCINTIRE, W. H. (1935), *Soil Sci.*, 39, 359.

SIELING, D. H. (1941), *J. am. Soc. Agron.*, 33, 34.

SIMONSON, C. H., and AXELY, J. H. (1963), *Proc. Soil Sci. Soc. Am.*, 27, 26.
SMITH, H. W., MOODIE, C. D., *et al.* (1966), *Soil Sci.*, 102, 94.
STREET, N. (1963), Ibid., 95, 367.
TEDROW, J. C. F., and GILLAM, W. S. (1941), Ibid., 51, 223.
THOMPSON, H. S. (1850), *J. Roy. Agric. Soc.*, 11, 68.
WAY, J. T. (1850), Ibid., 11, 313.
— (1852), Ibid., 13, 123.
YAALON, D. H., *et al.* (1960). *Neth. J. agric. Sci.*, 10, 217.

Chapter 8

AL-ABBAS, H., and BARBER, S. A. (1964), *Soil Sci.*, 97, 103.
ALBRECHT, W. A., and SMITH, G. E. (1952), *Trans. Int. Soc. Soil Sci. Comm. II and IV*, 1, 119.
ALLAWAY, W. H. (1945), *Soil Sci.*, 59, 207.
ALTEN, F., WEILAND, H., and KNIPPENBERG, E. (1933), *Biochem. Z.*, 265, 85.
— and WERNER, W. (1960), *Trans. 7th int. Congr. Soil Sci.*, 2, 260.
ÅSLANDER, A. (1952), *Soil Sci.*, 74, 181.
BARNES, H. D. (1918), *J. South Afr. Chem. Inst.*, 11, 67.
BARROWS, H. L., and SIMPSON, E. L. (1962), *Proc. Soil Sci. Soc. Am.*, 26, 443.
BARSHAD, I. (1949), *Am. Miner.*, 34, 675.
— (1960), *Science*, 131, 988.
BERG, R. (1927), *Z. analyt. Chem.*, 71, 23.
BLACK, I. A., and SMITH, E. (1950), *J. Sci. Fd. Agric.*, 7, 201.
BLUME, J. M., and SMITH, D. (1954), *Soil Sci.*, 77, 9.
BOULD, C. (1963), *J. Sci. Fd. Agric.*, 14, 710.
BOWER, C. A., REITEMEIER, R. F., and FIREMAN, M. (1952), *Soil Sci.*, 73, 251.
BRADY, N. C., and COLWEL, W. E. (1945), *J. am. Soc. agron.*, 37, 429.
CHENERY, E. M. (1964), *Analyst*, 89, 365.
CHENG, K. L., MELSTED, S. W., and BRAY, R. H. (1953), *Soil Sci.*, 75, 37.
COLWEL, W. E., and BRADY, N. C. (1945), *J. am. Soc. agron.*, 37, 413.
CORNFIELD, A. H., and POLLARD, A. G. (1951), *J. Sci. Fd. Agric.*, 2, 135.
EGNÉR, H., and RIEHM, H. (1955), *Landw. Forsch.*, 6.
ELLIS, G. H. (1938), *Ind. Eng. Chem. A.E.*, 10, 112.
GÄRTEL, W. (1962), *Landw. Forsch. Sonderh.*, 16, 133.
GEISSLER, T., and KURNOTH, P. (1960), *Wiss. Z. Humboldt Univ. Berlin nathnaturwiss Reihe.*, 9, 795.
GÖHLER, F., and BECKER, M. (1961), *Albrecht-Thaer Arch.*, 5, 241.
GOLDSTEIN, D., and STARK-MAYER, C. (1958), *Anal. chim. Acta*, 19, 437.
GRAD, W. (1962), *Roczn. Nauk. rol.*, 86A, 645.
GRAHAM, E. R., POWELL, S., and CARTER, M. (1956). *Miss. agric. Exp. Stn. Res. Bull.*, 607.
HALL, R. J., GRAY, G. A., and FLYNN, L. R. (1966), *Analyst*, 91, 102.
HARRISON, G. E., and RAYMOND, W. H. A. (1953), Ibid., 78, 528.
HEAGY, F. C. (1948), *Can. J. Res.*, 26E, 295.
HISSINK, D. J. (1923), *Soil Sci.*, 15, 269.
HOFFMAN, W. E., and SCHROEDER, D. (1960), *Trans. 7th int. Congr. Soil Sci.*, 2, 253.
HOGG, D. E. (1962), *N.Z. J. Soil Sci.*, 5, 64.
HOVLAND, D., and CALDWELL, A. E. (1960), *Soil Sci.*, 89, 92.
HUNTER, J. G. (1950), *Analyst*, 75, 91.
JACKSON, M. L. (1958), *Soil Chemical Analysis*. Constable, Lond.
JACOBY, E. (1961), *Pl. Soil*, 15, 74.
KERR, J. W. R. (1960), *Analyst*, 85, 867.
KING, H. G. C., *et al.* (1967), Ibid., 92, 695.

LACY, J. (1965), *Analyst*, 90, 65.

LATTES, F. R. (1962), *C.R. Acad. Sci. Paris*, 254, 922.

LUNDEGARDH, H. (1941), *Lantbr. Högsk. Ann.*, 9, 127.

MANN, C. K., and YOE, J. H. (1957), *Anal. chim. Acta*, 16, 155.

MATTOCKS, A. M., and HERNANDEZ, H. R. (1950), *J. am. pharm. Ass.*, 39, 519.

MCCOLLOCH, R. C., BINGHAM, F. T., and ALDRICH, D. G. (1957), *Proc. Soil Sci. Soc. Am.*, 21, 85.

METSON, A. J. (1961), *New Zealand C.S.I.R. Soils Bull. No. 12*.

METZGER, W. H. (1929), *Soil Sci.*, 27, 305.

MIKKELSEN, D. S., and TOTH, S. J. (1947), *J. am. Soc. agron.*, 39, 165.

—, —, and PRINCE, A. L. (1948), *Soil Sci.*, 66, 385.

MITCHELL, R. L. (1964), *Comm. Bur. Soils Tech. Comm.*, 44A.

— and ROBERTSON, I. M. (1936), *J. Soc. Chem. Ind.*, 55, 269T.

MOORE, C. W. E. (1961), *Aust. J. Bot.*, 9, 92.

MYTTENAERE, C. (1963), *Ann. Physiol. Veg. Univ. Brux.*, 8, 47.

ORLOVA, L. P. (1961), *Pochvovedenie*, 12, 110.

PEASLEE, D. E. (1964), *Soil Sci.*, 97, 248.

— (1966), *Proc. Soil Sci. Soc. Am.*, 30, 443.

PEREIRA, R. S. (1944), *Rev. Brasil biol.*, 4, 263.

PURVIS, E. R., and HANNA, W. J. (1949), *Soil Sci.*, 67, 29.

REITH, J. W. S. (1963), *J. Sci. Fd. Agric.*, 14, 417.

REITEMEIER, R. F. (1943), *Ind. Eng. Chem. A. E.*, 15, 393.

RICH, C. I. (1965), *Monog. am. Soc. Agron.*, 9, 849.

RUSSELL, E. W. (1961), *Soil conditions and plant growth*. 9th edn., Longmans.

SAALBACH, E., and JUDEL, G. K. (1961), *Z. PflErnähr Düng.*, 95, 23.

SALMON, R. C. (1963), *J. Sci. Fd. Agric.*, 14, 605.

SCHACHTSCHABEL, P. (1954), *Z. PflErnähr Düng.*, 67, 9.

SCHOUWENBURG, J. C. VAN, and WALINGA, I. (1962), *Neth. J. agric. Sci.*, 10, 154.

SCHROEDER, W. E., and ZAHIROLESLAM, S. (1963), *Z. PflErnähr Düng.*, 100, 207.

—, —, and HOFFMAN, W. E. (1963), Ibid., 100, 215.

SCHWARZENBACH, G., et al. (1946), *Helv. chim. acta*, 29, 811.

SCOTT, R. O., and URE, A. M. (1958), *Analyst*, 561.

SEMENOVA, F. I. (1962), *R. Zh. (biol.)*, 7D, 189.

SENDROY, J. (1942), *J. biol. Chem.*, 144, 243.

SNELL, F. D., and SNELL, C. T. (1959), *Colorimetric methods of analysis*. 3rd edn., 2A, 545. D. van Nostrand.

STAHLBERG, S. (1960), *Acta agric. scand.*, 10, 205.

SWANSON, R. A., HOVLAND, D., and FINE, L. O. (1966), *Soil Sci.*, 102, 244.

THOMAS, J. G. (1953), *J. Soil Sci.*, 4, 238.

TRUOG, E., et al. (1947), *Soil Sci.*, 63, 1.

URBACH, C. (1932), *Biochem. Z.*, 252, 74.

VOELKER, A. (1871), *J. chem. Soc.*, 24, 276.

VOGEL, A. I. (1962), *A textbook of quantitative inorganic analysis*. 3rd edn., Longmans.

WACKER, W. E. C., IIDA, C., and FUWA, K. (1964), *Nature, Lond.*, 202, 659.

WADLEIGH, G. H., GAUCH, H. G., and KOLISCH, M. (1951), *Soil Sci.*, 72, 275.

WELTE, E., and WERNER, W. (1963), *J. Sci. Fd. Agric.*, 14, 180.

—, —, and NIEDERBUDDE, E. A. (1960), *Trans. 7th int. Congr. Soil Sci.*, 2, 246.

WEYBREW, J. A., MATRONE, G., and BAXLEY, H. M. (1948), *Anal. Chem.*, 20, 759.

WILLIAMS, T. R., et al. (1966), *J. Sci. Fd. Agric.*, 17, 344.

WILLSON, A. E. (1951), *Anal. Chem.*, 21, 754.

YIEN, C. H., and CHESNIN, L. (1952), *Proc. Soil Sci. Soc. Am.*, 17, 240.

YOUNG, H. Y., and GILL, R. F. (1951), *Anal. Chem.*, 23, 751.

YOUNG, A., and SWEET, T. R. (1955), Ibid., 27, 418.

ZUGRAVESCU, P. (1961), *Rev. chim. Bucharest*, 12, 86.

Chapter 9

ALESHIN, S. N., and BOLDȲREV, A. I. (1962), *Pochvovedenie*, 2, 114.

AYRES, A. S., TAKAHASHI, M., and KANECHIRO, Y. (1946), *Proc. Soil Sci. Soc. Am.*, 11, 175.

BARBER, S. A. (1960), *Trans. 7th int. Congr. Soil Sci.*, 3, 435.

BARBER, T. E., and MATTHEWS, B. C. (1962), *Can. J. Soil Sci.*, 42, 266.

BECKETT, P. H. T. (1964), *J. Soil Sci.*, 15, 1 and 9.

— (1965), *Agrochimica*, 9, 150.

BENETT, A. C., and REED, R. M. (1965), *Proc. Soil Sci. Soc. Am.*, 29, 192.

BLACK, C. A. (1957), *Soil Plant Relationships*. John Wiley, New York.

BLANCHET, R., STUDER, R., and CHAUMNONT, C. (1962), *Ann. agron. Paris*, 13, 175.

BOLT, G. H., SUMNER, M. E., and KAMPHORST, A. (1963), *Proc. Soil Sci. Soc. Am.*, 17, 294.

BOWER, C. A. (1960), *Trans. 7th int. Congr. Soil Sci.*, 16.

— and HATCHER, J. T. (1962), *Soil Sci.*, 93, 235.

— and WADLEIGH, C. H. (1948), *Proc. Soil Sci. Soc. Am.*, 13, 218.

BOYNTON, D., and BURRELL, A. B. (1944), *Soil Sci.*, 58, 441.

BRAY, R. H., and DE TURK, E. E. (1939), *Proc. Soil Sci. Soc. Am.*, 3, 101.

CHEW, W. C., WONG, C. N., and HVANG, Y. (1962), *Acta pedol. sin.*, 10, 98.

CROWTHER, E. M. (1947), *Brit. Sugar-beet rev.*, 16, 19.

DE TURK, E. E., WOOD, C. K., and BRAY, R. H. (1943), *Soil Sci.*, 55, 1.

DOWDY, R. H., and HUTCHESON, T. B. (1963), *Proc. Soil Sci., Soc. Am.*, 27, 31.

EISENMAN, G., RUDI, D. O., and CASBY, J. U. (1957), *Science*, 126, 831.

EVANS, C. E., and SIMON, R. H. (1949), *Proc. Soil Sci. Soc. Am.*, 14, 126.

FEHRENBACHER, J. B., WILDING, L. P., and BEAVERS, A. H. (1963), Ibid., 27, 152.

FIELDES, M., and SWINDALE, C. D. (1954), *N. N. J. Sci. Tech.*, B36, 140.

FRAPS, G. S. (1929), *Texas agric. Exp. Stn. Bull.*, 391.

GOUVEIA, D. H. G., GOUVEIA, J., and VIOLANTE, M. T. (1960), *Bol. Soc. Estud. Moçambique*, 29, 10.

GUGGENHEIM, E. A. (1950), *Thermodynamics, an advanced treatise.* North Holland Pub. Co.

HEINMANN, H. (1958), *4th Congr. int. Potash Inst., Madrid*, p. 212.

HERRERA, J. M. A. (1958), Ibid., p. 62.

HSI, C. F., and CHAO, C. (1962), *Acta pedol. sin.*, 10, 235.

HUMBERT, R. P. (1958), *4th Congr. int. Potash Inst., Madrid*, p. 327.

HUNZINKER, R. R. (1958), Thesis, Iowa State College.

JACKSON, M. L. (1958), *Soil chemical analysis.* Constable, London.

KOGAN, R. M., NIKIFOROV, M. V., and FRIDMAN, SH. D. (1961), *Pochvovedenie*, 8, 92.

KOTER, Z., and WACHWOLA, M. (1962), *Acta agrobot.*, 11, 131.

LENGYEL, B., and BLUM, E. (1934), *Trans. Faraday Soc.*, 30, 461.

MATTHEWS, B. C., and BECKETT, P. H. T. (1962), *J. agric. Sci.*, 58, 59.

MATTOCK, G., and UNCLES, R. (1964), *Analyst*, 89, 380.

MCLEAN, A. J. (1961), *Can. J. Soil Sci.*, 41, 1926.

MERWIN, H. D., and PEECH, M. (1950), *Proc. Soil Sci. Soc. Am.*, 15, 125.

METSON, A. J. (1961), *New Zealand C.S.I.R., Soils Bull. No. 12.*

MILCHERA, M. (1962), *Izv. tsent. nauchno, iszled. Inst. Pochvozn. Agrotekh. 'Puskarov'*, 5, 137.

MILFORD, M. H. (1962), *Diss. Abstr.*, 23, 1854.

MITCHELL, R. L. (1964), *Comm. Bur. Soils Tech. Comm.*, 44A.

MORANI, V., and FORTINI, S. (1962), *Agrochimica*, 7, 29.

MOSS, P. (1963), *Pl. Soil*, 18, 99.

MUNK, H. (1962), *Landw. Forsch.*, 15, 145.

MURACA, R. F., and BONSACK, J. P. (1955), *Chemist-Analyst*, 44, 38.

NEEB, K. H., and GEBAUR, W. (1958), *Z. anal. Chem.*, 162, 167.

NOWOSIELS, O. (1962), *Roczn. glebozn.*, 12, 269.

OLSON, R. V. (1953), *Proc. Soil Sci. Soc. Am.*, 17, 20.

ORLOV, D. S., and TSIKURIA, N. N. (1962), *Vest. mosk. gos. Univ. Ser. biol. Pochv.*, 2, 57.

PARKS, R. Q., *et al.* (1943), *Ind. Eng. Chem. A. E.*, 15, 527.

PORTNOY, H. D., THOMAS, L. M., and GURDJIAN, E. S. (1962), *Tantala*, 9, 119.

RATHJE, W. (1961), *Z. PflErnähr. Düng.*, 94, 174.

REDIES, M., and VIMPANY, I. (1966), *Nature, Lond.*, 210, 1078.

RICHARDS, G. E., and MCLEAN, E. O. (1963), Ibid., 95, 308.

ROUSE, R. D., and BERTRAMSON, B. R. (1950), *Proc. Soil Sci. Soc. Am.*, 14, 113.

SCHULTE, E. E., and COREY, R. B. (1963), Ibid., 27, 358.

SCOTT, A. D., and BATES, T. E. (1962), Ibid., 26, 209.

— and HANWAY, J. J. (1960), *Trans. 7th int. Congr. Soil Sci.*, 3, 72.

— and REED, M. G. (1963), *Proc. Soil Sci. Soc. Am.*, 24, 326.

SEMB, G., and ØIEN, A. (1961), *Meld. Norg. Landbrhøisk*, 40, 19.

SHOHL, A. T., and BENNETT, H. B. (1928), *J. biol. Chem.*, 78, 643.

SKOBETS, V. D., ABARBARCHUK, I. L., and SKOBETS, E. M. (1962), *Nauch. Dokl. vÿssh. Shk. Biol. Nauki.*, 3, 189.

SMITH, J. L. (1881), *Am. J. Arts.*, 1, 3.

SNELL, F. D., and SNELL, C. T. (1949), *Colorimetric methods of analysis*. 3rd edn., D. van Nostrand, 2, 576.

SOILEAU, J. M., JACKSON, W. A., and MCCRACKEN, R. J. (1964), *J. Soil Sci.*, 15, 117.

SUNDERMAN, F. W., JNR., and SUNDERMAN, F. W. (1958), *Am. J. Clin. Pathol.*, 29, 95.

TINKER, P. B. (1964), *J. Soil Sci.*, 15, 24 and 35.

— (1966), *Trans. int. Soc. Soil Sci., Comm. II and IV*, 223.

VOGEL, A. I. (1962), *A textbook of quantitative inorganic analysis*. 3rd edn., Longmans.

VRIES, P. DE (1961), *Inst. Biol. scheik. Onderz. LandbGewassen Jaarb.*, 107.

VYSKREBENTENSEVA, E. I. (1963), *Fiziol. Rast.*, 10, 307.

WALLACE, T. (1958), *4th Congr. int. Potash Inst., Madrid*, p. 145.

WALSH, T., and CLARKE, E. J. (1942), *Eire Dept. Agr. J.*, 39, 316.

WARREN, R. G., and COOKE, G. W. (1962), *J. agric. Sci.*, 59, 269.

WEBSTER, G. C., and VARNER, J. E. (1954), *Arch. biochem. biophys.*, 52, 22.

WILDING, C. P. (1963), *Diss. Abstr.*, 23, 2642.

WILLIAMS, E. G. (1962), *Trans. Meeting Comm. IV and V int. Soc. Soil Sci.*, N.Z., p. 820.

WILLIAMS, W. O. (1941), *Proc. am. Soc. hort. Sci.*, 39, 47.

WOODRUFF, C. M. (1955), *Proc. Soil Sci. Soc. Am.*, 19, 36.

— and MCINTOSH, J. L. (1960), *Trans. 7th int. Congr. Soil Sci.*, 3, 80.

ZAZVORKA, M. (1959), *Sborn. vys. Skol. Zemed. Praha.*, 15, 73.

Chapter 10

ACQUAYE, D. K., and CUNNINGHAM, R. K. (1965), *Trop. Agric. Trin.*, 42, 281.

ADAMS, C. I., and SPAULDING, G. H. (1955), *Anal. Chem.*, 27, 1003.

ALLISON, F. E., DOETSCH, J. H., and ROLLER, E. M. (1951), *Soil Sci.*, 72, 187.

— and ROLLER, E. M. (1955), Ibid., 80, 349.

ALVES, J. A., and ALVES, E. L. N. (1952), *Melhoramento*, 5, 77.

ARNOLD, P. W. (1954), *J. Soil Sci.*, 5, 116.

ASHRAF, M., BHATTY, M. K., and SHAH, R. A. (1960), *Pak. J. Sci. Ind. Res.*, 3, 1.

ASHTON, F. L. (1936) *J. agric. Sci.*, 26, 239.

ASSOC. OF OFFIC. AGRIC. CHEMISTS, WASHINGTON (1955), *Official methods of analysis*, 8th edn.

AUDUS, L. J. (1946). *Nature, Lond.*, 158, 419.

BAL, D. V. (1925), *J. agric. Sci.*, 15, 454.

BARSHAD, I. (1951), *Soil Sci.*, 72, 361.

BEET, A. E. (1954), *J. Appl. Chem.*, 4, 373.

— (1955), *Nature, Lond.*, 175, 513.

BELCHER, R., and GODBERT, A. L. (1941), *J. Soc. Chem. Ind.*, 60, 196T.

BIRCH, H. F. (1958), *Pl. Soil*, 10, 9.

— (1959), Ibid., 11, 262.

BLUE, W. G., and ENO, C. F. (1952), *Florida Soil Sci. Soc. Proc.*, 12, 157.

BOSWELL, F. C., RICHER, A. C. and CASIDA, L. E. (1962), *Proc. Soil Sci. Soc. Am.*, 26, 254.

BOUAT, A., and CROUZET, C. (1965), *Annls. agron.*, 16, 107.

BOWER, C. A. (1950), *Soil Sci.*, 70, 375.

BRADSTREET, R. B. (1965), *The Kjeldahl method for organic nitrogen*. Academic Press, N.Y. and Lond.

BRATTON, A. C., and MARSHALL, E. K. (1939), *J. biol. Chem.*, 128, 537.

BREMNER, J. M. (1959), *J. agric. Sci.*, 52, 147.

— (1960), Ibid., 55, 11.

— (1965), *Monog. am. Soc. Agron.*, 9,

— and SHAW, K. (1955), *J. agric. Sci.*, 46, 320.

— and — (1958), Ibid., 51, 22.

— and HARADA, T. (1959), Ibid., 52, 137.

— and EDWARDS, A. P. (1965), *Proc. Soil Sci. Soc. Am.*, 29, 504.

BUDIN, W. (1914), *J. agric. Sci.*, 6, 452.

CAMPBELL, W. R., and HANNA, M. I. (1937), *J. biol. Chem.*, 119, 1.

CHAKRABORTY, S. P., and SEN GUPTA, S. P. (1960), *Nature, Lond.*, 184, 2033.

CHAPMAN, H. D. (1965), *Monog. am. Soc., Agron.*, 9, 891.

— and LIEBIG, G. F. (1952), *Proc. Soil Sci. Soc. Am.*, 16, 276.

CHASE, F. E. (1948), *Sci. Agric.*, 28, 315.

CHENG, H. H., and BREMNER, J. M. (1965), *Monog. am. Soc. Agron.*, 9, 1287.

CHIBNALL, A. C., REES, M. W., and WILLIAMS, E. F. (1943), *Biochem. J.*, 37, 354.

CLARKE, E. P. J. (1941), *J. Ass. off. agric. Chem., Washington*, 24, 641.

CLARKE, A. L., and JENNINGS, A. C. (1965), *J. agric. Fd. Chem.*, 13, 174.

CLEMENT, C. R., and WILLIAMS, T. E. (1962), *J. Soil Sci.*, 13, 82.

CONWAY, E. J. (1947), *Microdiffusion analysis and volumetric error*. 2nd edn., Crosby Lockwood, Lond.

COPE, W. C. (1916), *J. Ind. eng. Chem.*, 8, 592.

CORNFIELD, A. H. (1960), *Nature, Lond.*, 187, 260.

CSAKY, T. Z. (1948), *Acta chem. Scand.*, 2, 450.

DABIN, B. (1965), *Cah. Pédol. ORSTOM*, 3, 335.

DAVIES, E. B., and COUP, M. R., *et al.* (1940), *N.Z. J. Sci. Tech.*, A21, 348.

DAVISSON, B. S., and PARSONS, J. T. (1919), *J. Ind. Eng. Chem.*, 11, 306.

DHARIVAL, A. P. S., and STEVENSON, F. J. (1958), *Soil Sci.*, 48, 467.

DUMAS, J. B. A. (1831), *Ann. chim. phys.* (Paris), (2), 47, 198.

DYCK, A. W. J., and MCKIBBIN, R. R. (1935), *Can. J. Res.*, B.13, 264.

EAGLE, D. J., and MATTHEWS, B. C. (1958), *Can. J. Soil Sci.*, 38, 161.

EMMERT, E. M. (1929), *J. Ass. off. agric. Chem. Washington*, 12, 240.

— (1934), *Soil Sci.*, 38, 139.

ENGLISH, F. L. (1947), *Anal. Chem.*, 19, 850.

ENSMINGER, L. E., and GIESEKING, J. E. (1939), *Soil Sci.*, 48, 467.

FERRY, P., and BLANCHERE, H. (1957), *Annls. Inst. Natn. Rech. Agron. Paris*, A8, 111.

FRAPS, G. S., and STERGES, A. J. (1947), *Texas agric. Exp. Stn. Bull.*, No. 693.

FRENEY, J. R. (1964), *J. agric. Sci.*, 63, 297.

FRIEDRICH, A., KÜHAAS, E., and SCHNÜRCH, R. (1933), *Z. physiol. Chem.*, 216, 68.

GASSER, J. K. R. (1958), *Nature, Lond.*, 181, 1334.

GRABARAOV, P. G., and VISHERSKAYA, B. N. (1960), *Izv. Akad. Nauk. Kazakh SSR Ser. Bot. Pochvoved.*, 2, 55.

GREENWOOD, D. J. (1962), *Pl. Soil*, 17, 365.

— and LEES, H. (1959), Ibid., 11, 87.

— and — (1956), Ibid., 7, 253.

— and — (1960), Ibid., 14, 360.

GRIESS, P. (1879), *Chem. Ber.*, 12, 426.

GUNNING, J. W. (1889), *Z. anal. Chem.*, 28, 188.

HAAG, E., and DALPHIN, C. (1943), *Arch. sci. phys. et nat.*, 25, 148.

HARMSEN, G. W., and VAN SCHREVEN, D. A. (1955), *Adv. Agron.*, 7, 299.

HARPSTEAD, M. I., and BRAGE, B. L. (1958), *Proc. Soil Sci. Soc. Am.*, 26, 326.

HART, M. G. R., and ROBERTS, E. H. (1961), *Nature, Lond.*, 189, 598.

HAUCK, R. D., MELSTED, S. W., and YANKWICH, P. E. (1958), *Soil Sci.*, 86, 287.

HESSE, P. R. (1957), *Pl. Soil*, 9, 86.

HILLER, A., PLAZIN, J., and VAN SLYKE, D. D. (1948), *J. biol. Chem.*, 176, 1401.

HOSKINS, J. L. (1944), *Analyst*, 69, 271.

HUNTER, A. S., and CARTER, D. L. (1965), *Soil Sci.*, 100, 112.

ILOSVAY, M. L. (1889), *Bull. Soc. Chim.*, 2, 388.

JACKSON, M. L. (1958), *Soil chemical analysis.* Constable, London.

— and CHANG, S. C. (1947), *J. am. Soc. Agron.*, 39, 623.

JENNY, H. (1930), *Missouri agric. Exp. Stn. Res. Bull.*, 152.

KEENEY, D. R. and BREMNER, J. M. (1966), *Nature, Lond.*, 211, 892.

— and — (1967), *Soil Sci.*, 104, 358.

KIRK, P. L. (1936), *Ind. Eng. Chem. A. E.*, 8, 223.

— (1947), *Adv. Protein Chem.*, 3, 139.

— (1950), *Analyt. Chem.*, 22, 354.

KJELDAHL, J. (1883), *Z. anal. Chem.*, 22, 366.

KOCH, F. C., and MCMEEKIN, T. L. (1924), *J. am. Chem. Soc.*, 46, 2066.

KRESGE, C. B., and MERKLE, F. G. (1957), *Proc. Soil Sci. Soc. Am.*, 21, 516.

LAKE, G. R., *et al.* (1951), *Anal. Chem.*, 23, 1635.

LAURO, M. F. (1931), *Ind. Eng. Chem. A. E.*, 3, 401.

LEES, H., and QUASTEL, J. H. (1944), *Chem. Ind.*, 26, 238.

— and — (1946), *Biochem. J.*, 40, 803.

LEGGETT, G. E., and MOODIE, C. D. (1962), *Proc. Soil Sci. Soc. Am.*, 26, 160.

LEWIS, D. G. (1961), *J. Sci. Fd. Agric.*, 12, 735.

MA, T. S., and ZUAZAGA, G. (1942), *Ind. Eng. Chem. A. E.*, 14, 280.

MACURA, J. (1960), *Trans. 7th int. Congr. Soil Sci.*, 2, 664.

— and MÁLEK, I. (1958), *Nature, Lond.*, 182, 1796.

MADER, D. L., and HOYLE, M. C. (1964), *Soil Sci.*, 98, 295.

MÄKITIE, O. (1963), *Ann. agric. fenn.*, 2, 159.

MARAGHAN, J. T., and PESEK, T. (1963), *Proc. Soil Sci. Soc. Am.*, 27, 361.

MARCALI, K., and RIEMAN, W. (1946), *Ind. Eng. Chem. A. E.*, 18, 709.

— and — (1948), *Anal. Chem.*, 20, 381.

MARKHAM, R. (1942), *Biochem. J.*, 36, 790.

MATTHEWS, D. J. (1920), *J. agric. Sci.*, 10, 72.

MATTSON, S., and KOUTLER-ANDERSSEN, E. (1943), *Lantbrükshögskolans Ann.*, 11, 107.

MCCLEAN, W., and ROBINSON, G. W. (1924), *J. agric. Sci.*, 14, 548.

MCKENZIE, H. A., and WALLACE, H. S. (1954), *Aus. J. Chem.*, 7, 55.

MCLEAN, A. A. (1964), *Nature, Lond.*, 203, 1307.

METSON, A. J. (1961), *New Zealand C.S.I.R. Soils Bull.*, No. 12.

MIDDLETON, G., and STUCKEY, R. E. (1951), *J. Pharm. Pharmacol.*, 3, 829.

MONTGOMERY, H. A. C., and DYMOCK, J. F. (1961), *Analyst*, 86, 414.

MORRILL, L. G., and DAWSON, J. E. (1964), *Proc. Soil Sci. Soc. Am.*, 28, 710.

MURNEEK, A. E., and HEINTZE, P. H. (1937), *Res. Bull. Mo. agric. Exp. Stn.*, No. 261.

NAGAI, K., and HIRAYA, C. (1962), *J. Sci. Soil Tokyo*, 33, 303.

NOMMIK, H. (1957), *Acta agr. Scand.*, 7, 395.

— and NILSSON, K. O. (1963), Ibid., 13, 205.

ÖBRINK, K. J. (1955), *Biochem. J.*, 59, 134.

OLSEN, C. (1929a), *C. R. Trav. Lab. Carlsberg*, 17, No. 3.

— (1929b), Ibid., 17, No. 15.

OLSON, R. A. (1960), *Trans. 7th int. Congr. Soil Sci.*, 2, 463.

PANGANIBAN, E. H. (1925), *J. am. Soc. Agron.*, 17, 1.

PAPENDICK, R. I., and PARR, J. F. (1965), *Soil Sci.*, 100, 182.

PARNAS, J. K., and WAGNER, J. (1931), *Biochem. Ztg.*, 125, 253.

PARR, J. F., and REUSZER, H. W. (1959), *Proc. Soil Sci. Soc. Am.*, 23, 214.

PATEL, S. M., and SREENIVASAN, A. (1948), *Analyt. Chem.*, 29, 63.

PIPER, C. S. (1942), *Soil and Plant Analysis*, Univ. of Adelaide.

POE, C. F., and NALDER, M. E. (1935), *Ind. Eng. Chem. A. E.*, 7, 189.

PRASAD, R. (1965), *Pl. Soil*, 23, 261.

PURVIS, E. R., and LEO, M. W. M. (1961), *J. Agr. Fd. Chem.*, 9, 15.

QURESHI, M. I. N., *et al.* (1962), *Pak. J. sci. industr. Res.*, 5, 159.

RICHARD, T. A., *et al.* (1960), *Trans. 7th int. Congr. Soil Sci.*, 2, 28.

RICHARDSON, H. L. (1938), *J. agric. Sci.*, 28, 73.

RIDER, B. F. and MELLON, M. G. (1946), *Ind. Eng. Chem. A. E.*, 18, 97.

ROBINSON, J. B. D. (1964), *East Afr. agric. for. Res. Org. Ann. Rep.*, p. 44.

— (1967a), *Pl. Soil*, 27, 53.

— (1967b), *J. Soil Sci.*, 18, 109.

— and GACOKA, P. (1962), *J. Soil Sci.*, 13, 133.

RODRIGUES, G. (1954), Ibid., 5, 264.

ROLLER, E. M., and MCKAIG, N. (1939), *Soil Sci.*, 47, 397.

ROSS, D. J. (1964), *Nature, Lond.*, 204, 503.

SALT, H. B. (1953), *Analyst*, 74, 4.

SAUNDER, D. H., ELLIS, B. S., and HALL, A. (1957), *J. Soil Sci.*, 8, 301.

SCHACHTSCHABEL, P. (1960), *Trans. 7th int. Congr. Soil Sci.*, 2, 22.

SCHREVEN, VAN, D. A. (1963), *Pl. Soil*, 28, 163.

SHEERS, E. H., and COLE, M. S. (1953), *Analyt. Chem.*, 25, 535.

SHEPARD, D. E., and JACOBS, M. B. (1951), *J. am. Pharm. Assoc.*, 40, 154.

SHER, I. M. (1955), *Anal. Chem.*, 27, 831.

SHINN, M. B. (1941), *Ind. Eng. Chem., A. E.*, 13, 33.

SKYRING, G. W., CAREY, B. J., and SKERMAN, V. B. D. (1961), *Soil Sci.*, 91, 388.

SNELL, F. D., and SNELL, C. T. (1949), *Colorimetric methods of analysis.* 3rd edn., D. van Nostrand.

SOHN, J. B., and PEECH, M. (1958), *Soil Sci.*, 85, 1.

SPERBER, J. I., and SYKES, B. J. (1964), *Pl. Soil*, 20, 127.

SPRENGELS, H. (1864), *Ann. de Physic. u Chemie*, 121, 188.

STANFORD, G., AYRES, A. S., and DOI, M. (1965), *Soil Sci.*, 99, 132.

— and HANWAY, J. (1955), *Proc. Soil Sci. Soc. Am.*, 19, 74.

STANLEY, F. A., and SMITH, G. E. (1956), Ibid., 20, 557.

STEWART, B. A., and PORTER, L. K. (1963), Ibid., 27, 41.

—, —, and BEARD, W. E. (1964), Ibid., 28, 366.

—, —, and CLARK, F. E. (1963), Ibid., 27, 377.

STEYERMARK, A., *et al.* (1958), *Analyt. Chem.*, 30, 1561.

SWAN, M. H., and ADAMS, M. L. (1956), *Anal. Chem.*, 28, 1630.

TRUOG, E., HULL, H. H., and SHIHATA, M. M. (1951), *Memo. Univ. Wisconsin*, Madison.

VAN SLYKE, D. D., and HILLER, A. (1933), *J. biol. Chem.*, 102, 499.

VESTERVALL, F. (1963), *K. Lantbr-Högsk. Ann.*, 29, 129.

VOGEL, A. I. (1962), *A textbook of quantitative inorganic analysis.* Longmans.

WALKLEY, A. (1935), *J. agric. Sci.*, 25, 598.

WALLACE, G. I., and NEAVE, S. L. (1927), *J. Bact.*, 14, 377.

WARING, S. A., and BREMNER, J. M. (1964a), *Nature, Lond.*, 201, 951.

— and — (1964b), Ibid., 202, 1141.

— and — (1964c), Ibid., 203, 819.

WHEELER, B. E. J. (1963), *Pl. Soil*, 12, 219.

WHITE, L. M., and LONG, M. C. (1951), *Anal. Chem.*, 23, 363.

WILFARTH, H. (1885), *Z. anal. Chem.*, 24, 455.

WILLITS, C. O., COE, M. R., and OGG, C. L. (1949), *J. Ass. off. agric. Chem. Washington*, 32, 118.

WINKLER, L. W. (1913), *Z. angew. Chem.*, 26, 231.

WOLF, B. (1947), *Anal. Chem.*, 19, 334.

YOUNG, J. L., and CATTANI, R. A. (1962), *Proc. Soil Sci. Soc. Am.*, 26, 147.

YUEN, S. H., and POLLARD, A. G. (1953), *J. Sci. Fd. Agric.*, 4, 490.

Chapter 11

ACHARD, F. K. (1786), *Crells. Chem. Ann.*, 2, 391.

ALLISON, L. E. (1935), *Soil Sci.*, 40, 311.

— (1960), *Proc. Soil Sci. Soc. Am.*, 24, 36.

ANDERSON, M. S., and BYERS, H. G. (1934), *Soil Sci.*, 38, 121.

APPELMAN, C. O. (1927), Ibid., 24, 241.

BALL, D. F. (1964), *J. Soil Sci.*, 15, 84.

BARTHOLOMEW, W. V., and BROADBENT, F. E. (1950), *J. agric. Sci.*, 14, 156.

BAUMANN, A. (1887), *Landw. VersStadt*, 33, 247.

BERTHELOT, M., and ANDRÉ, G. (1887), *Ann. Chem.*, 10, 368.

BHAUMIK, H. D., and CLARK, F. E. (1947), *Proc. Soil Sci. Soc. Am.*, 12, 234.

BIRCH, H. F. (1958), *Nature, Lond.*, 181, 788.

— (1959a), Ibid., 183, 1415.

— (1959b), *Pl. Soil*, 11, 262.

— and FRIEND, M. T. (1956), *Nature, Lond.*, 178, 500.

BLOM, L., EDELHAUSEN, L., and VAN-KREVELEN, D. W. (1957), *Fuel*, 36, 135.

BOYNTON, D., and REUTHER, W. (1938), *Proc. Soil Sci. Soc. Am.*, 3, 37.

BREMNER, J. M. (1949), *Analyst*, 74, 492.

— (1958), *J. Sci. Fd. Agric.*, 9, 528.

— (1965), *Monog. am. Soc. Agron.*, 9, 1238.

— and HARADA, T. (1959), *J. agric. Sci.*, 52, 137.

— and HO, C. L. (1961), *Agron. Abstr.*, p. 15.

— and LEES, H. (1949), *J. agric. Sci.*, 39, 247.

BRINK, R. H., DUBACH, P., and LYNCH, D. L. (1960), *Soil Sci.*, 89, 157.

BROOKS, J. D., and STERNHELL, S. (1957), *Aust. J. Appl. Sci.*, 8, 206.

BURGES, A. (1960), *Trans. 7th int. Congr. Soil Sci.*, 2, 91.

CAROLAN, R. (1948), *Soil Sci.*, 66, 241.

CHESHIRE, M. V., and MUNDIE, C. M. (1966), *J. Soil Sci.*, 17, 372.

CLARK, N. A., and OGG, C. L. (1942), *Soil Sci.*, 53, 27.

CLEMENT, C. R., and WILLIAMS, T. E. (1962), *J. Soil Sci.*, 13, 82.

COFFIN, D. E., and DE LONG, W. A. (1960), *Trans. 7th int. Congr. Soil Sci.*, 2, 91.

CORNFIELD, A. H. (1952), *J. Sci. Fd. Agric.*, 3, 154.

— (1961), *Pl. Soil*, 14, 90.

DAJI, J. A. (1932), *Biochem. J.*, 26, 1275.

DATTA, N. P., KHERA, M. S., and SAIMI, T. R. (1962), *J. Ind. Soc. Soil Sci.*, 10, 67.

DAVIES, E. B. (1950), *J. N.Z. Inst. Chem.*, 14, 86.

DEGTJAREFF, W. T. (1930), *Soil Sci.*, 29, 239.

DETMER, W. (1871), *Landw. VersStadt*, 14, 248.

DISCHE, Z. (1947), *J. biol. Chem.*, 167, 189.

— (1950), Ibid., 183, 489.

DODD, C. C., FOWDEN, L., and PEARSALL, W. H. (1953), *J. Soil Sci.*, 4, 69.

DOJARENKO, A. G. (1902), *Ann. Chem.*, 56, 311.

DORMAAR, J. F., and LYNCH, D. L. (1962), *Proc. Soil Sci. Soc. Am.*, 26, 251.

DUBACH, P., MEHTA, N. C., and DEUEL, H. (1961), *Z. PflErnähr. Düng.*, 95, 199.

— and — (1963), *Soils Fertil.*, 26, 293.

EKPETE, D. M., and CORNFIELD, A. H. (1965), *J. agric. Sci.*, 64, 205.

ELKAN, G. H., and MOORE, W. E. C. (1962), *Ecology*, 43, 775.

ELSON, L. A., and MORGAN, W. T. J. (1933), *Biochem. J.*, 27, 1824.

ENWEZOR, W. O., and CORNFIELD, A. H. (1965), *J. Sci. Fd. Agric.*, 16, 277.

FERRARI, G., and DELL'AGNOLA, G. (1963), *Soil Sci.*, 96, 418.

FORSYTH, W. C. G. (1950), *Biochem. J.*, 46, 141.

— and FRASER, G. K. (1947), *Nature, Lond.*, 160, 607.

FREYTAG, H. E., and IGEL, H. (1964), *Zbl. Bakt.*, 117, Abt. II, 525.

FRITZ, J. S., YAMAMURA, S. S., and BRADFORD, E. C. (1959), *Anal. Chem.*, 31, 260.

GAINEY, P. L. (1919), *Soil Sci.*, 7, 293.

GRAHAM, E. R. (1948), Ibid., 65, 181.

GREENWOOD, D. J. (1961), *Pl. Soil*, 14, 360.

— (1962), *Nature, Lond.*, 195, 161.

— and LEES, H. (1956), *Pl. Soil*, 7, 253.

— and — (1959), Ibid., 11, 87.

GRIFFITHS, E., and BIRCH, H. F. (1961), *Nature, Lond.*, 189, 424.

GUPTA, U. C., and SOWDEN, F. J. (1964), *Soil Sci.*, 97, 328.

HAYES, M. H. B., and MORTENSEN, J. L. (1963), *Proc. Soil Sci. Soc. Am.*, 27, 270.

HECK, A. F. (1929), *Soil Sci.*, 28, 225.

HELVEY, T. C. (1951), *J. Sci. Instr.*, 28, 354.

HORI, S., and OKUDA, A. (1961), *Soil and Plant Fd.*, 6, 170.

HUBACHER, M. H. (1949), *Anal. Chem.*, 21, 945.

IVARSON, K. C., and GUPTA, U. C. (1967), *Can. J. Soil Sci.*, 47, 74.

— and SOWDEN, F. J. (1962), *Soil Sci.*, 94, 245.

JACKSON, M. L. (1958), *Soil chemical analysis.* Constable, London.

JACQUIN, F. (1961), *Bull. Ecole nat. supér. agron. Nancy*, 3, 106.

JOHNSON, D. D., and GUENZI, W. D. (1963), *Proc. Soil Sci. Soc. Am.*, 27, 663.

JUDD, B. I., and WELDON, M. D. (1939), *J. am. Soc. Agron.*, 31, 217.

KEELING, P. S. (1962), *Clay Miner. Bull.*, 28, 155.

KOJIMA, R. T. (1947), *Soil Sci.*, 64, 157.

KONONOVA, M. M., and TITOVA, N. A. (1961), *Pochvovedenie*, 11, 81.

LEES, H. (1949), *Pl. Soil*, 2, 123.

— (1950), *Nature, Lond.*, 166, 118.

LEIGHTY, W. R., and SHOREY, E. C. (1930), *Soil Sci.*, 30, 257.

LITTLE, I. P., HAYDOCK, K. P., and REEVE, R. (1962), *C.S.I.R.O. Div. Soils Divl. Rep. 3/62*, 6.

LUNT, H. A. (1931), *Soil Sci.*, 32, 27.

LYNCH, D. L., HEARNS, E. E., and COTNOIR, L. J. (1957), *Proc. Soil Sci. Soc. Am.*, 21, 160.

MACKENZIE, A. F., and DAWSON, J. E. (1962), *J. Soil Sci.*, 13, 160.

MACURA, J. (1960), *Trans. 7th int. Congr. Soil Sci.*, 2, 664.

MARKERT, S. (1962), *Albrecht-Thaer.-Arch.*, 6, 523.

MARSH, F. W. (1928), *Soil Sci.*, 25, 253.

MCCREADY, R. M., and HASSID, W. T. (1942), *Ind. Eng. Chem. A. E.*, 14, 525.

METSON, A. J. (1961), *New Zealand C.S.I.R. Soils Bull. No. 12*.

MEYER, J. A., HILGER, F., and PEETERS, A. (1959), *Proc. 3rd Inter. Afr. Soils Conf.*, 513.

MITCHELL, R. L. (1932), *J. am. Soc. Agron.*, 24, 256.

MONNIER, G., TURE, L., and JEANSON-LUUSINGANG, G. (1962), *Ann. agron. Paris*, 13, 55.

MOON, F. E., and ABOU-RAYA, A. K. (1952), *J. Sci. Fd. Agric.*, 3, 407.

MORRIS, D. L. (1948), *Science*, 107, 254.

MORRISON, R. I. (1963), *J. Soil Sci.*, 14, 201.

MORTENSEN, J. L. (1960), *Trans. 7th int. Congr. Soil Sci.*, 2, 98.

NELLER, J. R. (1918), *Soil Sci.*, 5, 225.

PARSONS, J. W., and TINSLEY, J. (1961), Ibid., 92, 40.

PAUL, E. A., and SCHMIDT, E. L. (1960), *Proc. Soil Sci. Soc. Am.*, 24, 195.

PAULI, F. W., and GROBLER, J. H. (1961), *S. afr. J. agr. Sci.*, 4, 157.

— (1965), *Pl. Soil*, 22, 337.

PEECH, M., DEAN, L. A., and REED, J. (1947), *U.S. Dep. Agric. Circ. 757.*

POMMER, A. M., and BREGER, I. A. (1960), *Geochem. cosmochim. Acta*, 20, 35.

POSNER, A. M. (1966), *J. Soil Sci.*, 17, 65.

PURVIS, E. R., and HIGSON, G. E. (1939), *Ind. Eng. Chem. A. E.*, 11, 19.

PUTNAM, H. D., and SCHMIDT, E. L. (1959), *Soil Sci.*, 87, 22.

QUINN, J. G., and SALOMON, M. (1964), *Proc. Soil Sci. Soc. Am.*, 28, 456.

RATHER, J. B. (1917), *Ark. Exp. Stn. Bull.*, 140.

READ, J. W., and RIDGELL, R. H. (1922), *Soil Sci.*, 13, 1.

RICHER, A., and MASSON, P. (1964), *Ann. agron. Paris*, 15, 619.

ROBINSON, W. O. (1927), *J. agric. Res.*, 34, 339.

ROSS, D. J. (1964), *Nature, Lond.*, 204, 503.

ROULET, N., *et al.* (1963), *Z. PflErnähr. Düng.*, 101, 210.

ROVIRA, A. D. (1953), *Nature, Lond.*, 172, 29.

SAUNDER, D. H., and GRANT, P. M. (1962), *Trans. Jnt. Mtg. int. Soc. Soil Sci., Comm. IV*, p. 235.

SCHOLLENBERGER, C. J. (1927), *Soil Sci.*, 24, 65.

— (1931), Ibid., 31, 483.

— (1935), Ibid., 59, 53.

SCHNITZER, M., and DESJARDINS, J. G. (1962), *Proc. Soil Sci. Soc. Am.*, 26, 362.

— and GUPTA, U. C. (1965), Ibid., 29, 274.

— and SKINNER, S. I. M. (1965), Ibid., 29, 400.

— and — (1966), *Soil Sci.*, 101, 120.

SHAW, K. (1959), *J. Soil Sci.*, 10, 316.

SHEWAN, J. M. (1938), *J. agric. Sci.*, 28, 324.

SHOREY, E. C., and LANTHROP, E. C. (1910), *J. am. Chem. Soc.*, 32, 1680.

SIMONART, P., BATISTIC, L., and MAYANDON, J. (1967), *Pl. Soil*, 27, 153.

SKYRING, G. W., and THOMPSON, J. P. (1966), *Pl. Soil*, 24, 289.

SMITH, H. W., and WELDON, M. D. (1940), *Proc. Soil Sci. Soc. Am.*, 5, 177.

SOWDEN, F. J. (1959), *Soil Sci.*, 88, 138.

— and DEUEL, H. (1961), Ibid., 91, 44.

— and IVARSON, K. C. (1962), Ibid., 94, 340.

SPRENGEL, C. (1826), *Kastners Arch. ges. Naturlehre (Nürenburg)*, 8, 145.

STANFORD, G., and HANWAY, J. (1955), *Proc. Soil Sci. Soc. Am.*, 19, 74.

STEVENSON, I. L. (1956), *Pl. Soil*, 8, 170.

STEWART, B. A., PORTER, L. K. and BEARD, W. E. (1964), *Proc. Soil Sci. Soc. Am.*, 28, 366.

STOTZKY, G. (1960), *Can. J. Microbiol.*, 6, 439.

SWABY, R. J., and LADD, J. N. (1962), *Trans. Jnt. Meeting inter. Soc. Soil Sci., Comm. IV*, p. 197.

— and PASSEY, B. I. (1953), *Aust. J. agric. Res.*, 4, 334.

THOMAS, R. L., and LYNCH, D. L. (1961), *Soil Sci.*, 91, 312.

TINSLEY, J. (1950), *Trans. 4th int. Congr. Soil Sci.*, 1, 161.

— and SALAM, A. (1961), *J. Soil Sci.*, 12, 259.

TIURIN, I. V. (1931), *Pedology*, 5–6, 36.

TRACEY, M. V. (1950), *Biochem. J.*, 47, 433.

VANDECAVEYE, S. C., and KATZNELSON, H. (1938), *Soil Sci.*, 46, 139.

VAN SLYKE, D. D. (1911), *J. biol. Chem.*, 10, 15.

— and FOLCH, J. (1940), Ibid., 136, 509.

— *et al.* (1941), Ibid., 141, 627.

VAQUELIN, C. (1797), *Ann. Chim.*, 21, 39.

WAKSMAN, S. A., and STARKEY, R. L. (1924), *Soil Sci.*, 17, 141.

— and STEVENS, K. R. (1928), Ibid., 26, 113 and 239.

WALDRON, A. C., and MORTENSEN, J. L. (1961), *Proc. Soil Sci. Soc. Am.*, 25, 29.

WALKLEY, A. (1935), *J. agric. Sci.*, 25, 598.

— (1947), *Soil Sci.*, 63, 251.

WALKLEY, A., and BLACK, I. A. (1934), *Soil Sci.*, 37, 29.

WARBURG, O. (1926), *Uber den Stoffwechsel der Tumoren.* Trans. F. Dickens (1930). Constable, London.

WHITEHEAD, D. C., and TINSLEY, J. (1963), *J. Sci. Fd. Agric.*, 14, 849.

—, and — (1964), *Soil Sci.*, 97, 34.

WISON, B. D., and STAKER, E. V. (1932), *J. am. Soc. Agron.*, 24, 477.

WRIGHT, J. R., and SCHNITZNER, M. (1959), *Can. J. Soil Sci.*, 39, 44.

YOUNG, J. L. (1962), *Soil Sci.*, 93, 397.

— (1964), Ibid., 98, 133.

— and LINDBECK, M. R. (1964), *Proc. Soil Sci. Soc. Am.*, 28, 377.

YUAN, T. L. (1964), *Soil Sci.*, 98, 133.

Chapter 12

ADAMS, A. P., BARTHOLOMEW, W. V., and CLARK, F. E. (1954), *Proc. Soil Sci. Soc. Am.*, 18, 40.

AMMON, R., and HINSBERG, K. (1936), *Z. physiol. Chem.*, 239, 207.

ANDERSON, C. A., and BLACK, C. A. (1965), *Proc. Soil Sci. Am.*, 29, 255.

ANDERSON, G. (1956), *J. Sci. Fd. Agric.*, 7, 437.

— (1950), Ibid., 11, 497.

— (1962), *J. Soil Sci.*, 13, 225.

— and HANCE, R. J. (1963), *Pl. Soil*, 19, 296.

ARRIAGA E CUNHA, J. M. DE, *et al.* (1960), *Agron. lusit.*, 22, 171.

ASKINAZI, D. L., *et al.* (1963), *Pochvovedenie*, 5, 6.

ÅSLYNG, H. C. (1954), *Kong. Vet.-og Landbohøjskole Arsskrift*, 1, 50.

— (1964), *Acta agric. scand.*, 14, 260.

ASSOC. OFF. AGRIC. CHEM. (1950), *Official methods of analysis.* Washington.

AVNIMELECH, Y., and HAGIN, J. (1965), *Proc. Soil Sci. Soc. Am.*, 29, 393.

BACON, A. (1950), *Analyst*, 75, 301.

BASS, G. B., and SIELING, D. H. (1950), *Soil Sci.*, 69, 269.

BECKWITH, R. S., and LITTLE, I. P. (1963), *J. Sci. Fd Agric.*, 14, 15.

BEHRENS, W. U. (1962), *Z. PflErnähr. Düng.*, 96, 35.

BIRCH, H. F. (1953), *J. agric. Sci.*, 42, 229 and 329.

— (1961), *Pl. Soil*, 15, 347.

— (1964), Ibid., 21, 391.

BLACK, C. A., and GORING, C. A. I. (1953), *Soil and fertilizer phosphorus*, ed. Pierre and Norman. Academic, N.Y.

—, PIERRE, W. H., and ALLAWAY, W. H. (1948), *Ann. Rep. Iowa agric. Res. Stn.*, 101.

BOLTZ, D. F., and MELLON, M. G. (1947), *Anal. Chem.*, 19, 873.

BOSWELL, G. W., and DE LONG, W. A. (1959), *Can. J. Soil Sci.*, 39, 20.

BOWER, C. A. (1945), *Soil Sci.*, 59, 277.

— (1949), *Iowa agric. Exp. Stn. Res. Bull.*, 362.

BRAY, R. H., and KURTZ, L. T. (1945), *Soil Sci.*, 59, 44.

BRELAND, H. L., and SIERRA, F. A. (1962), *Proc. Soil Sci. Soc. Am.*, 26, 348.

BROMFIELD, S. M. (1964), *Nature, Lond.*, 201, 321.

— (1967), *Aust. J. Soil Res.*, 5, 225.

BURD, J. S. (1948), *Soil Sci.*, 65, 227.

CALDWELL, A. G. (1955), Ph.D. Thesis, Iowa State College.

— and BLACK, C. A. (1958), *Proc. Soil Sci. Soc. Am.*, 22, 290.

CHANG, S. C., and CHU, W. K. (1961), *J. Soil Sci.*, 12, 286.

— and JACKSON, M. L. (1957), *Soil Sci.*, 84, 133.

— and JUO, S. R. (1963), Ibid., 95, 91.

— and LIAW, F. H. (1962), *Science*, 136, 386.

—, CHU, W. K., and ERH, K. T. (1966), *Soil Sci.*, 102, 44.

CHIANG, C. (1963), *J. Soil Sci.*, Tokyo, 34, 360.

COLWELL, J. D. (1965), *Chem. Ind.*, 21, 893.

COMMEN, P. K., *et al.* (1959), *Ind. J. agric. Sci.*, 29, 96.

COOKE, G. W. (1951), *J. Soil Sci.*, 2, 254.

COSGROVE, D. J. (1962), *C.S.I.R.O. Div. Plant Ind. Rep.*, 1961–2, p. 54.

— (1963), *Aust. J. Soil Res.*, 1, 203.

COTTON, R. H. (1945), *Ind. Eng. Chem. A. E.*, 17, 1945.

DAHNKE, W. C., MALCOM, J. L., and MENENDEZ, M. E. (1964), *Soil Sci.*, 98, 33.

DAS, S. (1930), Ibid., 30, 33.

DATTA, N. P., and KAMARTH, M. B. (1959), *Ind. J. agric. Sci.*, 29, 87.

— and SRIVASTAVA, S. C. (1959), *Ind. J. agric. Sci.*, 29, 104,

DEAN, L. A. (1937), *Proc. Soil Sci. Soc. Am.*, 2, 223.

DENIGES, G. (1920), *C. R. Acad. Sci. Paris*, 171, 802.

DICKMAN, S. R., and DE TURK, E. E. (1938), *Soil Sci.*, 45, 29.

DORMAAR, J. F. (1964), *Can. J. Soil Sci.*, 44, 265.

— and WEBSTER, G. R. (1963), Ibid., 43, 35.

DOUGHTY, J. L. (1935), *Soil Sci.*, 40, 191.

DUPUIS, M. (1950), *Ann. agron. ser. A.*, 10.

DUVAL, L. (1962), *Ann. agron. Paris*, 13, 469.

DYER, B. (1894), *Trans. chem. Soc.*, 65, 115.

EGNÉR, H., and RIEHM, H. (1955), *Landw. Forsch.*, 6.

EID, M. T., BLACK, C. A., and KEMPTHORNE, O. (1951), *Soil Sci.*, 71, 361.

FIFE, C. V. (1959), Ibid., 87, 13 and 83.

— (1962), Ibid., 93, 113.

FISHER, R. A., and THOMAS, R. P. (1935), *J. am. Soc. Agron.*, 27, 863.

FRAPS, G. S. (1906), Ibid., 28, 823.

— (1911), *Texas agric. Exp. Stn. Bull.*, 136.

FREI, E., PEYER K., and SCHULTZE, E. (1964), *Schweiz. landw. Forsch.*, 3, 318.

FRIED, M., and DEAN, L. A. (1952), *Soil Sci.*, 73, 263.

FRIEND, M. T., and BIRCH, H. F. (1960), *J. agric. Sci.*, 54, 341.

GARDNER, R., and KELLEY, O. J. (1940), *Soil Sci.*, 50, 91.

GASSER, J. K. R. (1956), *VIe Congr. int. Sci. Sol. Rapp. C.*, 479.

GHANI, M. O. (1943a), *Ind. J. agric. Sci.*, 13, 29.

— (1943b), Ibid., 13, 562.

— and ISLAM, A. (1957), *Soil Sci.*, 84, 445.

GORING, C. A. I. (1955), *Pl. Soil*, 6, 17.

HANCE, R. J., and ANDERSON, G. (1962), *J. Soil Sci.*, 13, 225.

— and — (1963a), *Soil Sci.*, 96, 94.

— and — (1963b), Ibid., 96, 157.

HANLEY, K. (1962), *Irish J. agric. Res.*, 1, 192.

HANNAPEL, R. J., FULLER, W. H., and FOX, R. H. (1964), *Soil Sci.*, 97, 421.

HARRAP, F. E. G. (1963), *J. Soil Sci.*, 14, 82.

HARRIS, C. I., and WARREN, G. F. (1962), *Proc. Soil Sci. Soc. Am.*, 26, 381.

HESSE, P. R. (1962), *Nature, Lond.*, 193, 295.

— (1963), *Pl. Soil*, 19, 205.

HIBBARD, P. C. (1937), *Soil Sci.*, 39, 337.

HOLMAN, W. M. (1936), *J. Roy. Soc. N.S.W.*, 70, 264.

HOPKINS, C. G., and PETTIT, J. A. (1908), *Illinois Exp. Stn. Bull. No. 123*, p. 187.

JACKSON, M. L. (1958), *Soil Chemical Analysis*. Constable, London.

JORET, G., and HERBERT, J. (1955), *Ann. Agron. ser. A.*, 233.

KAILA, A. (1962), *Maataloust. Aikakausk.*, 34, 187.

— and VIRTANEN, O. (1955), Ibid., 27, 104.

KURTZ, L. T., and ARNOLD, C. Y. (1946), *Univ. Illinois Dep. Agric. Mimco.*, AG-1306.

LACY, J. (1965), *Analyst*, 90, 65.

LARSEN, J. E., WARREN, G. F., and LANGSTON, R. (1959), *Proc. Soil Sci. Soc. Am.*, 23, 438.

LARSEN, S. (1965), *J. Soil Sci.*, 16, 275.

LARSEN, S., and COURT, M. N. (1960), *Trans. 7th int. Congr. Soil Sci.*, 2, 413.

— and SUTTON, C. D. (1963), *Pl. Soil*, 18, 77.

— and WIDDOWSON, A. E. (1964), *Nature, Lond.*, 203, 942.

LAVERTY, J. C. (1963), *Proc. Soil Sci. Soc. Am.*, 27, 360.

LEGG, J. O., and BLACK, C. A. (1955), Ibid., 19, 139.

LEMARE, P. H. (1960), *Trans. 7th int. Congr. Soil Sci.*, 3, 600.

LEVINE, H., ROWE, J. J., and GRIMALDI, F. S. (1954), *Science*, 119, 327.

LIPMAN, J. G., and CONYBEARE, A. B. (1936), *New Jersey agric. Exp. Stn. Bull.*, 607.

LORENZ, N. V. (1901), *Landw. Versuchs*, 55, 183.

MARTIN, J. K. (1964), *N.Z. J. agric. Res.*, 7, 723.

MATTINGLY, G. E. G. (1957), *Soils Fertil.*, 20, 59.

— and PINKERTON, A. (1961), *J. Sci. Fd. Agric.*, 12, 772.

MATTSON, S. and KARLSSON, N. (1938), *Ann. agric. Coll. Sweden*, 6, 109.

MCCALL, W. W., DAVIS, J. F., and LAWTON, K. (1956), *Proc. Soil Sci. Soc. Am.*, 20, 81.

MCDONALD, J. A. (1933), *Trop. Agric. Trin.*, 10, 108.

MEHLICH, A., FRED, E. B., and TRUOG, E. (1935), *Trans. 3rd int. Congr. Soil Sci.*, 1, 169.

MEHTA, N. N. (1951), Thesis, Iowa State College.

MEHTA, N. C., LEGG, I. O., and GORING, C. A. I. (1954), *Proc. Soil Sci. Soc. Am.*, 18, 443.

METSON, A. J. (1961), *N.Z. C.S.I.R. Soils Bull.*, No. 12.

MITSCHERLICH, E. A. (1923), *Landw. Jahrb.*, 58, 601.

MOORTHY, B. R., and SUBRAMANIAN, T. R. (1960), *Trans. 7th int. Congr. Soil Sci.*, 3, 590.

MURPHY, J., and RILEY, J. P. (1962), *Anal. chim. Acta*, 27, 31.

NELSON, W. L., MEHLICH, A., and WINTERS, E. (1953), *Agron.*, 4, 153.

NEUBAUER, H. (1932), *Z. PflErnähr. Düng.*, 27, 505.

NIKLAS, A., *et al.* (1927), Ibid., 3A, 136.

OLSEN, S. R., COLE, C. V., *et al.* (1954), *U.S. Dep. Agric. Circ.*, 939.

— and WATANABE, F. S. (1957), *Proc. Soil Sci. Soc. Am.*, 21, 144.

—, —, and COLE, C. V. (1960), *Soil Sci.*, 90, 44.

OSMUND (1887), *Bul. soc. chim. biol.*, 47, 745.

PARTON, D. J. (1963), *J. Soil Sci.*, 14, 167.

PEARSON, R. W. (1940), *Ind. Eng. Chem. A. E.*, 12, 198.

PEECH, M., *et al.* (1947), *U.S. Dep. Agric. Circ.*, 757.

PETERSEN, G. W., and COREY, R. B. (1966), *Proc. Soil Sci. Soc. Am.*, 30, 563.

PIERRE, W. H., and PARKER, F. W. (1927), *Soil Sci.*, 24, 119.

PIPER, C. S. (1942), *Soil and Plant Analysis*. Univ. of Adelaide.

PONS, W. A., and GUTHRIE, J. D. (1946), *Ind. Eng. Chem. A. E.*, 18, 184.

POTTER, R. S., and BENTON, T. H. (1916), *Soil Sci.*, 2, 291.

PRATT, R. F., and GARBER, M. J. (1964), *Proc. Soil Sci. Soc. Am.*, 28, 23.

PURI, M. N., and ASGHAR, A. C. (1936), *Soil Sci.*, 42, 39.

ROBINSON, R. J. (1941), *Ind. Eng. Chem. A.E.*, 13, 465.

RUSSELL, E. W. (1961), *Soil Conditions and Plant Growth*, 9th edn. Longmans.

SAKADI, J., KRAMER, M., and THAMM, F. (1965), *Agrokém. Talajt.*, 14, 75.

SAUNDER, D. H. (1956), *Soil Sci.*, 82, 457.

— and METELERKAMP, H. R. (1962), *Int. Soil Conf. N.Z.*, 847.

SAUNDERS, W. M. H., and WILLIAMS, E. G. (1955), *J. Soil Sci.*, 6, 254.

SAXENA, S. N. (1964), *J. Ind. Soc. Soil Sci.*, 12, 137.

SEMB, G., and UHLEN, G. (1954), *Acta agric. scand.*, 5, 44.

SCHMOEGER, M. (1897), *Biedermanns Centbl.*, 26, 579.

SCHOFIELD, R. K. (1935), *Trans. 3rd int. Congr. Soil Sci.*, 2, 37.

— (1955), *Soils Fertil.*, 18, 373.

SCHOLLENBERGER, C. J. (1918), *Soil Sci.*, 6, 365.

SHEARD, R. W., and CALDWELL, A. G. (1955), *Can. J. agric. Sci.*, 35, 36.
SMITH, A. N. (1965a), *Agrochimica*, 9, 162.
— (1965b), *Pl. Soil*, 22, 314.
SMITH, D. H., and CLARK, F. E. (1951), *Soil Sci.*, 72, 353.
SMITH, F. B., BROWN, P. E., and SCHLOTS, F. E. (1932), *J. am. Soc. Agron.*, 24, 452.
SMITH, F. W., ELLIS, B. G., and GRAVA, J. (1957), *Proc. Soil Sci. Soc. Am.*, 21, 400.
SNELL, F. D., and SNELL, C. T. (1949), *Colorimetric methods of analysis*, 3rd edn.
    D. van Nostrand, N.Y.
SOKOLOV, D. F. (1948), *Pochvovedenie*, 502.
STELLY, M., and RICAUD, R. (1960), *Trans. 7th int. Congr. Soil Sci.*, 3, 60.
STEWART, R. (1910), *Illinois Exp. Stn. Bull.*, 145, 91.
— (1932), *Imp. Bur. Soil Sci. Tech. Comm.*, 25.
STRZEMIENSKI, K. (1955), *N.Z. J. Sci. Tech.*, B37, 243.
SUTHERLAND, W. H., and BLACK, C. A. (1959), *Pl. Soil*, 10, 356.
TALIBUDEEN, O. (1957), *J. Soil Sci.*, 8, 86.
— (1958), Ibid., 9, 120.
THOMPSON, E. J., OLIVEIRA, A. L. F., *et al.* (1960), *Pl. Soil*, 13, 28.
THOMPSON, L. M., BLACK, C. A., and ZOELLNER, J. A. (1954), *Soil Sci.*, 77, 185.
TOMBESI, L., and CALÉ, M. T. (1962), Ibid., 17, 137.
TRUOG, E. (1930), *J. am. Soc. Agron.*, 22, 874.
— and MEYER, A. H. (1929), *Ind. Eng. Chem. A. E.*, 1, 136.
TURNER, R. C., and RICE, H. M. (1954), *Soil Sci.*, 74, 141.
VAN DIEST, A. (1963), Ibid., 96, 261.
— and BLACK, C. A. (1959), Ibid., 87, 145.
VÁRALLAY, G. (1934), *Z. PflErnähr. Düng. Bodenk.*, 34A, 218.
VINCENT, V. (1937), *C. R. 17th Congr. Chim. Ind. Paris*, 2, 861.
VOGEL, A. I. (1962), *A textbook of quantitative inorganic analysis.* 3rd edn., Longmans.
VOLK, V. V., and MCLEAN, E. O. (1963), *Proc. Soil Sci. Soc. Am.*, 27, 53.
WALKER, T. V., and ADAMS, A. F. R. (1958), *Soil Sci.*, 85, 307.
WARREN, R. G., and COOKE, G. W. (1962), *J. agric. Sci.*, 59, 269,
WATANABE, F. S., and OLSEN, S. R. (1962), *Soil Sci.*, 93, 183.
— and — (1965), *Proc. Soil Sci. Soc. Am.*, 29, 677,
WEIR, C. C. (1962), *Trop. Agric. Trin.*, 30, 67.
— and SOPER, R. J. (1963), *J. Soil Sci.*, 14, 256.
WHITE, R. E., and BECKETT, P. H. T. (1964), *Pl. Soil*, 20, 1.
WIKLANDER, L. (1950), *LantbrHögsk. Medd.*, 17, 407.
WILD, A. (1964), *Nature, Lond.*, 203, 326.
WILLIAMS, C. H. (1950), *J. agric. Sci.*, 40, 233 and 243.
WILLIAMS, E. G., and KNIGHT, A. H. (1963), *J. Sci. Fd. Agric.*, 14, 555.
WILLIAMS, R. (1937), Ibid., 27, 259.
WILSON, H. N. (1951), *Analyst*, 76, 65.
— (1954), Ibid., 79, 535.
WINOGRADSKY, S., and ZIEMIECKA, J. (1928), *Ann. Inst. Pasteur*, 42, 36.
WOODS, J. T., and MELLON, M. G. (1941), *Ind. Eng. Chem. A. E.*, 13, 762.
WRENSHALL, C. L., and DYER, W. J. (1939), *Can J. Res.*, 17B, 199.
— and MCKIBBIN, R. R. (1937), Ibid., 15B, 475.
YUAN, T. L., ROBERTSON, W. K., and NELLER, J. R. (1960), *Proc. Soil Sci. Soc. Am.*, 24,
    447.

Chapter 13

ADERIKHI, P. G. (1960), *Trans. 7th int. Congr. Soil Sci.*, 2, 281.
ALLPORT, N. L. (1933), *Quart. J. Pharm. and Pharmacol.*, 6, 431.
ANDERSEN, L. (1953), *Acta chem. scand.*, 7, 689.
ANDERSON, A. J., and SPENCER, D. (1950), *Aust. J. Sci. Res.*, B3, 431.
BAERNSTEIN, H. D. (1932), *J. Biol. Chem.*, 97, 663.

BAERNSTEIN, H. D. (1936), Ibid., 116, 25.

BARDSLEY, C. E., and LANCASTER, J. D. (1960), *Proc. Soil Sci. Soc. Soc. Am.*, 24, 265.

— and — (1965), *Monog. am. Soc. Agron.*, 9, 1102.

BLOCK, R. J., and BOLLING, D. (1951), *The amino-acid composition of proteins and foods.* 2nd edn. C. C. Thomas, Illinois.

BLOOMFIELD, C. (1962), *Analyst*, 87, 586.

BOND, R. D. (1955), *Chem. and Ind.*, 30, 941.

BOWER, C. A., and HUSS, R. B. (1948), *Soil Sci.*, 66, 199.

BREMNER, J. M. (1950), *Biochem. J.*, 47, 538.

BURIEL, M. F., and JIMENEZ, G. S. (1955), *An. Edafol. Fisiol. Veg.*, 14, 103.

BUTLIN, K. R. (1953), *5th int. Congr. Microb.*, Abstr., 183.

BUTTERS, B., and CHENERY, E. M. (1959), *Analyst*, 84, 239.

CANTINO, E. C. (1946), *Soil Sci.*, 61, 361.

CHAO, T. T., HARWARD, M. E., and FANG, S. C. (1962), *Proc. Soil Sci. Soc. Am.*, 26, 27.

CHAPMAN, H. D., and PRATT, P. F. (1961), *Methods of analysis of soils, plants and waters.* Univ. Calif.

CHAUDHRY, I. A., and CORNFIELD, A. H. (1966), *Analyst*, 91, 528.

CHESNIN, L., and YIEN, C. H. (1961), *Proc. Soil Sci. Soc. Am.*, 15, 149.

DEAN, G. A. (1966), *Analyst*, 91, 530.

DE LONG, W. A., and LOWE, L. R. (1962), *Can. J. Soil Sci.*, 42, 223.

— and — (1963), Ibid., 43, 151.

DENT, C. E. (1947), *Biochem. J.*, 41, 240.

ENSMINGER, L. E. (1954), *Proc. Soil Sci. Soc. Am.*, 18, 259.

EVANS, C. A., and ROST, C. C. (1945), *Soil Sci.*, 59, 125.

FIELD, E., and OLDACH, C. S. (1946), *Ind. Eng. Chem. A. E.*, 18, 665.

FISKE, C. H. (1921), *J. Biol. Chem.*, 80, 623.

FLEMMING, R. (1930), *Biochem. J.*, 24, 965.

FOULK, C. W., and BAWDEN, A. T. (1926), *J. am. Chem. Soc.*, 48, 2045.

FRENEY, J. R. (1958), *Soil Sci.*, 86, 241.

— (1961), *Aust. J. agric. Res.*, 12, 424.

—, BARROW, N. J., and SPENCER, K. (1962), *Pl. Soil*, 17, 295.

GILBERT, F. A. (1951), *Botan. Rev.*, 17, 671.

GLEEN, H., and QUASTEL, J. H. (1953), *Appl. Microbiol.*, 1, 70.

GRIGG, J. L. (1953), *Analyst*, 78, 470.

HARMSEN, G. W., QUISPEL, A., and OTZEN, D. (1954), *Pl. Soil*, 5, 324.

HART, M. G. R. (1959), Ibid., 11, 215.

— (1961), *Analyst*, 86, 472.

— (1962), *Pl. Soil*, 17, 87.

— (1963), Ibid., 19, 106.

HESSE, P. R. (1957a), Ibid., 9, 86.

— (1957b), *Hydrobiologia*, 11, 29.

— (1957c), *Analyst*, 82, 710.

— (1958), *Hydrobiologia*, 11, 171.

— (1961a), *Pl. Soil*, 14, 249.

— (1961b), Ibid., 14, 335.

JENSEN, H. L., and SORENSEN, H. (1952), *Acta agric. scand.*, 2, 295.

JOHNSON, C. M., and NISHITA, H. (1952), *Anal. Chem.*, 24, 736.

JOUIS, E., and LECACHEUX, M. T. (1962), *Ann. agron. Paris*, 13, 483.

KAHN, B. S., and LIEBOFF, S. L. (1928), *J. biol. Chem.*, 80, 623.

KILMER, V. J., and NEARPASS, D. C. (1960), *Proc. Soil Sci. Soc. Am.*, 24, 337.

KITCHENER, J. A., LIBERMAN, A., and SPRATT, D. A. (1952), *Analyst*, 76, 509.

KOYAMA, T., and SUGARAWA, K. (1953), *J. Earth Sci. Nagoya Univ.*, 1, 24.

LARSEN, R. P., ROSS, L. E., and INGBER, N. M. (1959), *Anal. Chem.*, 31, 1596.

LITTLE, R. C. (1953), *J. Sci. Fd. Agric.*, 4, 336.

— (1957), Ibid., 8, 271.

17*

LITTLE, R. C. (1958), *J. Sci. Fd. Agric.*, 9, 273.

LIU, M., and THOMAS, G. W. (1961), *Nature, Lond.*, 192, 384.

LOWE, L. E., and DE LONG, W. A. (1961), *Can. J. Soil Sci.*, 41, 141.

MANN, H. V. (1955), *J. Soil Sci.*, 6, 241.

MASSOUMI, A., and CORNFIELD, A. H. (1963), *Analyst*, 88, 321.

MCCARTHY, T. E., and SULLIVAN, M. X. (1941), *J. Biol. Chem.*, 141, 871.

METSON, A. J. (1961), *N.Z. C.S.I.R. Soils Bull. No. 12.*

MONAND, P. (1953), *Soc. Chim. France, Mem. 218*, 1063.

MOORMAN, F. R. (1961), *Researches on acid–sulphate soils and their amelioration by liming*, p. 3. Min. of Rural Affairs Agron. Library, Saigon.

MORRIS, H. E. (1948), *Anal. Chem.*, 20, 1037.

MORTIMER, C. H. (1941), *J. Ecol.*, 29, 317.

MUNGER, J. R., NIPPLER, R. W., and INGOLS, R. S. (1950), *Anal. Chem.*, 22, 1455.

ORY, H. A., WARREN, V. L., and WILLIAMS, H. B. (1957), *Analyst*, 82, 189.

PARR INSTRUMENT CO. (1934), *Bull. No. 113.*

QUASTEL, J. H., HEWITT, E. J., and NICHOLAS, D. J. D. (1948), *J. Agric. Sci.*, 38, 315.

— and SCHOLFIELD, P. G. (1949), *Nature, Lond.*, 164, 1068.

REVOL, L., and FERRAND, M. (1935), *Bull. Soc. chim. biol.*, 17, 1451.

ROBINSON, W. O. (1945), *Soil Sci.*, 59, 11.

ROMM, I. I. (1944), *J. app. Chem. (USSR)*, 17, 188.

SAALBACH, E., KESSEN, G., and JUDEL, G. K. (1962), *Landw. Forsch.*, 15, 6.

SHEEN, R. T., KAHLER, H. L., and ROSS, E. M. (1935), *Ind. Eng. Chem. A. E.*, 7, 262.

SHOREY, E. C. (1913). *U.S. Dep. Agric. Bur. Soils Bull.*, 88.

SMITTENBURG, J., HARMSEN, G. W., et al. (1951), *Pl. Soil*, 3, 353.

SNELL, F. D., and SNELL, C. T. (1949), *Colorimetric methods of analysis*, Vol. II, 3rd edn. D. van Nostrand.

SOWDEN, F. J. (1955), *Soil Sci.*, 80, 181.

STARKEY, R. L. (1935), *J. gen. Physiol.*, 18, 325.

STEINBERGS, A. (1953), *Analyst*, 78, 47.

—, IISMAA, C., et al. (1962), *Anal. chim. Acta*, 27, 158.

STEVENSON, F. J. (1954), *Proc. Soil Sci. Soc. Am.*, 18, 373.

SUGARAWA, K., KOYAMA, T., and KOZAWA, A. (1954), *J. Earth Sci. Nagoya Univ.*, 2, 1.

SWANSON, C. O., and LATSHAW, W. C. (1922), *Soil Sci.*, 14, 421.

THOMAS, M. D., et al. (1950), *Ind. Eng. Chem.*, 42, 2231.

TOMLINSON, T. E. (1957), *Emp. J. exp. Agric.*, 25, 108.

VALLENTYNE, J. R. (1952), *Science*, 116, 667.

VASSEL, B. (1941), *J. biol. Chem.*, 140, 323.

VICKERY, H. B., and WHITE, A. (1933), Ibid., 99, 701.

VOGEL, A. I. (1962), *A textbook of quantitative inorganic analysis*, 3rd edn. Longmans.

WILLIAMS, C. H., and STEINBERGS, A. (1959), *Aust. J. Agric. Res.*, 10, 340.

— and — (1962), *Pl. Soil*, 17, 279.

WILLIAMS, H. W. (1955), *Cornell Univ. agric. exp. Stn. Mem.*, 337.

YOMEDA, S. (1961), *Sci. Rep. Fac. Agric. Okayama Univ.*, 17, 39.

Chapter 14

ABUTALYBOV, M. G., and ALIEV, D. (1961), *Izv. Akad. Nauk. azerb. SSR. Ser. biol. med. Nauk.*, 5, 31.

ALEXANDER, L. T., and CADY, J. G. (1962), *U.S. Dep. Agric. Tech. Bull.*, No. 1282.

ALLISON, L. E., and SCARSETH, G. D. (1942), *J. am. Soc. Agron.*, 34, 616.

ALTEN, F., and WEILAND, H. (1933), *Z. PflErnähr. Düng.*, 30A, 193.

ATKINSON, H. J., and WRIGHT, J. R. (1957), *Soil Sci.*, 84, 1.

BAO, H. M., LIU, C. K., and YÜ, T. J. (1964), *Acta pedol. sin.*, 12, 216.

BLOOMFIELD, C. (1956), *6th int. Congr. Sci. Sol. Rapp. B.*, 427.

BOWER, C. A., and TRUOG, E. (1940), *Proc. Soil Sci. Soc. Am.*, 5, 86.

BREMNER, J. M., *et al.* (1946), *Nature, Lond.*, 158, 790.

BROMFIELD, S. M. (1958), *Pl. Soil*, 9, 325.

— and WILLIAMS, E. G. (1963), *J. Soil Sci.*, 14, 346.

CASIDA, L. E., and SANTORO, T. (1961), *Soil Sci.*, 92, 287.

CHAMBERLAIN, G. T. (1956), *Ann. Rep. E. Afr. agric. for. Res. Org.*, p. 32.

CHAMURA, S. (1962), *Proc. Crop Sci. Soc. Japan*, 30, 350.

CHENERY, E. M. (1948), *Analyst*, 73, 501.

— (1949), *J. Soil Sci.*, 2, 97.

COFFIN, D. E. (1963), *Can. J. Soil Sci.*, 43, 7.

COLEMAN, N. T., *et al.* (1959), *Proc. Soil Sci. Soc. Am.*, 23, 146.

CORNFIELD, A. H., and POLLARD, A. G. (1950), *J. Sci. Fd. Agric.*, 1, 107.

COULSON, C. B., *et al.* (1960), *J. Soil Sci.*, 11, 20 and 30.

CRAFT, C. H., and MAKEPEACE, C. R. (1945), *Ind. eng. Chem. A. E.*, 17, 206.

DANIELS, R. B., *et al.* (1962), *Proc. Soil Sci. Soc. Am.*, 26, 75.

DEAN, J. A., and LADY, J. H. (1953), *Analyt. Chem.*, 25, 947.

DE ENDREDY, A. S. (1962), *Clay Miner. Bull. No. 5*, p. 209.

DESHPANDE, T. L., *et al.* (1964), *Nature, Lond.*, 201, 107.

DION, H. G. (1944), *Soil Sci.*, 58, 411.

DROSDOFF, M., and TRUOG, E. (1935), *J. amer. Soc. Agron.*, 27, 312.

— and NIKIFOROFF, C. C. (1940), *Soil Sci.*, 49, 333.

DUCHAFOUR, P. (1963), *C. R. Acad. Sci. Paris*, 256, 2657.

FRINK, C. R., and PEECH, M. (1962), *Soil Sci.*, 93, 317.

FUJIMOTO, C. K., and SHERMAN, G. D. (1946), *Proc. Soil Sci. Soc. Am.*, 10, 107.

— and — (1948), *J. amer. Soc. Agron.*, 40, 527.

GALABUTSKAYA, E., and GOVAROVA, R. (1934), *Min. Suir*, 9, 27.

GOTO, K., *et al.* (1958), *Bull. Chem. Soc. Japan*, 31, 783.

GRAMMS, G., and KAGELMANN, K. (1962), *Z. landw. Vers. u. UntersWes.*, 8, 141.

GROVES, A. W. (1951), *Silicate Analysis*, 2nd edn. Allen & Unwin.

HACKETT, C. (1962), *Nature, Lond.*, 195, 471.

HALDANE, A. D. (1956), *Soil Sci.*, 82, 483.

HAMMETT, L. P., and SOTTERY, C. T. (1925), *J. Am. Chem. Soc.*, 47, 142.

HARRY, R. G. (1931), *Chem. Ind.*, 50, 796.

HEM, J. D., and CROPPER, W. H. (1959), *U.S. Geol. Surv. Wat. Supp.*, Pap. 1459-A.

— (1960), Ibid., Pap. 1459-B.

HESSE, P. R. (1963), *Pl. Soil*, 19, 205.

HILL, U. T. (1956), *Anal. Chem.*, 28, 1419.

HOFF, D. J., and MEDERSKI, H. J. (1948), *Proc. Soil Sci. Soc. Am.*, 22, 129.

HSU, P. H., and RENNIE, D. A. (1962), *Can. J. Soil Sci.*, 42, 197.

— (1963), *Soil Sci.*, 96, 230.

IGNATIEFF, V. (1941), Ibid., 51, 249.

IRI, H., *et al.* (1957), *Soil and Pl. Fd.*, 3, 36.

IYER, C. R., and RAJAGOPALAN, R. (1936), *J. Ind. Inst. Sci.*, 19A, 56.

JACKSON, M. L. (1958), *Soil chemical analysis*. Constable.

JEFFERIES, C. D. (1941), *Soil Sci.*, 52, 451.

— (1945), *Proc. Soil Sci. Soc. Am.*, 11, 211.

— and JOHNSON, L. (1961), *Soil Sci.*, 92, 402.

JEFFERY, J. W. O. (1961), *J. Soil Sci.*, 12, 172.

JOHANSSON, A. (1962), *St. JordbrFörs. Medd.*, 132, 25.

JOHNSON, M. O. (1916), *Hawaii agric. Exp. Stn. Bull.*, 51, 1.

JONES, L. H. P., and LEEPER, G. W. (1950), *Science*, 11, 463.

— and — (1951), *Pl. Soil*, 3, 141.

JONES, L. H. and THURMAN, D. A. (1957), Ibid., 9, 131.

KILMER, J. (1960), *Proc. Soil Sci. Soc. Am.*, 24, 420.

KOLTHOFF, I. M., and SANDELL, E. B. (1945), *Textbook of quantitative inorganic analysis*, rev. edn. Macmillan, N.Y.

KORBL, J., and PŘIBIL, R. (1956), *Chemist-Analyst*, 45, 162.

KRUPTSKII, N. K., *et al.* (1961), *Pochvovedenie*, 10, 93.

LAMM, C. G., and MAURY, M. (1962), *Acta agric. scand.*, 12, 9.

LEEPER, G. W. (1947), *Soil Sci.*, 63, 79.

LE RICHE, H. H., and WEIR, A. H. (1963), *J. Soil Sci.*, 14, 225.

LIGON, W. S., and PIERRE, W. H. (1932), *Soil Sci.*, 34, 145 and 307.

LUTZ, J. F. (1937), *Proc. Soil Sci. Soc. Am.*, 1, 43.

MACKENZIE, R. C. (1954), *J. Soil Sci.*, 5, 167.

— (1957), *The differential thermal investigation of clays.* Pub. Miner. Soc. (Clay Miner. Group), London.

MARTIN, A. E., and REEVE, R. (1958), *J. Soil Sci.*, 9, 89.

MATTSON, S. (1931), *Soil Sci.*, 31, 311.

MAYR, C., and GEBAUER, A. (1938), *Z. anal. Chem.*, 113, 189.

MCCOOL, M. M. (1934), *Boyce Thompson Inst. Contrib.*, 6, 147.

MCINTYRE, D. S. (1956), *J. Soil Sci.*, 7, 302.

MCLEAN, E. O., *et al.* (1959), *Proc. Soil Sci. Soc. Am.*, 23, 289.

MEHLIG, J. P. (1939), *Ind. Eng. Chem. A. E.*, 11, 274.

MITSUCHI, M., and OYAMA, M. (1963), *J. Sci. Soil Tokyo*, 34, 23.

MUKHERJEE, J. N., *et al.* (1947), *J. Colloid Sci.*, 2, 247.

NORRISH, K., and TAYLOR, R. M. (1961), *J. Soil Sci.*, 12, 294.

OADES, J. M. (1963), *Soils Fertil.*, 26, 69.

— and TOWNSEND, W. M. (1963), *J. Soil Sci.*, 14, 134.

PADDICK, H. E. (1948), *Proc. Soil Sci. Soc. Am.*, 13, 197.

PAGE, A. L., and BINGHAM, F. T. (1962), Ibid., 26, 351.

PAGE, E. R., *et al.* (1962), *Pl. Soil*, 16, 238 and 247.

PARKER, D. T. (1962), *Agron. J.*, 54, 303.

PASSIOURA, J. B., and LEEPER, G. W. (1963a), *Agrochimica*, 8, 81.

— and — (1963b), *Nature, Lond.*, 200, 29.

PAVER, H., and MARSHALL, C. E. (1934), *J. Soc. chem. Ind.*, 53, 750.

PEECH, M., *et al.* (1947), *U.S. Dep. Agric. Circ.*, 757.

PIERRE, W. H., and STUART, A. D. (1933), *Soil Sci.*, 36, 211.

PLUCKNETT, D. C., and SHERMAN, D. G. (1963), *Proc. Soil Sci. Soc. Am.*, 27, 39.

PRATT, P. F. (1961), Ibid., 25, 467.

— and BAIR, F. L. (1961), *Soil Sci.*, 91, 359.

PRITCHARD, D. T. (1967), *Analyst*, 92, 103.

PURDY, W. C., and HUME, D. N. (1955), *Anal. Chem.*, 27, 256.

RANDALL, P. J., and VOSE, P. B. (1963), *Plant Physiol.*, 38, 403.

RAUPACH, M. (1963), *Aust. J. Soil Res.*, 1, 28.

RICHARDS, M. B. (1930), *Analyst*, 55, 554.

ROBICHET, O. (1957), *Ann. agron. Paris*, 8, 257.

ROOKSBY, H. P. (1961), *The X-ray identification and crystal structures of clay minerals.* Pub. Miner. Soc. (Clay Miner. Group), London.

RUBIN, B. A., *et al.* (1962), *Fiziol. Rast.*, 9, 657.

RUSSELL, E. W. (1961), *Soil conditions and plant growth*, 9th edn. Longmans.

SHEN, M. J., and RICH, C. I. (1962), *Proc. Soil Sci. Soc. Am.*, 26, 33.

SHERMAN, G. D., and HARMER, P. M. (1943), Ibid., 7, 398.

SIDERIS, C. P. (1937), *Ind. Eng. Chem. A. E.*, 9, 445.

— (1940), Ibid., 12, 307.

SIVARAJASINGHAM, S., *et al.* (1962), *Adv. Agron.*, 14, 1.

SNELL, F. D., and SNELL, C. T. (1949), *Colorimetric methods of analysis*, vol. II, 3rd edn. D. Van Nostrand, London.

STRAFFORD, N., and WYATT, P. F. (1947), *Analyst*, 72, 54.

TAMM, O. (1922), *Meddn. St. SkosforskInst.*, 19, 385.

THOMAS, G. W. (1960), *7th int. Congr. Soil Sci.*, 2, 364.

— and COLEMAN, N. T. (1964), *Soil Sci.*, 97, 229.

TIMONIN, M. I. (1947), *Proc. Soil Sci. Soc. Am.*, 11, 284.
TOTEV, T., and KULEV, I. (1965), *Pochvovedenie*, 1, 90.
TRUOG, E., *et al.* (1936), *Proc. Soil Sci. Soc. Am.*, 1, 101.
VLAMIS, J. (1953), *Soil Sci.*, 75, 383.
VOGEL, A. I. (1962), *A textbook of quantitative inorganic analysis.* 3rd edn., Longmans.
WAEGEMAN, S. G., and HENRY, S. (1954), *5th int. Congr. Soil Sci.*, 2, 384.
WALKER, J. L., and SHERMAN, G. D. (1962), *Soil Sci.*, 93, 325.
WEIR, C. C., and MILLER, M. H. (1962), *Can. J. Soil Sci.*, 42, 105.
WILLARD, H. H., and GREATHOUSE, L. H. (1917), *J. am. Chem. Soc.*, 39, 2366.
WILLIAMS, E. G., *et al.* (1958), *J. Sci. Fd. Agric.*, 9, 551.
YOE, J. H., and ARMSTRONG, A. R. (1947), *Anal. Chem.*, 19, 100.
YUAN, T. L., and FISKELL, J. G. A. (1959), *Proc. Soil Sci. Soc. Am.*, 23, 202.

Chapter 15

BIRCH, H. F. (1955), *Ann. Rep. E. Afr. agric. for. Res. Org.*, 1954–5, 17.
BUNTING, W. E. (1944), *Ind. Eng. Chem. A. E.*, 16, 574.
DIENERT, F., and WANDENBULKE, F. (1923), *C. R.*, 176, 1478.
ETTORI, J. (1936), Ibid., 202, 852.
FISHER, R. A. (1929), *J. agric. Sci.*, 19, 132.
HALLSWORTH, E. G., and WARING, H. D. (1964), *J. Soil Sci.*, 15, 158.
HINES, E., and BOLTZ, D. F. (1952), *Anal. Chem.*, 24, 947.
JACKSON, M. L. (1958), *Soil chemical Analysis.* Constable.
JONES, L. H. P., and HANDRECK, K. A. (1963), *Nature, Lond.*, 198, 852.
KAHLER, H. L. (1941), *Ind. Eng. Chem. A. E.*, 13, 536.
KNUDSON, H. W., JUDAY, C., and MELOCHE, V. W. (1940), Ibid., 12, 270.
MCKEAGUE, J. A. (1962), *Diss. Abstr.*, 23, 2.
MILLER, R. W. (1963), *Agron. Abstr.*, 55th Ann., p. 23.
NIKIFOROFF, C. C., and ALEXANDER, C. T. (1942), *Soil Sci.*, 53, 157.
RUSSELL, E. W. (1961), *Soil conditions and plant growth*, 9th edn. Longmans.
SCHWARTZ, M. C. (1934), *Ind. Eng. Chem. A. E.*, 6, 364.
SILVERMAN, L. (1948), *Chemist-Analyst*, 37, 62.
SNELL, F. D., and SNELL, C. T. (1949), *Colorimetric methods of analysis*, vol. II. D. Van
    Nostrand.
STRAUB, F. G., and GRABOWSKI, H. A. (1944), *Ind. Eng. Chem. A. E.*, 16, 574.
STRUNTZ, H. (1957), *Min. Tab. Akad. Verlag*, Leipzig.
UCHIYAMA, N., and ONIKURA, Y. (1955), *J. Sci. Soil Tokyo*, 25, 291.
VOGEL, A. I. (1961), *A textbook of quantitative inorganic analysis*, 3rd edn., Longmans.
WEIHRICH, R., and SCHWARTZ, W. (1941), *Arch. Eisenhuttenw.*, 14, 501.
WEISSLER, A. (1945), *Ind. Eng. Chem. A. E.*, 17, 695.
WELLER, A. (1882), *Ber.*, 15, 2592.
WOODS, J. T., and MELLON, M. G. (1941), *Ind. Eng. Chem. A. E.*, 13, 760.

Chapter 16

AKEMAN, H. J. (1948), *J. Ind. Hyg. Toxicol.*, 30, 300.
ALDRICH, D. G., VANESLOW, A. P., and BRADFORD, G. R. (1951), *Soil Sci.*, 71, 291.
ALEXANDER, J. D., BEAVERS, A. H., and JOHNSON, P. R. (1962), *Proc. Soil Sci. Soc. Am.*,
    26, 189.
ALMOND, H. (1953), *Anal. Chem.*, 25, 1766.
ANDERSEN, A. J. (1963), *Soil Sci.*, 95, 52.
ANDRUS, S. (1955), *Analyst*, 80, 514.
ANTIPOV-KARATAEV, I. N. (1947), *Pedology*, 652.
ARNON, D. I. (1959), *Nature, Lond.*, 184, 10.
ATAMANENKO, N. N., *et al.* (1962), *Visn. sil-hospod. Naukȳ*, 5, 108.
BAKER, J., and BURNS, W. T. (1964), *Can. J. Soil Sci.*, 44, 361.

BARSHAD, I. (1949), *Anal. Chem.*, 21, 1148.

BAUER, I., and LINDSAY, W. L. (1963), *Agron. Abstr.*

BERGER, K. C., and TRUOG, E. (1939), *Ind. Eng. Chem. A.E.*, 11, 540.

BERGH, H. (1948), *K. NorskeVidensk. Selskabs. Skrifter*, 1945, No. 3.

BERTRAND, D., and WOLF, A. (1961), *C. R. Acad. Sci. Paris*, 253, 1342.

BINGHAM, F. T., PAGE, A. L., and SIMS, J. R. (1963), *Proc. Soil Sci. Soc. Am.*, 27, 167.

BINGLEY, J. B. (1963), *J. agric. Fd. Chem.*, 11, 130.

BLUNDY, P. D. (1958), *Analyst*, 83, 555.

BOAWN, L. C. (1962), *Proc. Soil Sci. Soc. Am.*, 26, 208.

BOBKO, E. V., and MATVEEVA, T. V. (1936), *Zh. Prikl. Khim.*, 9, 532.

BOGG, H. M., and ALBEN, A. O. (1936), *Ind. Eng. Chem. A. E.*, 8, 97.

BORST-PAUWELS, G. W. F. H. (1961), *Pl. Soil*, 14, 377.

BOWDEN, P. (1964), *Analyst*, 89, 771.

BOWEN, H. J. M., and CAWSE, P. A. (1962), *Nature, Lond.*, 194, 399.

BRADFORD, G. R. (1963), *Soil Sci.*, 96, 77,

BRADFORD, G. R., and PRATT, P. F. (1961), Ibid., 91, 189.

BRANDT, W. W., and DUSWALT, A. A. (1958), *Anal. Chem.*, 30, 1120.

BRECKENRIDGE, J. G., et al. (1939), *Can. J. Res.*, 17, 258.

BREWER, R. F. (1965), *Monog. am. Soc. Agron.*, 9, 1135.

BRINKLEY, F. (1948), *J. biol. Chem.*, 173, 403.

BROWN, A. L., et al. (1962), *Proc. Soil Sci. Soc. Am.*, 26, 167.

CANNING, R. G., and DIXON, R. (1955), *Anal. Chem.*, 27, 877.

CANNON, H. L. (1963), *Soil Sci.*, 96, 196.

CHAMBERLAIN, G. T. (1956), *E. Afr. agric. for. Res. Org. Ann. Rep.*, 1955–56.

CHENG, K. L. (1956), *Anal. Chem.*, 28, 1738.

CHENG, K. L., and BRAY, R. H. (1953), Ibid., 25, 655.

COLEMAN, N. T., et al. (1963), *Proc. Soil Sci. Soc. Am.*, 27, 290.

COLWELL, W. E. (1946), *Soil Sci.*, 62, 43.

CONWAY, E. J. (1950), *Microdiffusion analysis and volumetric error*, 3rd edn. D. Van Nostrand.

CRONHEIM, G. (1942), *Ind. Eng. Chem. A. E.*, 14, 445.

DAVID, D. J. (1962), *Analyst*, 87, 576.

DEAN, J. A., and THOMPSON, C. (1955), *Anal. Chem.*, 27, 42.

— et al. (1957), *J. Ass. off. agr. Chem.*, 40, 949.

DROUINEAU, G., and MAZOYER, R. (1962), *Ann. agron. Paris*, 13, 31.

EATON, F. M. (1953), *U.S. Dep. Agric. Tech. Bull.*, 448, 131.

ELLIS, G. H., and THOMPSON, J. F. (1945), *Ind. Eng. Chem. A. E.*, 17, 254.

— et al. (1949), *Anal. Chem.*, 21, 1345.

ENNIS, M. T. (1962), *Irish J. agric. Res.*, 1, 139 and 147.

ESSINGTON, E., et al. (1962), *Soil Sci.*, 94, 96.

EVE, D. J. (1955), M.Sc. thesis, Rhodes Univ.

EVERETT, C. F. (1962), *Diss. Abstr.*, 23, 1851.

FAILYER, G. H., et al. (1908), *U.S. Dep. Agric. Bur. Soils Bull.*, 54.

FINE, L. O. (1965), *Monog. am. Soc. Agron.*, 9, 1117.

FISCHER, H. (1922), *Z. anal. Chem.*, 73, 45.

FISKELL, J. G. A., and WESTGATE, P. J. (1955), *Fla. State Hort. Soc. Proc.*, 68, 192.

FRIEND, M. T. (1958), *Ann. Rep. E. Afr. agric. for. Res. Org.*, p. 27.

GALLEGO, A. R., and JOLIN, B. T. (1960), *Ciencias*, 25, 983.

GODFREY, P. R., et al. (1951), *Anal. Chem.*, 23, 1850.

GOLDSCHMIDT, V. M. (1954), *Geochemistry*. Clarendon, Oxford.

GRAHAM-BRYCE, I. J. (1963), *J. Soil Sci.*, 14, 188.

GREEN, D. E. (1948), *Anal. Chem.*, 20, 370.

GREEN, V. (1952), *Adv. Agron.*, 4, 147.

GREGERS-HANSEN, B. (1964), *Nature, Lond.*, 201, 738.

GRIGG, J. L. (1953), *Analyst*, 78, 470.

GRIMALDI, F. S., and WHITE, C. E. (1953), *Anal. Chem.*, 25, 1886.

GUPTA, U. C., and MACKAY, D. C. (1966), *Soil Sci.*, 101, 93.

HAAS, A. R. C. (1929), *Botan. Gaz.*, 87, 630.

HAGHIRI, F. (1962), *Agron. J.*, 54, 278.

HAMILTON, J. W., and BEATH, O. A. (1963), Ibid., 55, 528.

HARDY, F. (1948), *Trop. Agric.*, 25, 68.

HATCHER, J. T., and WILCOX, L. V. (1950), *Anal. Chem.*, 22, 567.

HEALY, W. B., *et al.* (1961), *Soil Sci.*, 92, 359.

HEMPLER, K. (1963), *Landw. Forsch.*, 16, 1.

HENKENS, C. H. (1961), *Landbouwk. Tijdschr.'s Grav.*, 74, 16.

HENRIKSEN, A., and JENSEN, H. L. (1958), *Acta agric. scand.*, 8, 441.

HESSE, P. R. (1962), *W. Afr. Rice Res. Stn. Ann. Rep.*

HIATT, A. J., *et al.* (1963), *Agron. J.*, 55, 284.

HIBBARD, P. L. (1940), *Hilgardia*, 13, 1.

HOGAN, K. G., and BREEN, J. N. (1963), *N.Z. J. Sci.*, 6, 535.

HOLDEN, E. R., and BONTOYAN, W. R. (1962), *J. Ass. off. agric. Chem.*, 45, 455.

HOLMES, R. S. (1945), *Soil Sci.*, 59, 77.

HOSTE, J., and GILLIS, J. (1955), *Anal. chim. Acta*, 12, 158.

HUBBARD, D. M., and SCOTT, E. W. (1943), *J. am. Chem. Soc.*, 65, 2390.

HYBBINETTE, A. E., and SANDELL, E. B. (1942), *Ind. Eng. Chem. A. E.*, 14, 715.

INDOVINA, R. (1935), *Biochem. Z.*, 275, 286.

JACKSON, M. L. (1958), *Soil chemical analysis.* Constable.

JEFFERIES, C. D. (1951), *Soil Sci.*, 71, 287.

JOHNSON, C. M., *et al.* (1958), *J. agric. Fd. Chem.*, 6, 114.

JOHNSON, W. C., and SIMMONS, M. (1946), *Analyst*, 71, 554.

KANEHIRO, Y. (1964), *Diss. Abstr.*, 25, 2683.

KARIM, A. Q. M. B., and DERAZ, O. (1961), *Soil Sci.*, 92, 408.

KIDSON, E. B. (1938), *J. Soc. Chem. Ind.*, 57, 95.

KIVZNETSOV, V. I. (1948), *Zh. analit. Khim.*, 3, 295.

KLEIN, A. K. (1941), *J. Ass. off. agric. Chem.*, 24, 363.

KOLTHOFF, I. M. (1928), *J. am. Chem. Soc.*, 50, 393.

LAMBERT, J. L., *et al.* (1951), *Anal. Chem.*, 23, 1101.

LE RICHE, H. H., and WEIR, A. H. (1963), *J. Soil Sci.*, 14, 225.

LOWE, R. H., and MASSEY, H. F. (1965), *Soil Sci.*, 100, 238.

LUNDBLAD, K., *et al.* (1949), *Pl. Soil*, 1, 375.

LYMAN, C., and DEAN, L. A. (1942), *Soil Sci.*, 54, 315.

MARMOY, F. B. (1939), *J. Soc. Chem. Ind.*, 58, 275.

MARTENS, D. C., *et al.* (1966). *Proc. Soil Sci. Soc. Am.*, 30, 67.

MCDONALD, A. J., and STANTON, R. E. (1962), *Analyst*, 87, 600.

MCHARGUE, J. S., *et al.* (1932), *Ind. Eng. Chem. A.E.*, 4, 214.

MEGREGIAN, S. (1954), *Anal. Chem.*, 26, 1161.

MENZEL, R. G., and JACKSON, M. L. (1950), *Proc. Soil Sci. Soc. Am.*, 15, 122.

MIDDLETON, K. R., and WESTGARTH, D. R. (1964), *Soil Sci.*, 97, 221.

MILBOURN, G. M. (1960), *J. agric. Sci.*, 55, 273.

MILTON, R. F. (1949), *Analyst*, 74, 54.

MITCHELL, R. L. (1956), *Comm. Bur. Soils Tech. Comm.*, 44A.

MITSUI, S., *et al.* (1960), *J. Sci. Soil Tokyo*, 31, 321.

MORTIMORE, D. M., and ROMANS, P. A. (1952), *J. Opt. Soc. Am.*, 42, 673.

NAFTEL, J. A. (1939), *Ind. Eng. Chem. A. E.*, 11, 407.

NAKAMURA, M. T., and SHERMAN, C. D. (1961), *Hawaii agric. Exp. Stn. Tech. Bull. 45.*

NELSON, J. L., *et al.* (1959), *Soil Sci.*, 88, 275.

NG, S. K., and BLOOMFIELD, C. (1962), *Pl. Soil*, 16, 108.

NICHOLL, W. E. (1953), *Can. J. Chem.*, 31, 145.

NISHITA, H., *et al.* (1961), *Agric. Fd. Chem.*, 9, 101.

NORRIS, J. F., and FAY, H. (1896), *Am. Chem. J.*, 18, 703.

NORTH, A. A. (1956), *Analyst*, 81, 660.

OERTL, J. J. (1962), *Soil Sci.*, 94, 214.

— and KOHL, H. C. (1961), Ibid., 92, 243.

OLSON, R. V., and BERGER, K. C. (1946), *Proc. Soil Sci. Soc. Am.*, 11, 216.

PATASSY, F. Z. (1965), *Pl. Soil*, 22, 395.

PETRETIC, G. J. (1951), *Anal. Chem.*, 23, 1183.

PICKARD, J. A., and MARTIN, J. T. (1963), *J. Sci. Fd. Agric.*, 14, 706.

POLLEY, D., and MILLER, V. L. (1955), *Anal. Chem.*, 27, 1162.

POLUEKTOV, N. S. (1943), *Trudy Vses. Konferentsii Anal. Khim.*, 2, 393.

POMETUN, E., and BOYAROVA, V. I. (1961), *Ah. Anal. Khim.*, 16, 103.

POPEA, F., and GUTMAN, M. (1961), *Acad. Rep. pop. Rom. Stud. cerc. chim.*, 9, 673.

— and JEMANEAUNU, M. (1960), *Studii. Cerc. Chim.*, 8, 607.

POTRATZ, H. A., and ROSEN, J. M. (1949), *Anal. Chem.*, 21, 1276.

PRATT, P. F., and BRADFORD, G. R. (1958), *Proc. Soil Sci. Soc. Am.*, 22, 399.

PROSKURYAKOVA, G. F. (1964), *Trudy sverdlovsk. sel'.-khoz. Inst.*, 11, 488.

PUIG, A. C., et al. (1960), *An. Soc. Fis. Quim.*, 56A, 121.

PURUSHOTTAM, D. (1960), *J. scient. ind. Res.*, 19B, 449.

PURVIS, E. R., and PETERSON, N. K. (1956), *Soil Sci.*, 81, 223.

RÁB, F. (1962), *Sb. vys. Sk. zemed. Brne.*, 2A, 127.

REISENAUER, H. M., et al. (1962), *Proc. Soil Sci. Soc. Am.*, 26, 23.

ROBINSON, W. O., et al. (1934), *Ind. Eng. Chem. A. E.*, 6, 274.

ROMNEY, E. M., et al. (1959), *Soil Sci.*, 87, 150.

ROSCHACH, H. (1961), *Z. PflErnähr. Düng.*, 94, 134.

RUSH, R. M., and YOE, J. H. (1954), *Anal. Chem.*, 26, 1345.

SANDELL, E. B. (1937), *Ind. Eng. Chem. A. E.*, 9, 464.

— (1939), Ibid., 11, 364.

— (1940), Ibid., 12, 674.

— (1950), *Colorimetric determination of traces of metals.* 2nd edn. Inter. Sci., N.Y.

SARMA, T. P., and RAGHAVA-RAO, B. S. V. (1955), *J. scient. ind. Res.*, 14B, 450.

SCHARRER, K., and GOTTSCHALL, R. (1935), *Z. PflErnähr. Düng.*, 39, 178.

SCHNEIDER, W. A., and SANDELL, E. B. (1954), *Mikrochim. Acta*, 263.

SCHOFIELD, R. K., and GRAHAM-BRYCE, I. J. (1960), *Nature, Lond.*, 188, 1048.

SCHOLL, W. (1962), *Landw. Forsch. Sonderh.*, 16, 138.

SHAW, W. M. (1954), *Anal. Chem.*, 26, 1212.

SHOEMAKER, C. E. (1955), Ibid., 27, 552.

SHRIFT, A. (1964), *Nature, Lond.*, 201, 1304.

SINGH, S. S., and KANWAR, J. S. (1963), *J. Ind. Soc. Soil Sci.*, 11, 283.

SMALES, A. A. (1951), *Analyst*, 76, 348.

SMALL, H. G., and MCCANTS, C. B. (1961), *Proc. Soil Sci. Soc. Am.*, 25, 346.

SNELL, F. D., and SNELL, C. T. (1949), *Colorimetric methods of analysis*, 3rd edn. D. Van Nostrand.

SORTEBERG, A. (1962), *Soil Sci.*, 94, 80.

SPECHT, R. C., and MACINTIRE, W. H. (1961), *Soil Sci.*, 93, 172.

STANFIELD, K. E. (1935), *Ind. Eng. Chem. A. E.*, 7, 273.

STANTON, R. E., and MCDONALD, A. J. (1962a), *Trans. Inst. Min. Metall.*, 71, 517.

— and — (1962b), *Bull. Inst. Metall.*, 667, 511.

— and — (1964), *Analyst*, 89, 767.

— and — (1966), Ibid., 91, 775.

STEENBJERG, F., and BOKEN, E. (1948), *Tidsskr. PlAvl.*, 52, 375.

STEWART, I., and LEONARD, C. D. (1963), *Soil Sci.*, 95, 149.

STEWART, J. A., and BERGER, K. C. (1965), Ibid., 100, 244.

STOCK, A. (1938), *Svensk. kem. Tidskr.*, 50, 242.

STOUT, P. R., and JOHNSON, C. M. (1965), *Monog. am. Soc. Agron.*, 9, 1124.

STRMESKI, G. (1959), *Zborn. kmet. gozd.*, 6, 63.

SUNDERASAN, M., and SANKAR DAS, M. (1955), *Analyst*, 80, 697.

SWAINE, D. J. (1962), *Comm. Bur. Soils Tech. Comm.*, No. 52.

TELEP, G., and BOLTZ, D. F. (1951), *Anal. Chem.*, 23, 901.

THOMPSON, C. E., and LAKIN, H. W. (1957), *Geol. Surv. Bull.*, 209, 1036.

TOBIA, S. K., and HANNA, A. S. (1961), *Soil Sci.*, 92, 123.

URONE, P. F., and ANDERS, H. K. (1950), *Anal. Chem.*, 22, 1317.

VASILEVSKAYA, A. E., and SHCHERBAKOV, V. P. (1963), *Pochvovedenie*, 10, 96.

VERDIER, E. T., *et al.* (1957), *J. Agric. Fd. Chem.*, 5, 354.

VIRO, P. J. (1955a), *Soil Sci.*, 79, 459.

— (1955b), Ibid., 80, 69.

VOGEL, A. I. (1962), *A textbook of quantitative inorganic analysis*, 3rd edn. Longmans.

WALLACE, T. (1961), *The diagnosis of mineral deficiencies in plants by visual symptoms* 3rd edn. HMSO, London.

WARBURG, O. (1949), *Heavy metal prosthetic groups and enzyme action*. Oxford.

WARD, F. N. (1951), *U.S. Geol. Surv. Circ.*, 119.

— and LAKIN, H. W. (1954), *Anal. Chem.*, 26, 1168.

WEAR, J. I. (1965), *Monog. am. Soc. Agron.*, 9, 1059.

WEAR, J. I., and SOMMER, A. L. (1947), *Proc. Soil Sci. Soc. Am.*, 12, 143.

WELLS, N., and TAYLOR, N. H. (1960), *Trans. 7th int. Congr. Soil Sci.*, 2, 193.

WHETSTONE, R. R., *et al.* (1942), *U.S. Dep. Agric. Tech. Bull.*, No. 797.

WHISMAN, M., and ECCLESTON, B. H. (1955), *Anal. Chem.*, 27, 1861.

WILLARD, H. H., and HAHN, R. B. (1949), Ibid., 21, 293.

— and HORTON, C. A. (1950), Ibid., 22, 1194.

— and WINTER, O. B. (1933), *Ind. Eng. Chem. A. E.*, 5, 7.

WILLIAMS, C. H. (1955), *J. Sci. Fd. Agric.*, 6, 104.

WRIGHT, E. R., and MELLON, M. G. (1937), *Ind. Eng. Chem. A. E.*, 9, 375.

YATES, M. G., and HALLSWORTH, E. G. (1963), *Pl. Soil*, 19, 265.

ZAGRAN, V. D., and VLASOV, V. A. (1962), *Zh. anal. Khim.*, 17, 254.

Chapter 17

BANCROFT, W. D. (1892), *Z. Phys. Chem.*, 10, 837.

BIRCH, H. F. (1949), *Ann. Rep. E. Afr. agric. for. Res. Org.*, p. 32.

BLOOMFIELD, C. (1951), *J. Soil Sci.*, 2, 196.

— (1953), Ibid., 4, 5.

BROWN, L. A. (1934), *Soil Sci.*, 37, 65.

CLARK, W. M., and COHEN, B. (1923), *U.S. Pub. Health Rep.*, 38, 666.

—, —, *et al.* (1928), *Hyg. Lab. Bull.*, 151.

EVANS, U. R., and PRYOR, M. J. (1949), *J. Chem. Soc. Suppl.*, 157.

FARADAY, M. (1834), *Phil. Trans.*, p. 77.

GARRELS, R. M., and CHRIST, C. L. (1965), *Solutions, minerals and equilibria*. Harper, N.Y.

GILLESPIE, L. J. (1920), *Soil Sci.*, 9, 199.

GREENE, H. (1963), *J. Soil Sci.*, 14, 1.

GREENWOOD, D. J. (1961), *Pl. Soil*, 14, 360.

— and LEES, H. (1960), Ibid., 12, 69.

HART, M. G. R. (1959), Ibid., 11, 215.

HEINTZE, S. G. (1934), *J. agric. Sci.*, 24, 28.

HERTZNER, R. A. (1930), *Z. PflErnähr. Düng. Bodenk.*, 18A, 249.

HESSE, P. R. (1950), *Ann. Rep. E. Afr. agric. for. Res. Org.*, p. 50.

— (1957), *Hydrobiologia*, 11, 29.

INT. RICE RES. INST., PHILIPPINES (1963), *Ann. Rep.*

JEFFERY, J. W. O. (1960), *J. Soil Sci.*, 11, 140.

— (1961), Ibid., 12, 172.

LAMM, C. G. (1956), *Nature, Lond.*, 177, 620.

MCKEAGUE, J. A., and BENTLEY, C. F. (1960), *Can. J. Soil Sci.*, 40, 120.

MISRA, R. D. (1938), *J. Ecol.*, 36, 411.

MORTIMER, C. H. (1941), Ibid., 29, 280.
— (1942), Ibid., 30, 147.
PATRICK, W. H. (1960), *Trans. 7th int. Congr. Soil Sci.*, 2, 494.
PEARSALL, W. H. (1938), *J. Ecol.*, 26, 180.
— (1950), *Emp. J. exp. Agric.*, 18, 289.
— and MORTIMER, C. H. (1939), *J. Ecol.*, 27, 483.
PONNAMPERUMA, F. N., *et al.* (1967), *Soil Sci.*, 103, 374.
QUISPEL, A. (1947), *Soil Sci.*, 63, 265.
REMESOW, N. P. (1930), *Z. PflErnähr. Düng. Bodenk.*, 15A, 34.
ROBINSON, W. O. (1930), *Soil Sci.*, 30, 197.
SCOTT, A. D., and EVANS, D. D. (1955), *Proc. Soil Sci. Soc. Am.*, 19, 7.
SHAPIRO, R. E. (1958), *Soil Sci.*, 85, 267.
WILLIS, L. G. (1936), *Proc. Soil Sci. Soc. Am.*, 1, 291.

Chapter 18

ALLEN, C. A., *et al.* (1953), *J. Hyg.*, 51, 185.
ARMSTRONG, W. (1967), *J. Soil Sci.*, 18, 27.
BEACHER, B. L. (1956), *J. Soil Sci. Soc. Philippines*, 8.
BROGAN, J. C., *et al.* (1963), *Irish J. agric. Res.*, 2, 125.
CHAUDHRY, I. A., and CORNFIELD, A. H. (1966), *Pl. Soil*, 25, 474.
DUDAL, R. (1957), 1st S.E. Asian Soils Conf., Manila.
HART, M. G. R. (1961), *Analyst*, 86, 472.
HESSE, P. R. (1957), *Hydrobiologia*, 11, 29.
— (1958), Ibid., 11, 171.
Int. Rice Res. Inst. Philippines (1963), *Ann. Rep.*
JEFFERY, J. W. O. (1961), *W. Afr. Rice Res. Stn. Ann. Rep.*
— (1963), Ibid., p. 21.
JOFFE, J. S. (1935), *Soil Sci.*, 39, 391.
METZGER, W. H. (1930), *Soil Sci.*, 29, 251.
MORAGHAN, J. T. (1963), *Proc. Soil Sci. Soc. Am.*, 27, 361.
MORTIMER, C. H. (1941), *J. Ecol.*, 29, 280.
NG SIEW KEE and BLOOMFIELD, C. (1962), *Pl. Soil*, 16, 108.
PONNAMPERUMA, F. N. (1955), Thesis, Cornell Univ.
— *et al.* (1966), *Soil Sci.*, 101, 421.
ROBINSON, W. O. (1930), Ibid., 30, 197.
SCHOLLENBERGER, C. J. (1928), Ibid., 25, 357.
SCOTT, A. D., and EVANS, D. D. (1955), *Proc. Soil Sci. Soc. Am.*, 19, 12.
SHAPIRO, R. E. (1958), *Soil Sci.*, 85, 267.
SUBRAHMANYAN, V. (1927), *J. agric. Res.*, 17, 429.
TURNER, R. C. (1958), *Soil Sci.*, 86, 32.
West Afr. Rice Res. Stn. (*1961*) *Ann. Rep.*
WHITNEY, R. S., and GARDNER, R. (1943), *Soil Sci.*, 55, 127.
YAALON, D. H. (1957), *Pl. Soil*, 8, 275.

Index